# GETTING STARTED WITH PRO/ENGINEER™

ROBERT RIZZA
Department of Mechanical Engineering and Applied Mechanics
NORTH DAKOTA STATE UNIVERSITY
FARGO, NORTH DAKOTA

PRENTICE HALL
Upper Saddle River, New Jersey 07458

Library of Congress Cataloging-in-Publication Data

Rizza, Robert
Getting started with Pro/Engineer / by Robert Rizza.
p. cm.
ISBN 0-13-040212-5
1. Engineering design—Data processing. 2. Pro/ENGINEER.
3. Computer-aided design. I. Title.
TA174 .R555 1999
620'.0042'028553042—dc21 99-35368
CIP

Editor-in-chief: *Marcia Horton*
Acquisitions editor: *Eric Svendsen*
Production editor: *Kim Rose/BookMasters*
Assistant managing editor: *Eileen Clark*
Executive managing editor: *Vince O'Brien*
Art director: *Jayne Conte*
Cover design: *Bruce Kenselaar*
Manufacturing buyer: *Beth Sturla*
Assistant vice president of production and manufacturing: *David W. Riccardi*





Printed in the United States of America

10 9 8 7 6 5 4 3 2 1

ISBN 0-13-040212-5

Prentice-Hall International (UK) Limited, *London*
Prentice-Hall of Australia Pty. Limited, *Sydney*
Prentice-Hall Canada Inc., *Toronto*
Prentice-Hall Hispanoamericana, S.A., *Mexico*
Prentice-Hall of India Private Limited, *New Delhi*
Prentice-Hall of Japan, Inc., *Tokyo*
Prentice-Hall (Singapore) Pte. Ltd., *Singapore*
Editora Prentice-Hall do Brasil, Ltda., *Rio de Janeiro*

# BRIEF CONTENTS

CHAPTER 1 • GETTING ACQUAINTED WITH THE PRO/E INTERFACE 1
CHAPTER 2 • CONSTRUCTING PARTS USING THE SKETCHER 24
CHAPTER 3 • CREATING BASE FEATURES WITH DATUM PLANES 53
CHAPTER 4 • ADDING HOLES TO BASE FEATURES 80
CHAPTER 5 • OPTIONS THAT REMOVE MATERIAL: CUT, SLOT, NECK, AND SHELL 101
CHAPTER 6 • OPTIONS THAT ADD MATERIAL: FLANGE, RIB, AND SHAFT 123
CHAPTER 7 • OPTIONS THAT SPEED UP MODEL CONSTRUCTION: PATTERN, COPY, GROUP, MIRROR GEOM, AND UDF 144
CHAPTER 8 • FILLETS, ROUNDS, AND CHAMFERS 185
CHAPTER 9 • DATUM POINTS, AXES, CURVES, AND COORDINATE SYSTEMS 200
CHAPTER 10 • THE REVOLVE OPTION 226
CHAPTER 11 • FEATURE CREATION WITH SWEEP 245
CHAPTER 12 • BLENDS 276
CHAPTER 13 • SOME OPTIONS FOR MANAGING FEATURES 293
CHAPTER 14 • COSMETIC FEATURES 311
CHAPTER 15 • QUILTS AND SOME OPTIONS FROM THE TWEAK MENU 322
CHAPTER 16 • THE DRAWING MODE 346
CHAPTER 17 • CREATING A SECTION 372
CHAPTER 18 • ADDING TOLERANCES TO A DRAWING 391
CHAPTER 19 • ASSEMBLIES AND WORKING DRAWINGS 404
CHAPTER 20 • ENGINEERING INFORMATION AND FILE TRANSFER 455

BIBLIOGRAPHY 467
INDEX 468

# CONTENTS


FOREWORD xi

ACKNOWLEDGMENTS xii

CHAPTER 1 • GETTING ACQUAINTED WITH THE PRO/E INTERFACE 1

INTRODUCTION AND OBJECTIVES 1

1.1 THE PRO/E INTERFACE 5

1.1.1 Tutorial # 1: Becoming Familiar with the Interface 6

1.2 PART FILES AND PRO/E 11

1.2.1 Tutorial # 2: Retrieving "BaseOrient" 11

1.3 CHANGING THE ORIENTATION OF A MODEL 14

1.3.1 Tutorial # 3: Changing the Orientation of "BaseOrient" 15

1.4 SOLID MODELS 16

1.4.1 Tutorial # 4: A Solid Model of "BaseOrient" 17

1.5 THE USE OF MAPKEYS 18

1.5.1 Tutorial # 5: Creating Mapkeys Interactively 19

1.6 PRINTING 20

1.6.1 Tutorial # 6: Printing a Copy of "BaseOrient" 21

1.7 SUMMARY AND STEPS FOR USING VARIOUS INTERFACE OPTIONS 21

1.8 ADDITIONAL EXERCISES 22

CHAPTER 2 • CONSTRUCTING PARTS USING THE SKETCHER 24

INTRODUCTION AND OBJECTIVES 24

2.1 PARTS AND THE SKETCHER 24

2.1.1 Tutorial # 7: A Vise Handle 25

2.2 SYMBOLS USED IN THE REGENERATION PROCESS 29

2.3 THE USE OF RELATIONS 31

2.4 CHANGING THE UNITS OF A PART 31

2.4.1 Tutorial # 8: A Hinge 32

2.5 THE SEC TOOLS OPTION 35

2.6 THE SKETCH VIEW OPTION 36

2.7 THE GEOM TOOLS OPTION 36

2.7.1 Tutorial # 9: A Tension Plate 36

2.8 THIN SECTIONS 39

2.8.1 Tutorial # 10: An Angle Bracket Using the Thin Option 39
2.9 CONIC SECTIONS 41
2.10 THE SKETCH MODE 42
2.10.1 Tutorial # 11: Creating and Using a Section 42
2.11 SUMMARY AND STEPS FOR USING THE SKETCHER 44
2.12 ADDITIONAL EXERCISES 47

CHAPTER 3 • CREATING BASE FEATURES WITH DATUM PLANES 53
INTRODUCTION AND OBJECTIVES 53
3.1 THE DEFAULT DATUM PLANES 54
3.1.1 Tutorial # 12: A Bushing 54
3.1.2 Tutorial # 13: A Shifter Fork 58
3.2 DATUM PLANES ADDED BY THE USER 65
3.2.1 Tutorial # 14: A Rod Support 67
3.3 START PART 69
3.3.1 Tutorial # 15: A Metric Start Part 71
3.4 SUMMARY AND STEPS FOR ADDING DATUMS 72
3.5 ADDITIONAL EXERCISES 73

CHAPTER 4 • ADDING HOLES TO BASE FEATURES 80
INTRODUCTION AND OBJECTIVES 80
4.1 ADDING HOLES IN PRO/E 80
4.1.1 Tutorial # 16: Adding Holes to the Angle Bracket 82
4.1.2 Tutorial # 17: A Coaxial Hole for the Bushing 84
4.1.3 Tutorial # 18: A Sketched Hole in a Plate 85
4.1.4 Tutorial # 19: A Pipe Flange with a Bolt Circle 87
4.1.5 Tutorial # 20: A Hole on the Inclined Plane of the Support 92
4.2 SUMMARY AND STEPS FOR ADDING HOLES 93
4.3 ADDITIONAL EXERCISES 94

CHAPTER 5 • OPTIONS THAT REMOVE MATERIAL: CUT, SLOT, NECK, AND SHELL 101
INTRODUCTION AND OBJECTIVES 101
5.1 ADDING A CUTOUT OR SLOT 101
5.1.1 Tutorial # 21: Using Cut to Remove Material from a U Bracket 102
5.1.2 Tutorial # 22: A Washer with a Cut 103
5.1.3 Tutorial # 23: An Axle with a Slot 106
5.2 ADDING A NECK 108
5.2.1 Tutorial # 24: An Adjustment Screw 108
5.3 USING THE SHELL OPTION 111

5.3.1 Tutorial # 25: An Emergency Light Cover 111
5.4 SUMMARY 115
5.4.1 Steps for Using the Cut or Slot Options 116
5.4.2 Steps for Adding a Neck to a Base Feature 116
5.4.3 Steps for Using the Shell Option 116
5.5 ADDITIONAL EXERCISES 117

CHAPTER 6 • OPTIONS THAT ADD MATERIAL: FLANGE, RIB, AND SHAFT 123
INTRODUCTION AND OBJECTIVES 123
6.1 FLANGES 123
6.1.1 Tutorial # 26: A Lamp Holder 124
6.2 RIBS 130
6.2.1 Tutorial # 27: A Connecting Arm 130
6.3 SHAFTS 134
6.3.1 Tutorial # 28: A Shaft for the U Bracket 134
6.4 SUMMARY 135
6.4.1 Steps for Adding a Flange 136
6.4.2 Steps for Adding a Rib 136
6.4.3 Steps for Adding a Shaft 136
6.5 ADDITIONAL EXERCISES 137

CHAPTER 7 • OPTIONS THAT SPEED UP MODEL CONSTRUCTION: PATTERN, COPY, GROUP, MIRROR GEOM, AND UDF 144
INTRODUCTION AND OBJECTIVES 144
7.1 USING THE PATTERN OPTION 145
7.1.1 Tutorial # 29: A Pattern with a Linear Increment 146
7.1.2 Tutorial # 30: A Pattern with a Radial Increment 147
7.1.3 Tutorial # 31: A Nut with a Rotational Pattern 148
7.1.4 Tutorial # 32: A Propeller 150
7.2 USING THE GROUP OPTION 157
7.2.1 Tutorial # 33: A Vibration Isolator Pad 160
7.3 THE COPY OPTION 162
7.3.1 Tutorial # 34: Using Copy and Copy Mirror 163
7.4 THE MIRROR GEOM OPTION 165
7.4.1 Tutorial # 35: Completing the Hinge with Mirror Geom 165
7.5 USER-DEFINED FEATURES 166
7.5.1 Tutorial # 36: Creating a UDF 168
7.5.2 Tutorial # 37: Adding a UDF to a Part 172
7.6 SUMMARY 175
7.6.1 Review and Steps for the Pattern Option 175
7.6.2 Review and Steps for the Group Option 177

7.6.3 Review and Steps for the Copy Option 177
7.6.4 Review and Steps for the Mirror Geom Option 178
7.6.5 Review and Steps for Creating and Placing a UDF 178
7.7 ADDITIONAL EXERCISES 179

CHAPTER 8 • FILLETS, ROUNDS, AND CHAMFERS 185
INTRODUCTION AND OBJECTIVES 185
8.1 THE USE OF ROUNDS AND FILLETS 186
8.1.1 Tutorial # 38: A Round and Fillet for the Angle Bracket 187
8.1.2 Tutorial # 39: Edge Rounds for the Connecting Arm 188
8.1.3 Tutorial # 40: A Round on a Circular Edge 188
8.1.4 Tutorial # 41: Edge Rounds for the Propeller Blades 189
8.1.5 Tutorial # 42: A Part with Surface-to-Surface Rounds 189
8.2 CHAMFERS 191
8.2.1 Tutorial # 43: Adding Chamfers to the Adjustment Screw and the Nut 191
8.3 SUMMARY 192
8.3.1 Steps for Adding a Round or Fillet 193
8.3.2 Steps for Adding a Chamfer 193
8.4 ADDITIONAL EXERCISES 193

CHAPTER 9 • DATUM POINTS, AXES, CURVES, AND COORDINATE SYSTEMS 200
INTRODUCTION AND OBJECTIVES 200
9.1 CONSTRUCTING A DATUM AXIS 200
9.1.1 Tutorial # 44: A Part Requiring a Datum Axis 201
9.2 DATUM COORDINATE SYSTEMS 203
9.2.1 Tutorial # 45: A Coordinate System for a Two-Dimensional Frame 204
9.3 DATUM POINTS 206
9.3.1 Tutorial # 46: Adding Datum Points to the Two-Dimensional Frame 209
9.4 DATUM CURVES 210
9.4.1 Tutorial # 47: Adding a Datum Curve to the Two-Dimensional Frame 211
9.4.2 Tutorial # 48: A King Post Truss 213
9.4.3 Tutorial # 49: Datum Curves for an Airfoil Section 215
9.5 SUMMARY 218
9.5.1 Steps for Creating a Datum Axis 218

9.5.2 Steps for Constructing a Datum Coordinate System 218
9.5.3 Steps for Adding Datum Points to a Model 219
9.5.4 Steps for Creating a Datum Curve 219
9.6 ADDITIONAL EXERCISES 219

CHAPTER 10 • THE REVOLVE OPTION 226
INTRODUCTION AND OBJECTIVES 226
10.1 REVOLVED SECTIONS 227
10.1.1 Tutorial # 50: A Pulley Wheel 227
10.1.2 Tutorial # 51: Creating a Mug Using a Revolved Thin 228
10.1.3 Tutorial # 52: A Parabolic Reflector 229
10.2 SUMMARY AND STEPS FOR USING THE REVOLVE OPTION 236
10.3 ADDITIONAL EXERCISES 236

CHAPTER 11 • FEATURE CREATION WITH SWEEP 245
INTRODUCTION AND OBJECTIVES 245
11.1 THE SWEEP OPTION 245
11.1.1 Tutorial # 53: A Simple Sweep 246
11.1.2 Tutorial # 54: A Gasket 249
11.2 HELICAL SWEEPS 255
11.2.1 Tutorial # 55: A Spring Constructed with a Helical Sweep 256
11.2.2 Tutorial # 56: Constructing Screw Threads Using a Helical Sweep 258
11.3 SWEEPS ALONG A DATUM CURVE 261
11.3.1 Tutorial # 57: Completing the Two-Dimensional Frame 262
11.3.2 Tutorial # 58: Completing the King Post Truss 262
11.4 SUMMARY AND STEPS FOR USING THE SWEEP OPTION 267
11.5 ADDITIONAL EXERCISES 268

CHAPTER 12 • BLENDS 276
INTRODUCTION AND OBJECTIVES 276
12.1 THE BLEND OPTION 277
12.1.1 Tutorial # 59: HVAC Takeoff 277
12.2 SWEPT BLENDS 282
12.2.1 Tutorial # 60: A Centrifugal Blower 282
12.3 SUMMARY AND STEPS FOR CREATING BLENDS 288
12.4 ADDITIONAL EXERCISES 289

CHAPTER 13 • SOME OPTIONS FOR MANAGING FEATURES 293
INTRODUCTION AND OBJECTIVES 293
13.1 THE REDEFINE OPTION 294
13.1.1 Tutorial # 61: Redefining a Feature 294
13.2 THE REORDER AND REROUTE OPTION 296
13.2.1 Tutorial # 62: An Example Using Reorder 296
13.3 LAYERS AND FEATURES 297
13.3.1 Tutorial # 63: Creating a Layer and Adding Features to the Layer 299
13.4 SUPPRESS AND RESUME 300
13.4.1 Tutorial # 64: Suppressing and Resuming Features 301
13.4.2 Tutorial # 65: Suppressing Features in a Layer 302
13.5 THE RESOLVE OPTION 302
13.5.1 Tutorial # 66: Resolving a Failed Regeneration 302
13.6 SUMMARY AND STEPS FOR USING THE FEATURE MANAGEMENT OPTIONS 306
13.7 ADDITIONAL EXERCISES 308

CHAPTER 14 • COSMETIC FEATURES 311
INTRODUCTION AND OBJECTIVES 311
14.1 PRO/E COSMETIC OPTIONS 311
14.1.1 Tutorial # 67: Cosmetic Threads for the Nut 314
14.1.2 Tutorial # 68: Cosmetic Text for the Cover Plate 316
14.2 SUMMARY AND STEPS FOR ADDING COSMETIC FEATURES 318
14.3 ADDITIONAL EXERCISES 318

CHAPTER 15 • QUILTS AND SOME OPTIONS FROM THE TWEAK MENU 322
INTRODUCTION AND OBJECTIVES 322
15.1 ADDING A DRAFT TO A BASE FEATURE 324
15.1.1 Tutorial # 69: A Draft for the Emergency Light Cover 325
15.2 EARS 325
15.2.1 Tutorial # 70: Ears for the Cover Plate 327
15.3 THE LIP OPTION 330
15.3.1 Tutorial # 71: Adding a Lip to a Part 330
15.4 QUILTS 334
15.4.1 Tutorial # 72: An Aircraft Wing 334
15.5 SUMMARY AND STEPS FOR THE TWEAK OPTIONS 340
15.6 ADDITIONAL EXERCISES 342

CHAPTER 16 • THE DRAWING MODE 346
INTRODUCTION AND OBJECTIVES 346

16.1 DRAWINGS AND FORMATS 346
16.1.1 Tutorial # 73: Creating a Format 348
16.1.2 Tutorial # 74: Creating a Drawing 352
16.2 ADDING DIMENSIONS AND TEXT TO A DRAWING 356
16.2.1 Tutorial # 75: Dimensions and Text for the Drawing of the Angle Bracket 359
16.3 AUXILIARY VIEWS AND PRO/E 364
16.3.1 Tutorial # 76: Creating Auxiliary Views in a Drawing 365
16.4 SUMMARY 367
16.4.1 Steps for Creating a Format 368
16.4.2 Steps for Creating and Adding Views to a Drawing 368
16.4.3 Steps for Adding Driven Dimensions to the Views 369
16.5 ADDITIONAL EXERCISES 370

CHAPTER 17 • CREATING A SECTION 372
INTRODUCTION AND OBJECTIVES 372
17.1 CREATING AND PLACING A SECTION 372
17.1.1 Tutorial # 77: A Full Section of the Angle Bracket 375
17.1.2 Tutorial # 78: An Offset Section of the Angle Bracket 378
17.1.3 Tutorial # 79: A Half Section of the Angle Bracket 381
17.1.4 Tutorial # 80: A Revolved Section of the Connecting Arm 383
17.2 SUMMARY AND STEPS FOR CREATING A SECTION 385
17.3 ADDITIONAL EXERCISES 386

CHAPTER 18 • ADDING TOLERANCES TO A DRAWING 391
INTRODUCTION AND OBJECTIVES 391
18.1 ADDING TRADITIONAL TOLERANCES TO A DRAWING 392
18.1.1 Tutorial # 81: Traditional Tolerances for the Drawing "AngleBracket" 394
18.2 GEOMETRIC DIMENSIONING AND TOLERANCING 395
18.2.1 Tutorial # 82: Creating a Drawing with Geometric Tolerances 397
18.3 SUMMARY AND STEPS FOR ADDING TOLERANCES TO A DRAWING 401
18.4 ADDITIONAL EXERCISES 403

CHAPTER 19 • ASSEMBLIES AND WORKING DRAWINGS 404
INTRODUCTION AND OBJECTIVES 404

19.1 CREATING AN ASSEMBLY FILE 405
19.2 CREATING PARAMETRIC ASSEMBLIES 407
19.2.1 Tutorial # 83: Assembling by Using Place Options 409
19.3 ASSEMBLIES USING DATUM PLANES 411
19.3.1 Tutorial # 84: An Assembly Using Datum Planes 411
19.4 CREATING PARTS IN THE ASSEMBLER 422
19.4.1 Tutorial # 85: An Axle for the Pulley Assembly 423
19.5 ADDITIONAL CONSTRAINTS 425
19.5.1 Tutorial # 86: Completing the Pulley Assembly 425
19.6 EXPLODED ASSEMBLIES 430
19.6.1 Tutorial # 87: An Exploded State for the Pulley Assembly 432
19.7 ADDING OFFSET LINES TO ASSEMBLIES 437
19.7.1 Tutorial # 88: Offset Lines for the Pulley Assembly 437
19.8 ADDING AN ASSEMBLY TO A DRAWING 438
19.8.1 Tutorial # 89: A Working Drawing of the Pulley Assembly 438
19.9 BILL OF MATERIAL (BOM ) AND BALLOONS 440
19.9.1 Tutorial #90: Adding a BOM to the Working Drawing 440
19.10 SUMMARY AND STEPS FOR DEALING WITH ASSEMBLIES 441
19.11 ADDITIONAL EXERCISES 444

CHAPTER 20 • ENGINEERING INFORMATION AND FILE TRANSFER 455
INTRODUCTION AND OBJECTIVES 455
20.1 DEFINING AND ASSIGNING MATERIAL PROPERTIES 456
20.1.1 Tutorial # 91: Entering the Density for a Model 457
20.2 MODEL ANALYSIS 457
20.2.1 Tutorial # 92: Mass Properties of the Angle Bracket 459
20.2.2 Tutorial # 93: Interference Between Two Parts 459
20.3 THE MEASURE OPTION 461
20.3.1 Tutorial # 94: Use of the Measure Option 461
20.4 IGES FILE TRANSFER 462
20.4.1 Tutorial # 95: An IGES File of the Angle Bracket 464
20.5 SUMMARY AND THE STEPS FOR THE OPTIONS 464
20.6 ADDITIONAL EXERCISES 465

BIBLIOGRAPHY 467
INDEX 468

# FOREWORD

This book originated from an introductory engineering graphics and computer-aided design (CAD) course taught at North Dakota State University (NDSU) for the past three years. The purpose of the course was to introduce the engineering student to sketching, visualization, and parametric modeling. Over the years, roughly 900 students have taken this course from several disciplines. Because of the variety of these students' interests and backgrounds, examples for the class and this book have been taken from different areas in the engineering sciences.

At NDSU, the CAD program is a cornerstone of the engineering program. CAD teaches the engineer to communicate design ideas graphically. This book was developed with that philosophy. While several chapters are devoted to describing and implementing different options in Pro/E for model construction, space is set aside for the techniques used in the representation of these models. For example, chapters 16, 17, and 18 are devoted to the creation of drawings, a traditional method for visualizing three-dimensional models. Furthermore, the use of Pro/E for creating working drawings containing assemblies is discussed in chapter 19. In addition, because CAD models are used in conjunction with other analysis and manufacturing software, chapter 20 has been devoted to the Pro/E options for obtaining engineering information and file transfer.

This book describes the options for Revision 20 and 21 of Pro/E, which have a Windows-style interface with buttons for easier navigation. Revision 20 and 21 are practically the same. The difference between the two, which affects this book, is the command **Analysis** in the pull-down menu bar of version 21. This command contains some of the options found in the **Info** command in version 20. Thus, for the options **Measure** and **Model Analysis,** a slight path difference exists for these options. In this book, where the two versions of Pro/E deviate, the difference is indicated by giving the equivalent version 21 path in parentheses. For example, in chapter 20, we have, "Choose **Info** and **Model Analysis** (**Analysis** and **Model Analysis** for Rev 21)." Besides this path difference, the tutorials are compatible for either version.

The chapters in this book are set in the following format. An introduction and objectives section is used to lay the groundwork for what is to follow. Then, the various sections ensue, describing the Pro/E software options. Tutorials are placed strategically within each chapter for immediate reinforcement in the use of the options. At the end of each chapter, a summary is provided, detailing the material discussed in the chapter. Furthermore, a numerated list of steps for each option is provided. These lists may be used as a reference.

Throughout the book, Pro/E options and commands are written in italic. The models used in the tutorials form two groups. One group contains models that are part of small projects used to reinforce the options being considered. The second group contains models from larger projects. These models from the larger projects are used at the end of chapter 19 to construct assemblies. Additional exercises are provided at the end of each chapter to further reinforce the options and concepts presented in that chapter. The models in these exercises were selected to challenge the student in the use of the options under consideration.

## ACKNOWLEDGMENTS

Any work, whatever its magnitude, is impossible without the support and encouragement of family, friends, and colleagues. I thank Eric Svendsen, my editor at Prentice Hall, for his support in getting this project on the right track and completed. Special thanks go to Keven Mueller, Prentice Hall representative, for inquiring about the manuscript in the first place.

Gratitude must go to the faculty of the Mechanical Engineering and Applied Mechanics (MEAM) Department at NDSU for their support while this project was being finished. In particular, Drs. Robert Pieri and Greg Gessel deserve many heartfelt thanks for listening to the ideas for this book and reviewing the manuscript. Bruce Horton deserves recognition for getting the software up and running.

Many of the models for the homework assignments were completed by Shaikh Rahman and Ken Jones, students in the MEAM department. Their hard work and timeliness are greatly appreciated.

Lastly, but in no way in the least, thanks go to my wife Barbara for her support during the long difficult times and for her help in editing the text.

R. R.

# CHAPTER 1

# GETTING ACQUAINTED WITH THE PRO/E INTERFACE

## INTRODUCTION AND OBJECTIVES

Pro/Engineer (Pro/E) is a software package developed and sold by Parametric Technology Corporation for the Computer-Aided Design and Drawing (CADD) market. It uses parametric or constraint-based modeling.

Parametric modeling allows the user to represent the dimensions of the object with parameters. For example, in traditional CADD packages, the block shown in Figure 1.1 would have been developed by entering the given dimensions. If the model dimensions were later changed, then the model would have to be redrawn, making the system inefficient. With a parametric modeler, the dimensions of the object are represented in the computer as parameters, as shown in Figure 1.2. Then, numbers are assigned to the parameters. If the dimensions of the object are changed, the model can be easily modified, because only the numerical values of the parameters need to be changed.

Pro/E is also a feature-based modeler. Feature-based modeling is a way to combine the commands needed to produce a common manufactured feature. For example, in traditional CADD packages, a hole in an object would have been produced by creating two circles and at least one line connecting the circles (see Figure 1.3), with the prescribed dimensions. In a software package such as Pro/E, the information to create the hole is grouped together in one feature. The circles and line(s) making up the hole would be represented and identified by Pro/E as a single feature, and the relationships among the individual parts making up the feature are maintained. Thus, if

FIGURE 1.1 *Block under consideration.*

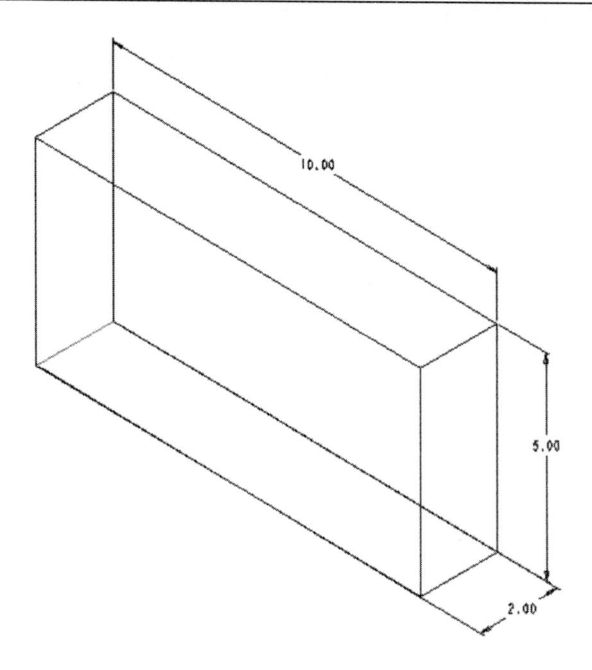


any changes are made to any part of the elements making up the hole, the entire feature is easily updated.

Non-constraint-based modelers are called primitive modelers. Such modelers use a finite set of geometric primitives to construct a three-dimensional

FIGURE 1.2 *Block represented by parameters.*

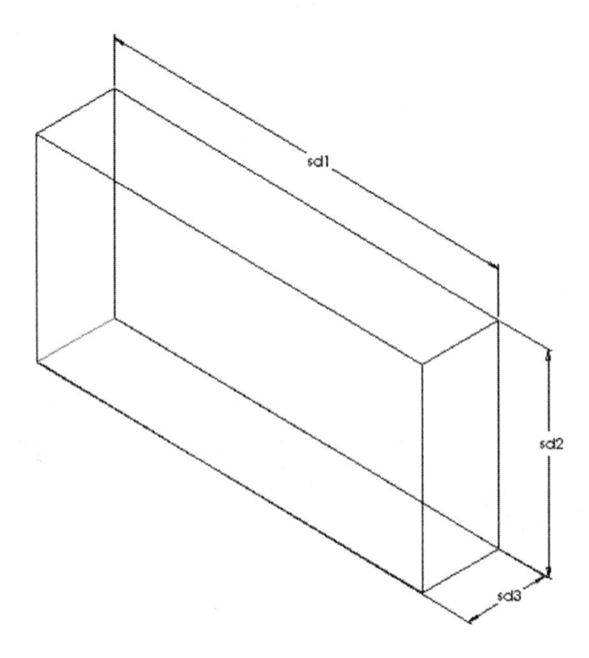

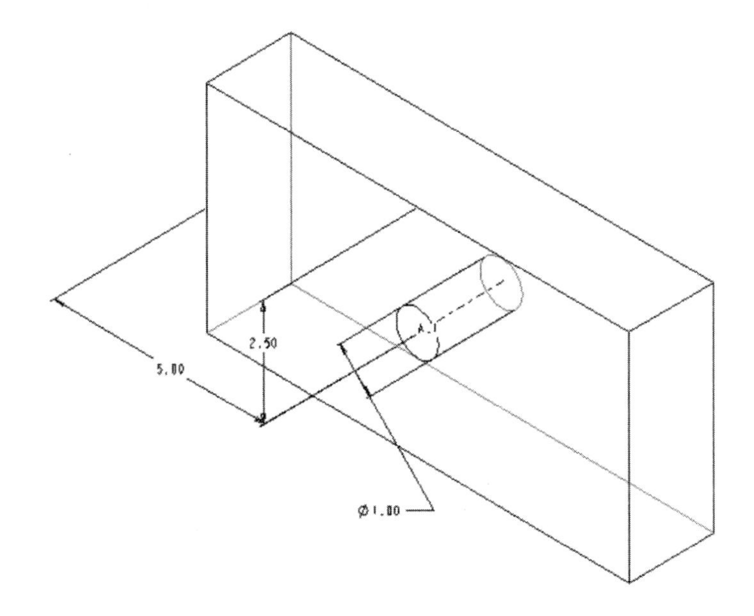


FIGURE 1.3 *Block with prescribed hole.*

model (see Figure 1.4). A three-dimensional primitive modeler constructs a model by combining these primitives, as in Figure 1.5.

In constraint-based modeling, modeling begins by creating a two-dimensional sketch. This sketch does not have to be accurate because it will be constrained by adding dimensions. When the generation process is performed, the dimensions constrain the sketch and the geometry is redefined using parametric equations. A three-dimensional model is generated from

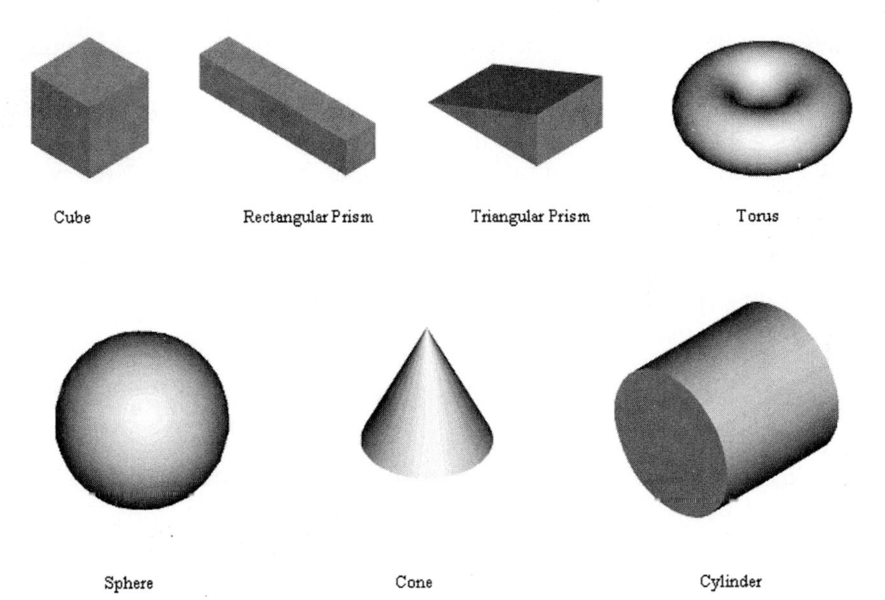


FIGURE 1.4 *Primitive modelers use a set of geometric shapes to construct models.*

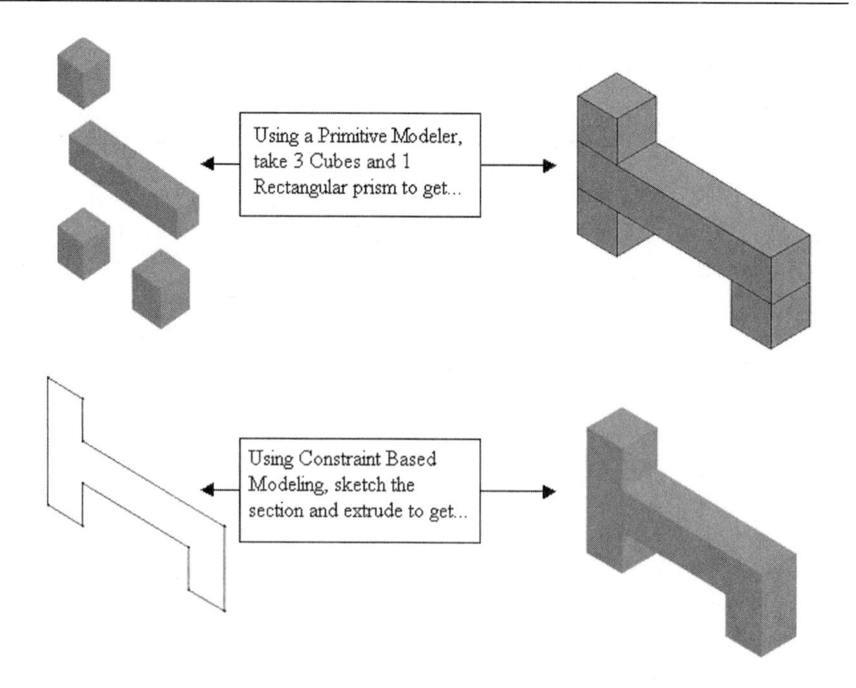


**FIGURE 1.5** *The difference between a Primitive and Constraint-Based Modeler.*

the sketch by extruding (as in Figure 1.5), revolving the sketch about an axis of revolution, or sweeping along a trajectory. This process often mimics the manufacturing process.

Constraint-based modelers are more intuitive because they capture the design intent and the dimensions are related to the geometry. A change in one of the dimensions will cause the model to change immediately. Any design flaw will be readily visible, because the model is automatically updated.

The approach in this book follows the method used in constraint-based modeling. That is, we begin with the options of Pro/E that are used to generate a sketch, and proceed to the options that are used to make the two-dimensional model, three-dimensional.

However, in this chapter, we concentrate on the Pro/E interface. The interface is the display, which is visible after the software has been loaded. We discuss the pertinent options that are used to create, retrieve, and print files. Therefore, the objectives of this chapter are the following:

1. The chapter will introduce the user to the Pro/E interface.
2. Through the various tutorials, the user will develop an ability to retrieve a part file, reorient the part, and print a hardcopy of the model.
3. The reader will attain some experience in customizing the display.

## 1.1 The Pro/E Interface

The actual loading and running of the software is system dependent. Before proceeding any further with this book, you should check the exact procedure for loading and running the software on your system.

The Pro/E interface is illustrated in Figure 1.6. In the figure, a snapshot has been taken of the software running in the Windows NT environment. This particular version has the PTC Application Manager that your system may not have available. For the sake of clarity, we have changed the default background color of the windows from blue to white. Notice that several windows are open. In particular, notice the window with the title:

**********Pro/ENGINEER**********

It is the main drawing window. Although we may add more windows from time to time to speed our work, most of your models will be displayed in this window.

Below the main window is a smaller window. It is the message window. All messages from Pro/E to the user are displayed in this window.

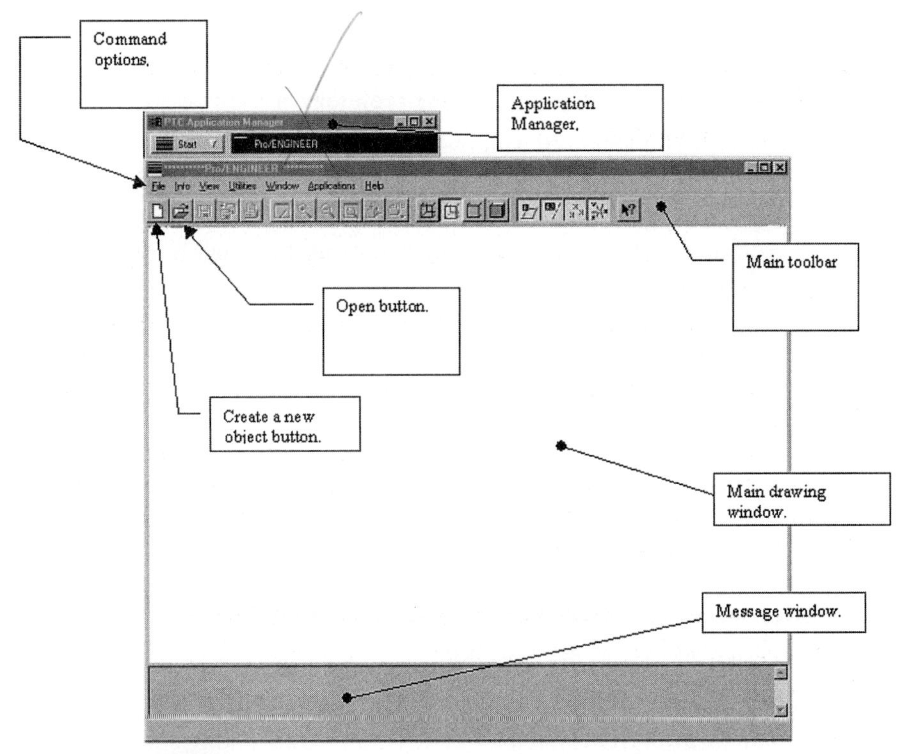


FIGURE 1.6 *The Pro/E interface running in a Windows NT environment.*

Across the top is a Microsoft Windows style toolbar containing various buttons. When the software is first loaded, the first two buttons are active and may be used to create a new file or open an existing file, respectively. Pro/E uses pull-down menus and icons for navigation. That is, you will activate various options by mouse clicks and pull-down menus. Depending on the options, "pop-up" menus and dialog boxes may also appear. These interfaces appear in order to enter all the information required to complete the requested task, and then disappear when all the appropriate information has been provided by the user.

In Figure 1.6, we can see the seven command options. Each of these command options, when activated will display a pull-down menu containing various options. The seven options are *File, Info, View, Misc, Utilities, Window, Applications,* and *Help.* The options are used in the following manner:

1. *File.* The pull-down menu contains options for creating, retrieving, printing, or saving a file in various configurations.
2. *Info.* The *Info* option allows the user to obtain information concerning the model in memory.
3. *View.* This command provides a series of options that control the display and orientation of the model on the screen.
4. *Utilities.* The *Utilities* command is a potpourri of options, which include changing the colors of the interface and features, defining and loading map keys, and setting preferences.
5. *Window.* This standard windows command allows the user to toggle between active windows.
6. *Applications.* For users with a license to applications, such as Pro/Mechanica, this command allows the user to quickly interface with the application.
7. *Help.* The seventh command allows the user to activate Pro/E's on-line help system. Please note that *Help* uses a web browser to display the help contents. Thus, such a browser is necessary in order to load *Help.*

The tutorials that follow are designed to illustrate the use of these options.

### 1.1.1 TUTORIAL # 1: BECOMING FAMILIAR WITH THE INTERFACE

Initiate the software using the specific commands for your system. After the program has loaded, compare your interface to the one shown in Figure 1.6. Identify the main window and the message window.

The options *File* and *Working Directory* allow the user to show and set the default directory. On systems where the user is assigned a specific folder, these options should be used to connect to the default directory immediately after launching the software.

Let us set the working directory. Click on *File* and then *Working Directory.* The *Select Working Directory* box will appear. This box is reproduced in Figure 1.7. Scroll down and find your assigned folder. Then select *OK.*

Let us examine some of the options under the *Utilities* command. Select *Utilities.* You will see several options listed in the pull-down menu. Among these are *Environment, MapKey, Preference, Colors,* and *Customize Screen.*

Select *Environment.* This option will load the *Environment* dialog box (Figure 1.8), which contains toggles for turning on and off features such as the datum planes, axes, coordinate system, and points. These features will often clutter your model; therefore, you may wish to turn off their display by using the toggles available. You can always turn the display back on as needed.

Try turning the display of the datum planes off. Click on this toggle and notice the message in the message window: "Datum planes will NOT be displayed." Select the toggle again. Pro/E should respond with "Datum planes will be displayed."

Notice that the default for the display of the model is *Trimetric.* Many users prefer *Isometric.* Select the toggle for *Isometric.* Pro/E will display the message: "The default view is Isometric." Select the option again and Pro/E will respond with "The default view is Trimetric."

Notice the remaining options in the *ENVIRONMENT* menu. Take some time and turn these options on and off. As you do so, take note of the messages sent by the software. After doing so, select *Done-Return* from the *ENVIRONMENT* menu.

Among the options of interest found under the *Utilities* command, is the *Colors* option. This option allows the user to make changes to the color scheme. It is quite possible that you will want to change the color defaults

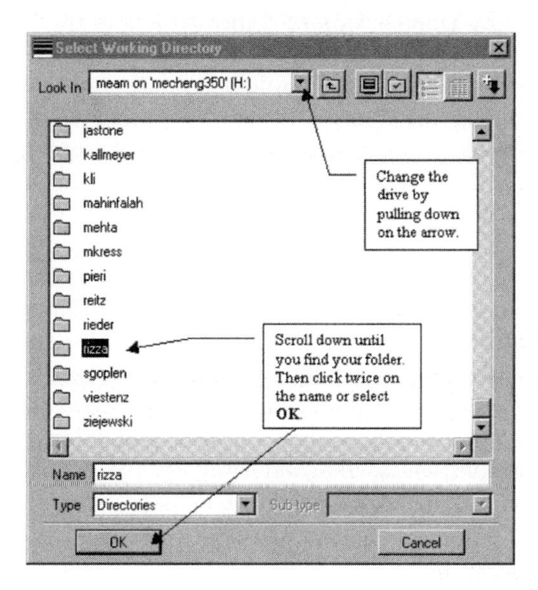


**FIGURE 1.7** *The* Select Working Directory *box is used to find and set the working directory.*

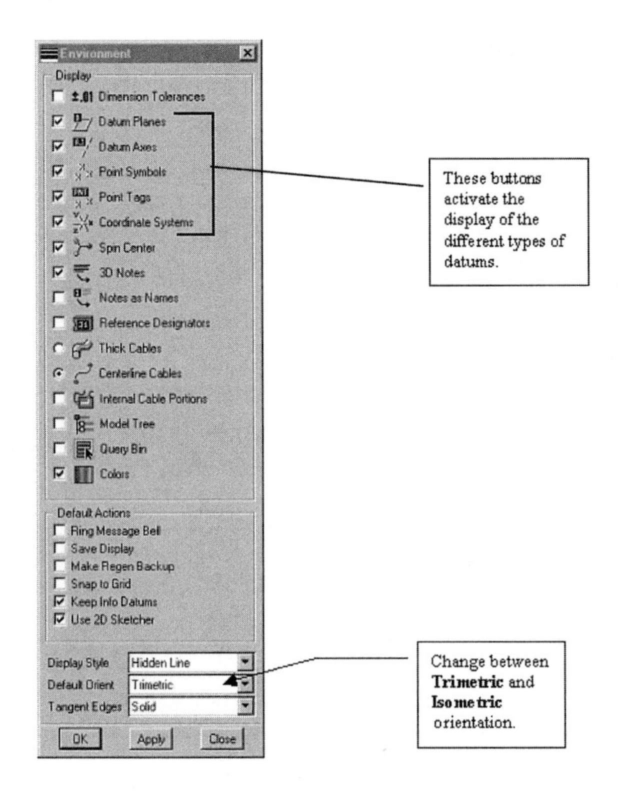


**FIGURE 1.8** *The* Environment *dialog box contains toggles for turning the display of certain features on and off. It also contains the field for changing the display orientation and style.*

for the features drawn by Pro/E, such as the datum planes. This change can be accomplished by the path *Utilities, Colors, Entity.*

Try this path. Choose *Utilities, Colors, Entity.* Note that the software has loaded the *Entity Colors* dialog box, which is reproduced in Figure 1.9a. The box is divided into two folders. The first folder is the *Datum* folder containing switches for modifying the colors of the various datum features. The second folder, *Geometry,* contains options for changing the color scheme of geometric entities. The colors for the various entities may be changed by selecting the button to the left of the entity. The *Set to Initial* button may be used to reinitialize the color scheme.

Choose the *Cancel* button. Now select *Utilities, Colors, System.* The software will load the *System Colors* dialog box illustrated in Figure 1.9b. The options contained within this box may be used to modify the system colors. With this option, you may customize the system display.

Pro/E has four built-in system color schemes that may be accessed by selecting the path *Utilities, Colors, System, Scheme.* The color schemes are *Black on White, White on Black, White on Green,* and the *Initial* setting. The *Default* setting is the color scheme visible when you first load the software. The option *Black on White* is used in this book.

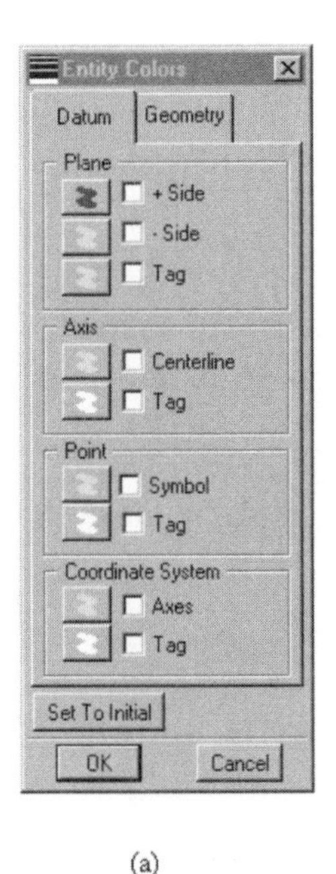


(a)

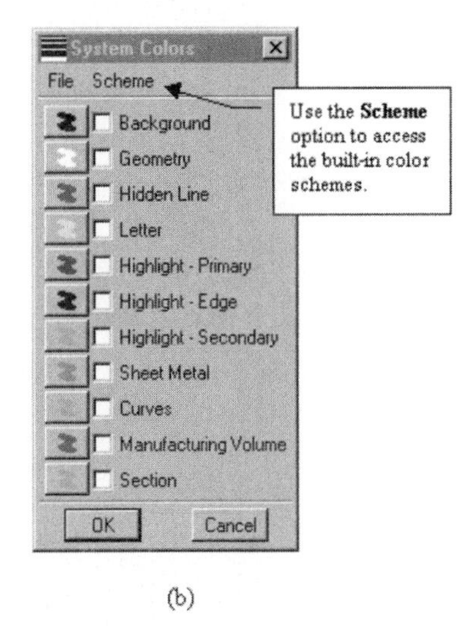


(b)

**FIGURE 1.9** *The color scheme for the system and the displayed entities may be modified by using the options found in the (a)* Entity Colors *dialog box and (b)* System Colors *dialog box.*

In general, be careful when you change any of the default settings. In this tutorial as well as those to follow, we will assume that you have *not* changed the default settings.

For some organizations, a configuration file is used. This file contains the options that control the operation of the software and interface. Assuming that you have been given the privilege to do so, you may edit this configuration file using the *Edit Config* option. After the configuration file has been edited, it may be loaded into memory using the *Load Config* option. The *Preferences* option displays a pull-down menu with the options *Load Config* and *Edit Config.*

You may wish to customize your display by adding additional icons to the toolbar or loading other toolbars that are not currently displayed. The *Utilities, Customize Screen* path may be used to achieve this goal.

Select *Utilities* and *Customize Screen.* The *Customize* box, reproduced in Figure 1.10, will become available. This box may be used to load built-in toolbars by clicking on the *Toolbars* folder and activating the toggle for the desired toolbar. In addition, the user may add icons to the toolbar by simply dragging the desired icon from the box to the toolbar.

We have found that it is helpful to have the *Default Orientation* icon on the toolbar. However, we do not wish to display the entire toolbar that contains this icon. The solution to this dilemma is to drag the desired icon onto the default toolbar.

In order to achieve this goal, click on *View* in the *Customize* box as shown in Figure 1.10. Several grayed icons will appear. These icons are gray because

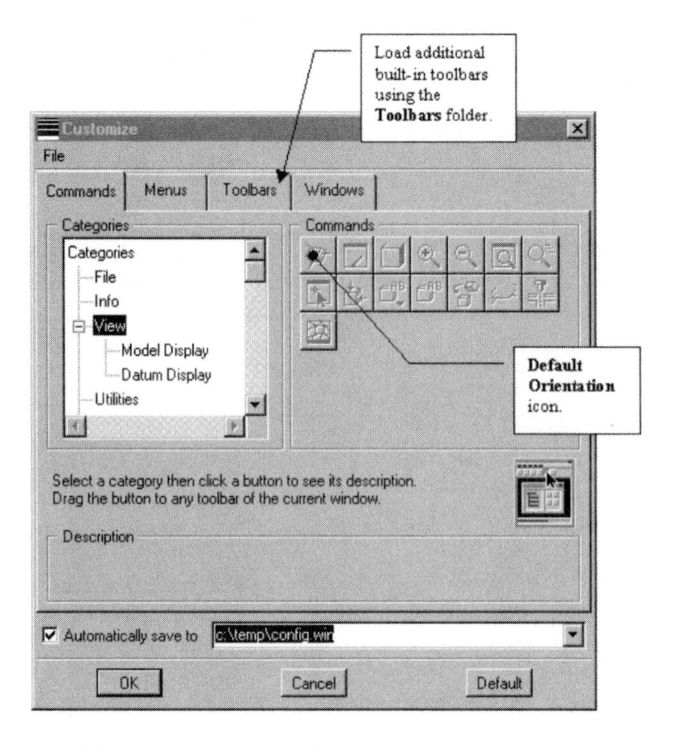


**FIGURE 1.10** *The commands available to the user may be customized by activating various built-in toolbars or by dragging desired icons to the toolbar currently displayed.*

a part is not active. However, one or more may be dragged onto the toolbar. The *Default Orientation* icon is the first icon from the left. Grab this icon and drag it onto the toolbar. Then select *OK.* If you wish, you may drag additional icons on to the toolbar using this approach.

The default icons are reproduced in Figure 1.11, along with a short description of their use. The description is also available by placing your mouse

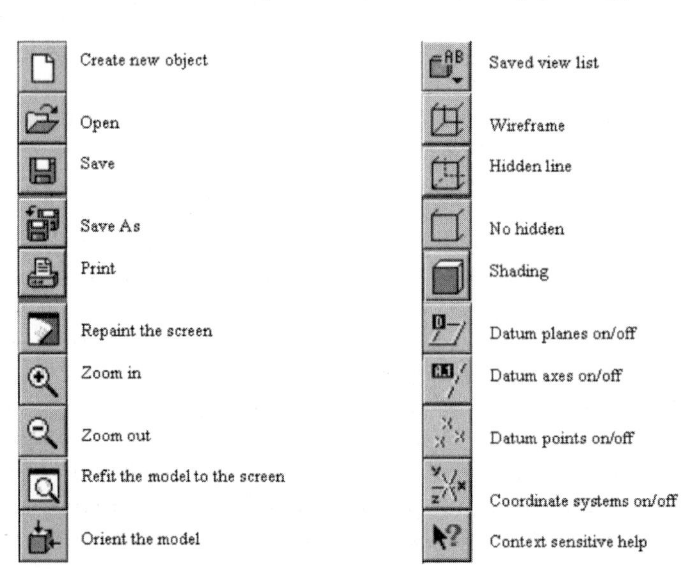


**FIGURE 1.11** *Several icons are available in the default toolbar.*

pointer over the icon. Try placing your mouse pointer over the first icon. In the message window, you should see: "Create new object."

## 1.2 PART FILES AND PRO/E

In Pro/E, part files are saved with the extension ".prt." Multiple versions of a part are allowed. The software will save each version with a version number appended to the end of the file name extension. For example, "BaseOrient.prt.1" is the first version of the part, "BaseOrient." In order to delete all but the current version from the disk use the path:

1. Select *File, Delete, Old Versions.*
2. Then, enter the name of the part.

The software will also delete *all* versions of a file on the disk. In this case, the path is:

1. Choose *File, Delete, All Versions.*
2. Then, enter the name of the part.

Pro/E loads all models into memory. Thus, it is possible that you will exhaust your computer's memory. It is a good idea to erase a model from memory after you are through working on it. In order to remove a model from memory use the sequence *File, Erase, Current.* If you wish to erase all the models except the currently displayed model, use the path *File, Erase, Not Displayed.*

### 1.2.1 TUTORIAL # 2: RETRIEVING "BASEORIENT"

We have bundled the part, "BaseOrient.prt" with this book. Before proceeding, you need to make sure that you have access to this part file.

Let us retrieve "BaseOrient." Make sure that your working directory is properly set. If not, follow the procedure:

1. Select *File* and *Working Directory.*
2. Scroll down the list of available folders or change the connection to the drive until you find the appropriate folder.
3. Double click on the name of the folder or click on the *OK* button.

Select *File* and *Open,* or use the *Open* icon. Scroll to find the file called "BaseOrient." In order to open the file, double click or highlight the name and select the *Open* button.

Pro/E will load the part and display the model in the main window. Let us also load the Model Tree for this part. Choose *Window* and *Model Tree.*

Pro/E normally draws the model tree window on top of the main window, which tends to get in the way of model creation. Therefore, you will need to drag the window off to the side or minimize it. In addition, the model tree window is usually too large and should be resized before moving. Consider Figure 1.12, resize the model tree and move it to the lower right-hand side of your screen.

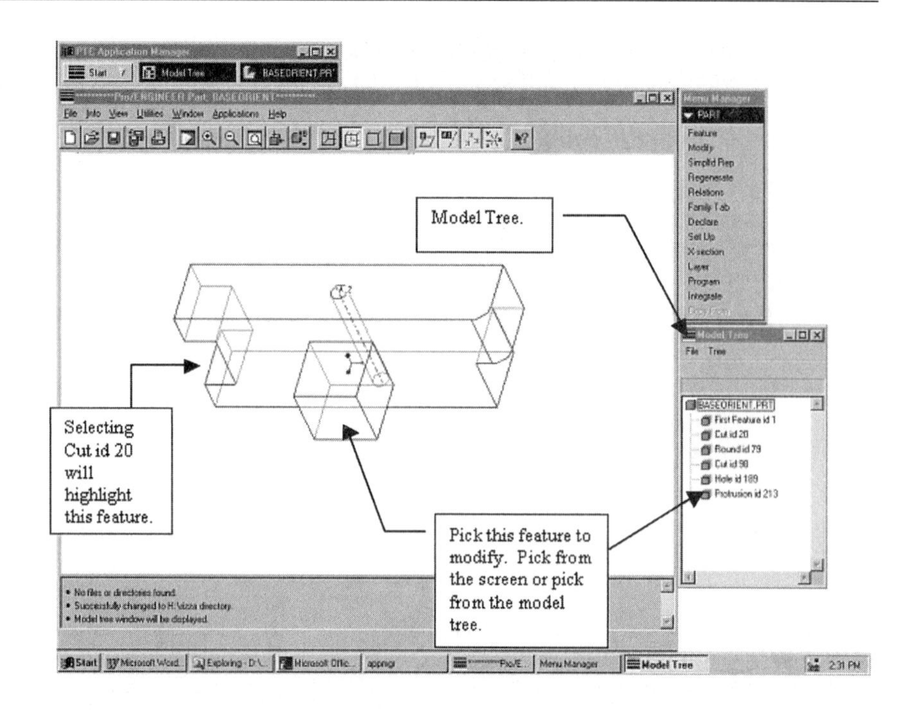


**FIGURE 1.12** *The part, "BaseOrient," loaded in Pro/E. Note the window contains the* Model Tree *that has been resized and moved to the lower right-hand side of the screen.*

Every part has a *Model Tree.* The model tree lists all the features in the model and the order in which they were created. The model tree can be used to select features on the model by clicking the element in the list. With your mouse, select the feature in the model tree list called "Cut id 20." Pro/E will highlight the feature (the default color is red), as shown in Figure 1.12.

Now, select *Modify* from the *PART* menu. Recall that Pro/E is a parametric modeler. That is, the models are constructed using parameters. We are going to change the value of one of these parameters and see what happens.

Select the box-shaped feature by either clicking on the feature with your mouse or by selecting the item in the model tree ("Protrusion id 213") as shown in Figure 1.12. Pro/E will display the dimensions of the feature as shown in Figure 1.13. Select the 6.00 dimension. Enter a value of 8.00. Hit *Return.* Then choose *Regenerate* from the *PART* menu.

Upon regeneration, the software will update the model. The updated model is shown in Figure 1.14.

Suppose we wanted to save this new version of the model, but under a different name. We can easily do this by using the *Save As* option. Go ahead and save the modified version using the procedure:

1. Select *File* and *Save As,* or the *Save As* icon.
2. In the space provided, enter the new name: "BaseOrient2."
3. Choose *OK.*

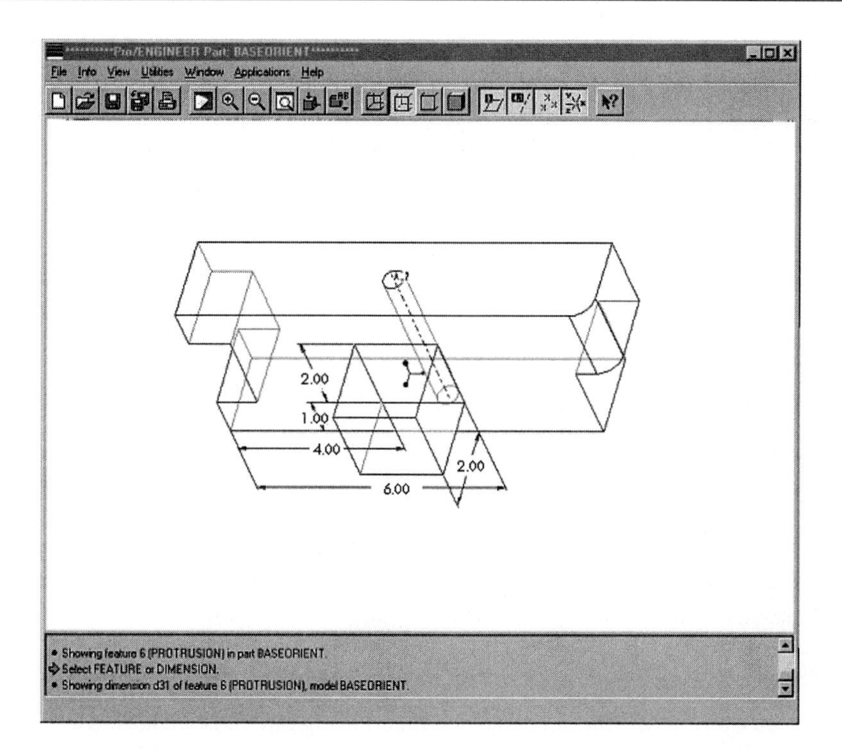


**FIGURE 1.13** *The main window with the "BaseOrient" part. The dimensions of the box-like feature ("Protrusion id 213") are displayed.*

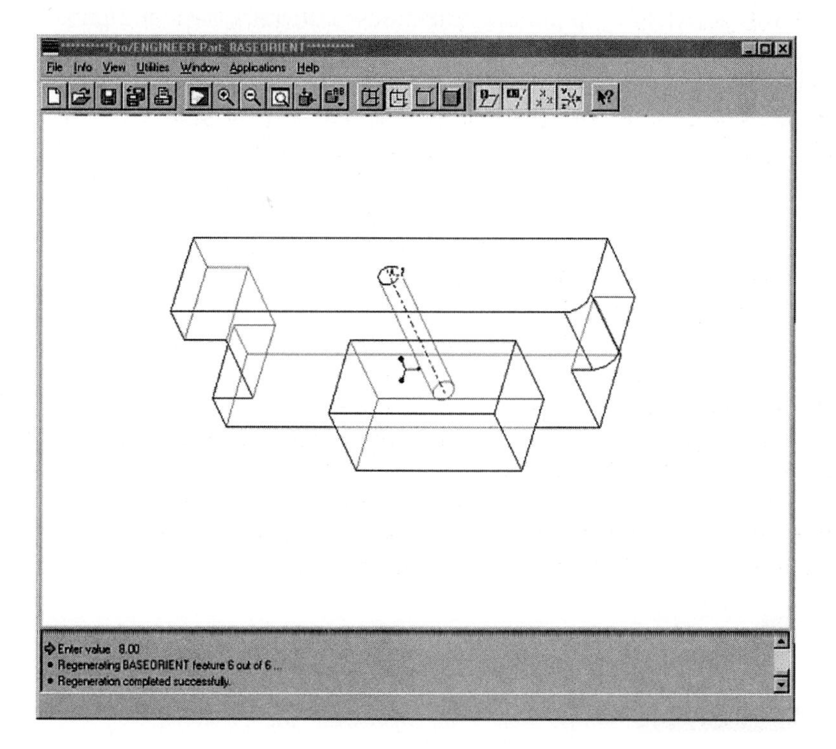


**FIGURE 1.14** *After modifying the 6.00 dimension to 8.00, the model changes in appearance.*

Notice that Pro/E does not automatically update the model in memory to the new name. By glancing at the top of the interface, you will see that the name of the model is "BaseOrient," not "BaseOrient2."

In order to see whether the model was properly saved, let us erase "Base-Orient" from memory and load "BaseOrient2." The sequence for accomplishing this task is:

1. Select *File, Erase, Current.*
2. Click on the *Yes* button.

In order to load "BaseOrient2," follow the steps:

1. Select *File* and *Open.*
2. Scroll down and find "BaseOrient2."
3. Double click on the file or hit *OK* after clicking once.

Does this part contain the modified protrusion? It should. If so, erase this part. In the next tutorial, we will retrieve "BaseOrient" again and manipulate its orientation.

## 1.3 CHANGING THE ORIENTATION OF A MODEL

The orientation of a model can be changed using one of two methods. If you have a three-button mouse, the orientation can be changed by using the Control (CTRL) key and one of the mouse buttons. Consider Figure 1.15. The Control key plus

1. The left mouse button allows user to *Zoom In* and *Zoom Out.*
2. The middle mouse button allows the user to *Rotate.*
3. The right mouse button allows the user to *Pan.*

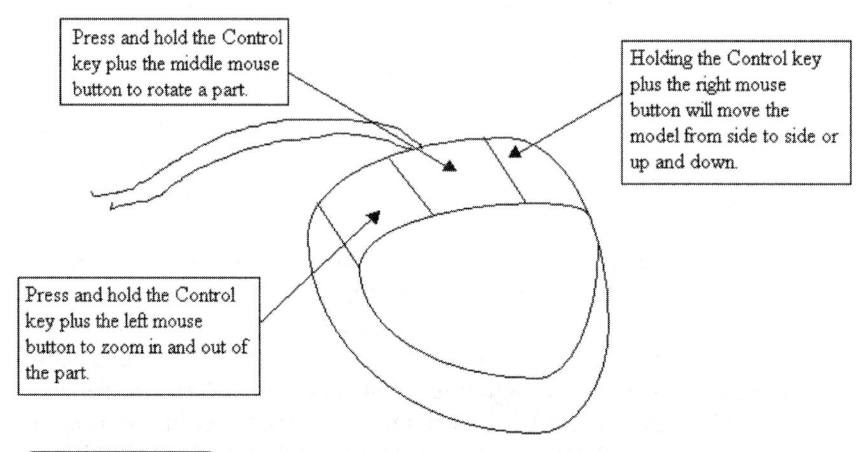


FIGURE 1.15 *Mouse showing the predefined button features.*

Furthermore, some of the icons shown in Figure 1.11 also may be used to change the orientation of the model. These icons are:

1. The *Zoom In* icon will allow the user to zoom in to a portion of the model. The user must select the region to zoom. To select an area, drag a window across the region of interest.
2. The *Zoom Out* icon performs the reverse operation of the *Zoom In* icon. All the user needs to do is click on the icon. Every time the icon is selected, the zoom out action is performed.
3. The *Refit* icon may be used to resize the model so that it fits into the window.
4. Finally, the *Orient* icon will launch the Orientation dialog box, which may be used to reorient the model by defining different views. This approach will form the basis of the next tutorial.

### 1.3.1 Tutorial # 3: Changing the Orientation of "BaseOrient"

Retrieve the part "BaseOrient." Use the path *File* and *Open.* Now, let us see how the mouse can be used to change the orientation. Press the Control key and select anywhere on the model with your mouse pointer. Keeping the left mouse button depressed move the mouse back and forth. The model should grow larger as you pull back and smaller as you push forward.

Now, let go of the left mouse button and depress the middle button. Again, move the mouse. This movement should rotate the model.

Finally, after letting go of the middle button, depress and hold the right button. Move the mouse to the right; the model should move to the right. Move the mouse to the left; the model should move to the left. If you push the mouse away from you, the model should move up. Move the mouse toward you; the model should move in the downward direction.

Try the *Zoom In* and *Zoom Out* icon buttons. Select *Zoom In* and drag the mouse across the model. After you release the mouse button, the software will zoom in on the chosen region of the model. Try clicking on the *Zoom Out* icon several times. After you are satisfied with the operation of these icons, choose the *Refit* icon. This option will reorient the model in the default orientation.

The model can also be reoriented by defining orthographic views. In defining such views, two orthogonal surfaces are needed. For models that do not have planar surfaces, datum planes can be used. We discuss datum planes in Chapter 3.

The part "BaseOrient" contains several planes. In the following paragraphs, we will use these planes to reorient the model.

This approach makes use of the *View* and *Orientation* commands, or the *Orient* icon. Consider the actions described in Figure 1.16. According to the figure, by properly defining the *"Top"* and *"Front"* the model can be reoriented. Notice that other possibilities present the ability to obtain the same orientation, because the model has multiple planar surfaces.

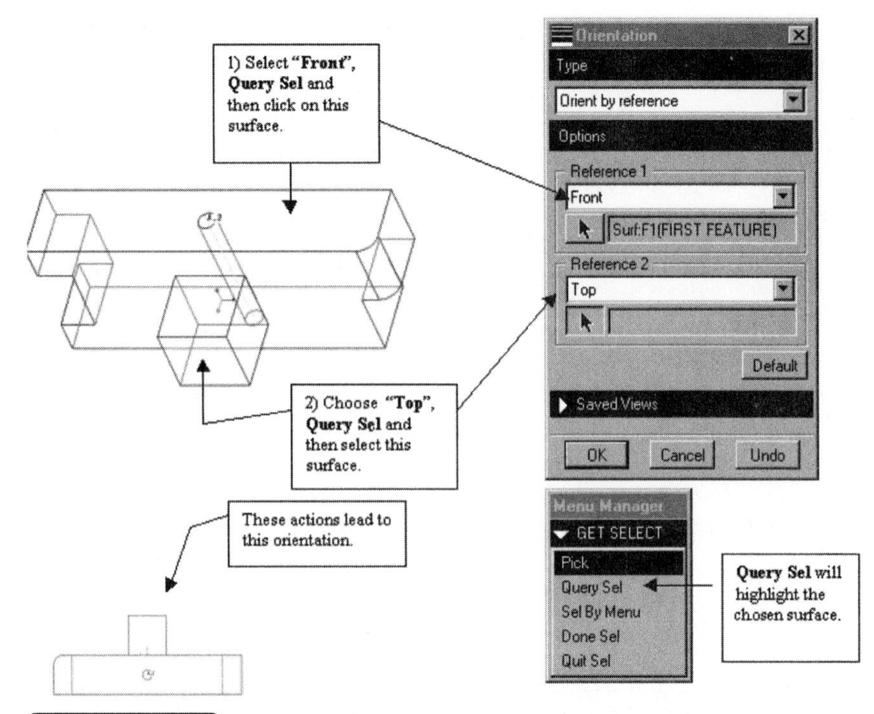


FIGURE 1.16 *The model can be reoriented by using appropriate selections from the* Orientation *dialog box.*

Perform the actions as indicated in the figure. Did the model reorient as shown?

It is possible to save the view. Click on *Save Views* in the *Orientation* dialog box (Figure 1.16). The software will list the defined views. Note that the only entry is *Default.* In the *Name* field, type in Front, which will then be the name of the view. Then click on the *Save* button. Select *OK.*

Let us see whether the view has been saved. First, restore the default view by selecting *View* and *Default.* Then, select the *Saved View List* icon (shown in Figure 1.11) or use the path *View* and *Saved Views.* The software will load a list containing the saved views. Select Front. Did the model reorient to the defined view?

## 1.4 SOLID MODELS

A solid model can be created for an active model by using the *View, Shade,* the *Shading* icon, or by selecting *View, Advanced, PhotoRender.* The option *Shade* provides a shaded solid model, while *PhotoRender* produces a solid model that is photo quality of the object. In this case, the model is displayed against a background.

The construction of the solid model is produced by using a virtual light source to illuminate the wireframe. The solid model that is produced depends on the location of this light source and its intensity. The options for the light source may be changed by using the *View, Model Setup,* and *Lights* options.

Because the *PhotoRender* option is graphics-intensive, it consumes a large amount of run time on the computer. It will not be used in this book. Instead, we will focus our attention on the *Shade* option.

## 1.4.1 TUTORIAL # 4: A SOLID MODEL OF "BASEORIENT"

Let us produce a shaded model of the part "BaseOrient." Select *View* and *Shade.* The software will use the default settings to create a shaded model of the part. Of course, the settings for the display can be changed.

In an assembly that is made up of multiple parts, it is advantageous to have each part a different color for greater visibility. Examine Figure 1.17a and 1.17b. Notice how the use of different colors leads to a more visible model.

The appearance of a shaded model may be changed by selecting a different color scheme from the palette. If the desired color scheme does not exist in the palette, it may be created and added to the palette.

Let us create a new color scheme, add it to the palette, and shade the model using the new color scheme. Select *View, Model Setup,* and *Color Appearances.* Note the default color scheme in the palette.

Choose *Add.* The software will load the *Appearance Editor.* The appearance of the shaded model is changed by using the options from the *Appearance Editor.* Both the palette and *Appearance Editor* are reproduced in Figure 1.18.

Let us change the color. Select the *Basic* folder and then the color button shown in Figure 1.18. Pro/E will load the *Color Editor.* The color is changed by increasing or decreasing the relative values of red (R), green (G), and blue (B). Move the sliders until the red is set to 69.00, the green to 145.00, and the blue to 179.00. Then select *OK.*

(a)

(b)

FIGURE 1.17 *The individual parts in the assembly are more visible in (b), because separate colors have been assigned to each part.*

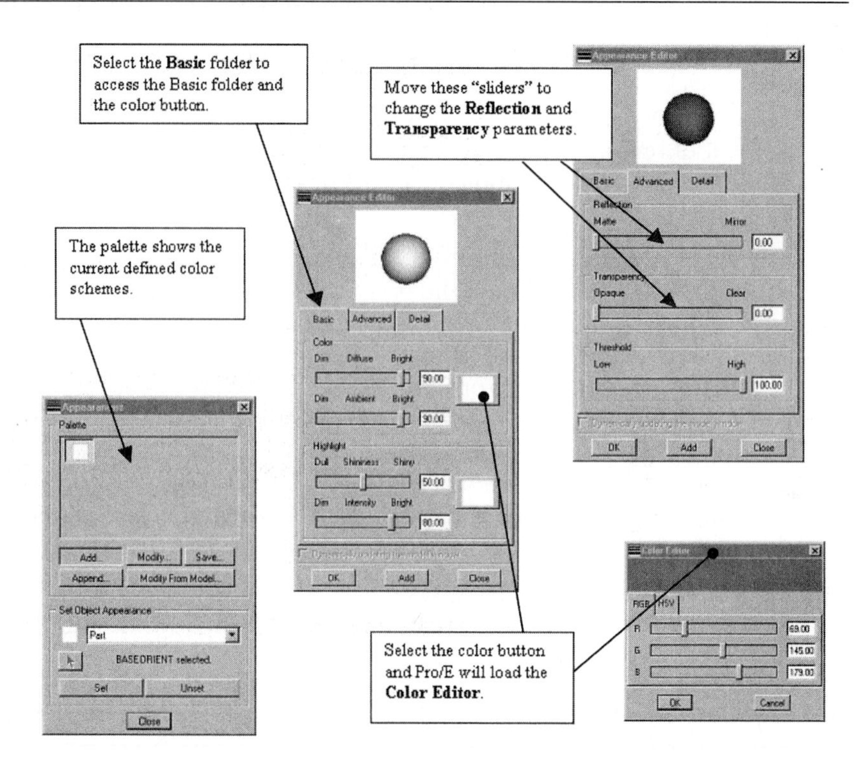


**FIGURE 1.18** *The color scheme for shading the model may be changed by using the options in the* Appearance *and* Color Editor. *The current available color schemes are shown in the palette.*

Change the *Reflection* from *Mirror* to *Matte,* by moving the slider shown in Figure 1.18 until the value is 0.00. Change the transparency from *Clear* to *Opaque* by moving the *Transparency* slider until the value is 0.00. Select *Add.* Pro/E will add the new color scheme to the palette.

After the color has been placed in the palette, it may be retrieved from the palette and used to redefine the shading scheme for the model. Click on the new color with your mouse. Select *Set* from the *Appearances* box. Click on the *Close* button. Notice, on the model, that the lines have changed color. Shade the model by selecting the *Shade* icon or the path *View* and *Shade.* Now choose *View, Model Setup,* and *Color Appearances.* Choose *UnSet* to restore the model to the original color scheme. Select *Close.*

## 1.5 THE USE OF MAPKEYS

Pro/E allows for the use of hot keys (*Mapkeys*) to help in navigating the menus. The default configuration file (Config.pro) supplied by PTC has numerous predefined *Mapkeys.* You might want to check this file for the appropriate *Mapkeys.*

The software also allows the user to define new keys or modify old ones in two different ways. The first approach is to edit the configuration file by using *Utilities, Preferences,* and *Edit Config.* Then, the commands for the keys are inserted into the configuration file. The format of the commands is:

Mapkey keyname #command; #command; . . .

In this statement, #command is the Pro/E command, and keyname is the name of the key to map. When using the function keys, you must use the form $Fn. In this case, the n is the number of the function key.

For example, the F1 key may be mapped to the *View, Environment, Isometric* commands by adding:

Mapkey $F1#View; #Environment; #Isometric; OK;

By defining this sequence of commands, every time the user hits the F1 key, the model on the screen will be reoriented to appear in *Isometric.*

The second approach is to add the *Mapkey* interactively by using the *Utilities* and *Mapkey* commands. This method is preferred, because it is more user friendly. This approach forms the basis of our next tutorial.

## 1.5.1 TUTORIAL # 5: CREATING MAPKEYS INTERACTIVELY

It is not necessary to have an active model when creating *Mapkey.* In this case, however, the model is needed to see the result of using the *Mapkey.* So, before proceeding any further, load the part "BaseOrient." The path is *File* and *Open.*

Now, let us map the keys "IS" so that depressing these keys will reorient the model in isometric orientation. Click on *Utilities* and *Mapkeys.* The software will load the *Mapkey* dialog box shown in Figure 1.19. Notice that the currently defined *Mapkeys* are listed under the heading "Mapkeys in Session." These *Mapkeys* may be modified by using the *Modify* button.

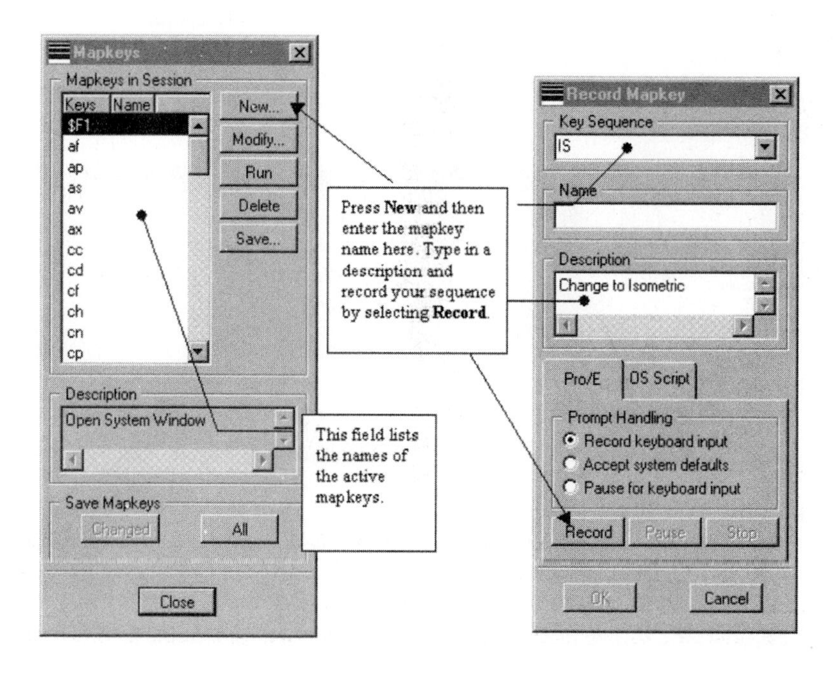


FIGURE 1.19 Mapkeys *are modified or created interactively by using the* Mapkey *and* Record Mapkey *dialog boxes, respectively.*

Select the *New* button. The software will load a new dialog box called the *Record Mapkey* box. It is the second box shown in Figure 1.19.

Type "IS" into the field provided. Also, add the text "Change to Isometric" in the description field. Then click on *Record.* Using your mouse, select the sequence *Utilities, Environment, Isometric,* and *OK.* Then select *Stop* and *OK* from the *Record Mapkey* dialog box.

Scroll through the list of *Mapkeys* in the *Mapkey* dialog box. You should see the letters "IS."

Now let us check to see if the new *Mapkey* works. If the model was reoriented during the process of creating the *Mapkey,* select *Utilities, Environment, Trimetric* and *OK.* This will reorient the model in *Trimetric.* Then type in the letters "IS." The model should reorient to an *Isometric* orientation.

## 1.6 PRINTING

In order to print (plot) in Pro/E you must have an active object. Therefore, if you want to print a certain model you must first retrieve it. Let us assume that the object is active.

In order to print, use the following steps:

1. Select *File* and *Print* or click on the *Print* icon.
2. Pro/E will load the *Print* box shown in Figure 1.20.

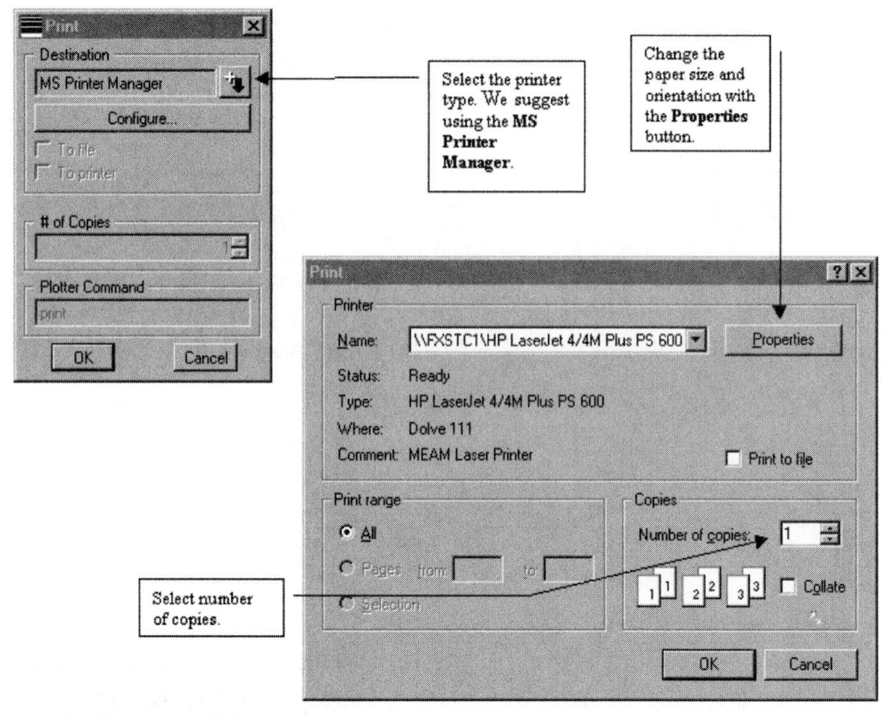


FIGURE 1.20 *The* Print *box is used to set the properties for the print job.*

3. Set your printer type in the *Destination* field. Because most organizations use PostScript printers, we suggest using the *MS Printer Manager.* This choice will use your system printer drivers.
4. Choose *OK.*
5. The software will load the *MS Print* box. Select the number of copies. If you wish to change the orientation or the paper size, use the *Properties* button.
6. Choose *OK.*

### 1.6.1 TUTORIAL # 6: PRINTING A COPY OF "BASEORIENT"

Print a copy of "BaseOrient" in default *Trimetric* orientation. Check to see whether your organization has special printer requirements. If you are connected to a PostScript printer, use the steps in section 1.6.

## 1.7 SUMMARY AND STEPS FOR USING VARIOUS INTERFACE OPTIONS

Our objectives in this chapter were to introduce the user to the Pro/E interface. We gained experience in customizing the interface by using the options contained within the *Utilities* command.

The first of these options was the *Environment* option, which is used to activate or deactivate the display of various features including datum features. The steps for accessing the *Environment* option are:

1. Select *Utilities* and *Environment.*
2. Turn the desired toggle on or off.
3. Choose *OK.*

We also spent some time investigating the methods and options for changing the colors of the display. The colors of the palette may be modified by using the *Utilities* and *Colors* options. In order to change the colors of the various entities, use the path *Utilities, Colors,* and *Entity.* In order to modify the color of the interface use the path *Utilities, Colors,* and *System.* The built-in system color schemes may be set by using the path *Utilities, Colors, System,* and *Scheme.*

In Pro/E, a part is saved in a part file with the extension ".prt." This part file may be loaded into memory by using the sequence:

1. Select *File* and *Open,* or the *Open* icon.
2. Scroll through the list of available files and highlight the desired one.
3. Double click or select the *Open* button.

It is quite possible that you may have been assigned your own work directory. The user may connect to the work directory. This approach saves time in searching for a particular file. In addition, all saved files will be placed in the work directory. The user may connect to the work directory by doing the following:

1. Select *File* and *Work Directory.*
2. Scroll through the list and find the appropriate folder. Change the drive designation if necessary.
3. Choose *OK.*

Because Pro/E is a feature-based modeler, it is a simple matter to change the dimensions of a part. The change is made by doing the following:

1. Select *Modify* from the *PART* menu.
2. Click on the desired dimension.
3. Enter the new value of the dimension.
4. Click on *Regenerate* in the *PART* menu.

A model may be reoriented by using the mouse button or by defining specific views of the model. These views may be saved. In order to define a view:

1. Select *View* and *Orientation,* or the *Orient* icon.
2. Choose the first reference. Use *Query Sel* to properly choose the reference.
3. Select the second reference again using *Query Sel.*
4. Click on the *OK* button.

In order to save the view:

1. Click on *Save Views* in the *Orientation* dialog box (Figure 1.16).
2. In the *Name* field, type in the name of the view.
3. Then click on the *Save* button.
4. Select *OK.*

## 1.8 ADDITIONAL EXERCISES

1.1 List the primitives shown in Figure 1.4 that may be used to create the three-dimensional model of the part "BaseOrient." Sketch how the primitives would be arranged to form the model.

1.2 List and sketch the extrusions that would be used to construct the part "BaseOrient" if a constraint-based modeler is used.

1.3 In Tutorial #3, the part "BaseOrient" was reoriented using the *"Front"* and *"Top"* options. List three more combinations that may be used to achieve the same orientation.

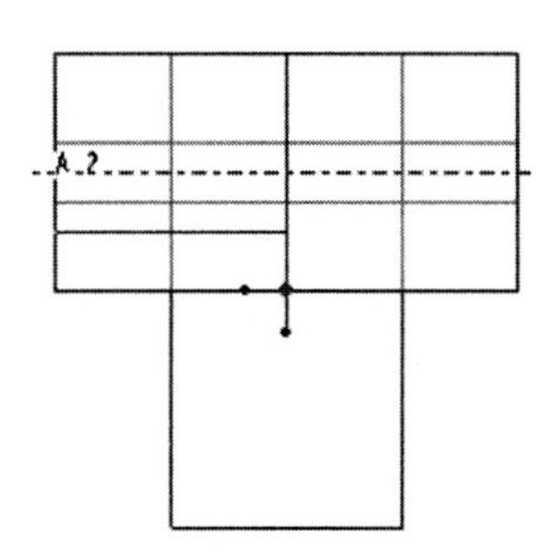


FIGURE 1.21 *Figure for exercise 1.3.*

1.4 Reorient the part "BaseOrient" so that the orientation in Figure 1.21 is obtained. Print out a copy of your work.

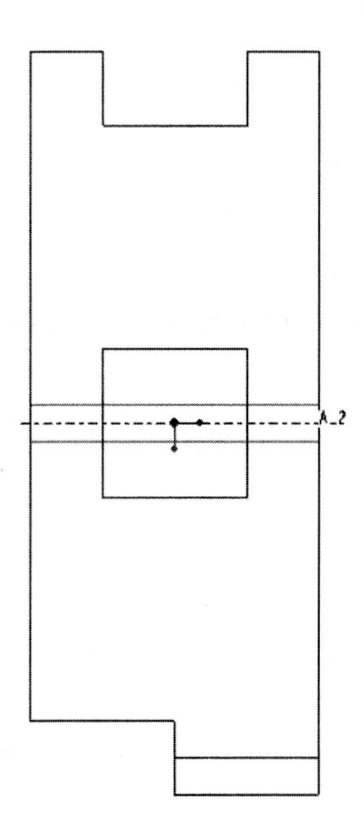


FIGURE 1.22 *Figure for exercise 1.4.*

1.5 Reorient the part "BaseOrient" so that the orientation in Figure 1.22 is obtained. Print out a copy of your work.
1.6 Modify the diameter of the hole in the part "BaseOrient" from 0.50 inches to 0.75 inches. Print out a copy of the result.
1.7 Change the 4.00 dimension to 4.50 in Figure 1.13. Print out a copy of the result.

# CHAPTER 2

# CONSTRUCTING PARTS USING THE SKETCHER

## INTRODUCTION AND OBJECTIVES

In Pro/E, features that are created from a section utilize the *Sketcher.* In the *Sketcher,* the user sketches the section using geometric entities such as lines, arcs, and circles. Then the section is dimensioned and regenerated to yield a two-dimensional parametric model. Upon successful regeneration, the section is made three-dimensional by extruding into a third depth, revolving about an axis or sweeping along a trajectory.

The objectives of this section are to:

1. Introduce the user to the *Sketcher* interface.
2. Construct some simple features using the *Sketcher.*
3. See how constraints and relations affect the regeneration process.
4. Introduce advance entities such as conic sections.
5. Create two-dimensional sections using the *Sketcher.*
6. Retrieve sections in the *Sketcher* when creating a part.

In this chapter, we will confine ourselves to the construction of solid or thin extrusions. Models created by using a revolution about an axis or a sweep along a trajectory will be treated in later chapters.

## 2.1 PARTS AND THE SKETCHER

The entry point into Pro/E for creating three-dimensional models is via the *PART* menu, as shown in Figure 1.12. The part can be constructed by drawing a two-dimensional sketch of the feature and then adding depth to the sketch or revolving around a centerline. For example, a protrusion is constructed by sketching a two-dimensional profile of the object in the *Sketcher* and then extending the sketch into the third dimension by adding depth.

In Pro/E, sketches can be either *Solid* or *Thin* (consider Figure 2.1). *Solid* is the default. The *Thin* option is used to create sections that have a small thickness. An object created with the *Solid* option must be closed. That is, the two-dimensional sketch must have an inside and an outside. *Thins* do not need to be closed. In the following sections, we will consider both solid and thin sections.

The interface of the *Sketcher* is shown in Figure 2.2. Note the three additional icons. These icons and their meanings are given in Figure 2.3. Figure 2.2 illustrates the section for the handle to be constructed in Tutorial # 7. Note the Sd0 parameter, which is the parameter name for the diameter of the circle. The circle was sketched using the *Sketch* option in the *SKETCHER* menu. The *Dimension* option was used to place the dimension. The software gave this dimension the name Sd0. When the *Regenerate* option is used, the software will use the spacing of the grid and sketch to determine a value for Sd0. The parameter may be modified by using the *Modify* option.

Consider Figure 2.4. Pro/E has predefined the mouse buttons for creating basic elements. The left mouse button will draw a line, the middle button a circle, and the right button an arc.

## 2.1.1 TUTORIAL # 7: A VISE HANDLE

If you have not already done so, load the software. If your system requires that you set the default directory, don't forget to do so. If so, use the path:

1. Select *File* from the Toolbar.
2. Choose *Working Directory.*
3. Scroll and find your folder.
4. Choose *OK.*

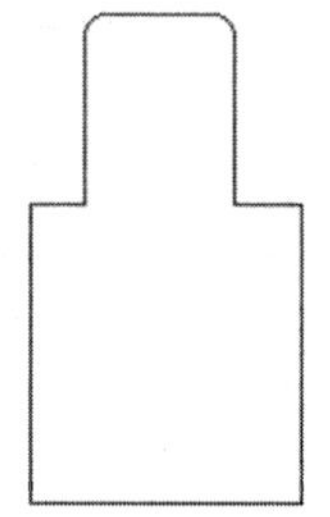

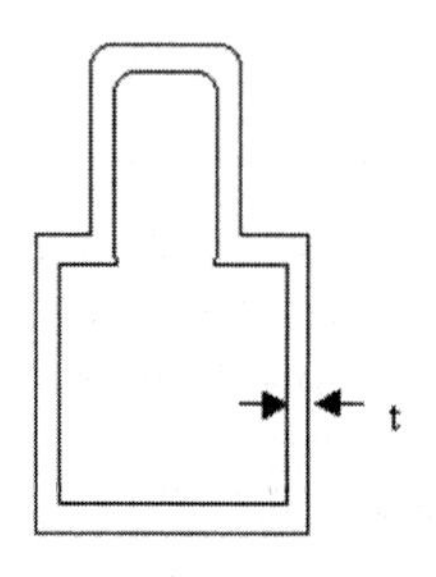


Thin section. The thickness is "t".

**FIGURE 2.1** *Both solid and thin sections may be constructed in the* Sketcher. *Thin sections have a constant thickness.*

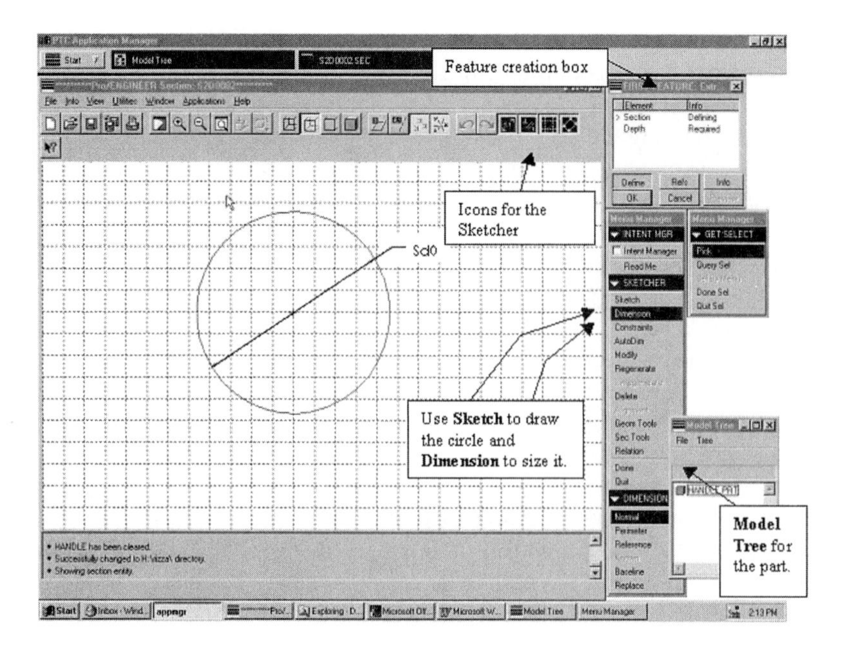


**FIGURE 2.2** *The section for the part "Handle" in the* Sketcher *has been drawn by using the* Sketch *option. A diameter dimension has been placed. Notice the variable sd0 that is the parameter for the diameter.*

Our first part is a handle for a vise. The figure illustrating the handle is given as Figure 2.3. The handle is simply a cylindrical protrusion created by drawing a circle with a diameter of 0.25 inches for the cross section and extruding (projecting) the section to a given depth of 6.00 inches.

Create the part by selecting *File* and *New.* Make sure that the file type is set to "*Part.*" Then enter "Handle." Click on the *OK* button.

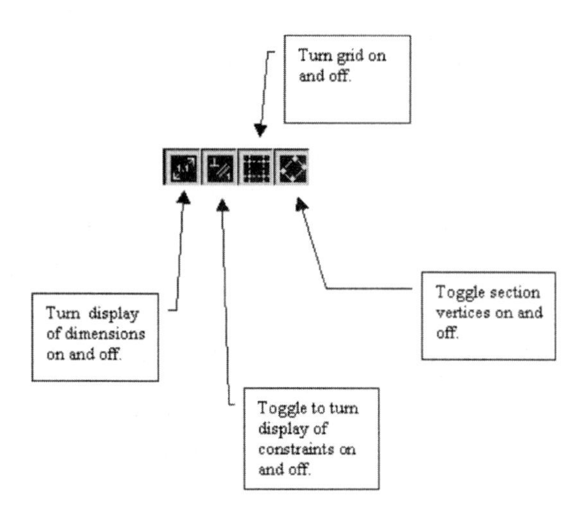


**FIGURE 2.3** *In default mode, the* Sketcher *contains four new buttons.*

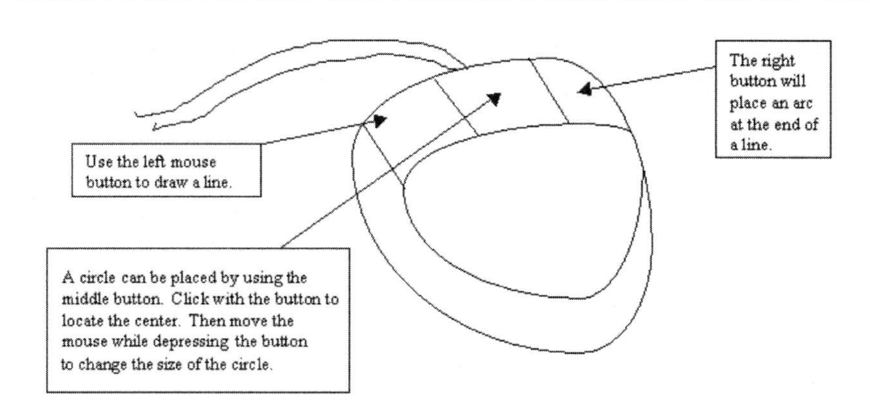


FIGURE 2.4 *The mouse with predefined drawing buttons.*

Pro/E will create a file named "handle.prt" and load it into memory. A *Model Tree* for the part will appear (shown in Figure 2.2). Notice that the first entry in the *Model Tree* is the name of the part. Resize this window and move it below the *PART* menu. Select *Feature.* Now choose *Create.* Choose *Protrusion.* Notice that the option *Extrude* is the default. We want to do an extrusion here. The other options, *Revolve, Sweep,* and *Blend,* will be considered in later tutorials.

Accept the defaults by clicking on *Done.* You should now be in the *Sketcher* mode. A grid with default spacing of horizontal and vertical lines should cover the main drawing window (Figure 2.2). Several new windows should be visible on the right-hand side of your screen.

The first of these new windows is the *Feature* creation dialog box shown in Figure 2.2. This dialog box lists all the elements required in the creation of the feature and their status. Notice the status on the two entries in the list. The section is being defined; that is, the program has entered the *Sketcher* where the user can sketch the section defining the feature. In addition, the depth of the feature is required. You can keep track of the necessary information needed to construct the feature by examining this dialog box.

After you have entered the required information, the status of the element will be changed. You can always change incorrect data by highlighting the element with the incorrect information and then clicking on *Define.* Pro/E will bring up the necessary windows or respond with the necessary queries in the message window so that you can change the incorrect information. The option *Refs* loads the associated references for the entire feature. Thus, you can keep checking how the feature is associated with other elements in the model. The button *Info* displays the information on the feature. The button *OK,* which is currently grayed because we have not entered all the required elements, is used upon completion of the feature to return to the *Part* mode. The option *Preview,* also currently grayed, is used to see the feature as it would be built with the defined elements. If the feature does not appear satisfactory, you can use *Define* to change some or all of the elements.

Now let us create the section and dimension it to proper value. Use the following steps to complete the task:

1. Use your middle mouse button and place a circle. Create the section so that it appears similar to the one shown in Figure 2.2.
2. Select *Dimension.* Click on two sides of the circle with the left mouse button and then place the dimension where desired by clicking with the middle button.
3. Let us associate a numerical value with the parameter Sd0. From the *SKETCHER* menu, select *Regenerate* and click.
4. Choose *Modify* and enter the value 0.25 as shown in Figure 2.5.

You may now choose *Done* from the *SKETCHER* menu. Notice in the feature creation box, that the status of the section has changed to "*Defined.*" Also, notice that the status of *Depth* has changed to "*Defining.*"

Enter the depth of 6.00 in the message window. Pro/E will extrude the circle to that depth. You may preview the construction before leaving the *Sketcher* by clicking on *Preview* in the feature creation box. Choose *OK* when you are satisfied with your work. Pro/E will return to the *Part* mode. Your handle should appear like the one shown in Figure 2.6, without the dimensions.

Congratulations! You have completed your first model using Pro/E! Save the part by selecting *File* and *Save.*

**FIGURE 2.5** *The parameter Sd0 may be modified to the proper dimension of 0.25 inches.*

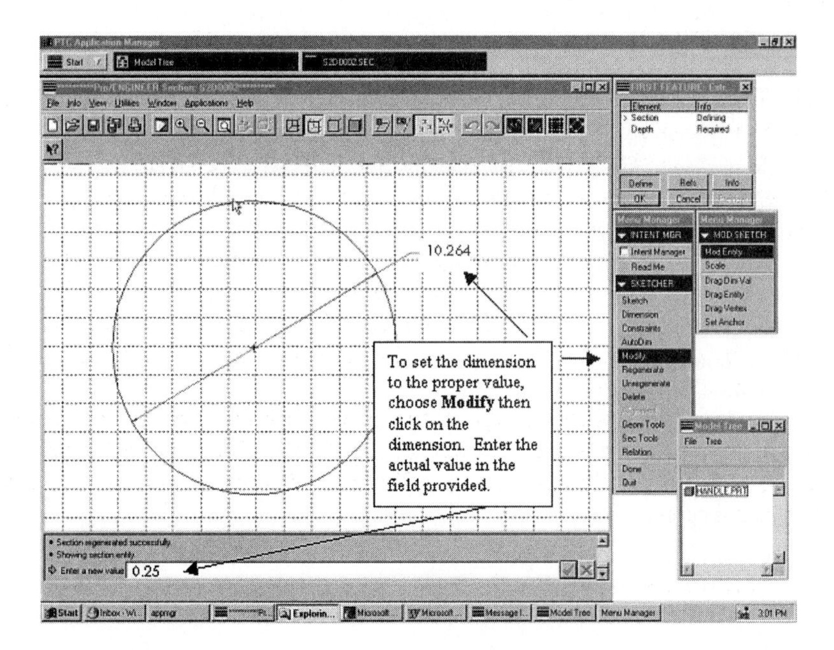

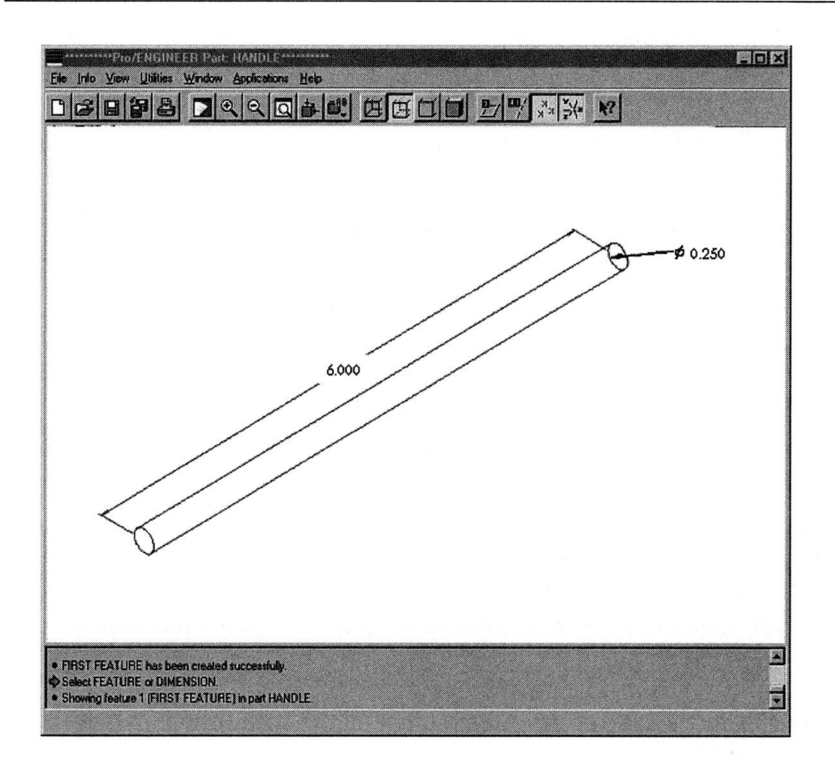


FIGURE 2.6 *The part "Handle" after extruding to a depth of 6.00 inches.*

## 2.2 SYMBOLS USED IN THE REGENERATION PROCESS

The regeneration phase is the most critical one in the sketching process. Depending on your sketch and dimensions, you may or may not get what you want. In addition, Pro/E may not always be able to generate the feature. Often the problem is a poorly dimensioned sketch.

It must be remembered that when you regenerate the sketch, Pro/E is attempting to interpret your rough sketch to produce the desired feature. If the feature fails to regenerate, it is because Pro/E is unable to interpret the sketch.

In the dimensioning of the sketch, all the pertinent elements must be located. If the sketch is underdimensioned, Pro/E will respond with "Underdimensioned section. Please add dimensions." Often Pro/E will notify you of which element is not dimensioned by changing its color. The default color in such a situation is red.

After a successful regeneration, Pro/E will display symbols that denote the various constraints used in interpreting the sketch drawn by the user. A listing of these constraints is given in Table 2.1. These constraints are based on several rules:

1. If you draw two arcs or circles with approximately the same radius, the arcs or circles are assumed to be of the same size.

TABLE 2.1 *SKETCHER* CONSTRAINT SYMBOLS

| SYMBOL | CONSTRAINT |
|---|---|
| "V" | Entity assumed vertical |
| "H" | Entity assumed horizontal |
| "L" with index | Entities with same index are equal length |
| "⊥" | Entities with perpendicular symbol are perpendicular |
| "//" | Entities with parallel symbol are parallel |
| Thick dashes | Points have equal coordinates |
| "T" | Entities are tangent |
| →← | Symmetric |
| "R" with index | Entities are of equal radii |
| —O— | Point entity |

2. If you draw a centerline, entities will be assumed symmetric about that centerline.
3. Lines approximately horizontal or vertical are assumed horizontal or vertical.
4. Entities will be assumed to be tangent if sketched approximately tangent.
5. Lines of roughly the same length will be assumed to be of the same length.
6. Point entities near other entities such as lines, arcs, and circles will be assumed to lie on that entity.
7. Endpoints of arcs and lines may be assumed to have the same coordinates.

Notice the "V" symbol in Figure 2.7. Pro/E is letting the user know that it is assuming the lines to be vertical. These symbols are based on the software's interpretation of the sketch drawn by the user.

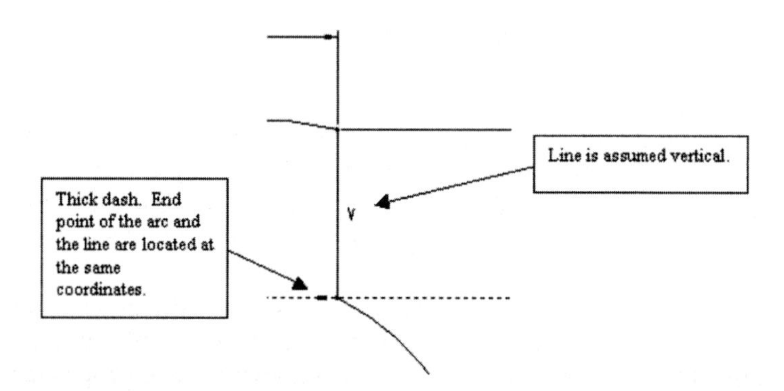


FIGURE 2.7 *The software uses symbols to convey its interpretation of the sketch. The symbol "V" in the figure indicates that the software is assuming a vertical line.*

Pro/E will assume that all lines close to the vertical (within an accepted tolerance) are vertical, and all lines that are close to horizontal are horizontal. Thus, you do not have to draw perfectly horizontal and straight lines. However, what should you do if you are trying to draw a line within the tolerance, say at 1°, and Pro/E interprets the line as being horizontal? In such a case, you must dimension the line using an angular dimension. The software will interpret the line as a nonhorizontal line. In order to give an angular dimension, click on both entities and then on the space between the entities defining the angle. Pro/E will place the angular dimension.

## 2.3 THE USE OF RELATIONS

The *Relation* option may be used to dimension a feature, which depends on another feature. It may also be used to dimension a feature with a given equation.

For example, suppose that parameter sd0 depends on sd1, such that sd0 is always one-half of sd1. We can ensure that sd0 is always one-half of sd1 by setting a relation between the two parameters by choosing the *Relation* option from the *SKETCHER* menu. The *RELATIONS* menu will appear as reproduced in Figure 2.8.

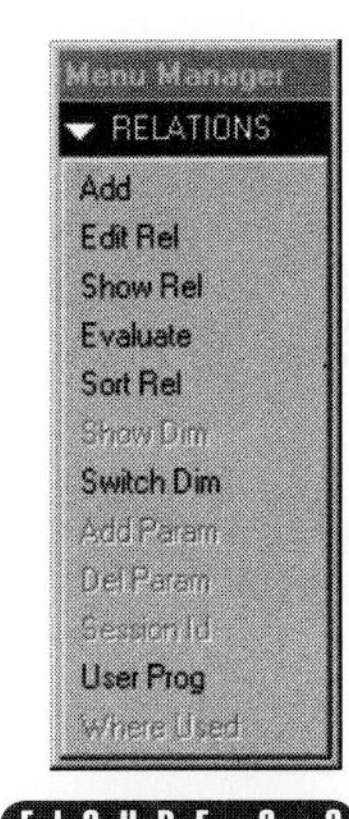


FIGURE 2.8 *The* Relations *menu may be used to add or edit relations between parameters.*

After selecting *Add* from the *RELATIONS* menu, the appropriate equation is entered. In the example considered here, the appropriate formula would be:

$$sd0 = sd1/2$$

Additional relations may be added. When all the relations are entered, you need to hit return twice in order to exit the *Add* relation mode.

Any relation can be edited using the *Edit Rel* option. Pro/E will load the Pro/Table editor with all defined relations. Edit the desired relation and then close the editor.

## 2.4 CHANGING THE UNITS OF A PART

Parameters describing the model, including the units, material properties, and the number of significant digits of the part, can be established by using the *Setup* option from the *PART* menu (Figure 1.7). Using the *Setup* option leads to the *PART SETUP* and *UNIT MGR* menus, shown in Figure 2.9.

The default unit for Pro/E is the inch. For parts with different units, the units of the model may be changed with the *Units* option from the *SETUP* menu. Selecting the *Units* option brings up the *UNIT MGR* menu shown in Figure 2.9, where the desired unit can be chosen by taking the *Principal Sys* option. Choosing the *Principal Sys* option brings up the dialog box shown in Figure 2.9.

Notice that material and mass properties can be defined by using the dialog box and the *PART SETUP* menu. The accuracy of the model can be changed by using the *Accuracy* option and giving the appropriate value.

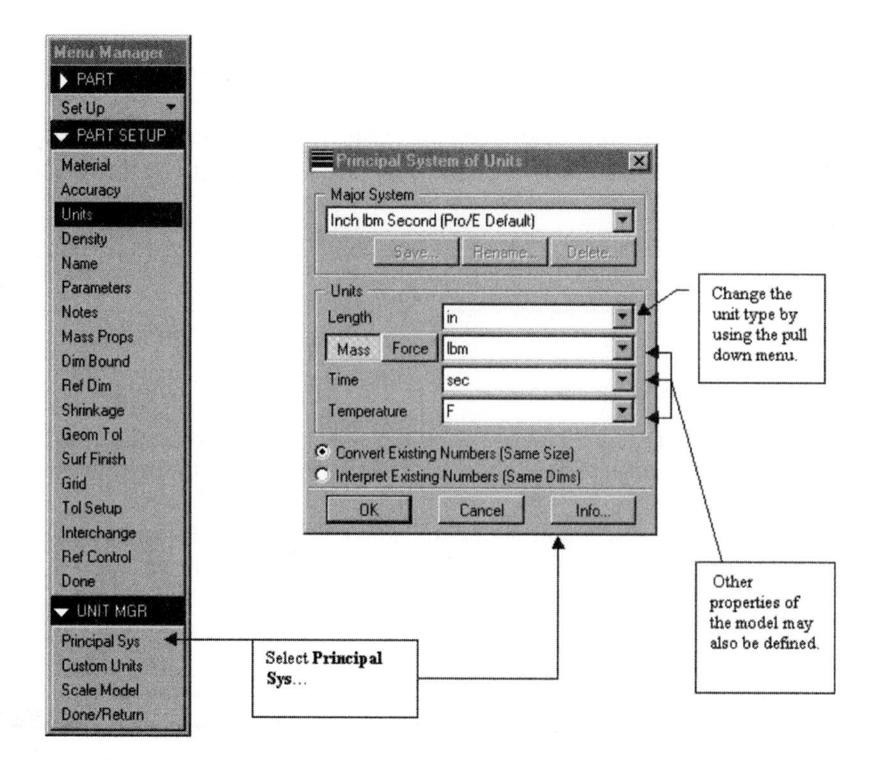


FIGURE 2.9 *The Pro/E* Set Up *menu and* Principal Sys *dialog box.*

## 2.4.1 TUTORIAL # 8: A HINGE

As we have discussed, relations may be used to define a dimension in terms of another. The part that we have chosen to illustrate the use of relations is one part of several from the assembly shown in Figure 2.10. The part under consideration is the hinge. Notice the teeth that form the barb. Upon assembly, the barbs bite into the end of the rectangular plate. It is in construction of the barb teeth that the *Relation* option is helpful.

Because of the symmetry, we will complete one-half of the model. Using a copying procedure called *Mirror Geom,* we complete the rest of the model in Chapter 7.

This part is metric, so we will need to change the default units. Let us begin by creating a part file called "Hinge." From the *PART* menu (Figure 1.7), select *Setup.* Then, from the *PART SETUP* menu, shown in Figure 2.9, select *Units.* The units of this part are millimeter, so choose the *Principal Sys* option. Then, change the *Length* setting from *in* to *mm.* Select *OK, Done/ Return,* and *Done.*

Now, choose *Feature, Create, Protrusion, Extrude, Solid,* and *Done.* Pro/E will open the *Sketcher.* You will need to sketch the geometry shown in Figure 2.11. Notice the parameters in the figure. Place the dimensions as shown in the figure but do not regenerate.

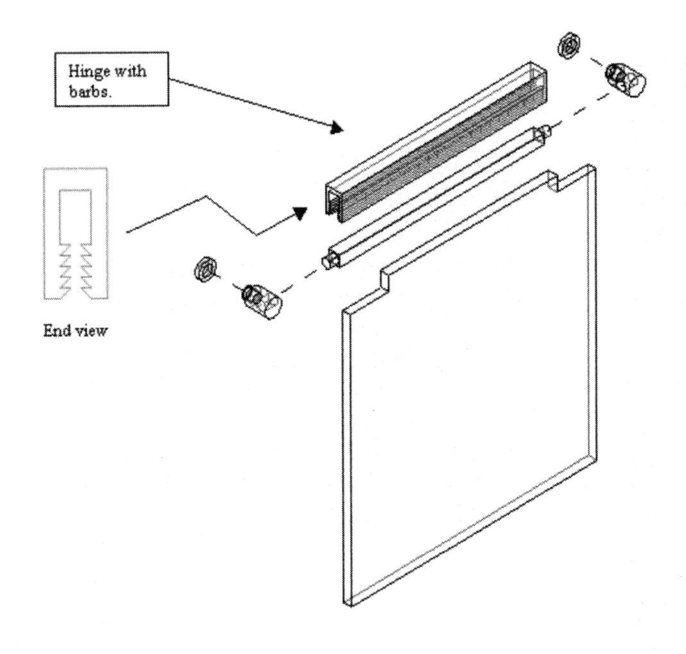


FIGURE 2.10 *The assembly contains the hinge to be constructed in this tutorial. The teeth of the barbs can be seen in the end view. Because the dimensions of all the teeth are the same, only one tooth needs to be dimensioned. The size of the rest of the teeth can be related to the first tooth by using the* Relation *option.*

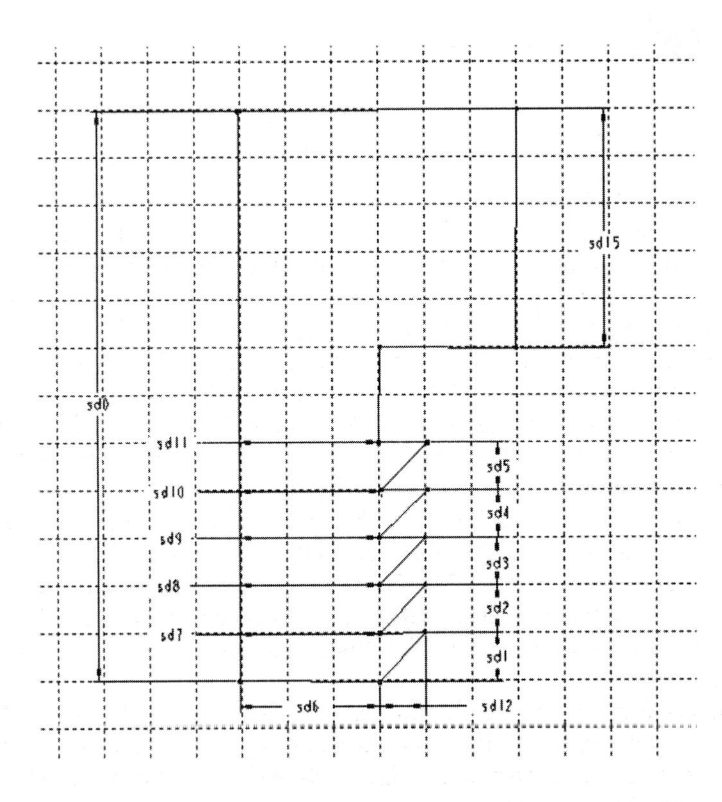


FIGURE 2.11 *The section needed to generate the hinge in Figure 2.10.*

This half of the part has five teeth. The parameters sd1, sd2, sd3, sd4, sd12, and sd5 define the height of the teeth. All of these parameters are of equal value. We can use the *Relation* option and set all these parameters equal to one of them, say, sd12. Then if sd12 were updated, the height of all the teeth would be automatically updated.

In order to set this relation, select *Relation* from the *SKETCHER* menu (Figure 2.2). From the *RELATION* menu, shown in Figure 2.8, choose *Add.* The option is set up so all the user has to do is enter the equation involving to the relation. In this case, the parameters are all the same, so all we want are equations: sd1 = sd12, sd2 = sd12, and so on.

Enter the first equation (sd1 = sd12), then hit *Return.* Notice that Pro/E allows the user to enter additional relations. Enter the next equation (sd2 = sd12). Type in the remaining equations.

We also need to set the parameters sd6, sd7, sd8, sd9, sd10, and sd11 as equal. These parameters define the width of the teeth. If we relate all the parameters to, say, sd6 we want to enter relations sd7 = sd6, sd8 = sd6, and so on. Enter these equations. When you are done, hit *Return* twice.

The sketch is now ready to be modified to the appropriate dimensions. *Regenerate* and *Modify* to the values shown in Figure 2.12. Note that of the parameters that were made equal, only sd12 (=1.000) and sd6 (=1.500) need be modified. Regenerate after modifying the dimensions. Extrude the part using the *Blind* option to a depth of 100 mm. The part should appear similar to Figure 2.13.

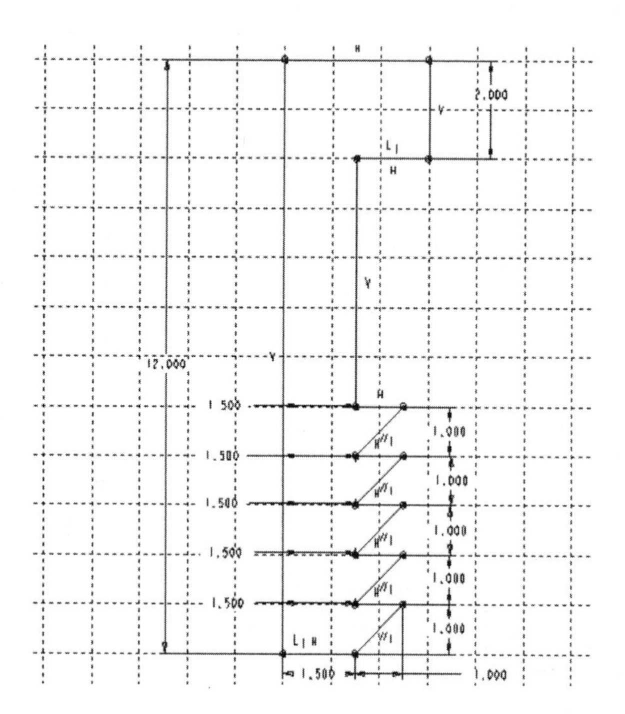


**FIGURE 2.12** *The section of the hinge with the appropriate dimensions needed for constructing the part.*

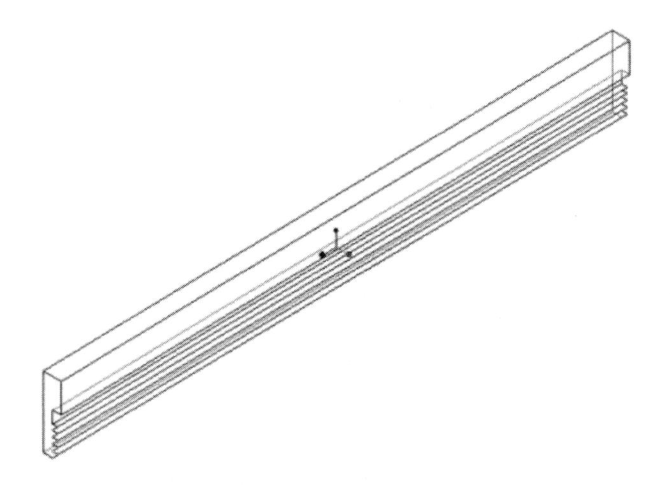

FIGURE 2.13 *The hinge after extruding to a depth of 100 mm.*

## 2.5 THE SEC TOOLS OPTION

The *Sec Tools* option may be used to define the look of the *Sketcher.* After clicking on *Sec Tools,* the *SEC ENVIRON* menu becomes active. This menu is reproduced in Figure 2.14. Through the *Sec Environ* option, the grid can be turned on, off, the grid spacing changed, the accuracy set, or the number of digits that are displayed changed.

If you ever want to plot the sketch while in the *Sketcher,* the *Interface* option can be used.

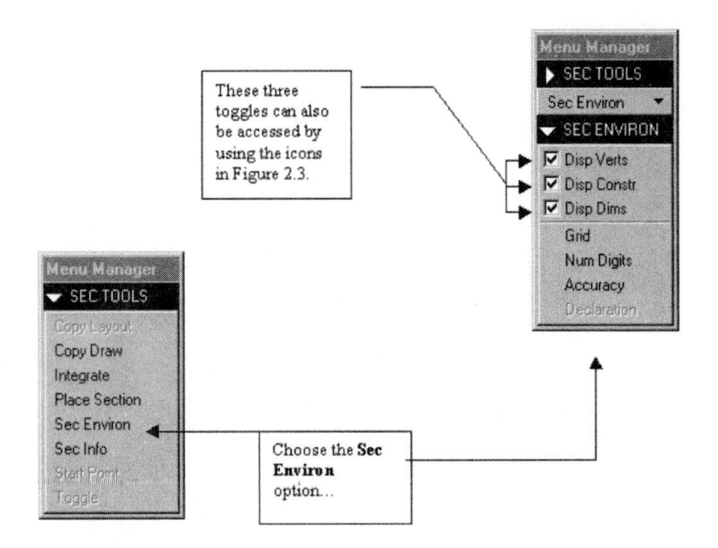


FIGURE 2.14 *The settings of the* Sketcher *can be changed by using the options from the* SEC TOOLS *and* SEC ENVIRON *menus.*

## 2.6 THE SKETCH VIEW OPTION

The *Sketcher* is designed to display the sketch in two dimensions. Usually this representation is sufficient for the creation of the sketch. However, in creating complicated parts, the user might want to reorient the part by rotating the sketch. This manipulation often reveals hidden features, which may be referenced.

In order to return the sketch to its original two-dimensional representation, the *Sketch View* option may be used. After the *Sketch View* option is selected, Pro/E reorients the sketch using the defined sketching and projection planes.

## 2.7 THE GEOM TOOLS OPTION

You may have noticed in the *SKETCHER* menu the option *Geom Tools.* The *Geom Tools* option is a series of built-in functions for advanced feature construction in the sketching mode. Among these options are *Intersect, Trim,* and *Divide.* The *GEOM TOOLS* menu is shown in Figure 2.15.

The *Geometric* tools can be used to create constructions that are more complicated, because the option *Intersect* obtains the intersection between two entities. *Trim* and *Divide* can be used to remove unwanted parts of the entity beyond a given limit. We have found that *Intersect* and *Delete* (from the *SKETCHER* menu, Figure 2.2) can be used for most constructions. Once the intersection has been obtained, the unwanted part of the entity (line, arc, etc.) can simply be deleted.

### 2.7.1 TUTORIAL # 9: A TENSION PLATE

Let us choose a new example to illustrate how the *Geom Tools* option can be used. Consider the tension plate shown in Figure 2.16. The tension plate is created by making a protrusion.

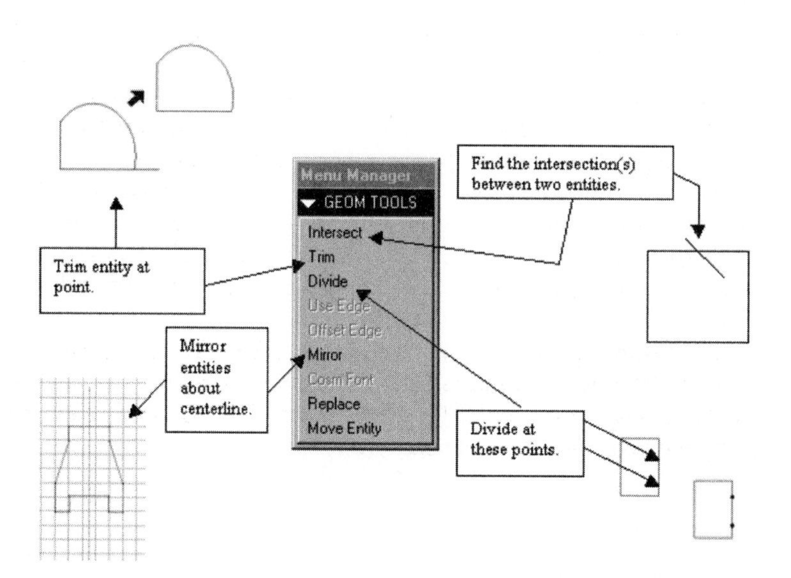


FIGURE 2.15 *The* GEOM TOOLS *menu and a description of the tools contained in the menu.*

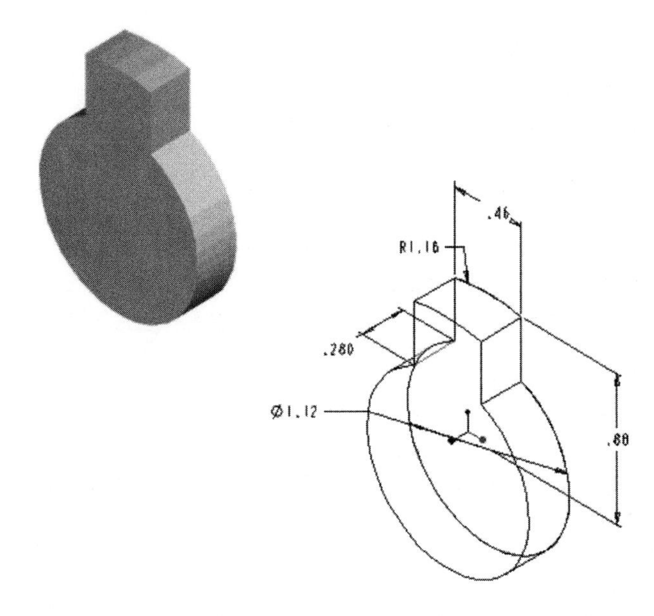


FIGURE 2.16 *The figures are showing a shaded and hidden model of the tension plate.*

Create a new part called "TensionPlate." Select *Feature, Create, Protrusion, Extrude, Solid,* and *Done.* In the *Sketcher,* begin by making the constructions shown in Figure 2.17. Draw a circle with the center of the circle on the intersection of a vertical *and* horizontal grid. This circle will become the lower half of the Tension Plate.

Now draw two vertical lines an equal distance from the center of the circle. Finally, draw a smaller circle with its center on the same vertical grid as the first circle.

You may wonder why an arc is not drawn between the two vertical lines instead of the second circle. The answer is that the center of the arc would be improperly located. The center would be placed aligned with the ends of the two vertical lines, which is not what we want in the case of the tension plate.

We essentially have our sketch; however, we need to remove the extra parts of the various entities. Choose *Geom Tools* and then *Intersect.* Then click near the intersection of the unwanted part of the entities. You will need to click on both parts of the intersection. After you are done obtaining the intersections, you should have something similar to the sketch in Figure 2.18. Notice that Pro/E has placed colored filled circles at the requested intersections. Now remove the unwanted part of the entities by using the *Delete* option.

Dimension the sketch as shown in Figure 2.19. Then *Regenerate.* Dimension with the proper dimensions and regenerate again. You should have a similar drawing to the one shown in Figure 2.19. Enter a depth of 0.28 to complete your model. Save your work.

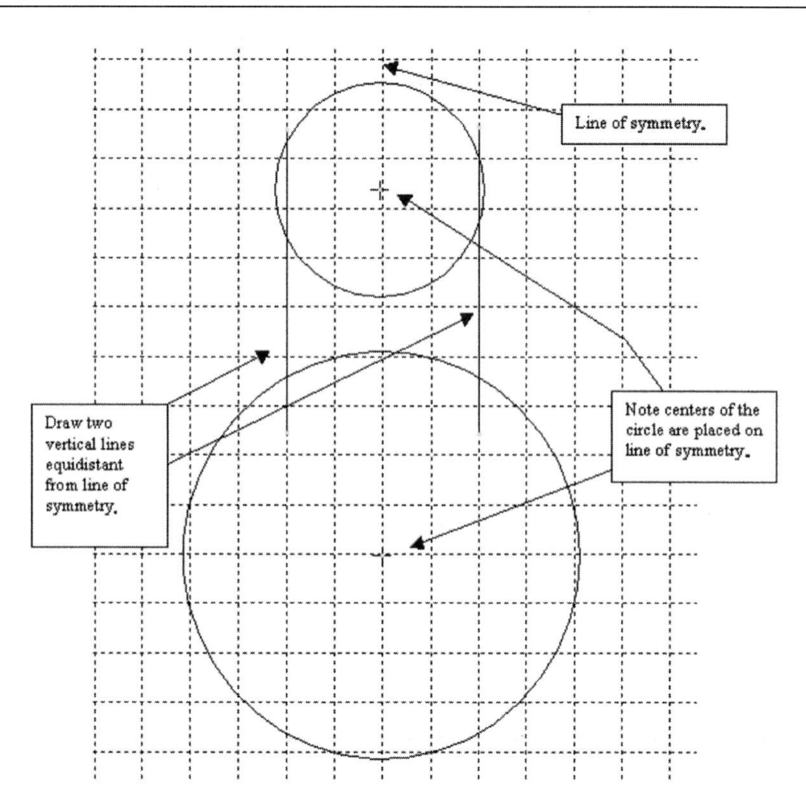


FIGURE 2.17 *Constructions needed to make the tension plate.*

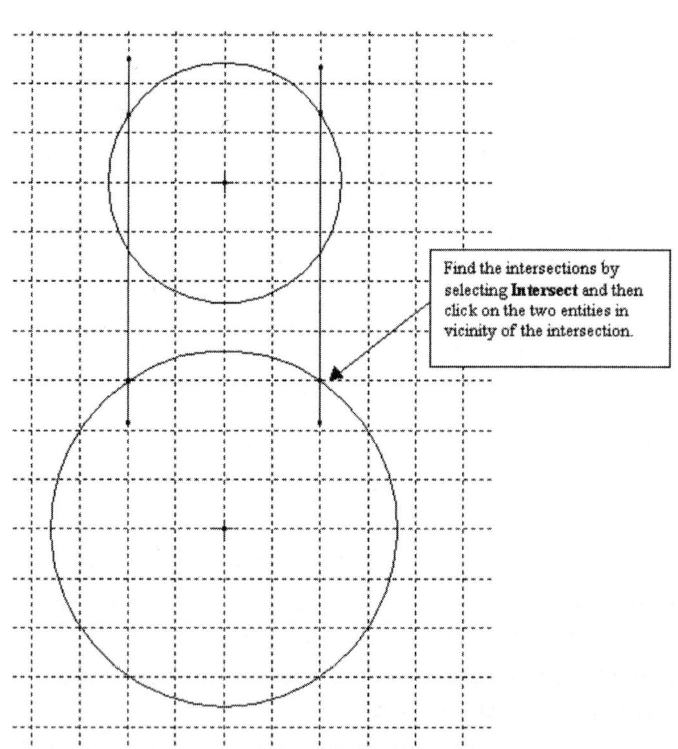


FIGURE 2.18 *Constructions with intersections shown. The intersections and endpoints of entities are shown by the software as solid black circles.*

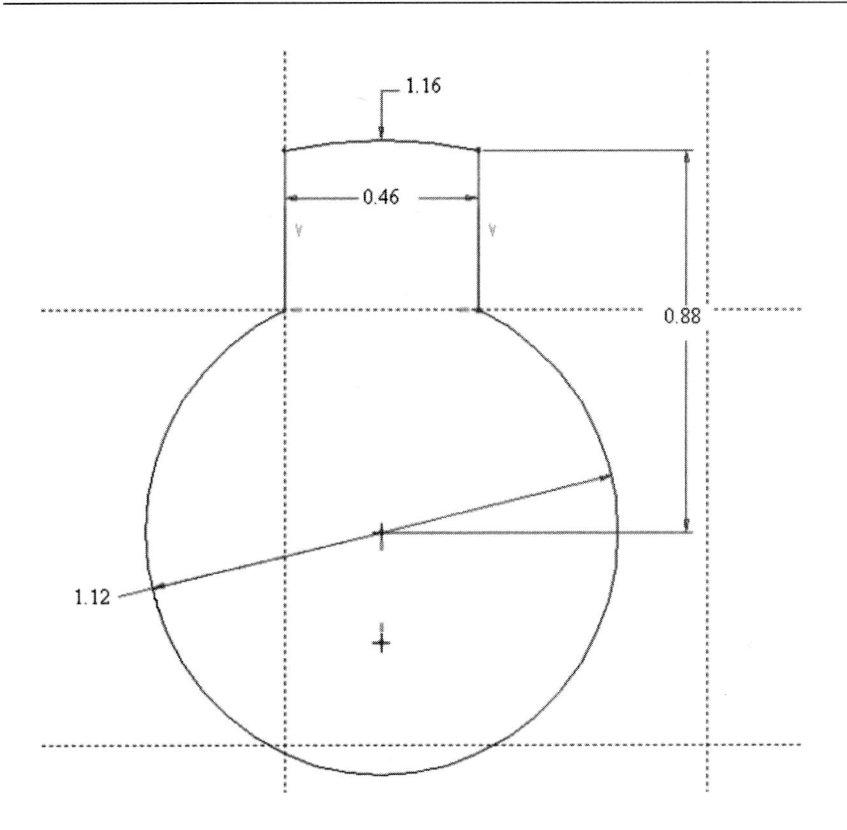


FIGURE 2.19 *Tension plate sketch with dimensions (after regenerating twice).*

## 2.8 THIN SECTIONS

As we saw in section 2.1, the geometry may have a constant thickness. In such cases, the *Thin* option is useful. When creating a thin section, the user must first sketch and then define in which direction the thickness is added. We refer to it as the fill direction.

An example of a model constructed using the *Thin* option is the bracket shown in Figure 2.20. The thickness of the bracket is 0.5; it can be constructed by drawing an "L" in the *Sketcher.*

### 2.8.1 TUTORIAL # 10: AN ANGLE BRACKET USING THE THIN OPTION

Construct the angle bracket by following the steps:

1. Create a new part by selecting *File* and *New.*
2. Make sure the file type is set to part and enter "AngleBracket." Click on the *OK* button.
3. Select *Feature, Create, Protrusion, Extrude, Thin,* and *Done.*
4. Sketch an "L" shape as shown in Figure 2.21. Then dimension the sketch as shown in the Figure 2.22. *Regenerate.*
5. Modify the dimensions to the values given in Figure 2.20. Then *Regenerate.* Choose *Done.*

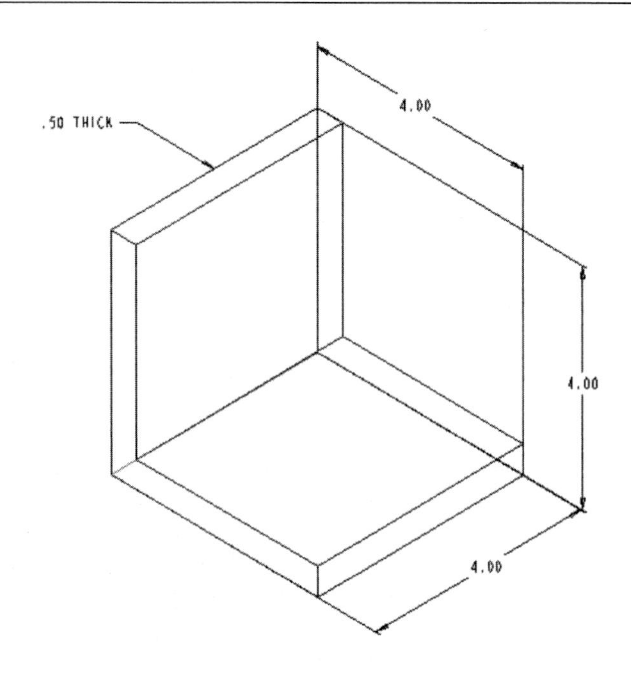


FIGURE 2.20 *A bracket constructed by using the* Thin *option.*

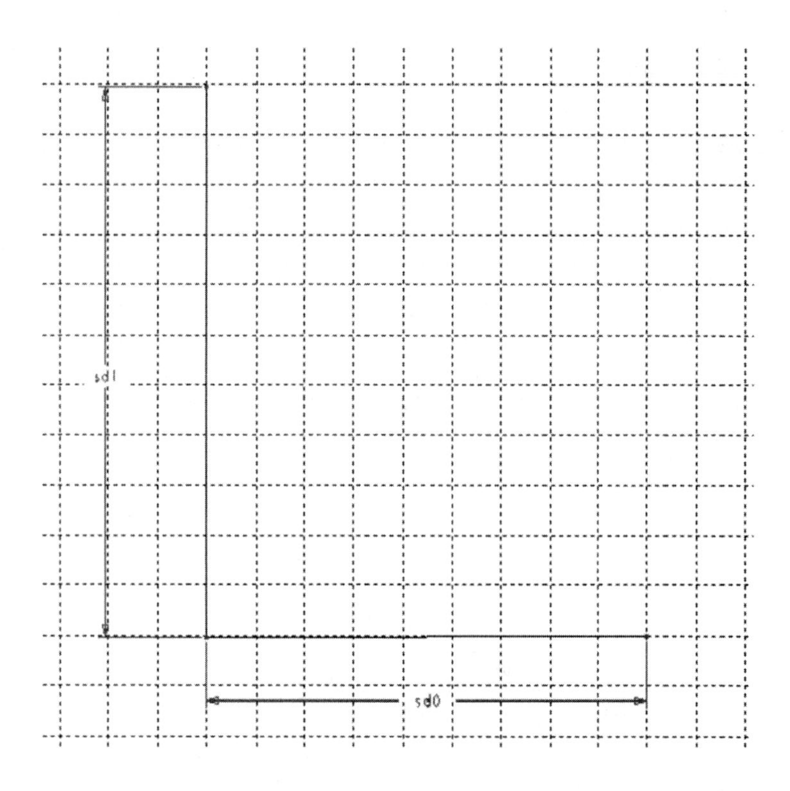


FIGURE 2.21 *"L" sketch in the* Sketcher *with parameter dimensions.*

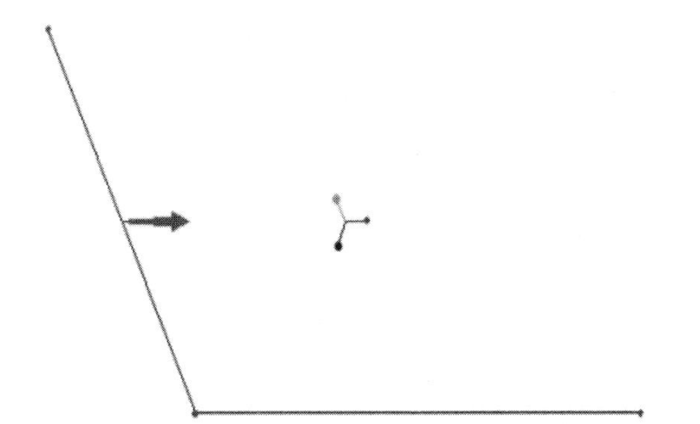

FIGURE 2.22 *The arrow indicating direction for thickness.*

You will see an image similar to the one shown in Figure 2.22 on your screen. Pro/E is looking for the direction to add the thickness. Notice the arrow indicating the fill direction. In this case, the direction of the arrow is correct as shown. You can flip the direction of the arrow by clicking on *Flip.* Notice that you can fill in both directions by choosing *Both.* Choose *OK* to accept the shown direction and then enter 0.50 for the thickness. Enter a depth of four and choose *Done.* Your part should look like the one shown in Figure 2.20.

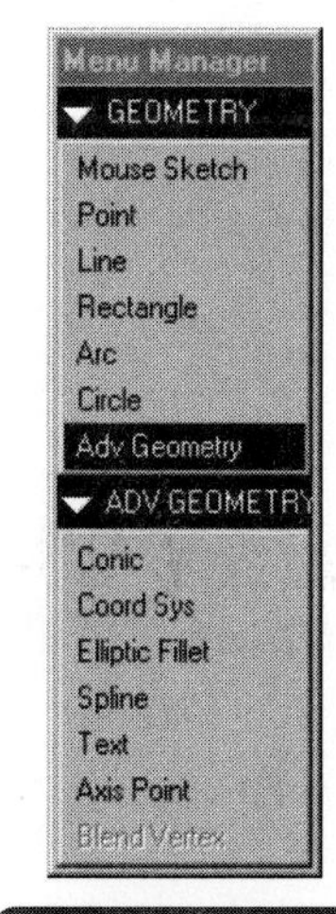


FIGURE 2.23 *Conic sections may be constructed by using the* ADVANCED GEOMETRY *menu.*

## 2.9 CONIC SECTIONS

A conic section can be created in the *Sketcher* by using the so-called *Advanced* (*Adv*) *Geometry* option from the *GEOMETRY* menu (Figure 2.23). Selecting the *Advanced Geometry* option leads to the *ADVANCED GEOMETRY* menu. This menu is reproduced in Figure 2.23. Among the options in this menu is the option *Conic.* A conic section can be an ellipse, hyperbola, or parabola.

In Pro/E, a conic is constructed by selecting the ends of the conic spline and a shoulder. The tangency angles of the endpoints must be prescribed by defining the angle with an angular dimension, an adjacent entity, or a centerline. This method is called the "rho" method. It is shown in Figure 2.24.

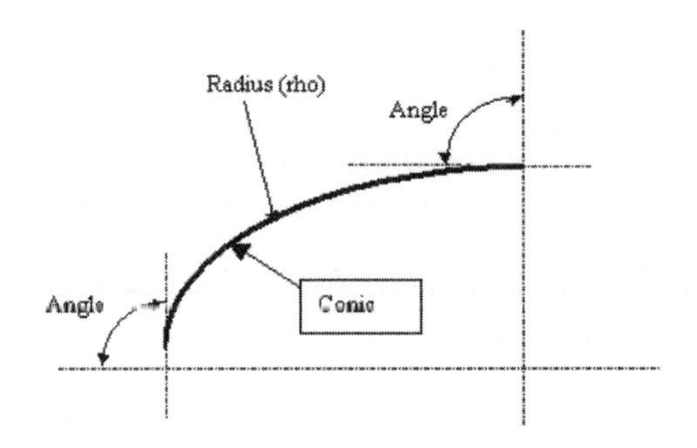


FIGURE 2.24 *The method for dimensioning a conic.*

In order to dimension a conic using the "rho" method, the value of rho, the radius, must be given. To enter rho, select the conic with the left mouse button and place the value with the middle button. Initially, Pro/E sets rho equal to 0.5. This value can be modified in the following manner:

1. $0.05 < \text{rho} < 0.5$ for an ellipse—a closed ellipse has a value (SQRT (2)-1), in this case enter the formula.
2. $\text{rho} = 0.5$, for a parabola.
3. $0.5 < \text{rho} < 0.95$ for a hyperbola.

In order to dimension the angle complete the following steps:

1. Select the conic, then the endpoint where the tangency is to be defined.
2. Pick the entity, centerline, or other conic to which the angle is to be defined.
3. Place the dimension by clicking on the screen, at the desired location, with your middle mouse button.

We are going to construct a conic section in chapter 10. The part, a parabolic reflector, makes use of the *Revolve* option. The methods and options discussed in this section will be revisited in section 10.1.3.

## 2.10 THE SKETCH MODE

The *Sketch* mode may be used to create two-dimensional sections. For sections that appear more than once in the same or different parts, it is convenient to create the section in the *Sketch* mode and then place the section in a library of sections. The section may be retrieved in the *Sketcher* when creating a part by using the *Place Sec* option.

When a section is loaded into the *Sketcher,* the user may change the scale as well as the orientation. The scale of the section is referenced to a point called the *Origin Point.* The *Origin Point* must be selected by the user.

Sections created in the *Sketch* mode are given the extension ".sec" by the software. The mode may be accessed by retrieving or creating a section. The *New* dialog box is used to create a section in a manner similar to creating a part. For the case of a section, this box is shown in Figure 2.25.

### 2.10.1 TUTORIAL # 11: CREATING AND USING A SECTION

As an example of how a section may be created, saved, and retrieved into the *Sketcher* during the construction of a part, let us create the section shown in Figure 2.26. Select *File* and *New* or use the *Create New* button in the toolbar. Choose the *Sketch* option, as shown in Figure 2.25 and enter the name "C_Section."

The software will open the *Sketcher.* Sketch the section shown in Figure 2.26. *Dimension, Regenerate,* and *Modify* to the values shown. *Regenerate*

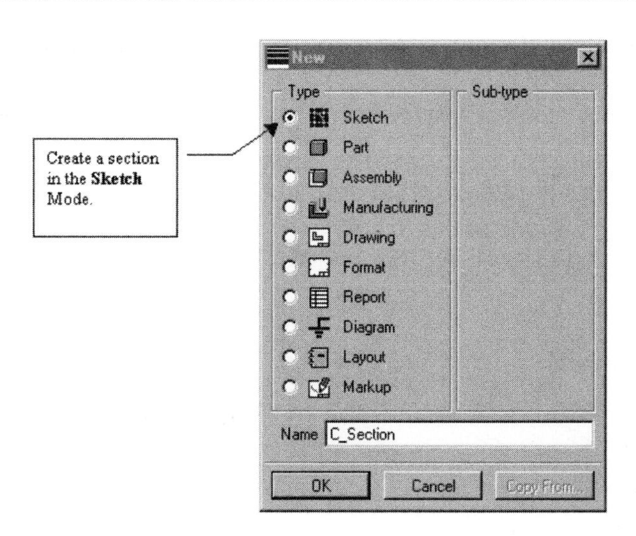


FIGURE 2.25 *Create a* Section *file by using the* New *box and the* Sketch *option.*

and select *Done.* Save the section by using *File* and *Save.* Then erase the sketch from memory.

Create a new file by selecting *File* and *New.* Choose the *Part* mode and give the part the name "C_Stiffener." Then, select *Feature, Create, Protrusion, Extrude, Thin,* and *Done.*

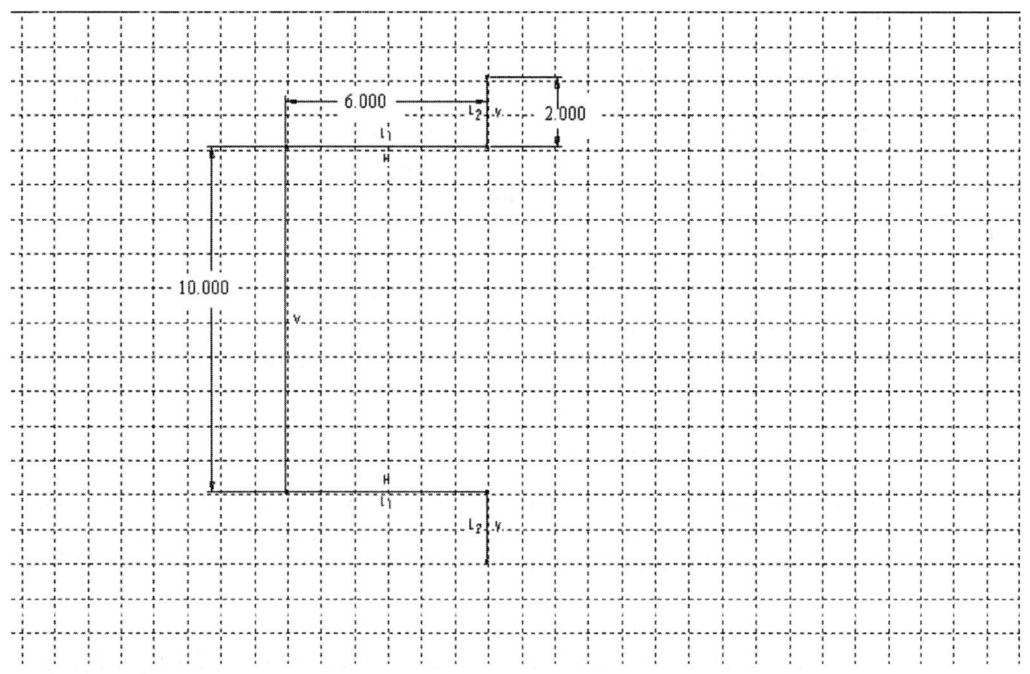


FIGURE 2.26 *A "C" section.*

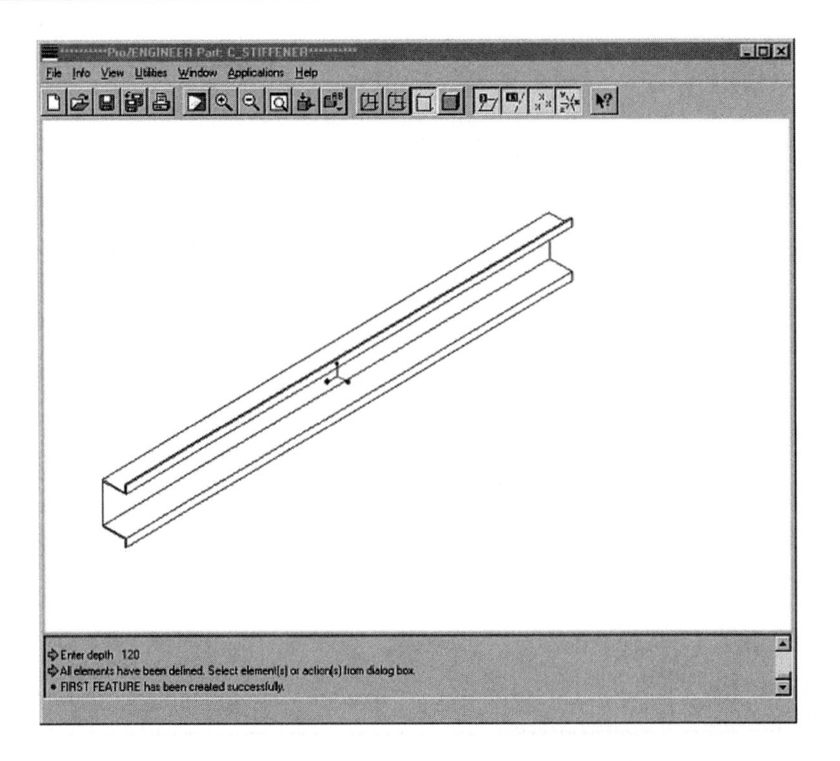


**FIGURE 2.27** *The stiffener created by placing a preexisting section in the* Sketcher.

We can now place the section "C_Section." Choose *Sec Tools* and *Place Section.* From the list, select "C_Section" and then choose the *Open* button. Because the part has no other features, the software will locate the section arbitrarily.

The dimensions of the section may be changed at this point. For this example, we can leave the dimensional values as they are. Choose *Regenerate* and *Done.* Fill the thin outward. Enter a value of 0.2 for the thickness. Then, enter a value of 120 for the depth. The model of the stiffener is shown in Figure 2.27.

## 2.11 SUMMARY AND STEPS FOR USING THE SKETCHER

In Pro/E, the *Sketcher* is used to draw and dimension the cross section of a feature. After the section has been regenerated into a parametric two-dimensional model, it may be made three-dimensional by extrusion, revolution about an axis, or swept along a trajectory. Therefore, it is important that the user of the software understand and be familiar with the *Sketcher.*

In this chapter, we concentrated on solid and thin sections that were extruded to form a three-dimensional feature. Solid sections are closed and have an inside and an outside. Thin sections, on the other hand, are open sections. Thin sections are used in creating features that have a constant thickness.

Sections are drawn in the *Sketcher* by using elementary entities such as lines, arcs, and circles. More advanced entities such as conic and spline sections are also available. In general, the procedure for using the *Sketcher* is:

1. Select *Feature* and *Create.*
2. Choose the type of feature. For example, *Protrusion.*
3. If a feature already exists in a model, then you will need to orient the model in the *Sketcher.* We will discuss this procedure more in chapter 3.
4. In the *Sketcher,* draw the geometry of the section.
5. *Dimension* the section.
6. *Regenerate* the section.
7. Modify the dimensions to the proper value.
8. *Regenerate* the section again.
9. Choose *Done.*
10. Define the depth of the section.
11. Choose *OK* or *Preview.*

The regeneration process is important in Pro/E. In translating the section from the sketch on the screen to a mathematical parametric model, the software makes several assumptions. These assumptions are listed in detail in section 2.2 and are called constraints. If a section fails to regenerate, most likely it is because the sketch conflicts with one or more of the constraints.

Because Pro/E is a parametric modeler, relations may be established between two or more dimensions. The relation is created simply in the *Sketcher* by the following steps:

1. Select *Relation* and *Add.*
2. Type in the formula expressing the relation.
3. Hit *Return.*
4. Enter additional relations, if desired, hitting return after each one has been entered.
5. In order to exit from this mode, hit *Return* one last time.

Notice that dimensions driven by a relation cannot be modified. A relation may be edited by using the *Relation* and *Edit Rel* sequence.

The default units for a Pro/E model are inches. The units may be changed by following this sequence:

1. Choose *Setup* from the *PART* menu.
2. Select *Units* and *Principal Sys.*
3. Change the units to the desired ones.
4. Click on *OK, Done/Return,* and *Done.*

Finally, the option *Geom Tools* may be used to construct complicated geometric sections in the *Sketcher.* This option may be used to find the intersection between two entities (*Intersect*), break an entity at discrete points (*Divide*), trim an entity at a point (*Trim*), or mirror one or more entities about a centerline (*Mirror*). In general, the options from the *GEOM TOOLS* menu may be accessed by doing the following:

In the *Sketcher,* select *Geom Tools.*

1. Choose the desired option from the *GEOM TOOLS* menu (Figure 2.15).
2. Provide the necessary information.

The *Sketcher* may be used to create a two-dimensional section in *Sketch* mode. In cases, where a section appears in multiple parts, it is convenient to use the *Sketcher* in the *Sketch* mode to create and store the section. Then, the section may be placed in each part by using the *Place Section* option.

A section may be created in *Sketch* mode by using the following sequence:

1. Select *File* and *New,* or use the *Create New Object* icon.
2. Choose *Sketch* and enter a name for the section.
3. Choose *OK.*

A section may be placed in the *Sketcher* by using the following procedure:

1. Create a part in the normal way.
2. Enter the *Sketcher.*
3. Select *Sec Tools* and *Place Section.*
4. Choose the desired section from the list.
5. If the part has no other sections, the software will place the section. Then, choose *Dimension, Regenerate,* and *Done.* If the part does have other features, the user may scale and rotate the new section. In this case, proceed to step six.
6. The software will open a new window containing the section. Enter the angle of rotation.
7. In the new window, select the origin for scaling.
8. In the new window, choose the point to grab and move the section.
9. Enter the scaling factor (enter one for no scaling).
10. If the scaling is not correct, use the right mouse button to switch between drag and scaling mode.
11. Place the section by clicking with your left mouse button.
12. *Dimension* the section to the part. *Regenerate* and select *Done.*

## 2.12 Additional Exercises

2.1 Construct the washer shown in Figure 2.28. Do not create the hole.

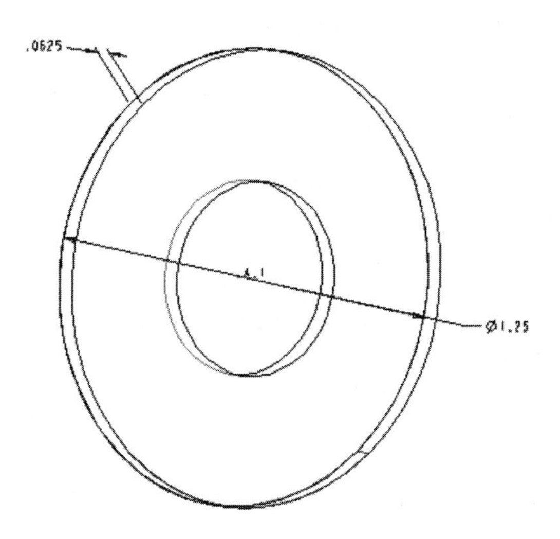


FIGURE 2.28 *Figure for exercise 2.1.*

2.2 Using Figure 2.29, create a model of the hold down. Do not construct the hole.

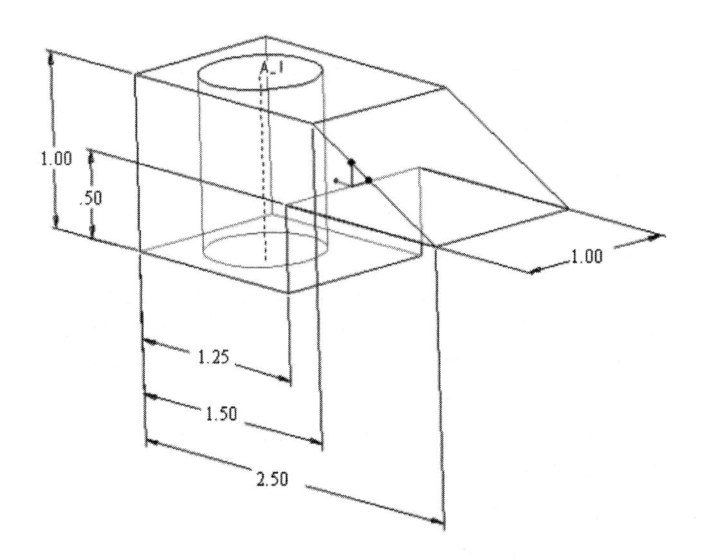


FIGURE 2.29 *Figure for exercise 2.2.*

2.3 Use the *Thin* option to construct a model of the retaining ring illustrated Figure 2.30.

**FIGURE 2.30** *Figure for exercise 2.3.*

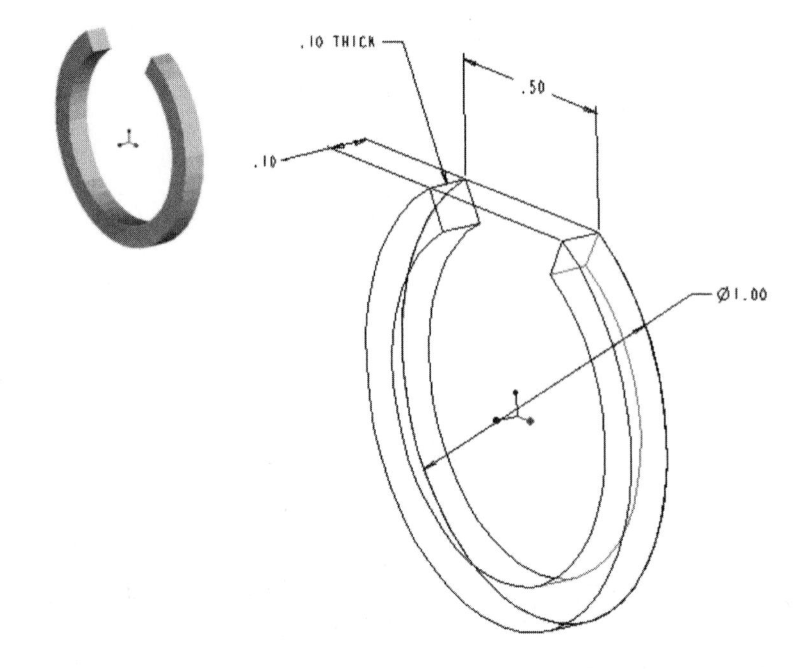


2.4 Create the lever arm using the geometry shown in Figure 2.31.

**FIGURE 2.31** *Figure for exercise 2.4.*

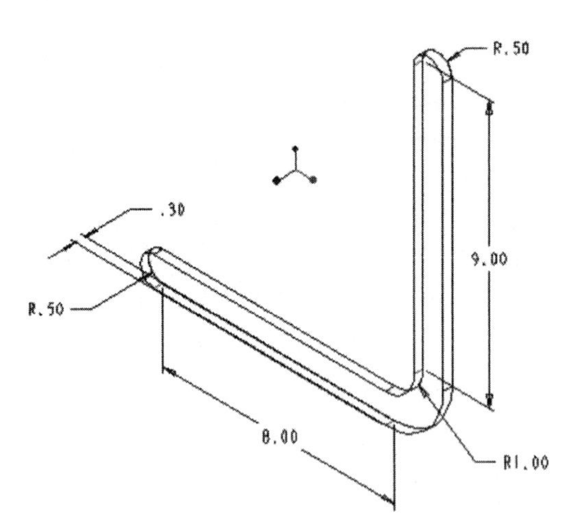

2.5 Construct the model of the cam shown in Figure 2.32.

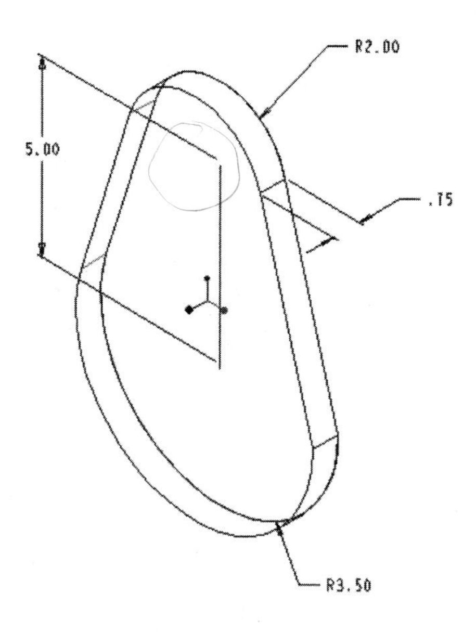


FIGURE 2.32 *Figure for exercise 2.5.*

2.6 The "S stiffener" shown in Figure 2.33 may be created using a single *Thin* protrusion. Construct a model of the stiffener.

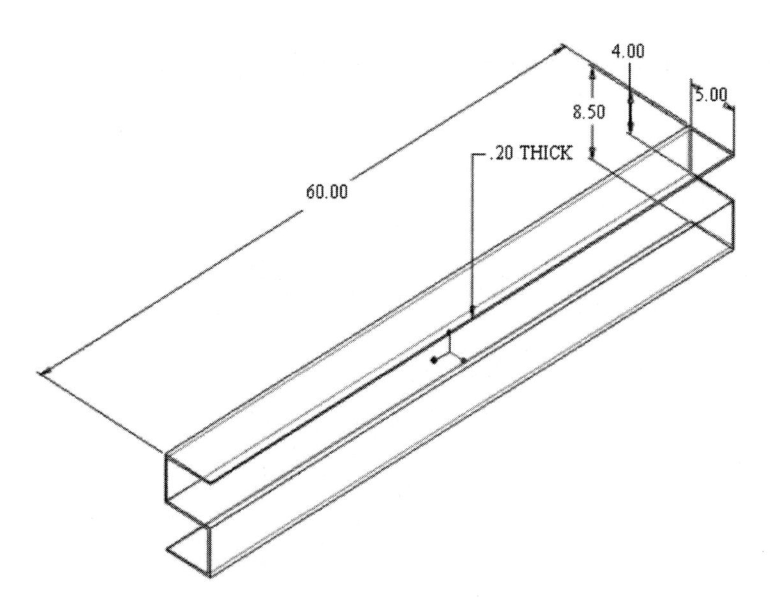


FIGURE 2.33 *Figure for exercise 2.6.*

2.7 The "T beam" has metric units. Based on Figure 2.34, create a model of the beam using the *Thin* option.

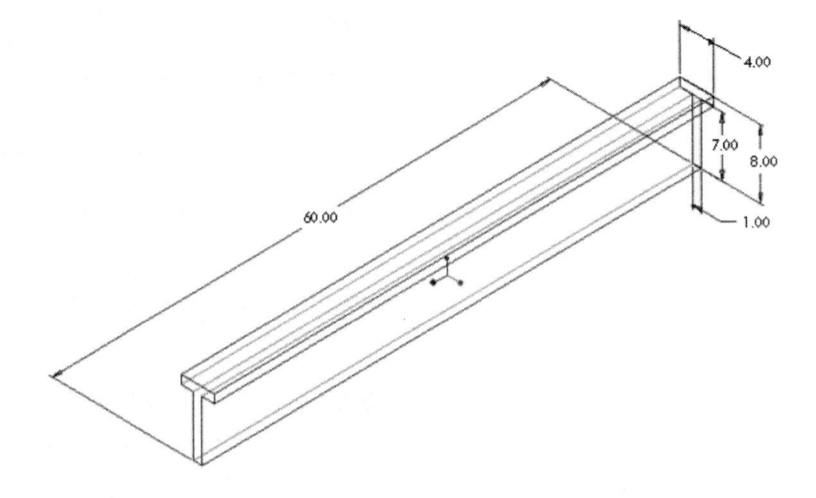


FIGURE 2.34 *Figure for exercise 2.7.*

2.8 Construct a model of the "I beam" using the geometry shown in Figure 2.35. Note that the units of this part are metric.

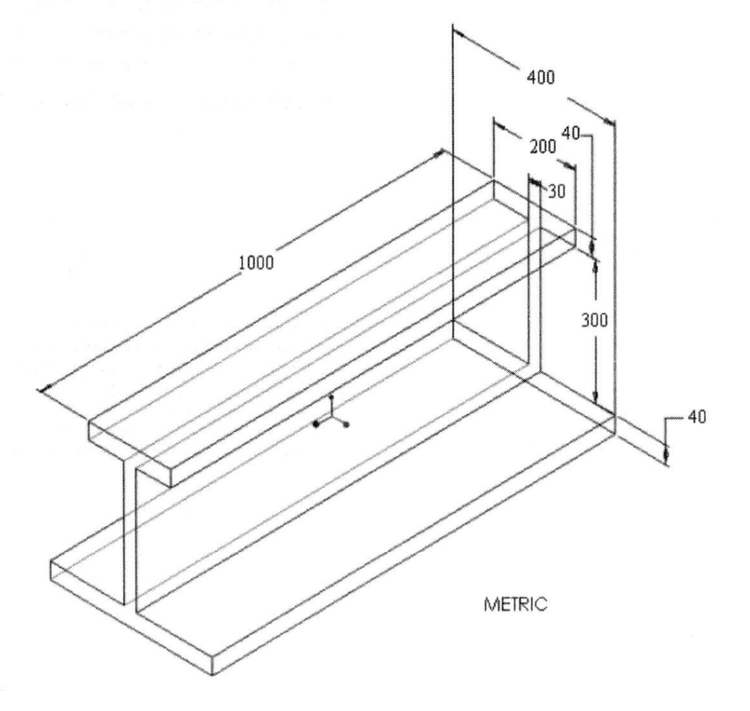


FIGURE 2.35 *Figure for exercise 2.8.*

2.9 Using the *Extrude* option, create the tension test specimen based on the geometry given in Figure 2.36.

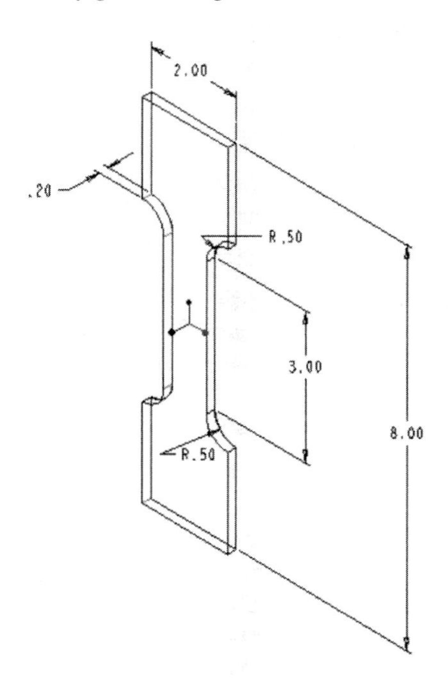


FIGURE 2.36 *Figure for exercise 2.9.*

2.10 Create a model of the gib head key illustrated in Figure 2.37.

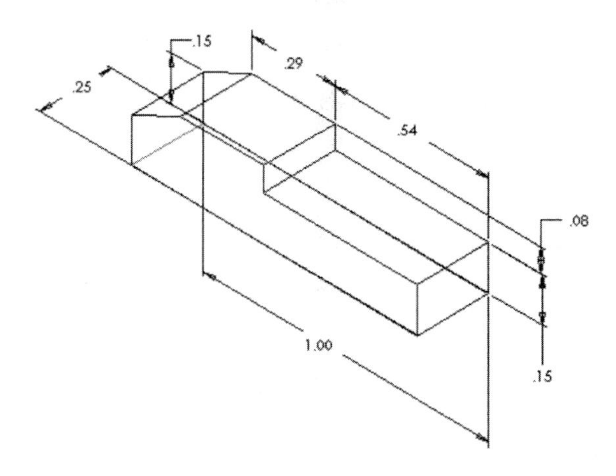


FIGURE 2.37 *Figure for exercise 2.10.*

2.11 The angled block, shown in Figure 2.38, may be constructed by using the *Extrude* option. Create a model of the block.

FIGURE 2.38 *Figure for exercise 2.11.*

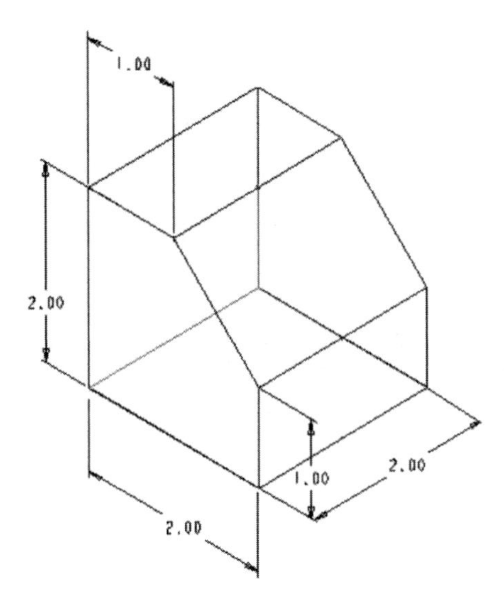


2.12 Use the *Thin* option to construct a model of the steel pin illustrated in Figure 2.39.

FIGURE 2.39 *Figure for exercise 2.12.*

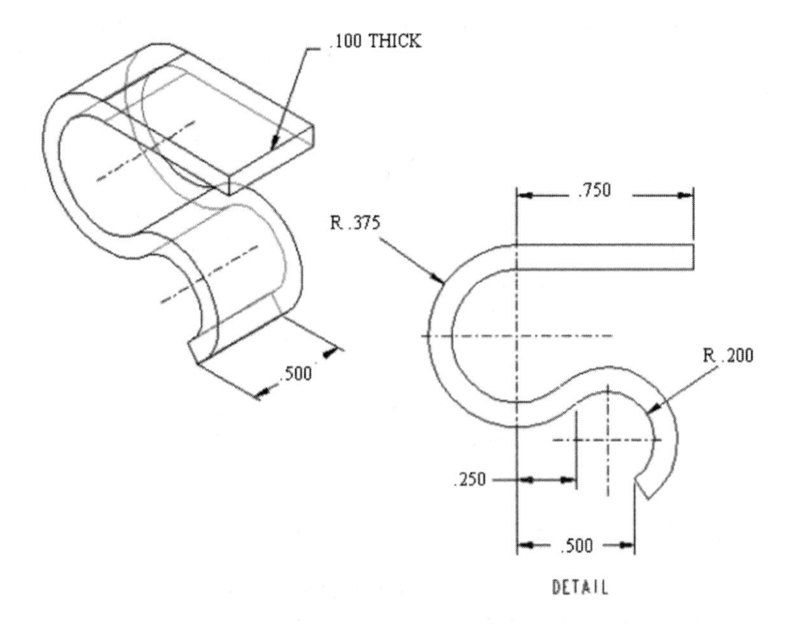

# CHAPTER 3

# CREATING BASE FEATURES WITH DATUM PLANES

## INTRODUCTION AND OBJECTIVES

In chapter 2, we used the *Sketcher* to create some simple base features. The objects are called base features because they do not contain any holes or cutouts. The objects were simple in the sense that a protrusion of a two-dimensional sketch into the third dimension was sufficient to create the basic shape of the object. However, for models where the base feature requires multiple protrusions, it is advantageous to use the built-in default datum planes. In addition, some objects, such as a sphere, contain no planar surfaces, which can be referenced for object orientation in the *Sketcher*. In this case, the datum planes can be used to orient the model as well as construct additional features.

Another advantage in using datum planes is that besides making model creation easier, datum planes allow assembly of parts to proceed in a straightforward manner. We will discuss assemblies and the role of datum planes in creating an assembly in chapter 20.

The default datum planes are not always sufficient for the construction of a model. Often, the user must create additional datum planes. In many cases, these additional planes are used as sketching planes. They are also used to reorient the model.

The objectives of this chapter are:

1. The user will understand the difference between default datum planes and user defined datum planes.
2. The user will be able to locate default or user-defined datum planes and be able to use these datum planes in the construction of one or more features.

## 3.1 THE DEFAULT DATUM PLANES

The default datum planes are shown in Figure 3.1. The three planes are mutually orthogonal. Notice the labels defining each plane. For example, plane 1 is referenced by the label DTM1. Often you will need to select one of the planes. The plane is selected by clicking on the label.

It must be remembered that when the default datum planes are used, not only must the sketch be dimensioned, but it also must be located relative to the datum planes. If other features exist, which have already been located with respect to the datum planes, then the feature under construction can be located with respect to the existing feature or features.

In Pro/E, the default datum planes must be added to a part *before* adding any additional features. After a feature has been added to the part, the user may construct and add user-defined datum planes. Thus, the rule to use when deciding whether to use datum planes is, when in doubt, use datum planes. The opinion of this author is to always use datum planes.

### 3.1.1 TUTORIAL # 12: A BUSHING

As an example for using the default datum planes, we take the bushing shown in Figure 3.2.

As can be seen from the figure, the bushing is made up of two cylinders. A model of the bushing can also be made by a revolution, as we will see in a later tutorial. The placement of the object with respect to the datum planes is something that you must consider before attempting to create the model on your computer screen. In this case, we have elected to place datum plane

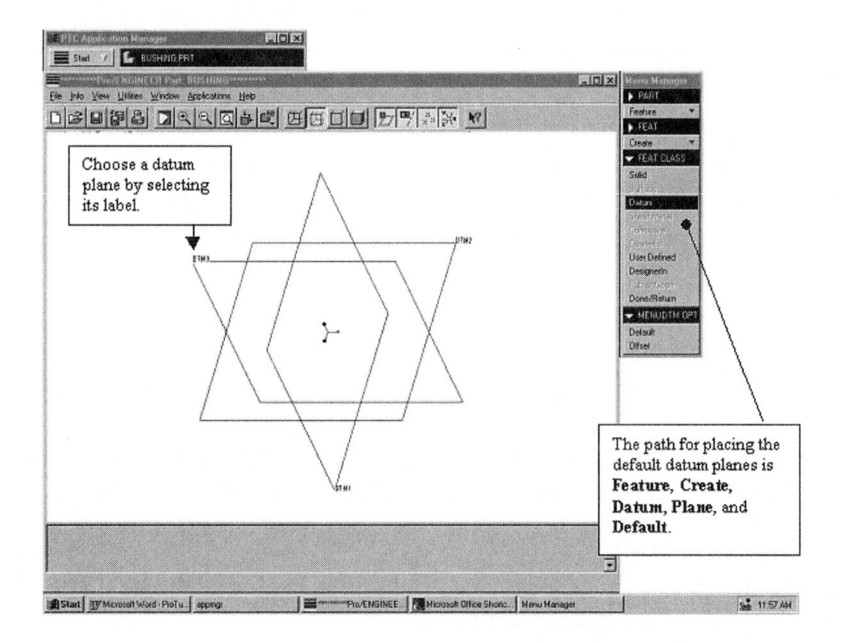


FIGURE 3.1 *The default datum planes.*

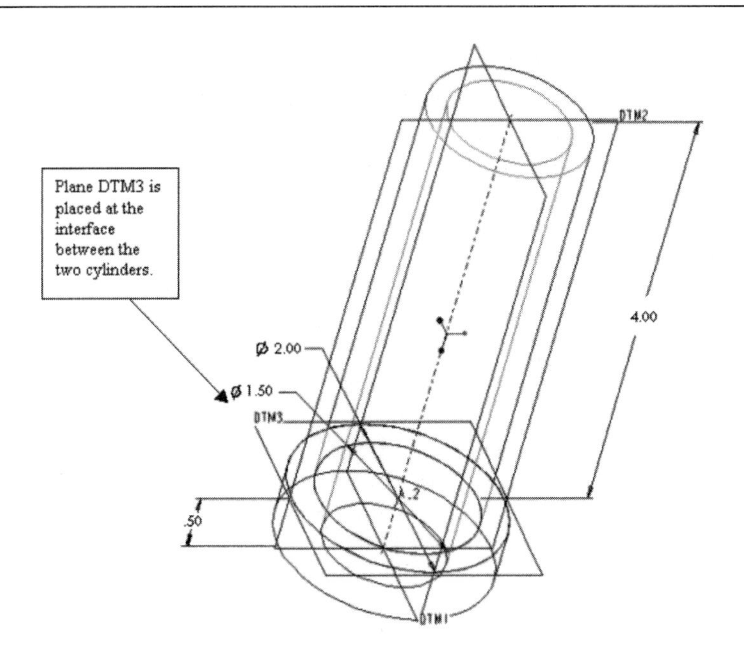


FIGURE 3.2 *The bushing to be constructed.*

# 3 (DTM3) at the intersection of the two cylinders. We will create the larger cylinder first as a protrusion using DTM3 as the sketching plane. Then, we will create the smaller diameter cylinder. Again, DTM3 will be used as the sketching plane to create this cylinder.

Whenever you are using the datum planes to create a feature, you must determine the orientation of the model before entering the *Sketcher.* The default datum planes can be used for this purpose. Because the planes are orthogonal (at ninety degrees) to each other, one plane can be chosen as the sketching plane and another orthogonal plane can chosen to define the orientation of the object in the *Sketcher.*

Because we have already chosen DTM3 as the sketching plane, we may define DTM2 as the top (or bottom) or DTM1 as a side (either right-hand or left-hand side). Either of these definitions will place DTM3 in the two-dimensional *Sketcher* window.

Let us proceed with the feature construction. Create a part called "BushingT3." Choose *Feature* and *Create.* Notice the *Datum* option in the *CREATE* menu (Figure 3.1).

Select *Datum.* You will see the *DATUM* menu appear, which is reproduced in Figure 3.3. Click on *Plane.* Then *Default.* The default datum planes should appear on your screen as in Figure 3.3. The datum planes are treated by Pro/E as features. Notice in the model tree the entries DTM1, DTM2, and DTM3.

Notice that Pro/E has returned to the *FEATURE* menu. Select *Create, Protrusion,* and *Done.* We will make a protrusion along one direction only, so accept *One Side* by selecting *Done.*

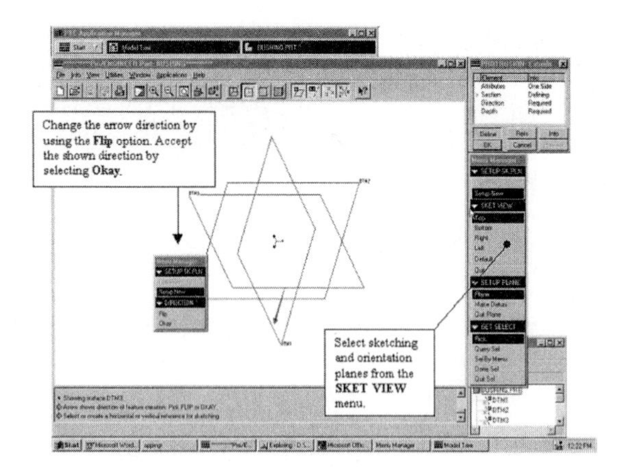


**FIGURE 3.3** *The default datum planes are loaded into the part as features. One of the planes can be selected as the sketching plane. Another of the planes is used as an orientation plane.*

Pro/E will inquire for the sketching plane. Click on the DTM3 label. A red arrow will appear indicating the default direction of the protrusion. We will be making the larger cylinder first, so accept the direction if the arrow is pointing *toward* you. Otherwise, use *Flip* to change the arrow direction.

The software will now ask for a vertical or horizontal surface to set the orientation of the datum planes. Select *Top* in the *SKETCH VIEW* menu (Figure 3.3). Then click on the label for DTM2. Pro/E will now enter the *Sketcher* mode. You should see a screen similar to the one shown in Figure 3.4. Notice the location of each datum plane label. Do their locations make sense?

Notice the thick lines indicating the edges of the DTM1 and DTM2 planes. The edges intersect at a point. We will place the center of our circle at this point. Then we will align the center of the circle with each edge. In doing so, we will fix the location of the circle with respect to the default datum planes. Because Pro/E always knows where the datum planes are, and the circle center has been aligned to the datum planes, the location of the circle center need not be dimensioned.

Sketch a circle using your middle mouse button. Place the center of the circle on the intersection of the edges of the datum planes. Your sketch should appear similar to the one shown in Figure 3.5. Choose *Alignment* from the *SKETCHER* menu (Figure 2.2). The *ALIGNMENT* menu will appear as shown in Figure 3.5. This menu contains options to *Align* or *Unalign* the features. If you click on *Unalign,* all the aligned features will appear in green. Thus, the *Unalign* option can be used to see what are the current aligned features.

Align the center of the circle to the horizontal edges by first clicking on the center of the circle and then on the horizontal edge of DTM2. Repeat to align the center with the vertical edge of DTM1. To see whether you have properly aligned the features, click on *Unalign.* The vertical edge of DTM1, the horizontal edge of DTM2, and the center of the circle should all appear in green.

Dimension the diameter of the circle. *Regenerate* and *Modify* to the correct diameter of 2.000. *Regenerate.* Then choose *Done.* Enter a depth

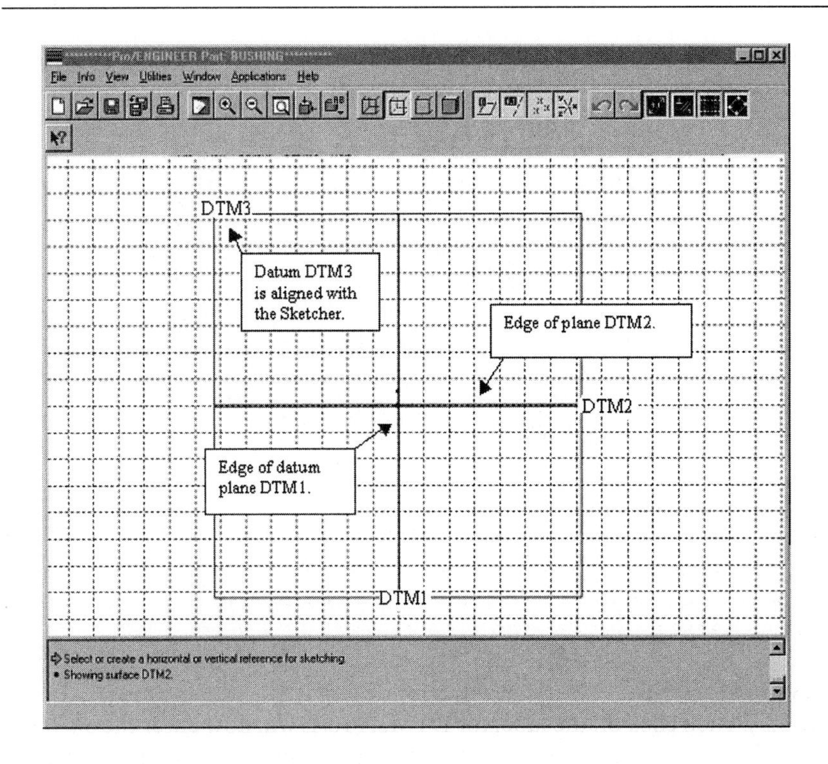


**FIGURE 3.4** *The* Sketcher *with the properly oriented datum planes.*

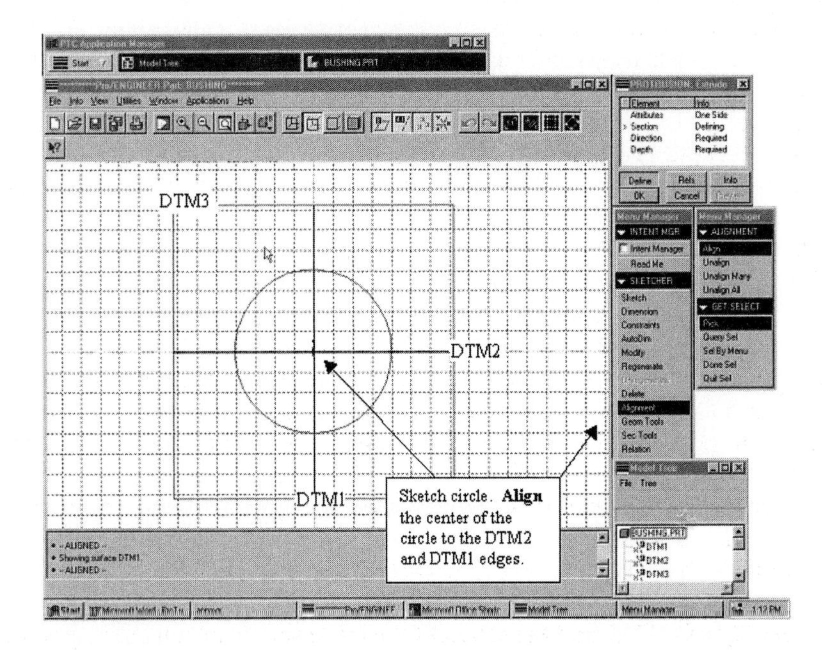


**FIGURE 3.5** *The* Align *option may be used to locate a feature with respect to another feature, including the default datum planes.*

of 0.5. Your feature should appear like the one shown in Figure 3.6 (choose default orientation).

We will now proceed to add the second cylinder, again choosing DTM3 as the default datum plane. Create the second protrusion. You will need to flip the direction on the arrow. Again, choose DTM2 as the "*Top*" to orient the model in the *Sketcher.*

In the *Sketcher,* draw a circle concentric to the first one. Again, place the center of the circle at the intersection of the two edges. *Align* this center with the edges. Size the circle to a diameter of 1.5. Enter a depth of 4.00. Your work should appear as in Figure 3.2. Save the bushing.

### 3.1.2 TUTORIAL # 13: A SHIFTER FORK

The base features for the shifter fork shown in Figure 3.7 is constructed in a straightforward manner if the default datum planes are used. The model is created by using multiple protrusions.

Let us construct this model. Create a part called "ShifterFork" and add the default datum planes. We begin by creating the "U" shaped protrusion. In order to construct this feature, follow the procedure:

1. Select *Feature, Create, Protrusion, Extrude, Thin,* and *Done.*
2. Choose *One Side* and *Done.*

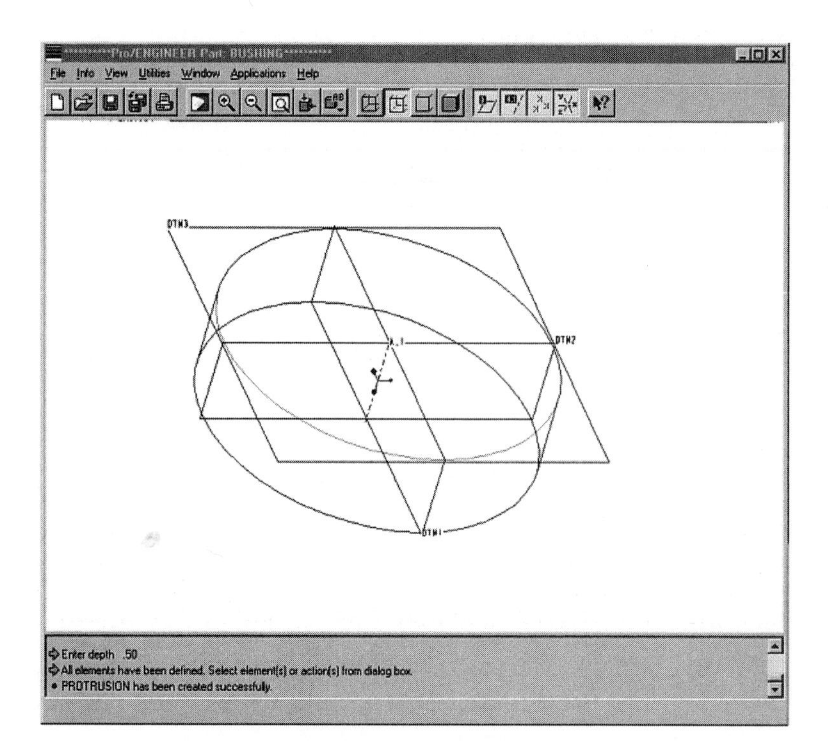


**FIGURE 3.6** *After the first protrusion, the bushing takes on this appearance.*

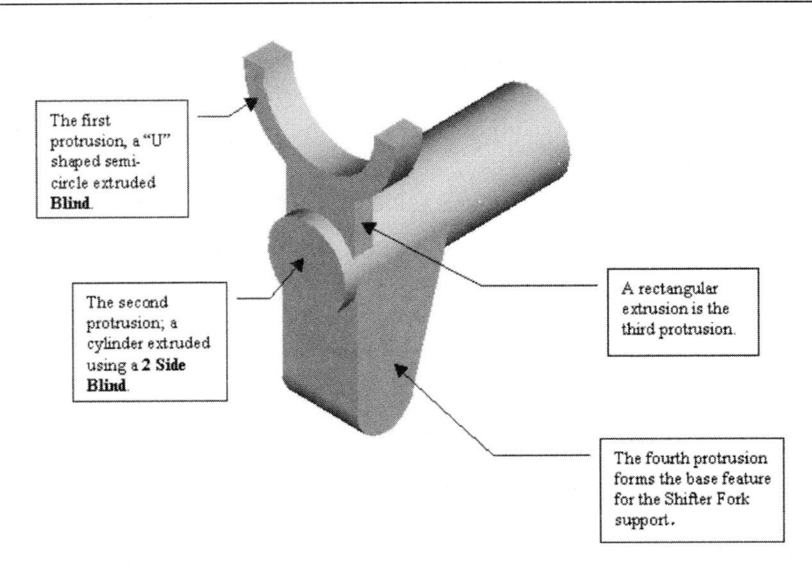


FIGURE 3.7 *The model of the shifter fork is constructed by using four protrusions. These protrusions are the base features.*

3. Pick DTM3 as the sketching plane and define DTM2 as the "*Top.*" It will orient the model as shown in Figure 3.8.
4. Make sure that the arrow points *into* the screen.
5. Sketch the semi-circle in the *Sketcher.*

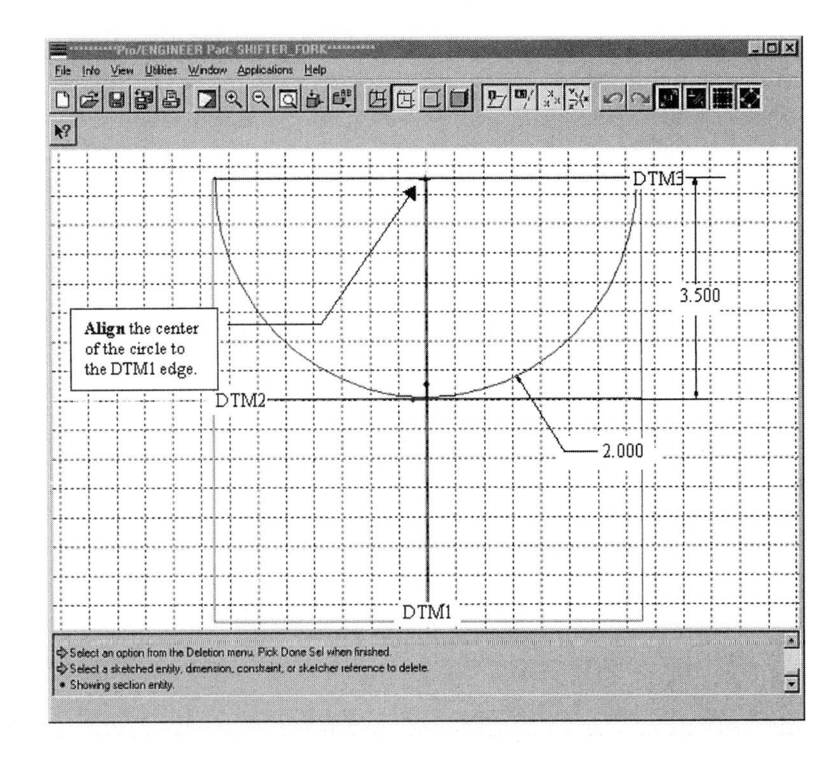


FIGURE 3.8 *Sketch a semi-circle for the first protrusion. Align the center to the DTM1 edge.*

6. *Dimension* the sketch. Align the center of the semi-circle to the DTM1 edge. *Regenerate* and *Modify* to values given in Figure 3.8. Choose *Done.*
7. The thickness is 0.500.
8. Extrude the model *Blind* to a value of 0.750.
9. Select *OK.*
10. The model after extrusion of the first protrusion is shown in Figure 3.9.

The circular cylinder, that is, the second protrusion, is constructed next. Because of the way we have chosen to orient the model with respect to datum plane DTM3, we need to use a *2 Side Blind* extrusion. Otherwise, one end of the cylinder will not extend ahead of the first protrusion.

The second protrusion is straightforward. You may wish to follow these steps:

1. Select *Feature, Create, Protrusion, Extrude, Solid,* and *Done.*
2. Use the *Both Sides* option.
3. Pick DTM3 as the sketching plane and DTM2 as the "*Top.*" This model will be oriented as in Figure 3.10.

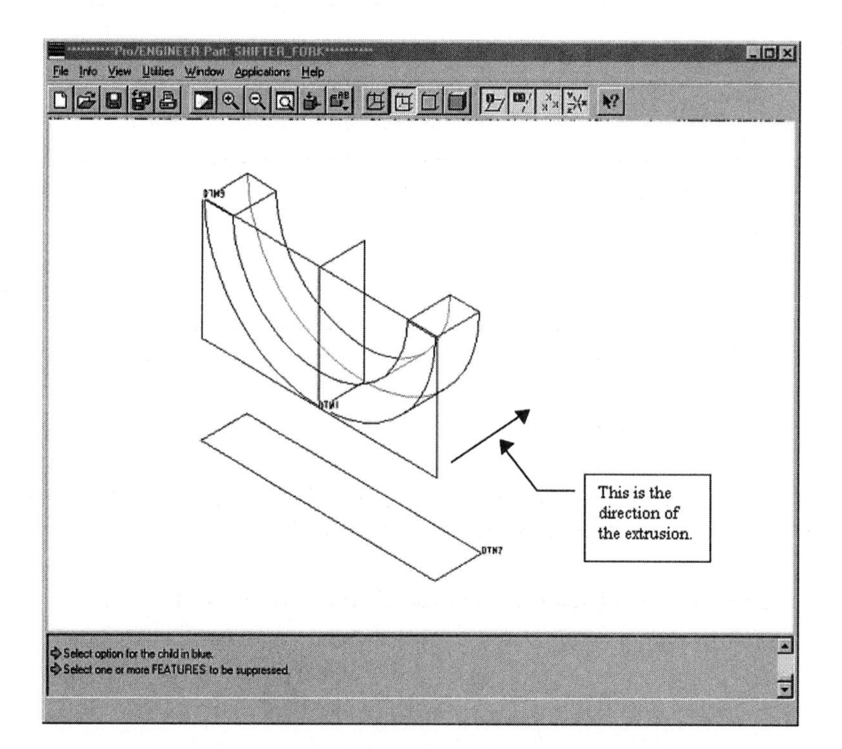


FIGURE 3.9 *After the first protrusion has been added, the model takes on the appearance illustrated in the figure.*

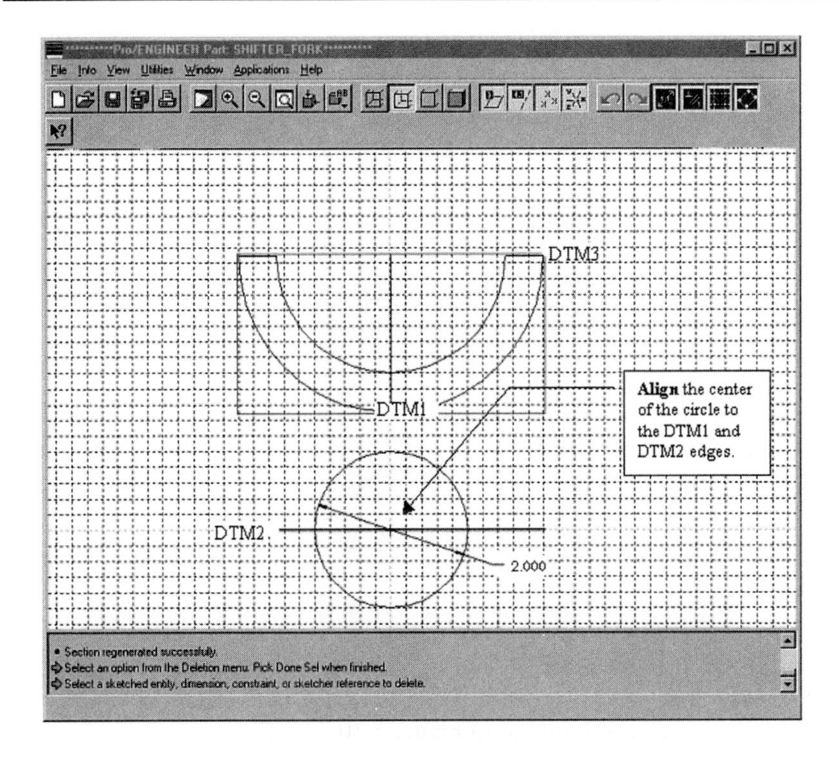


FIGURE 3.10 *Sketch a circle and align the center of the circle to the DTM1 and DTM2 datum plane edges.*

4. The direction of the extrusion doesn't really matter because of the *Both Sides* option. However, in order to obtain the same orientations as in this book, set the arrow to point *into* the screen.
5. Sketch a circle and align the center of the circle with the DTM1 and DTM2 edges.
6. *Dimension* the circle. *Regenerate* and *Modify* to 2.000 inches.
7. Choose *Done.*
8. Select *2 Side Blind* for the depth. In the direction shown in Figure 3.11, enter a value of 0.350. In the direction opposite to the one shown, enter a value of 6.00 − 0.350 = 5.650.
9. The model after the extrusion is shown in Figure 3.12.

The third protrusion may be constructed by following the same approach used for the first two protrusions. The approach has the following steps:

1. Select *Feature, Create, Protrusion, Extrude, Solid,* and *Done.*
2. Choose the *One Side* option.
3. Select DTM3 as the sketching plane and DTM2 as the "*Top.*" The model will be reoriented as in Figure 3.13.

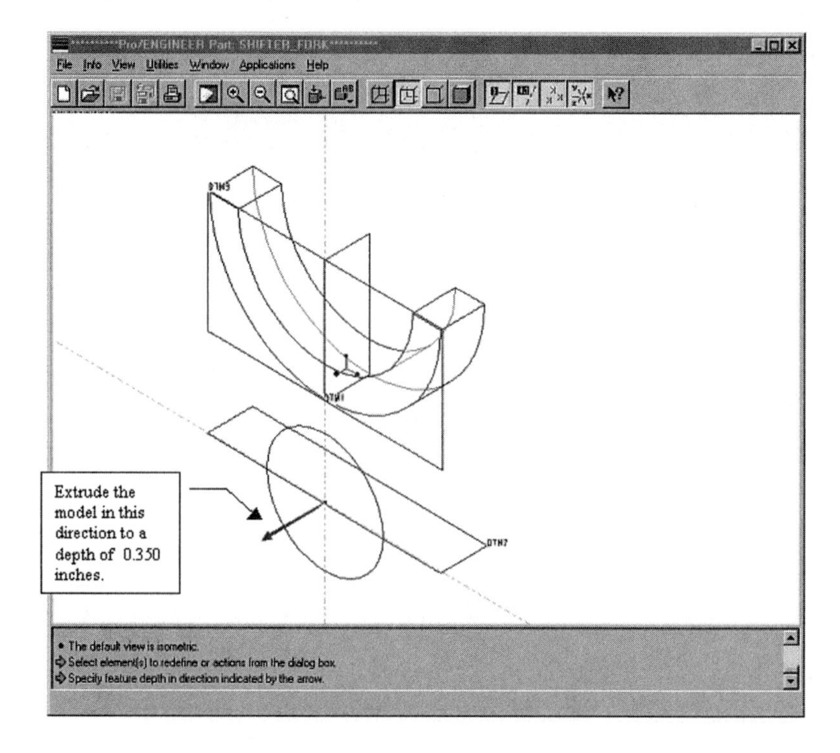


**FIGURE 3.11** *The second protrusion is constructed using a* 2 Side Blind, *which requires the user to enter the depth in both directions of the sketching plane.*

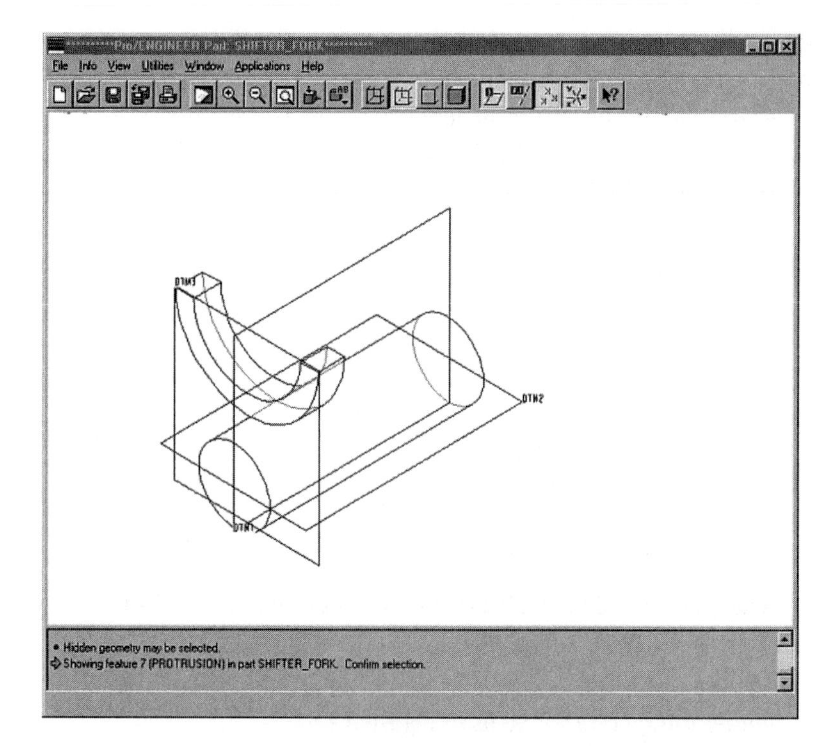


**FIGURE 3.12** *After the second protrusion, the model takes on the shape shown in the figure.*

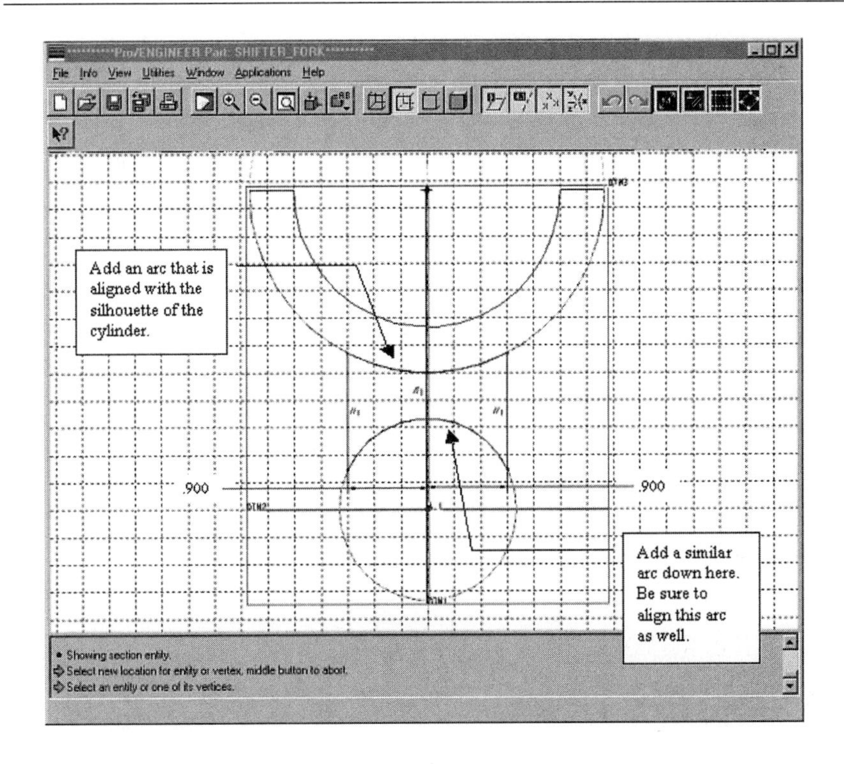


FIGURE 3.13 *The section for the third protrusion may be constructed by using two arcs and two vertical lines.*

4. For the direction of the extrusion, use the same as the one for the first protrusion.
5. Sketch the section shown in Figure 3.13. You may want to use two arcs aligned as shown in the figure.
6. *Dimension* and *Modify* the sketch to the dimensions shown.
7. *Regenerate* the model and select *Done.*
8. Extrude the feature to a *Blind* depth of 0.500.
9. Choose *OK.*
10. The model, with the third protrusion, is reproduced in Figure 3.14.

The fourth and final protrusion forms the base feature for the support of the fork. Because of the geometry of the section, we may use datum DTM1 as the sketching plane. Later, we can complete the section by using the *Cut* option to remove unwanted material from the feature.

Complete the final protrusion by following the sequence of steps:

1. Select *Feature, Create, Protrusion, Extrude, Solid,* and *Done.*
2. Choose the *Both Sides* option.
3. Select DTM1 as the sketching plane and DTM2 as the "*Top.*" The model will be reoriented as in Figure 3.15.

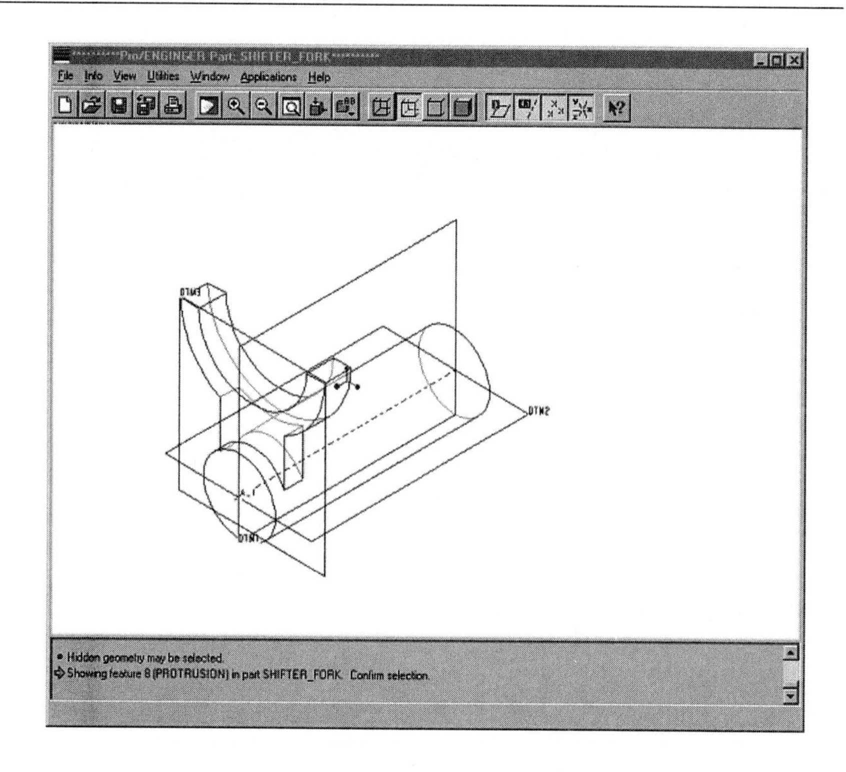


**FIGURE 3.14** *With the addition of the third protrusion, the effects of using the* 2 Side Blind *option for the second protrusion are readily visible.*

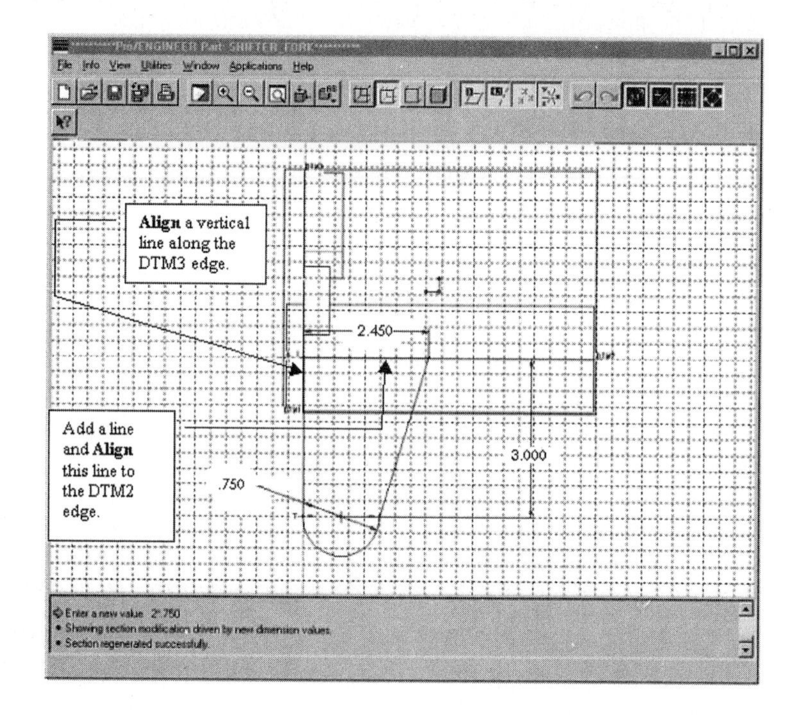


**FIGURE 3.15** *The geometry for the section of the fourth protrusion.*

4. The direction of the extrusion arrow really is immaterial. However, we have chosen the direction wherein the arrow points into the screen.
5. Sketch the section shown in Figure 3.15. Note that a line must be placed along the DTM2 edge. This line is required to close the section.
6. *Dimension, Regenerate,* and *Modify* to the appropriate values.
7. Choose *Done.*
8. Extrude *Blind* to a depth of 2.000 inches.
9. Choose *OK.*
10. The model is shown in Figure 3.16.

Save the model. It will be completed using the *Hole* and *Cut* options in a later chapter.

## 3.2 DATUM PLANES ADDED BY THE USER

Consider the model of the rod support shown in Figure 3.17. The support contains an inclined plane (surface A). For the most part, inclined or oblique

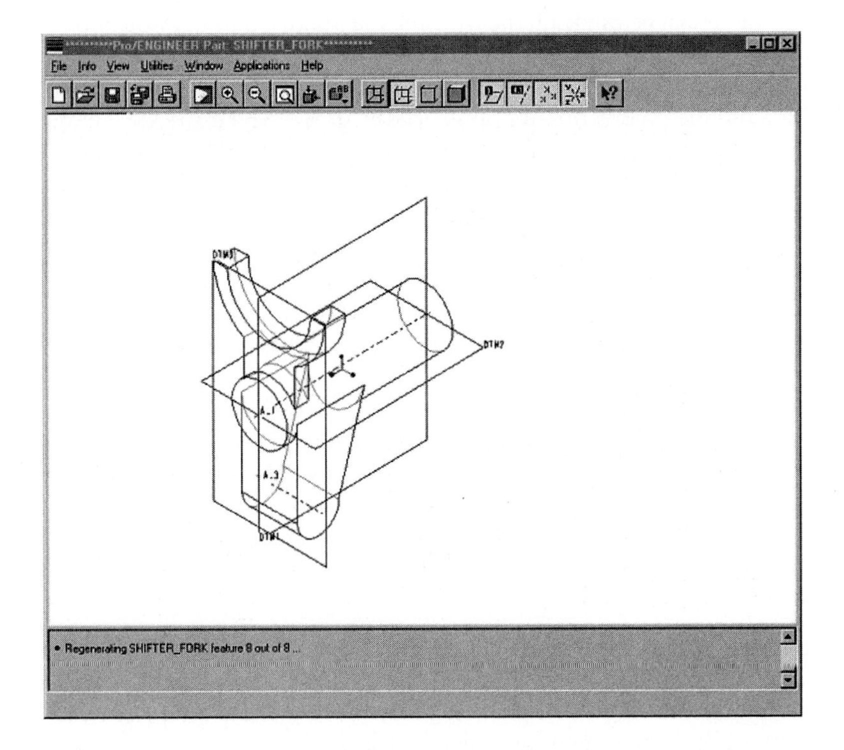


FIGURE 3.16 *A hidden line drawing of the model.*

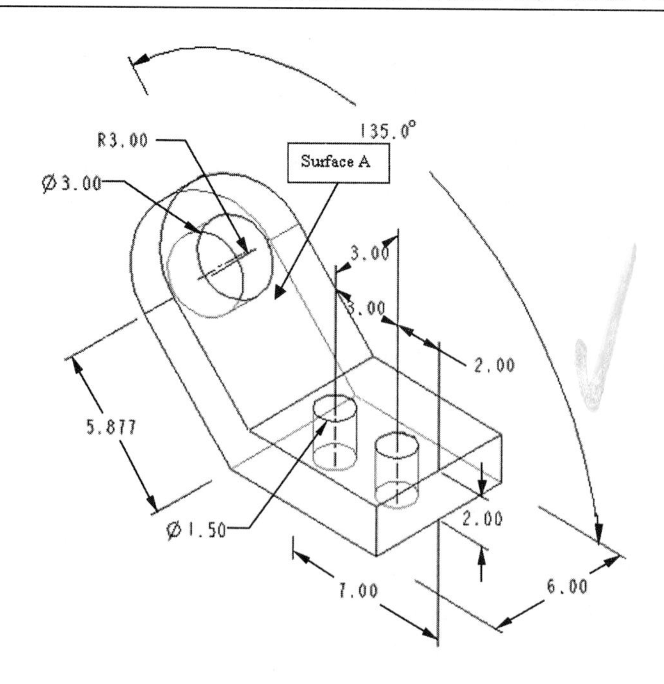


**FIGURE 3.17** *The model contains an inclined plane.*

planes are difficult to construct unless special sketching planes are created. These planes can be made on the fly using the *Make Datum* option or by constructing the plane ahead of time using the *Create* and *Datum* (Figure 3.1) options.

An example where the option is used is in the *SETUP SK PLN* menu (Figure 3.3). You may recall that this menu is used before entering the *Sketcher* to define the sketching plane and the orientation plane.

In using the *Make Datum* option, the user is presented with the *MAKE DATUM* menu. This menu is identical (except for its name) to the *DATUM SK PLN* menu shown in Figure 3.3. The options contained within the *DATUM PLANE* menu allow the user to create datum planes by properly choosing the constraints necessary to locate the plane with respect to the model or other existing datum planes.

In generating a datum plane, the user must be aware of these constraints. The datum plane must be properly constrained by choosing the appropriate combinations of constraining options. These options are *Through, Normal, Parallel, Offset, Angle, Tangent,* and *Blend Section.*

Some options will fully constrain the plane without any additional added constraints. Pro/E calls these stand-alone constraints. The stand-alone constraints are:

1. *Through/ Plane.* A datum is constructed coinciding with a plane.
2. *Offset/Plane.* A datum plane is constructed parallel to an existing plane at an increment provided by the user.

3. *Offset/Coord Sys.* The datum plane is created perpendicular to one of the axes of the coordinate system and offset from the origin.
4. *Blend Section.* This option creates a datum plane through the section that was used to create the feature. Pro/E handles multiple sections by asking for the section number before creating the datum plane.

### 3.2.1 Tutorial # 14: A Rod Support

Let us create the model shown in Figure 3.17. We will create an additional datum plane in order to construct surface A. The holes will be added in chapter 4.

Construct the base by using the procedure:

1. Create a part file called "RodSupport."
2. Add the default datum planes.
3. After the datum planes are added, select *Feature, Create, Protrusion, Extrude, Solid,* and *Done.*
4. Use the *Both Sides* option.
5. Select DTM3 as the sketching plane and DTM2 as the "*Top.*"
6. In the *Sketcher,* create the rectangular base of the part. Sketch a $7 \times 2$ inch rectangle with the DTM2 plane exactly between the 2 inch side.
7. *Dimension, Regenerate,* and *Modify.*
8. After successfully regenerating the sketch, extrude to a *Blind* depth of 6 inches. Your base should appear as in Figure 3.18.

In general, at least two datum planes must be created to sketch geometry on an inclined plane. One of these planes is to be used as the sketching plane, while the other is used to orient the part in the *Sketcher.* In this case, only one user-defined datum plane need be constructed, because any of the sides of the model can be used to orient the part (for example, Surface D in Figure 3.18).

The placement of these datum planes must be properly chosen. However, where should the first datum plane be located? One location would be through Line A in Figure 3.18 at 45° to the horizontal (135° to Surface C). In this case, after extruding the section, the model would appear as in Figure 3.19. Because the sketch was extruded perpendicular to the sketching plane, the protrusion does not pass fully through the base as it should.

The protrusion may be properly created by placing the sketching plane so that it passes through Line B instead of Line A as in Figure 3.18. The perpendicular direction of this new plane would pass through the base and so would the protrusion.

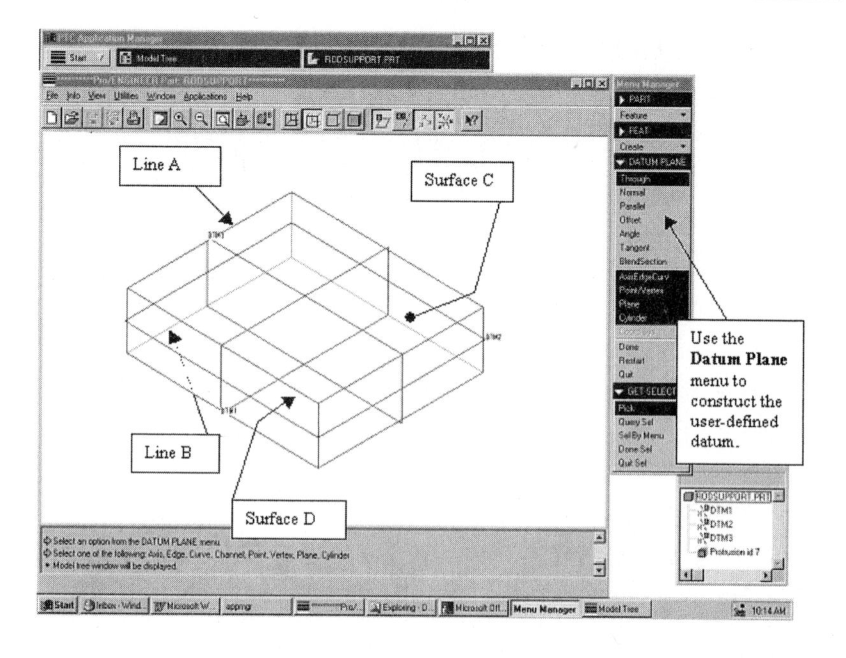


**FIGURE 3.18** *The base of the rod support after the first protrusion. The user-defined datum is located using the options from the* DATUM PLANE *menu. The location of this datum must be properly chosen.*

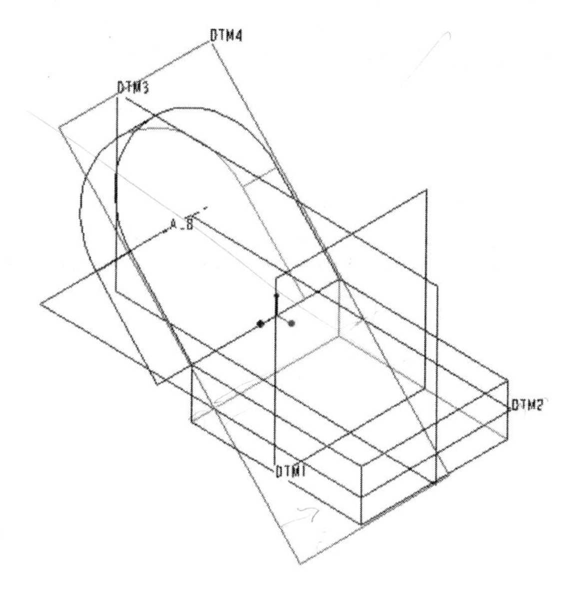


**FIGURE 3.19** *The inclined surface created by using a sketching datum plane through line A. Note the protrusions do not blend into one another.*

Create an inclined plane (DTM4) passing through Line B and at an angle of 45° to surface D. The steps for carrying out this task are:

1. Select *Feature, Create, Datum,* and *Plane.*
2. From the *DATUM PLANE* menu (Figure 3.18) choose *Through.*
3. Use *Query Sel* and pick Line B. *Accept.*

4. From the *DATUM PLANE* menu, select *Angle.*
5. Using *Query Sel,* pick Surface C. *Accept.*
6. At this point, the datum should be constrained and the options in the *DATUM PLANE* menu should be grayed. If they are not, choose *Restart* and proceed to step 2. Otherwise, select *Done.*
7. The value of the angular dimension is required. Select *Enter Value.* Note the direction of the arrow. If the arrow is pointing in the counterclockwise direction, enter 45. Otherwise, enter −45.
8. Hit *Return.*
9. Pro/E will create the datum. Is the datum the same as the one shown in Figure 3.20?

Sketch the appropriate section. Use DTM4 as the sketching plane and Surface D as "*Left.*" Construct the sketch shown in Figure 3.21 in the *Sketcher* and dimension. Be sure to align the features as shown in the figure. *Regenerate. Modify* the dimensions to the values 5.877 and 3.00.

Choose *Done* and extrude *Blind* to a depth of 2.00. Your part should appear similar to the one shown in Figure 3.17. In chapter 4, we will add the holes, so save your work. In adding the hole on the inclined surface, Surface A will be used as the placement surface.

## 3.3 START PART

Many engineering organizations utilizing Pro/E set up what is commonly referred to as a start part. The start part, an example of which is given in Figure 3.22, usually contains the default datum and settings for the units and

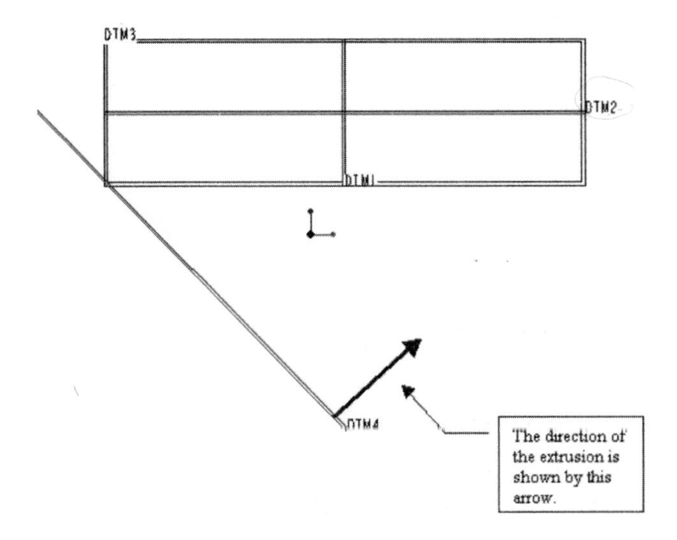


FIGURE 3.20 *The base after adding the user-defined datum plane DTM4. The model has been reoriented to show an edge view of the plane.*

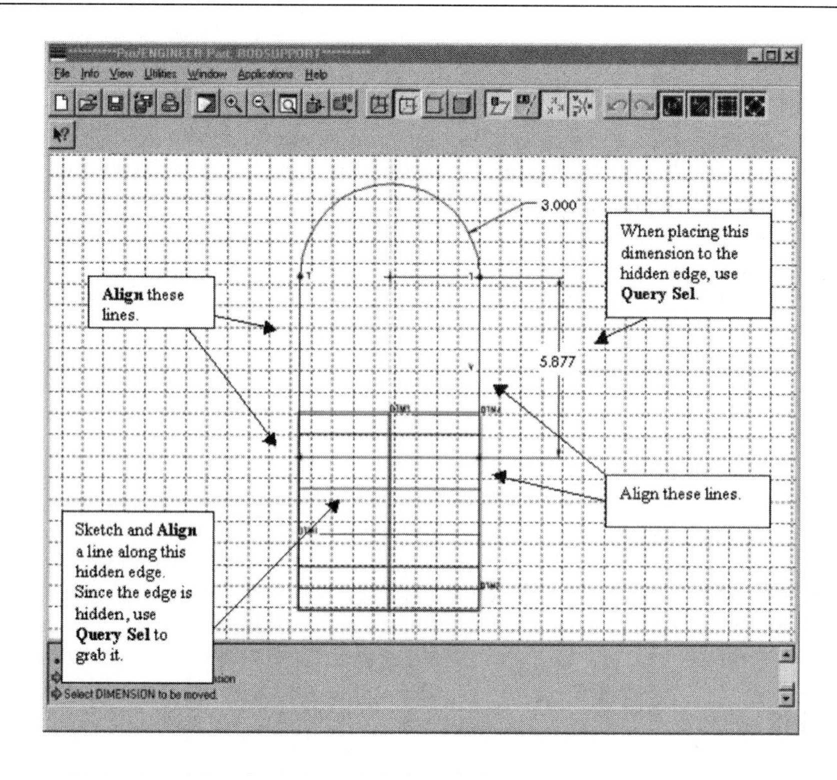


**FIGURE 3.21** *The part in the* Sketcher *with the geometry for the inclined plane. Align the features as shown. If you do not, you will need to add additional dimensions. When selecting the hidden edge, use* Query Sel.

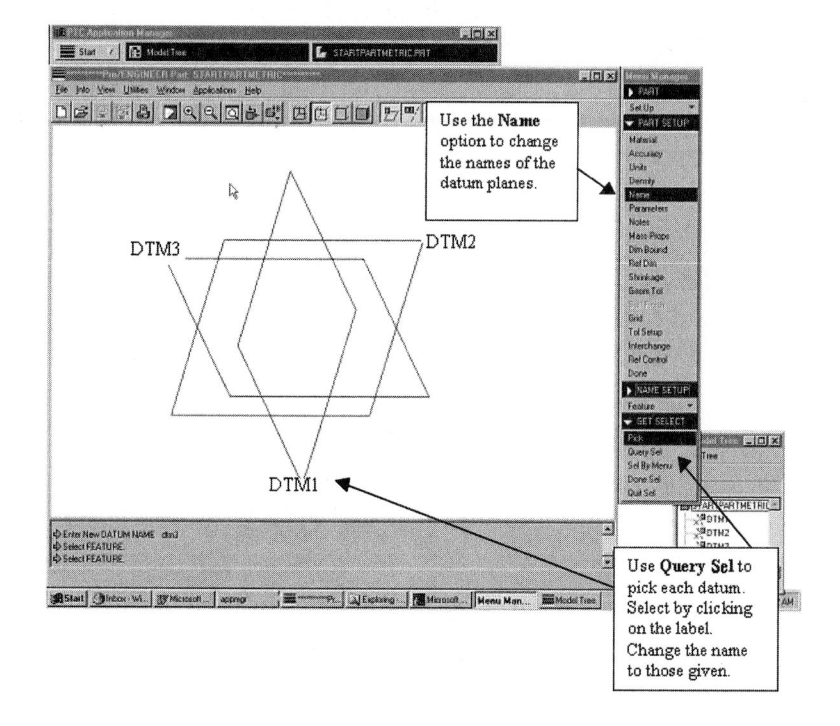


**FIGURE 3.22** *A start part can be constructed with predefined features such as the default datum planes. Often the names of the datum planes are changed.*

material properties. Many of these parts also have a coordinate system located at the intersection of the three default datum planes.

The purpose of the start part is to speed up part creation. Instead of creating a new part and loading the default datum planes for every new part, the start part can be loaded with datums in place. In the case of a metric part, a version of the start part with the metric units can be placed in the database for subsequent use.

Pro/E is bundled with start parts created by PTC. Check with your system administrator for the availability of these parts.

In this book, we assume that a start part is not being used. However, you may want to set up one or use the ones provided by PTC. In the next tutorial, we will examine how a simple start part can be created.

### 3.3.1 TUTORIAL # 15: A METRIC START PART

It is a simple matter to set up your own custom start part. Suppose that we want to set up a start part that contains the default datum planes and the units set to metric. Then, whenever a metric part with the default datum planes is needed, the start part can be loaded, the rest of the model completed, and the part saved under a new name using the *Save As* option.

For most parts, the model is oriented such that DTM3 is the front, DTM2 is the top and DTM1 is a side. Therefore, it might be advantageous to rename the datum planes.

Let us create the part shown in Figure 3.23. Begin by creating a part called "StartPartMetric."

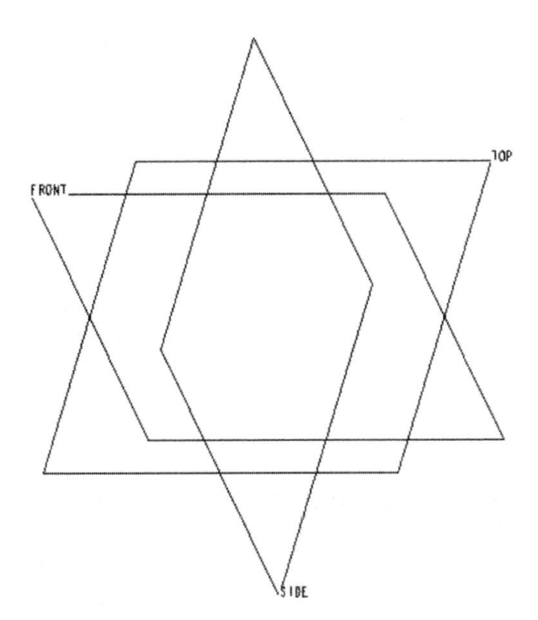


FIGURE 3.23 *The default datum planes may be renamed so that the orientation of the model has more significance. Typically, "Front," "Top," and "Side" as shown in the figure are used for this purpose.*

Select *Set Up* from the *PART* menu (Figure 1.4). Then choose *Units* from the *PART SETUP* menu shown in Figure 2.7. Because the millimeter is the standard unit for a metric part, select *Principal Sys* and change the *Length* parameter from in to mm. Click on *Done* and then *Done/Return.* Now the units of the part have been set.

Select *Feature, Create, Datum, Plane,* and *Default,* which will load the default datum. Choose *Done.* Now, we can rename the datum planes.

To change the name of the datum (or any feature for that matter) select *Set Up* and *Name* from the *PART SETUP* menu (Figure 2.7). It does not matter in what order the datums are renamed.

Suppose we start with DTM3. Choose *Query Sel* and, using your mouse, pick the DTM3 label. Choose *Accept.* Enter the new name: "Front." Now pick DTM2 and *Accept.* Enter the name: "Top." Finally, pick DTM1 using *Query Sel* and *Accept.* Change the name to "Side." Choose *Done Sel* and *Done.* The datum planes with their new names are shown in Figure 3.23.

Many start parts also have predefined coordinate systems. Such systems are discussed in chapter 9. Save this part. In chapter 9, we will add a coordinate system.

## 3.4 SUMMARY AND STEPS FOR ADDING DATUMS

For models with one or less planar surface, it is difficult to reorient the model in the *Sketcher.* Furthermore, a surface on the model suitable as a sketching plane may not exist. These difficulties may be overcome by using datum planes. Datums are features that may be renamed, modified, or deleted.

In Pro/E, datum planes come in two types. The first is the default datum planes that must be added to a part before any other feature. In general, the default datum planes may be added to a part by selecting: *Feature, Create, Datum, Plane,* and *Default.*

The second type is a user-defined datum. Such datums may be added to the model at any time, provided the user is able to locate the datum. This task is accomplished by using the options from the *DATUM PLANE* menu (Figure 3.18). The general steps for adding a user-defined datum to a model are as follows:

1. Select *Feature, Create, Datum,* and *Plane.*
2. Choose the constraint from the *DATUM PLANE* menu (Figure 3.18).
3. Pick the required feature from the model.
4. If the constraint is a stand-alone constraint, the datum may be placed by selecting *Done.* Otherwise, select additional constraints. When all the constraints have been chosen, select *Done.* Note that the entire procedure may be restarted at this point by using the *Restart* option.

5. Enter any additional information, such as an offset distance or angular dimension, and hit *Return.*
6. Pro/E will create the datum.

Most engineering departments use start parts. Start parts usually contain the default datum planes that have been renamed to names that are more meaningful. The names of a datum plane, or any feature, may be changed by using the following sequence:

1. Select *Set Up* and *Name.*
2. Click on the feature.
3. Enter the new name.

## 3.5 ADDITIONAL EXERCISES

3.1 Construct the model of the heat sink plate shown in Figure 3.24. Use user-defined datums.

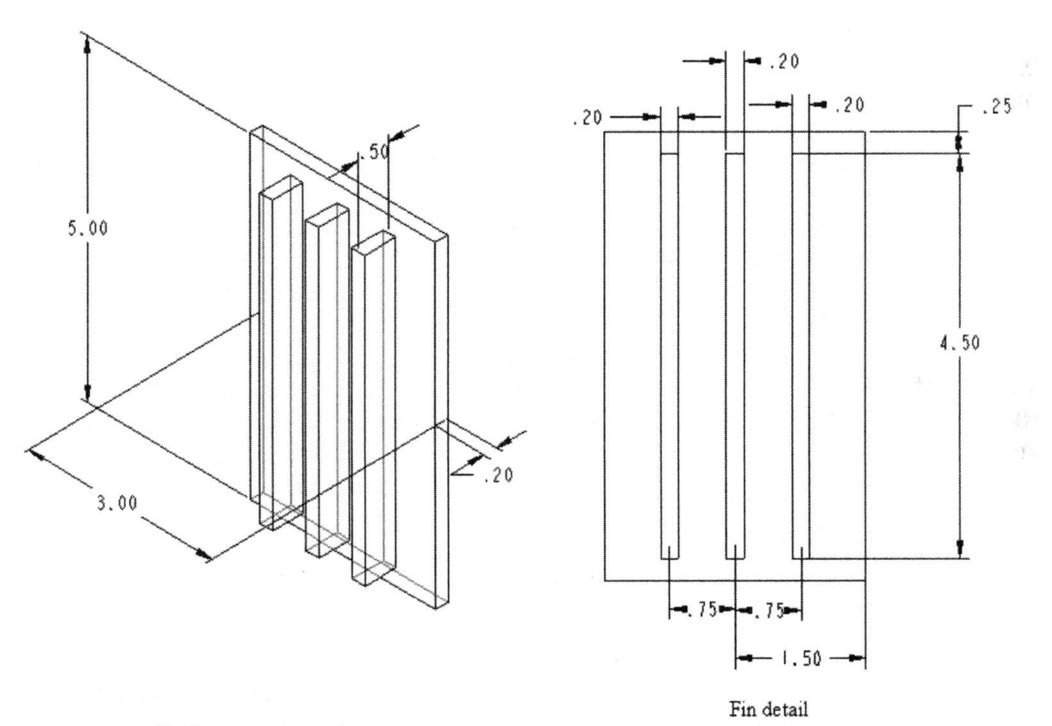


FIGURE 3.24 *Figure for exercise 3.1.*

3.2 Create the bushing shown in Figure 3.25. Do not create the hole.

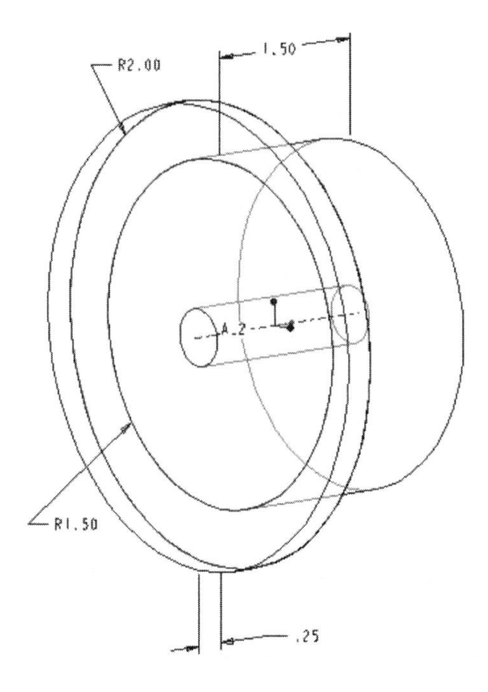


FIGURE 3.25 *Figure for exercise 3.2.*

3.3 Using the default datum planes, construct the axle shown Figure 3.26.

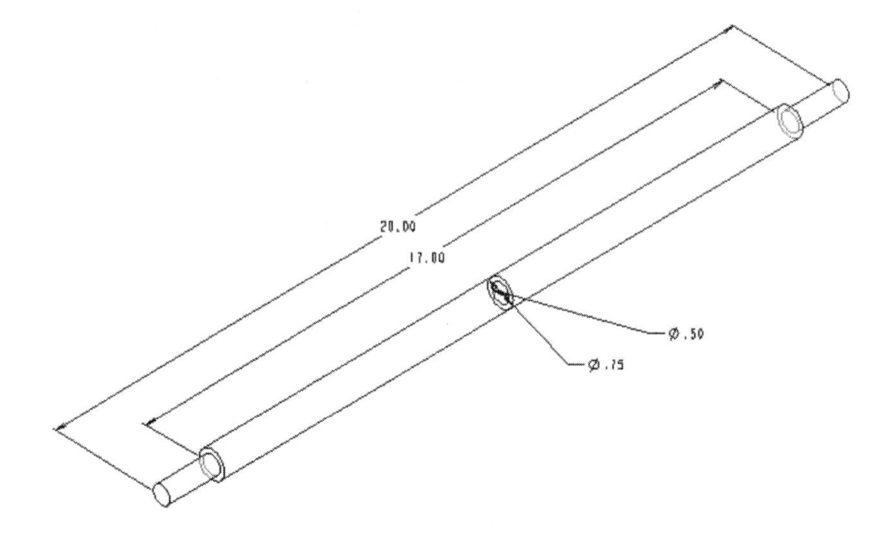


FIGURE 3.26 *Figure for exercise 3.3.*

3.4 The volume control dial illustrated in Figure 3.27 may be created using the default datum planes. Construct the model.

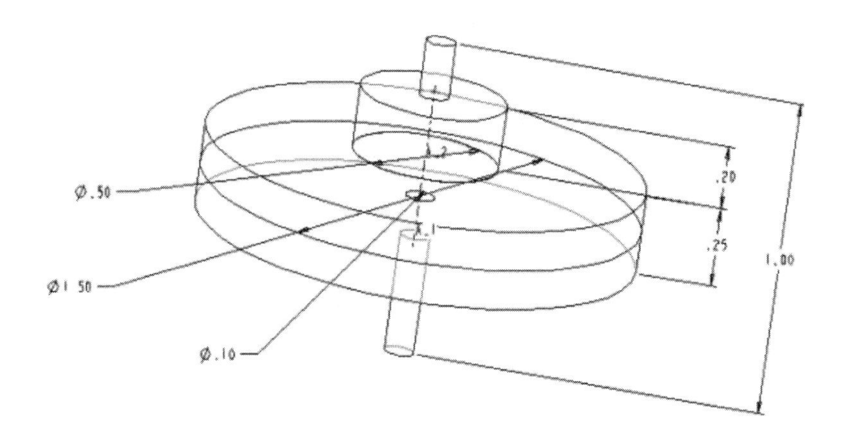


FIGURE 3.27 *Figure for exercise 3.4.*

3.5 Construct the base feature for a cement trowel as shown in Figure 3.28.

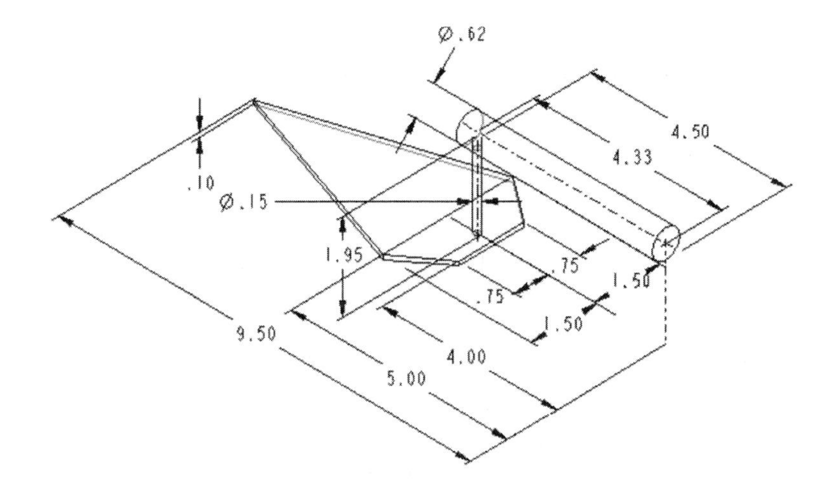


FIGURE 3.28 *Figure for exercise 3.5.*

3.6 Use the default datum planes in order to create the base features for a control handle as given in Figure in 3.29.

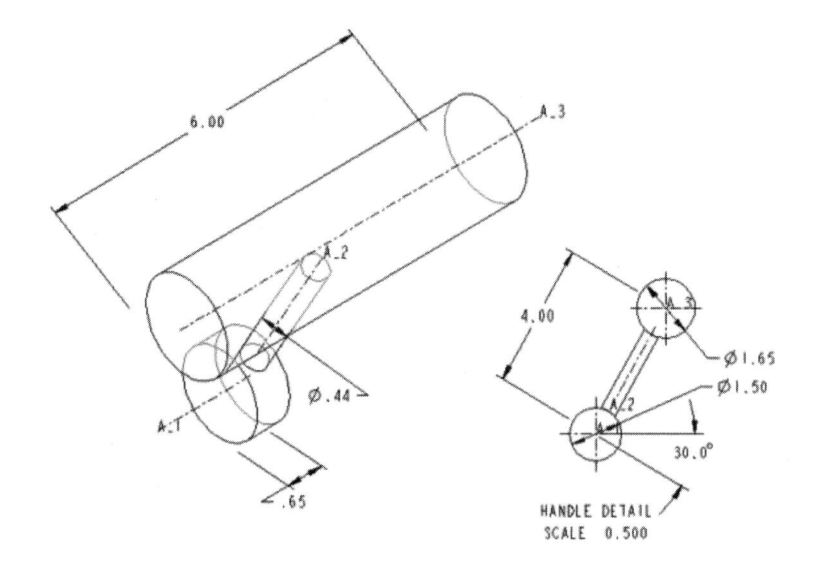


FIGURE 3.29 *Figure for exercise 3.6.*

3.7 The base features for an angled rod support may be created by using the default datum planes as well as an additional user-defined datum. Create the model, using Figure 3.30 as a guide.

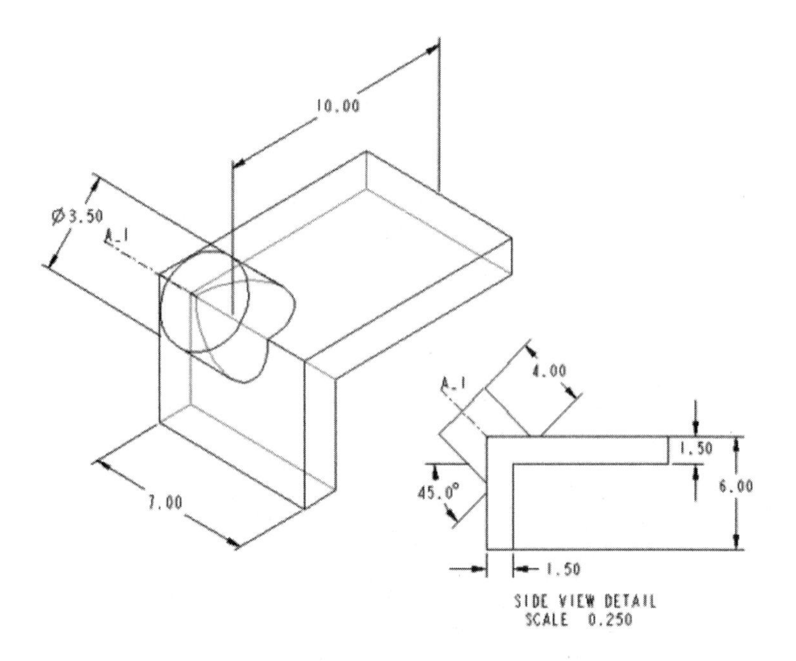


FIGURE 3.30 *Figure for exercise 3.7.*

3.8 Construct the features for the side guide as shown in Figure 3.31.

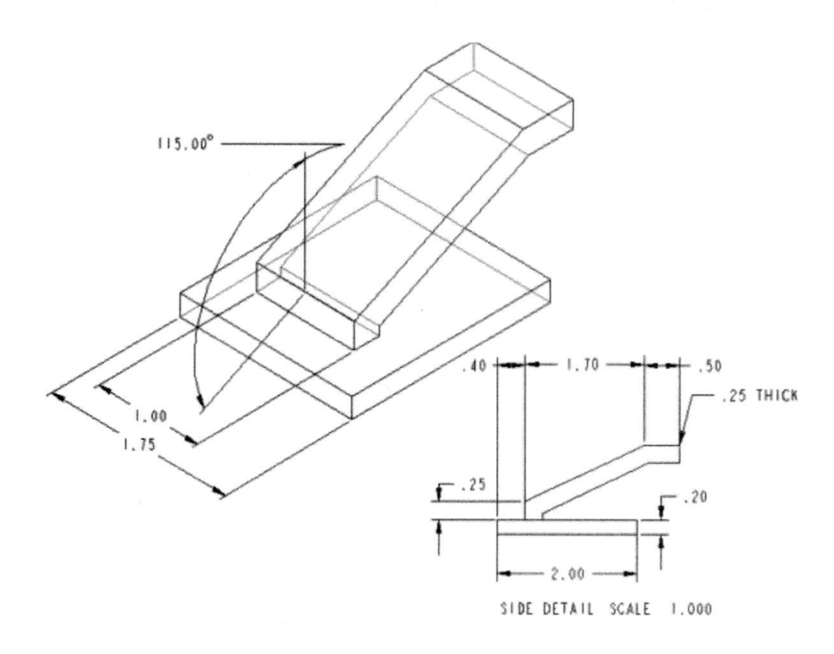


**FIGURE 3.31** *Figure for exercise 3.8.*

3.9 Use the geometry shown in Figure 3.32 to create the model of the offset bracket. Note an additional user-defined datum is required.

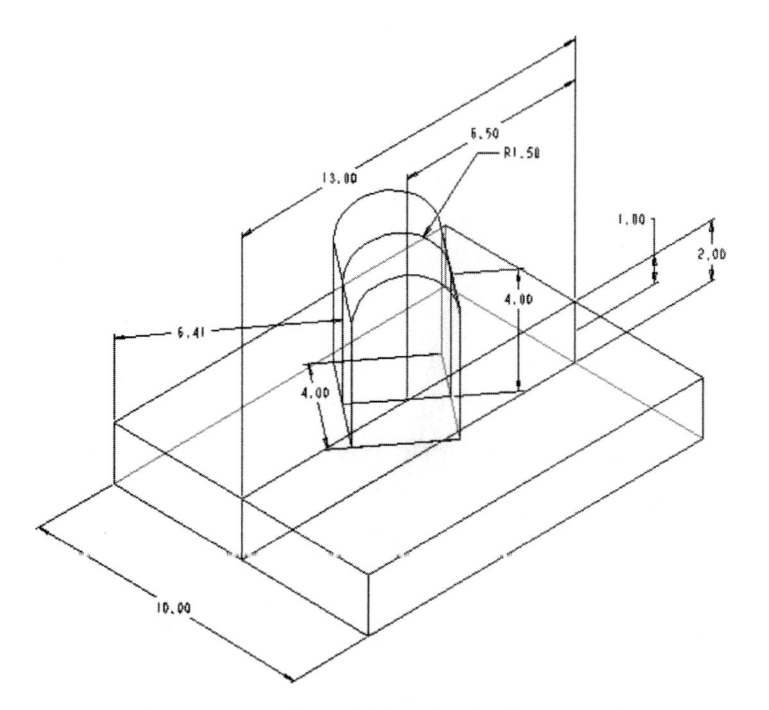


**FIGURE 3.32** *Figure for exercise 3.9.*

3.10 Construct the model of the side support reproduced in Figure 3.33. Hint: create a datum plane through the axis A2 and the lower edge of the rectangular base. Then project *Blind* 3.00.

FIGURE 3.33 *Figure for exercise 3.10.*

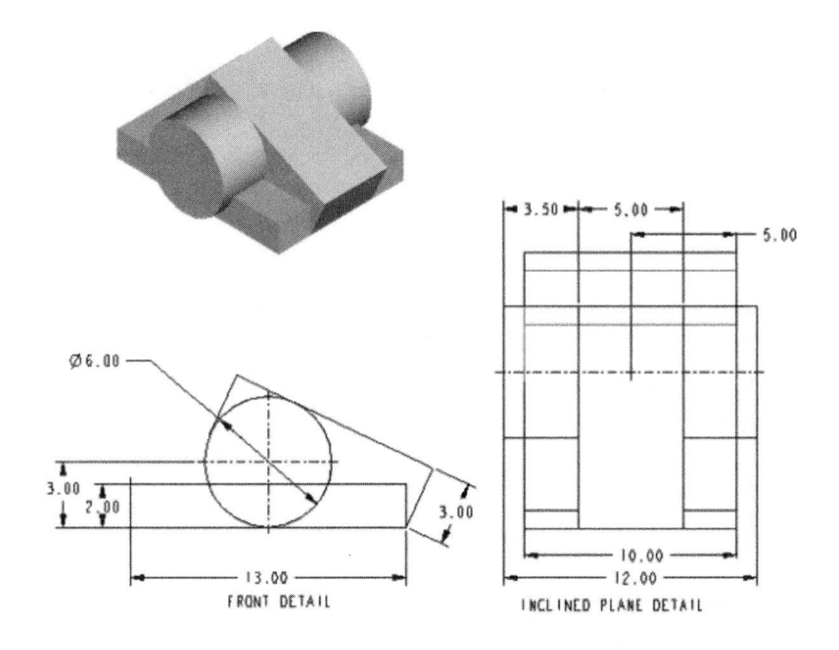


3.11 Create the model of the double side support as shown in Figure 3.34.

FIGURE 3.34 *Figure for exercise 3.11.*

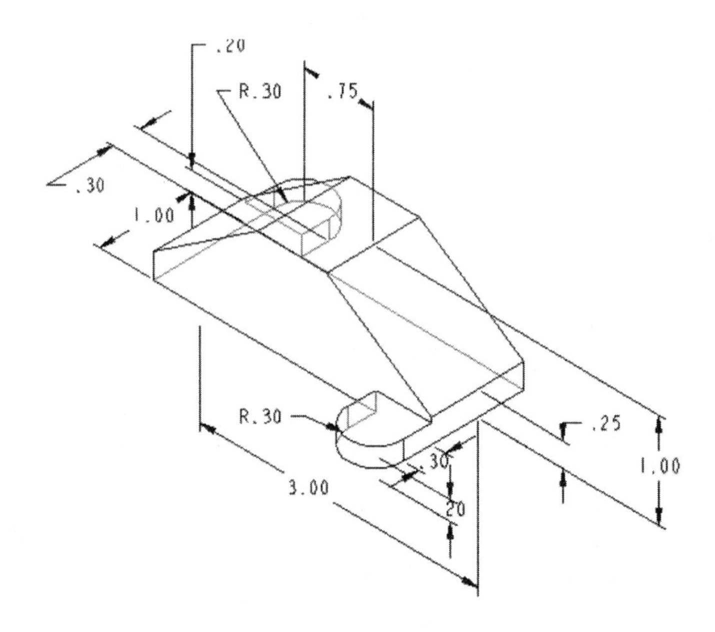

3.12 The slotted support is a part with metric units. Construct the model.

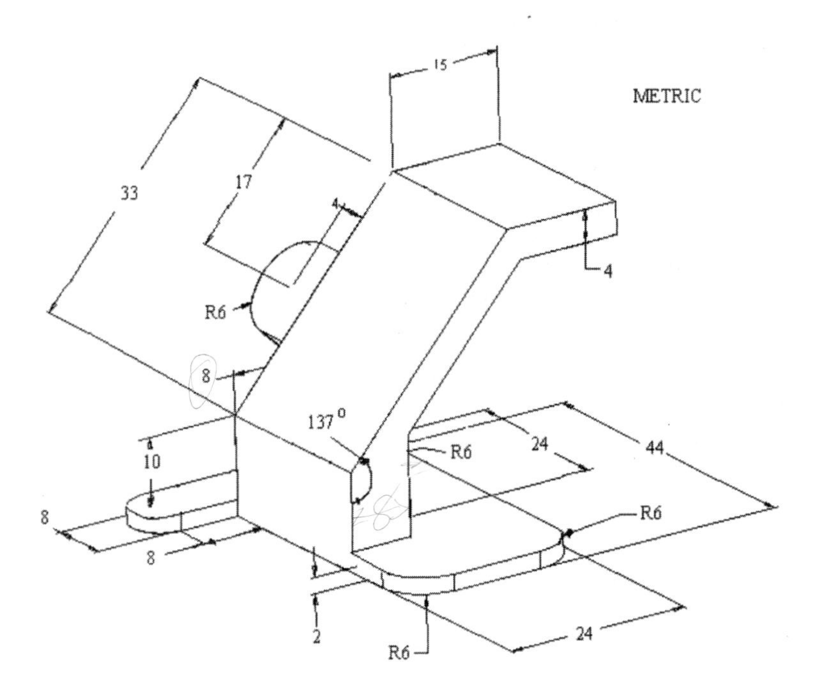


**FIGURE 3.35** *Figure for exercise 3.12.*

3.13 The clamp base may be created using a user-defined datum. Note that the units of the part are metric.

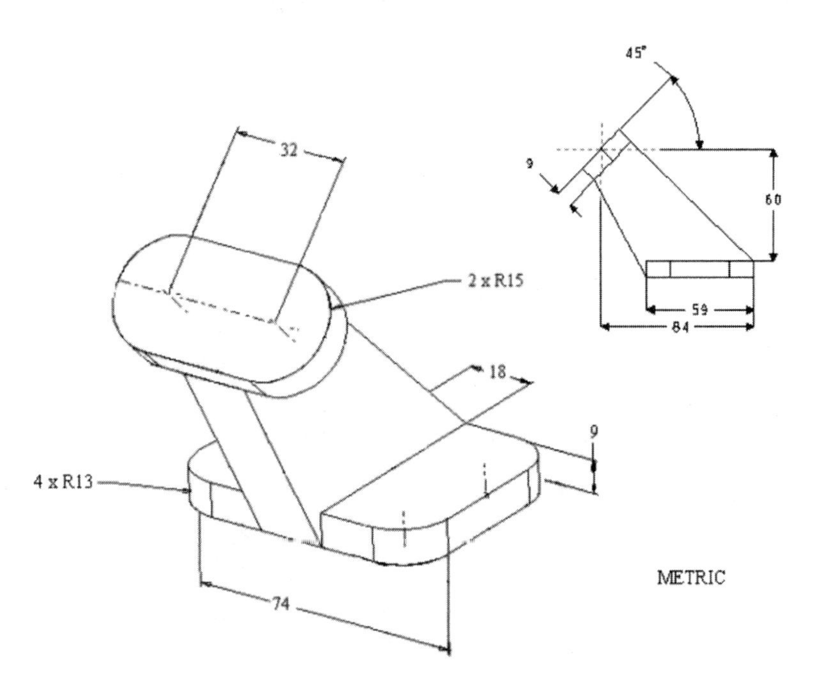


**FIGURE 3.36** *Figure for exercise 3.13.*

# CHAPTER 4

# Adding Holes to Base Features

## Introduction and Objectives

So far, all the models we have constructed are composed of protrusions. These base parts do not contain any holes. In this chapter, we will discuss how to create a hole. With Pro/E, holes can be placed by locating their centers with two linear dimensions along a known axis, at a point, or on a radial direction.

In creating a hole, the user must define three distinct aspects of the hole. The first is the geometry of the hole, that is, whether the hole has a constant diameter. The second is the location of the hole. The third aspect is the depth. These details, in creating a hole, will be considered in more detail in the sections that follow.

Thus, with these different aspects in mind, the objectives of this chapter become:

1. The reader will understand the different methods for locating a hole on a base feature.
2. The reader will know the difference between a *Straight* and *Sketched* hole.
3. The software user will be able to specify the depth of a hole.

## 4.1 Adding Holes in Pro/E

Pro/E allows the user to define a *Straight* or *Sketch* hole (Figure 4.1a). A sketched hole is user-defined with a varying diameter, whereas straight holes are holes with a constant diameter. *Sketched* holes are created by using the

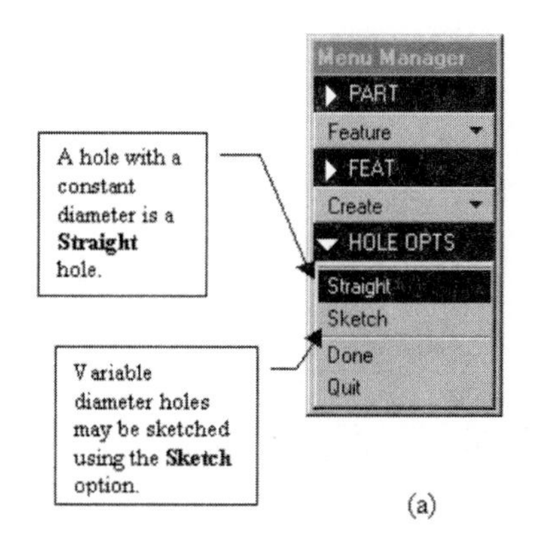


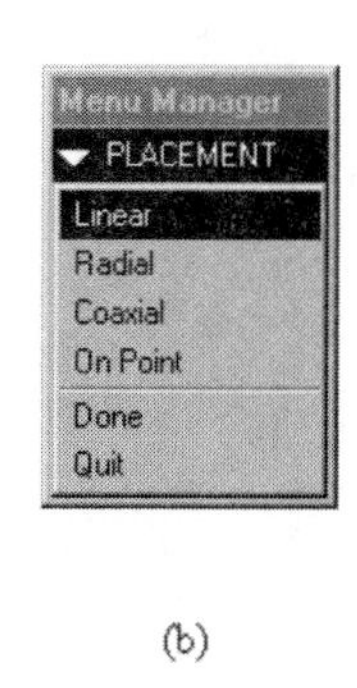


FIGURE 4.1 *(a) Hole type options are shown. (b) The Pro/E* Placement *options are illustrated. Pro/Engineer will need the depth of the hole. This information is provided by the user through the* SPEC TO *menu shown in Figure 4.2.*

*Sketcher.* Examples of sketched holes are counterbored and countersunk holes. The sketched hole is created by revolving the sketch around the axis of the hole. Sketched holes are always blind and one-sided.

Pro/E provides four options for placement of holes as seen in Figure 4.1b. The first of these methods is *Linear* placement, that is, the center of the circle is defined by linear dimensions from two boundaries or edges. In the second method, the *Radial* option, the hole is placed along a radius. This method is useful for placing holes on a bolt circle. The third way to place a hole is by using the *Coaxial* option. This option is convenient for holes along the center axis of a cylinder or shaft. Finally, the hole can be placed on a defined point, the *On Point* option.

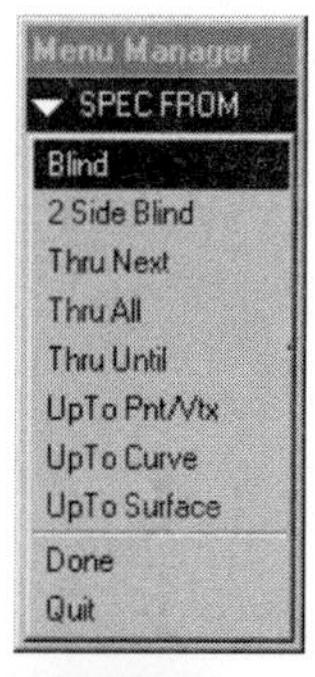


FIGURE 4.2 *The* SPEC TO *menu. This menu is used to define the depth of a* Straight *hole.*

The options from the *SPEC TO* menu are defined in the following manner:

1. *Blind.* Hole A is generated to a specific numerical depth (sd0 in Figure 4.3) provided by the user.
2. *2 Side Blind.* This option is available whenever the *Both Sides* option is used. The depth is specified from both sides of the placement plane.

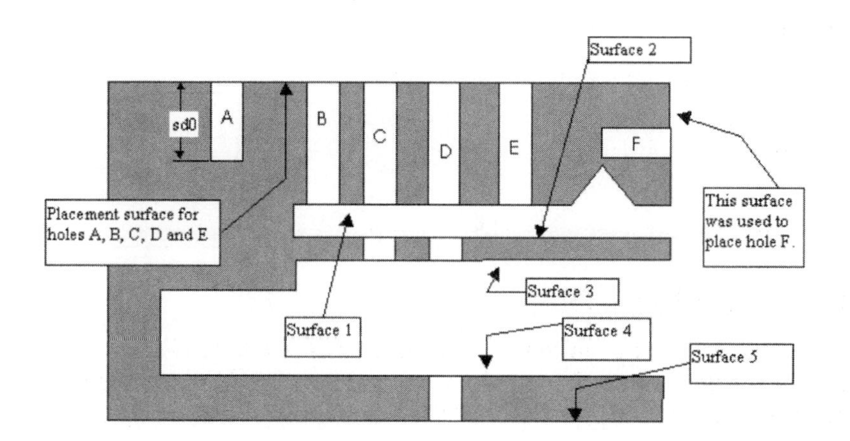


FIGURE 4.3 *A cross section of an object with holes in place using the options from the* SPEC TO *menu.*

3. *Thru Next.* The hole E is projected so that it passes and stops at the next surface (Surface 1 in Figure 4.3).
4. *Thru All.* For this option, Hole D is projected through all features and surfaces.
5. *Thru Until.* The hole (C in Figure 4.3) is created through all surfaces until Surface 3 is reached.
6. *UpTo Pnt/Vtx.* Creation of Hole F progresses until the desired point or vertex is reached.
7. *UpTo Curve.* The hole is projected up to a chosen curve.
8. *UpTo Surface.* The depth of Hole B is defined by a chosen surface or datum plane. The boundary, Surface 1 in this case, must be provided by the user.

### 4.1.1 TUTORIAL # 16: ADDING HOLES TO THE ANGLE BRACKET

In Pro/E, holes can be added to a base part by simply defining a new feature. As we saw in chapter 1, the software will define the circles and lines used to represent the hole as a single feature. For our example part, we take the angle bracket from chapter 2. Load the part. If you have not already created this part, refer to chapter 2 and construct the bracket before proceeding. The final version of the bracket with all the holes added is shown in Figure 4.4.

The holes in the angle bracket are straight type. Let us place one of these holes, say, Hole A, in Figure 4.4. Hole A and Hole B are identical except for their locations. For identical features, the Pro/E option *Pattern* may be used to create copies of the original feature. In this chapter, we will forgo the use

FIGURE 4.4 *The angle bracket with holes.*

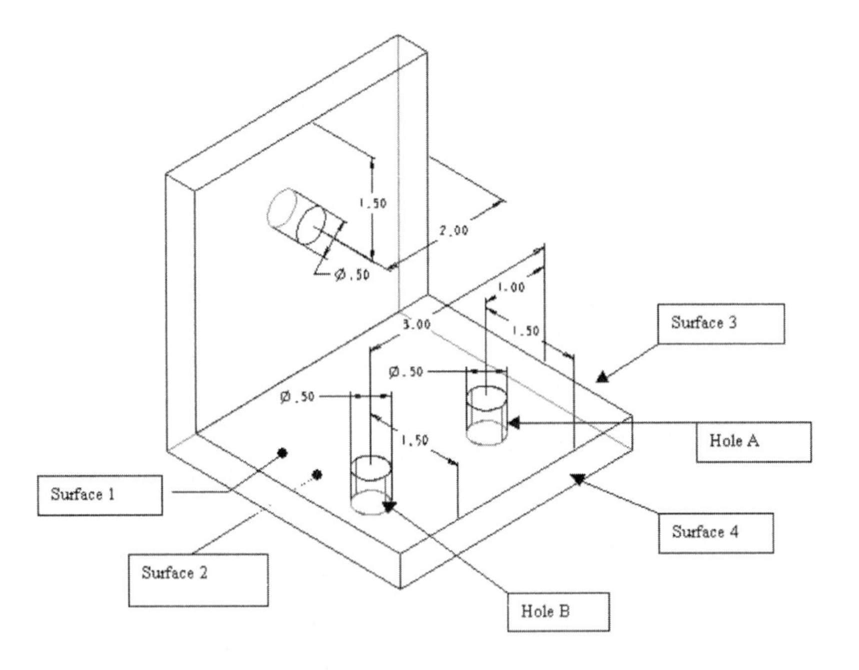

of the *Pattern* option and place Hole B using the same method used to place Hole A. The *Pattern* option will be discussed in chapter 6.

Select *Feature, Create,* and *Hole.* You will see a menu appear with the options shown in Figure 4.1a. Select *Straight.* Upon doing so, two more menus will appear. The first is a listing of the required elements for the hole and the status of the elements. This box, shown in Figure 4.5, is similar to the *Feature* creation dialog box obtained in the *Sketcher* mode (Figure 2.7). The second menu, shown in Figure 4.1b, is the *PLACEMENT* menu.

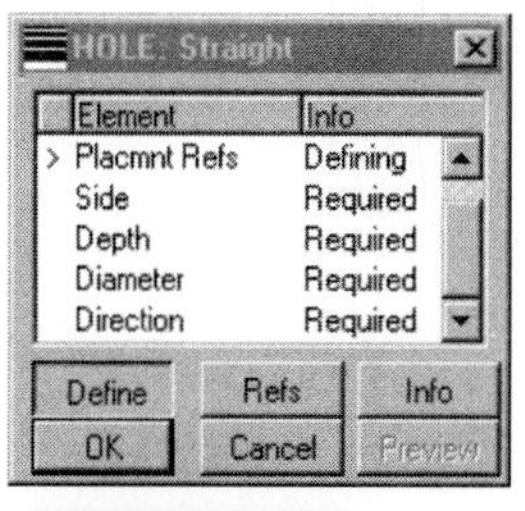


FIGURE 4.5 *Hole creation box.*

Let us place Hole A. We need to select a placement plane on which the hole will be located. Consider Figure 4.4 again. For Hole A, the placement plane can be either Surface 1 (and the hole created by projecting *down* from the surface) or surface 2 (and the hole created by projecting *up* from the surface).

Suppose we choose to place the hole on Surface 1 and project down to create the hole. Choose this surface when asked to do so. You may want to use the *Query Sel* option to select the proper surface.

Since we have selected the *Linear* option, two edges, axes, or planar surfaces need to be selected to locate the center of the circle. For the first surface, choose Surface 3 in Figure 4.4 (or any edge defining Surface 3) and enter a value of 1.00. Then select Surface 4 and enter a value of 1.50.

Because the hole is constructed by projection from one side of Surface 1 only, choose the *One Side* option when prompted to do so. Select *Done.*

Getting back to the placement of Hole A, we notice that since the hole goes entirely through the base feature, it can be constructed by using the *Thru All* option. Select *Thru All.* Enter a value of 0.5 for the diameter of the hole. Choose *OK* from the creation box (Figure 4.5) when it becomes active. Pro/E will place and extrude the hole. Your angle bracket should appear as in Figure 4.6.

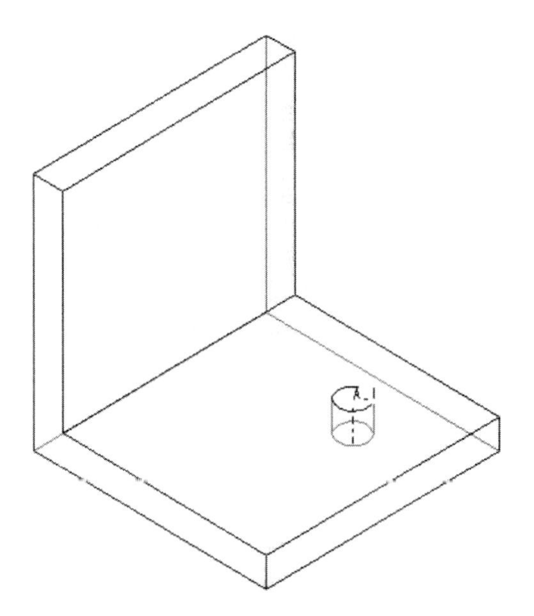

FIGURE 4.6 *The angle bracket after placing Hole A.*

Hole B can placed in the same manner as in placing Hole A. Again select *Create, Hole, Straight,* and *Done.* From Figure 4.4, select Surface 1 as the placement plane. Then use Surface 3 as the first planar surface. Enter a value of 3.00. Choose Surface 4 and enter a value of 1.5. Select *One Side* and *Done.* Choose *Thru All* from the *SPEC TO* menu and enter a value of 0.5 for the diameter of the hole. Finally, select *OK* from the creation box, or if you wish to preview placement of the hole, select the *Preview* option.

After placement of Hole B, the model should appear as in Figure 4.7. Now place the remaining hole using the appropriate dimensions and surfaces from Figure 4.1. Save the model to use in a subsequent section of this book.

### 4.1.2 TUTORIAL # 17: A COAXIAL HOLE FOR THE BUSHING

*Coaxial* holes are placed with the center of the hole along a known axis. Such holes are usually located along cylinders and shafts. Retrieve the bushing from chapter 3.

After retrieving the part, choose *Feature, Create, Hole, Straight,* and then *Coaxial.* Select *Done.* Pick the axis defined by the dotted line. You may need to use *Query Sel* to pick the proper axis. If chosen correctly, the dotted line will be highlighted in red. Choose the front of the larger diameter cylinder as the placement plane. Then, *One Side* and *Thru All.* For the diameter of the hole, enter 1.125, and then *Done.* The bushing should now look like the one reproduced in Figure 4.8. Save the part under the same file name as it was retrieved.

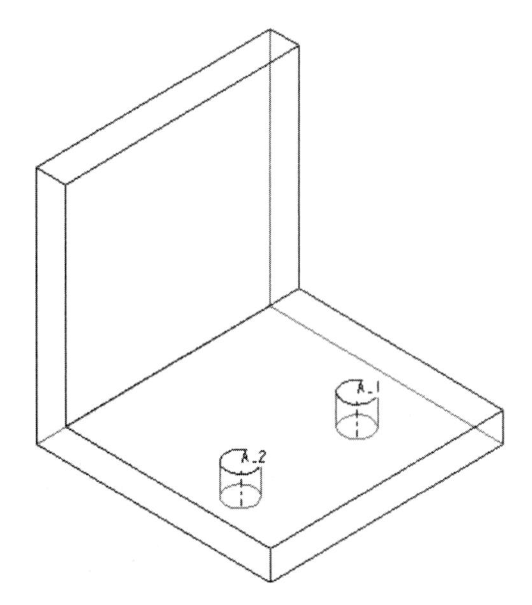


FIGURE 4.7 *The angle bracket with the second hole placed.*

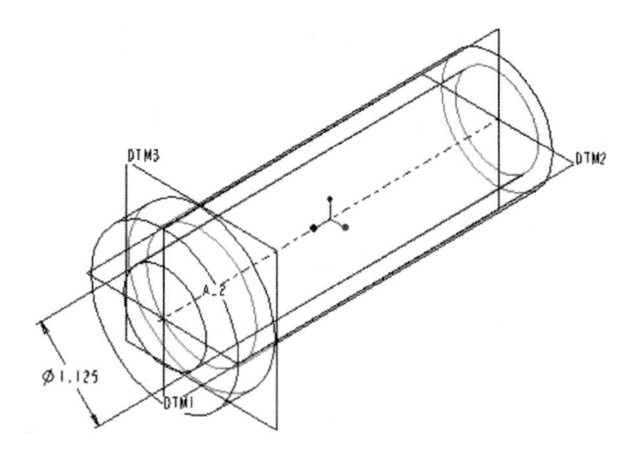


FIGURE 4.8 *The bushing from chapter 3 after adding the 1.125 diameter coaxial hole.*

## 4.1.3 Tutorial # 18: A Sketched Hole in a Plate

The *Sketched* option can be used to create a hole with a varying diameter. The variation in the diameter can be quite complex, because the *Sketcher* is used to define the section. Once the section has been defined, the hole is produced by using a 360° revolution of the section about the centerline. The user must define the centerline in the *Sketcher* by placing it at the appropriate distance from the section.

Create the simple base feature shown in Figure 4.9. Call the part "Plate-WithCounterboreHoles." We will place counterbore holes in the base feature shortly.

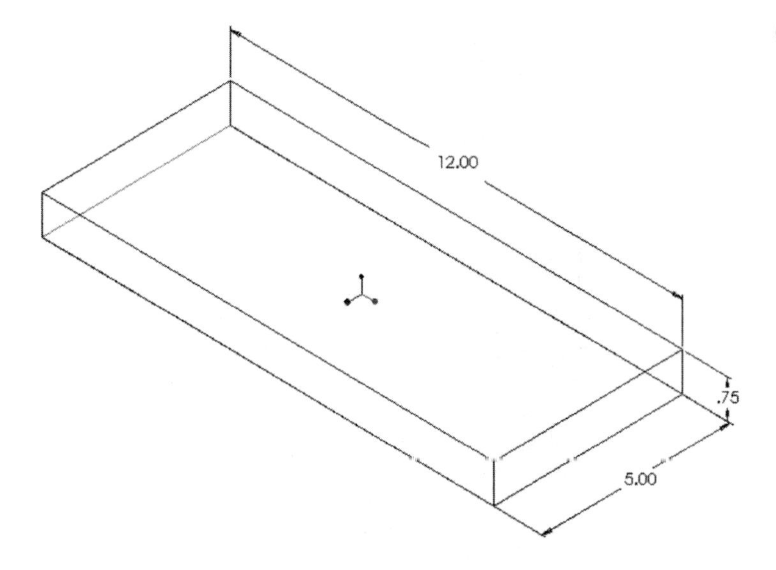


FIGURE 4.9 *The base feature to be used to illustrate how to create a sketched hole.*

The procedure for doing so requires these steps:

1. Add the default datum planes.
2. Select *Feature, Create, Extrude, Solid,* and *Done.*
3. Use the *One Side* option.
4. Select DTM2 as the sketching plane and DTM3 as the "*Bottom.*"
5. In the *Sketcher,* sketch a rectangle using the dimensions in Figure 4.9. It is a good idea to position the DTM1 and DTM3 edges so that they are the same distance from the edges of the plate.
6. *Regenerate* and *Modify* to the appropriate values.
7. Select *Done.*

After you have created the base feature, select *Feature, Create, Hole, Sketched,* and *Linear.* The *Sketcher* will appear. In the *Sketcher,* you will need to define the section of the hole. However, you must first define the centerline of the sketch. Select *Line* from the *GEOMETRY* menu (Figure 2.10). The *LINE TYPE* menu will appear. This menu is reproduced in Figure 4.10. Change the line type to *Centerline.* Then place a vertical centerline in the *Sketcher.* You will need to change the line type back to solid after placing the centerline by clicking on *Geometry.*

Proceed to reproduce the sketch defined in Figure 4.11. Notice that the sketch is closed by sketching a line coincident with the centerline. Dimension as shown and regenerate the sketch modifying the parameter to the dimensions given in Figure 4.12.

Sketched holes must be fully dimensioned. Thus, if the hole is to go through the entire part, the length of the hole must be dimensioned with the

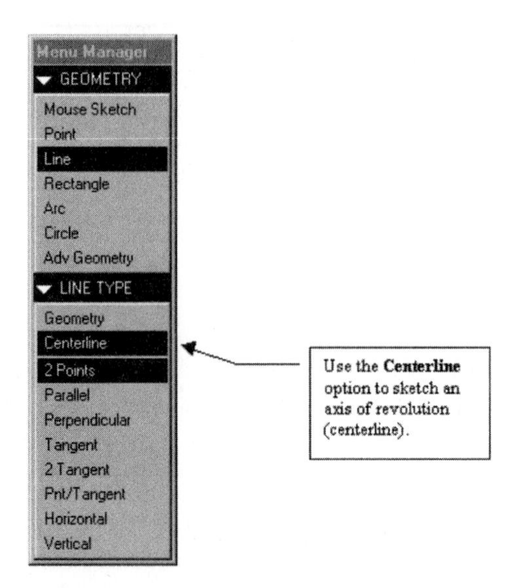


FIGURE 4.10 *The* LINE TYPE *menu is shown. A line can be set to centerline type by selecting* Centerline *from this menu.*

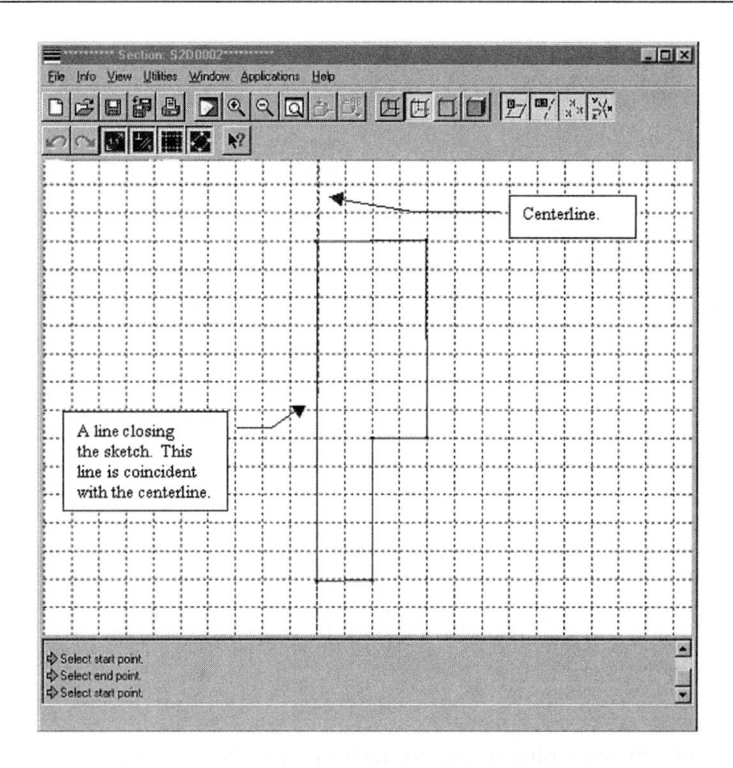


FIGURE 4.11 *Use this geometry to sketch the section for the counterbored hole.*

depth of the part. In such cases, a revolved cut may also be used. We will look at cuts in the next chapter.

Once the sketched section has been properly dimensioned, you will need to define the top of the base feature as the placement plane. Because we have chosen a linear sketched hole, two edges or planes can be used to place the hole. Consider Figure 4.13 and place the hole by using the edges of the base feature along with the given dimensions. The center of the hole is located 0.5 inches from the back and 0.75 inches from the side of the base feature.

Save this part for placement of additional sketched holes in a subsequent tutorial using the *Pattern* option.

### 4.1.4 Tutorial # 19: A Pipe Flange with a Bolt Circle

So far, all the holes that we placed were dimensioned either with respect to straight edges or along a known axis. Often, holes are placed at a radial position and a known angular position. Such a case occurs on bolt circles.

Figure 4.14 shows a pipe flange with a straight axial hole, as well as four radial holes. The base features for this part are constructed by using the default datum planes and two cylinders that intersect at the DTM2 plane, as shown in Figure 4.15. This method was discussed in chapter 3 for

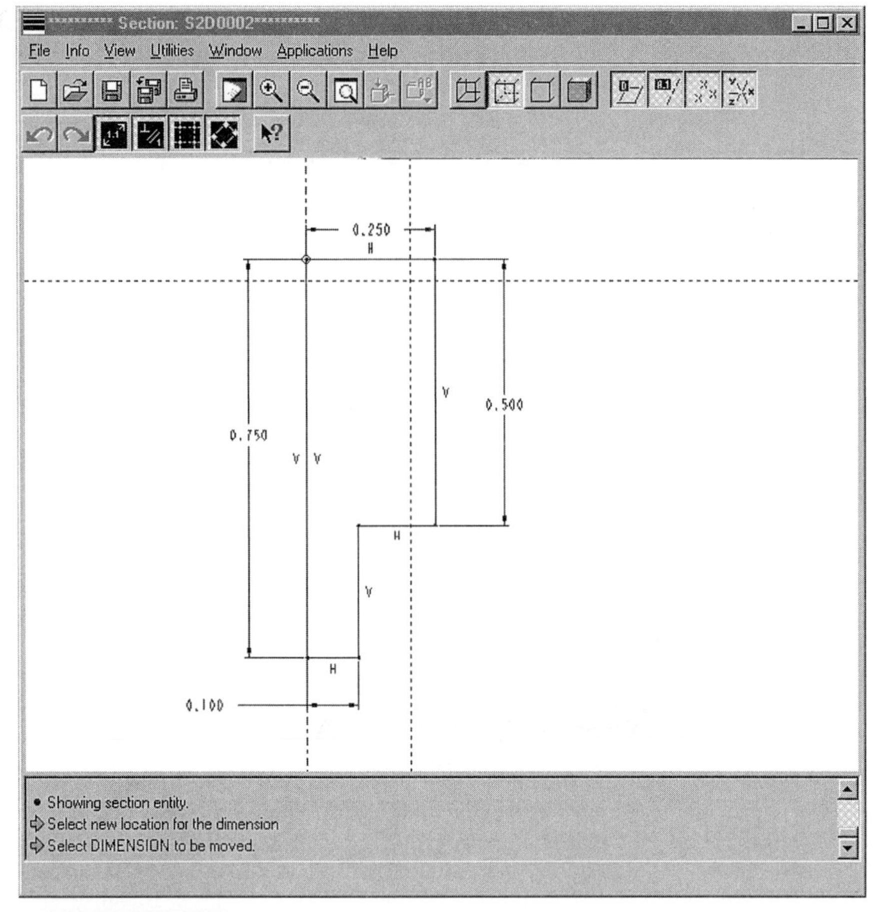


FIGURE 4.12 *Use these dimensions for the section.*

constructing the bushing. Construct the base features using the dimensions from Figure 4.15. Use the approach:

1. Create a part called "PipeFlange."
2. Add the default datum planes.
3. Select *Feature, Create, Protrusion, Extrude, Solid,* and *Done.*
4. Choose *One Side.*
5. Pick datum DTM2 as the sketching plane and DTM3 as "*Bottom.*"
6. Sketch a circle whose center is aligned with the DTM1 and DTM3 edges.
7. *Dimension* and *Modify* the circle to a diameter of 2.50.
8. Select *Done* and extrude *Blind* to a depth of 0.25. Extrude the disk "downward," that is, in the direction opposite to the cylinder in Figure 4.15.

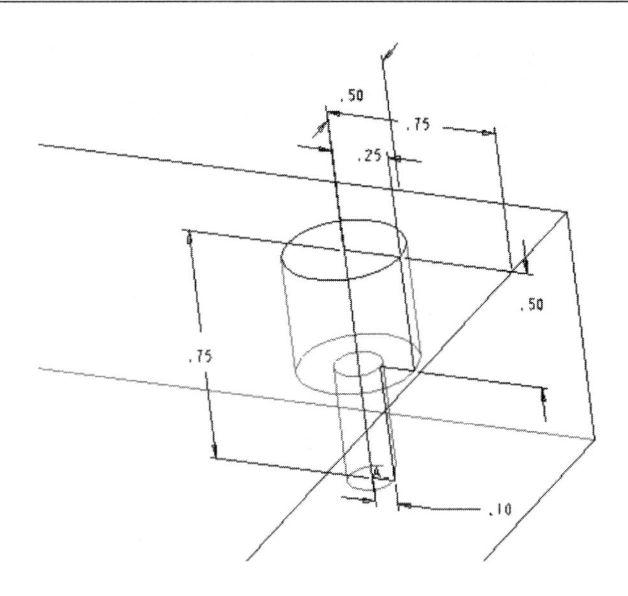


FIGURE 4.13 *Corner of the plate with dimensions locating the counterbored hole.*

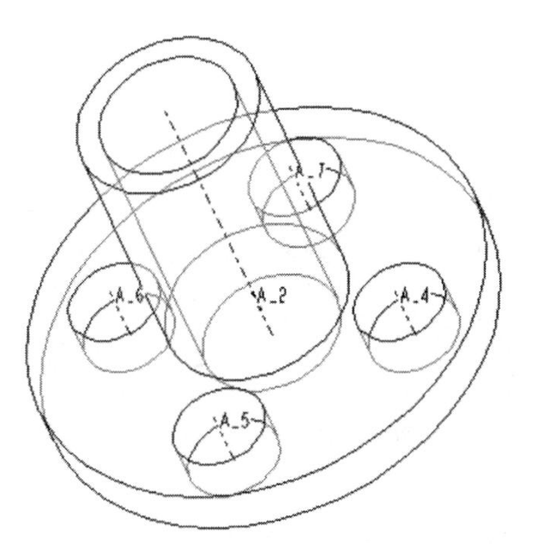


FIGURE 4.14 *A pipe flange with four equally spaced radial holes.*

9. Choose *OK* from the *Feature Creation* box.
10. Select *Feature, Create, Protrusion, Extrude, Solid,* and *Done.*
11. Again, choose *One Side.*
12. Use Surface 1 shown in Figure 4.15 as the sketching plane. Select DTM3 as "*Bottom.*"
13. Sketch a circle in the *Sketcher* whose center is aligned with the edges of the DTM1 and DTM2 datum planes.
14. *Dimension* and *Modify* the circle to a diameter of 1.00 inches.
15. Choose *Done.* Extrude the cylinder, using the *Blind* option in the direction opposite to the disk to a depth of 1.50.
16. Select *OK.*

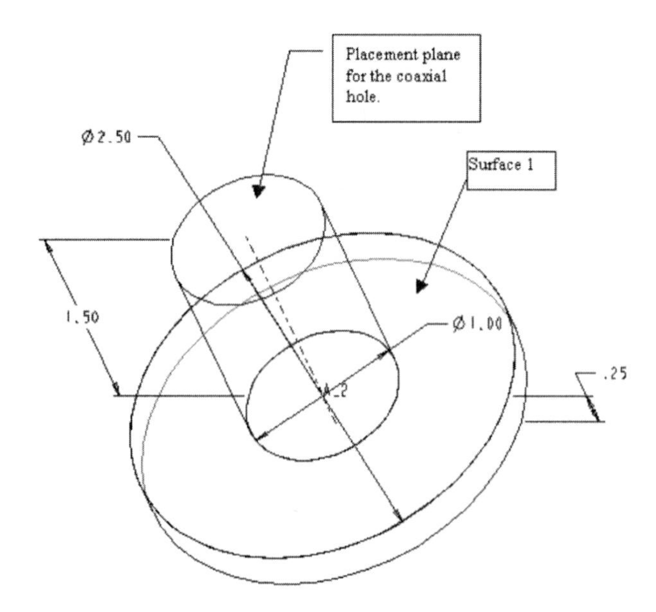


FIGURE 4.15 *The base features for the pipe flange and their associated dimensions.*

Add the coaxial hole by using the method outlined in the previous section of this chapter:

1. Select *Feature, Create, Hole, Straight,* and *Coaxial.*
2. Choose *Done.* Use axis A_2 from Figure 4.15 as the axis for placing the hole.
3. Pick the placement plane shown in Figure 4.15.
4. Use the *Thru All* option to specify the depth of the hole.
5. The hole has a diameter of 1.00. Enter this value for the size of the hole.
6. The pipe flange with the added coaxial hole is shown in Figure 4.16.

In order to place the radial holes, a radial dimension must be given with respect to an axis. In this part, the desired axis is the one that runs through the cylinder (A_2 in Figure 4.15). If this particular axis did not already exist, then an axis would have to be created with the command *Make Datum, Axis.* Axis construction using this method will be discussed in a subsequent chapter.

Select *Create, Hole, Straight,* and *Radial.* Then click on *Done.* Pick the top of the disk (Surface 1 in Figure 4.15) as the placement plane. Pick the axis through the center of the cylinder. You may need to use *Query Sel* to grab the axis.

The holes are placed by giving a radial distance from the reference axis and an angular position relative to a plane. For this part, we can use one of the default datum planes, which is perpendicular to the top of the disk, to define the angular position. If such a plane did not exist, then a plane would

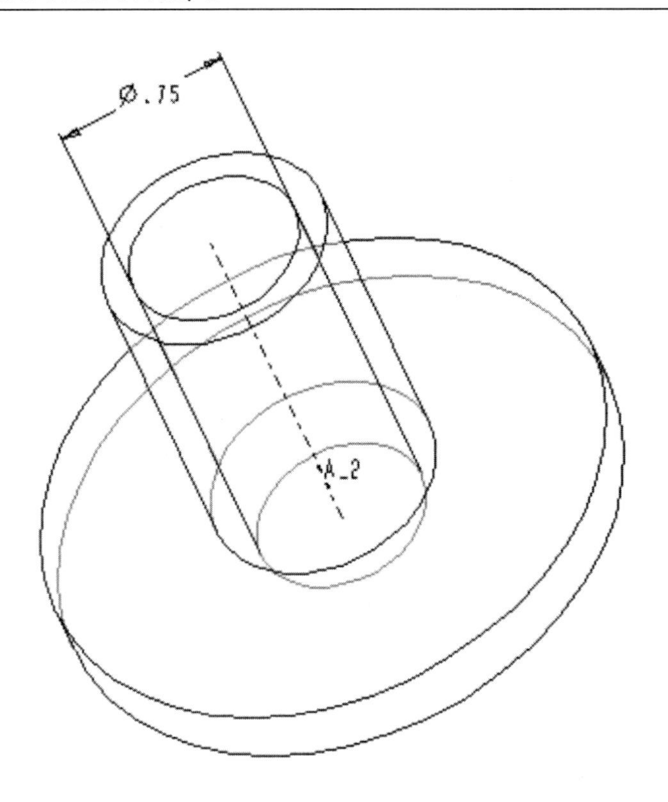


FIGURE 4.16 *The pipe flange with added coaxial hole.*

have to be created using the *Make Datum* option. Select a proper plane. For example, we selected DTM2 and entered a value of 0° for the angle for the part shown in Figure 4.17.

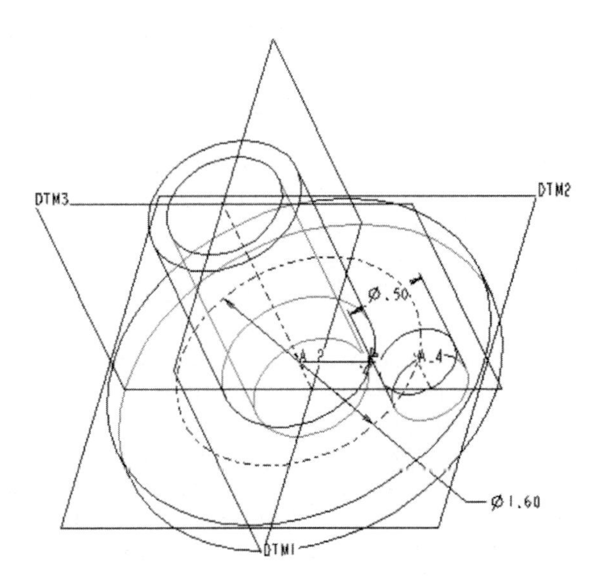


FIGURE 4.17 *The pipe flange with one radial hole.*

The bolt circle can either be defined by giving the diameter (*Diameter* option), the radius (*Radial* option), or a *Linear* position. For the problem under consideration, the diameter of the bolt circle is 1.6. Therefore, select *Diameter* and enter the value 1.6. Enter a diameter of 0.5 for the hole. This will place the first hole.

To place the remaining holes, you can repeat the previous steps three more times changing the placement angle, or you can make a pattern of the first hole. Because repeating the process for the additional holes is tedious, let us construct the remaining holes using the pattern option. At this point, save the part. We will add the remaining holes using the *Pattern* option in chapter 6.

### 4.1.5 TUTORIAL # 20: A HOLE ON THE INCLINED PLANE OF THE ROD SUPPORT

After an inclined plane has been created, it is a small matter to locate a hole on the plane. Usually, datum planes have been created to construct the inclined plane. These datum planes can be used to place the hole.

Recall the model of the rod support created in chapter 3. The model is shown in Figure 3.7. Here we are interested in the 3.00 diameter hole. The other two holes can be placed using the *Linear* option as discussed earlier. Take a few minutes and place these holes using Figure 3.14 for the pertinent information and the *Linear* option. After placement of these holes, your model should appear similar to the one shown in Figure 4.18.

So how do we place the remaining hole? In general, holes can be placed on inclined surfaces by referencing the user-defined datum or datums used in constructing the surface.

In this model, the sides of the support could be used to position the hole. Since the depth of the model is 6 units, then the hole can be placed from either side at a distance of 3 units. Also, notice that DTM3 runs through the model separating its depth into equal parts. We also can use this datum to locate the hole, placing the hole a distance of 0 units from datum DTM3.

It so happens, that the center of the hole is located at the center of the round protrusion (axis A_9 in Figure 4.18). Therefore, for this model, we can use the *Coaxial* option and the axis A_9 to place the hole.

For this model, using the coaxial option is the easiest and fastest method. So, let us create the hole using this approach.

Select *Feature, Create,* and *Hole.* Then choose *Straight, Done, Coaxial,* and *Done.* For the placement plane, select Surface A in Figure 4.18. Then choose the axis on your model corresponding to A_9 in Figure 4.18.

Use the *One Side* option and from the *SPEC TO* menu, select the *Thru All* option. The diameter of the hole is 3.00.

Choose *OK* from the creation box to place the hole. The model should appear as in Figure 4.19.

The rod support illustrates several points. The placement of the model to the datum planes should be carefully thought out so that the datum planes can be fully utilized to construct the base feature. In addition, the location of

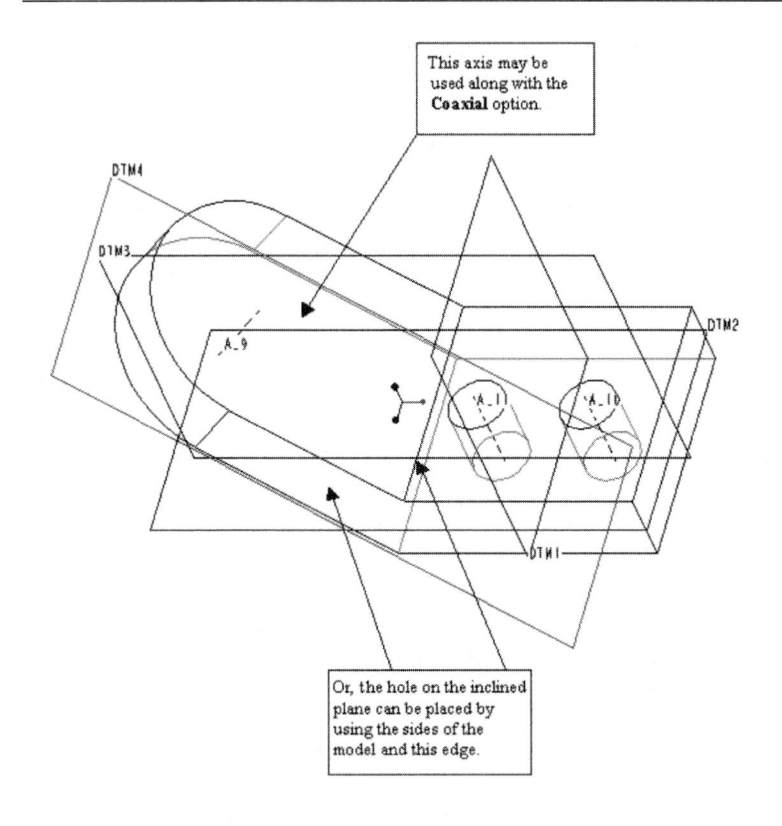


FIGURE 4.18 *The rod support with two of the three holes placed.*

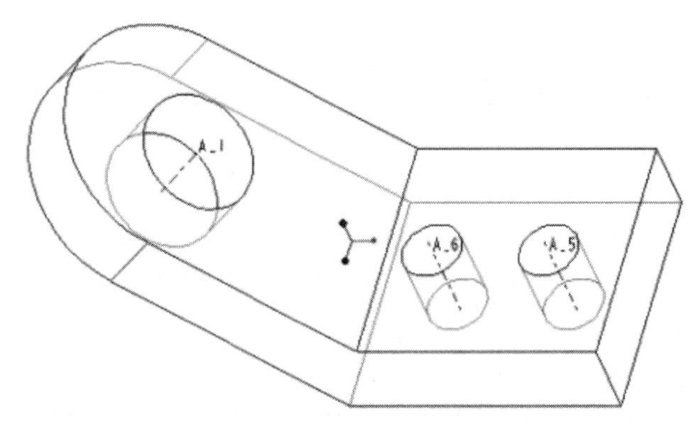


FIGURE 4.19 *The model of the rod support after adding the hole on the inclined plane.*

default, or user-defined datum can be exploited to also place additional features, if their placement is carefully considered ahead of time.

## 4.2 SUMMARY AND STEPS FOR ADDING HOLES

In this chapter, we examined the options available in Pro/E for placing a hole on a model. Holes in Pro/E are of two types. The first type is a *Straight* hole, which is a hole with a constant diameter. The second type is a *Sketch* hole. Sketch holes are used for holes with variable diameters.

In Pro/E, holes are placed using one of four possible options. These options are *Linear, Radial, Coaxial,* and *On Point.* The *Linear* option is used to place a hole when the linear location of the center of the hole is known relative to two surfaces or edges. The *Radial* option may be used to create a hole along a radial direction such as in a bolt circle. If the hole extends along a known axis, then the *Coaxial* option may be used. The *On Point* option allows the user to place a hole at a point.

In general, a hole may be placed on a model by using the following steps:

1. Select *Feature, Create,* and *Hole.*
2. Choose either a *Straight* or *Sketch* hole.
3. Select the placement option, that is, choose from *Linear, Radial, Coaxial,* and *On Point.*
4. Select the placement plane.
5. For the *Linear* option, select the two edges or surfaces. After each selection, enter the linear dimensional value. For the *Radial* option, pick the axis of revolution on the model and then enter the radial dimension from that axis. If you are using the *Coaxial* option, select the axis. For the *On Point* option, simply pick the desired point on the model.
6. If the type of hole is *Sketch* type, draw the geometry of the hole using the *Sketcher.* Do not forget to place a centerline. Then *Dimension* and *Modify* to the appropriate values.
7. If the hole is of *Straight* type, enter the diameter of the hole. Then choose the option from the *SPEC TO* menu (Figure 4.2) for specifying the depth.
8. Choose *Done.*

## 4.3 ADDITIONAL EXERCISES

The following problems require the placement of various types of holes. In some cases, we have chosen problems from previous chapters. In such cases, their source is indicated in parentheses.

4.1 Add the hole to the washer shown in Figure 4.20 (exercise 2.1).

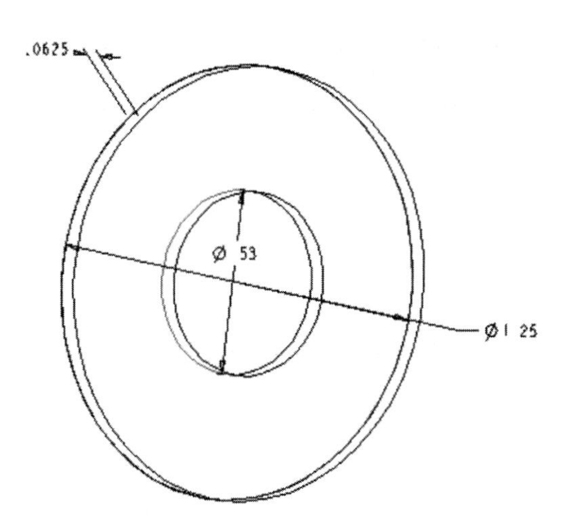


FIGURE 4.20 *Figure for exercise 4.1.*

4.2 Using Figure 4.21, complete the model of the hold down (exercise 2.2).

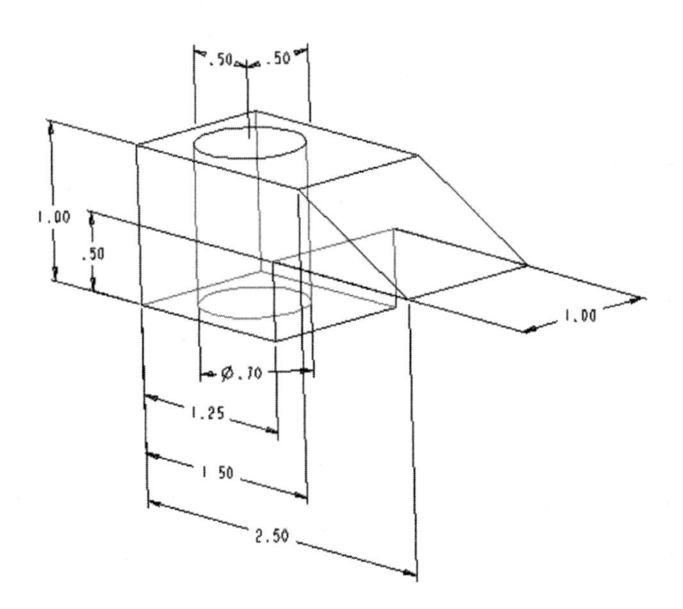


FIGURE 4.21 *Figure exercise 4.2.*

4.3 Add the hole to the bushing from exercise 3.2, shown in Figure 4.22.

FIGURE 4.22 *Figure for exercise 4.3.*

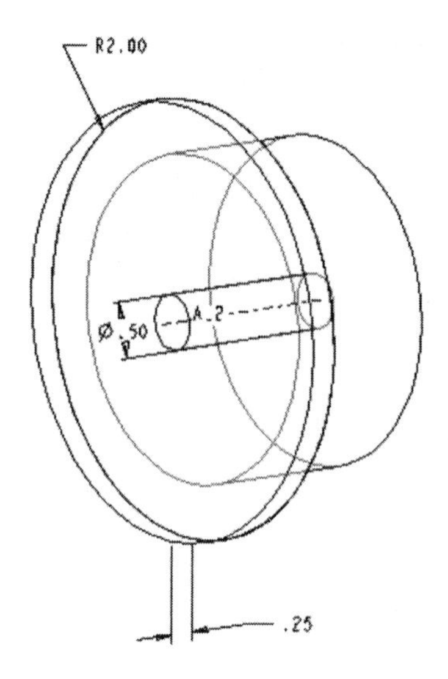


4.4 Create the model of the bumper using Figure 4.23.

FIGURE 4.23 *Figure for exercise 4.4.*

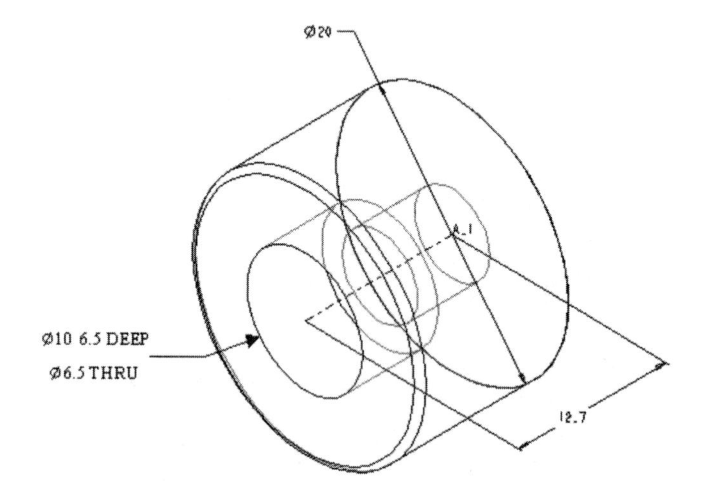

4.5 Complete the model of the angled rod support from exercise 3.7. Use Figure 4.24 for reference.

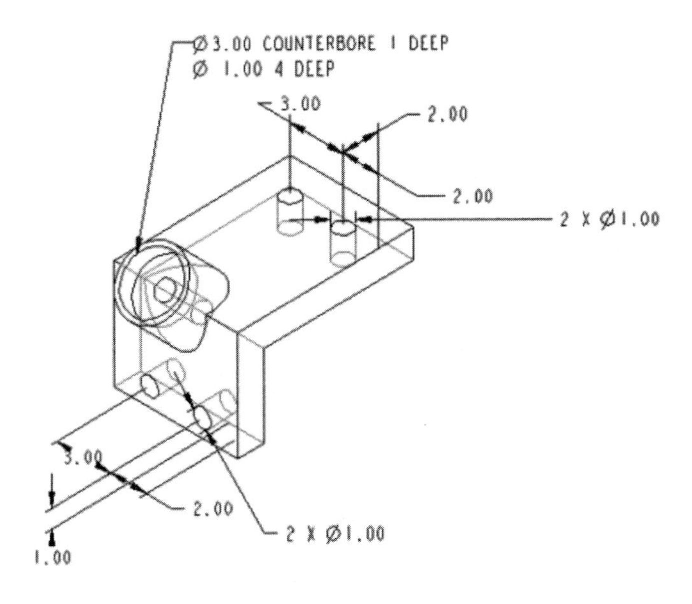


FIGURE 4.24 *Figure for exercise 4.5.*

4.6 Add the holes shown in Figure 4.25 to the model of the offset bracket (exercise 3.9).

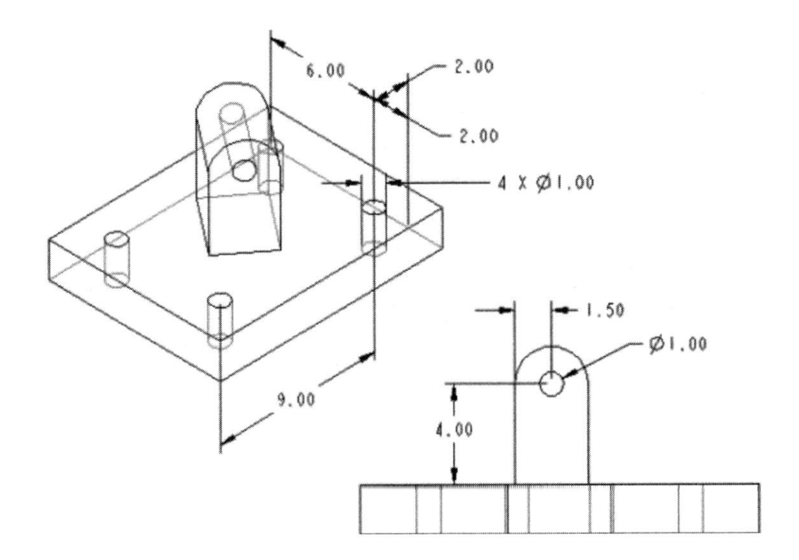


FIGURE 4.25 *Figure for exercise 4.6.*

4.7 For the model of the side support constructed in exercise 3.10, add the holes as shown in Figure 4.26.

FIGURE 4.26 *Figure for exercise 4.7.*

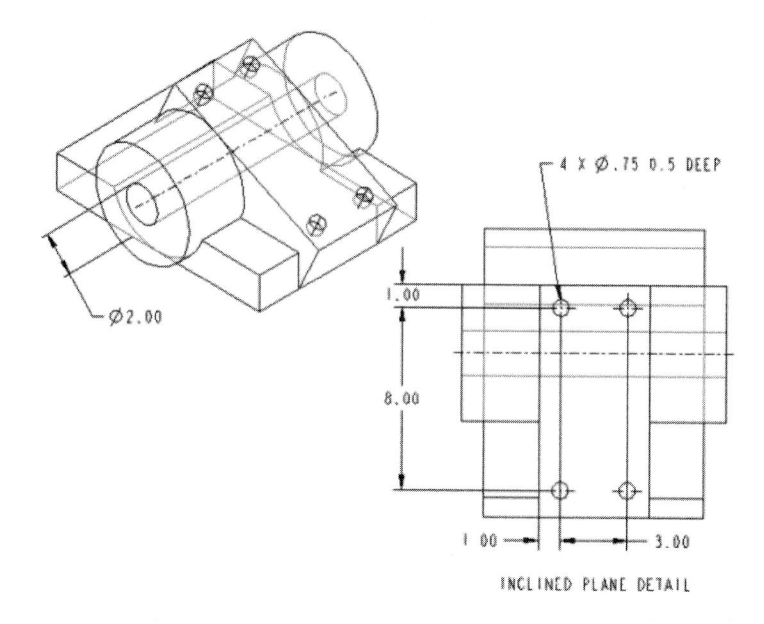


4.8 Complete the model of the control handle from exercise 3.6. Use the geometry shown in Figure 4.27.

FIGURE 4.27 *Figure for exercise 4.8.*

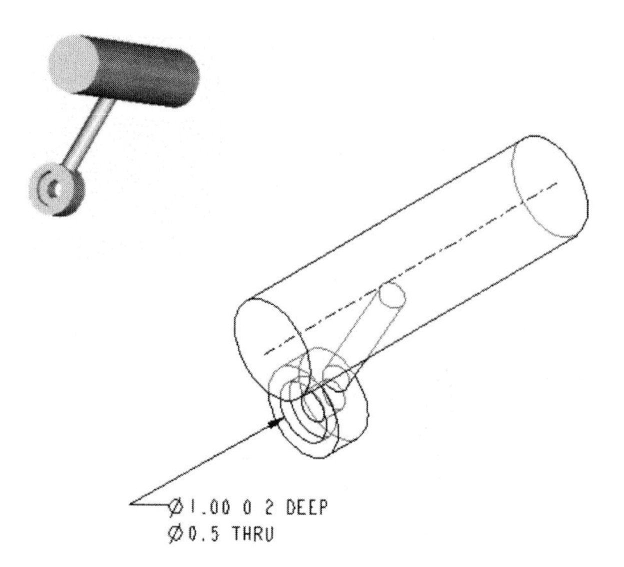

4.9 Add the counterbored hole to the model of the angled block from exercise 2.11. Use the geometry shown in Figure 4.28.

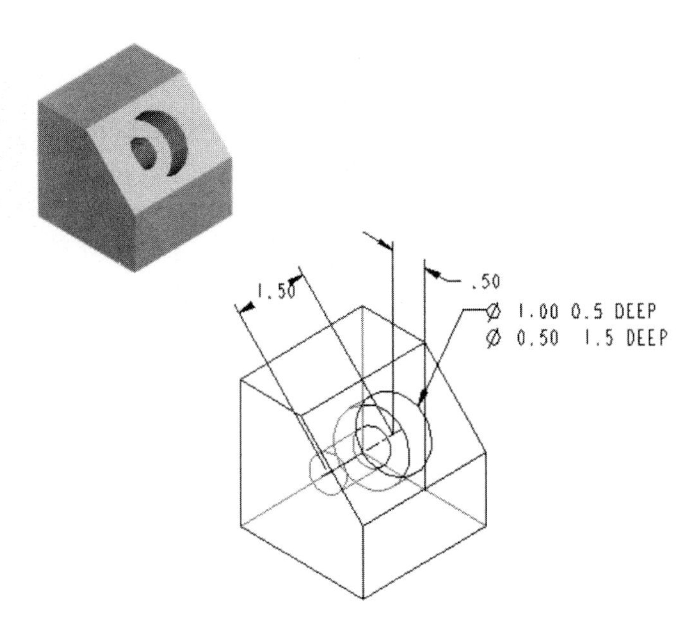


FIGURE 4.28 *Figure for exercise 4.9.*

4.10 Add the four holes to the model of the side guide from exercise 3.8. Use Figure 4.29 as a reference.

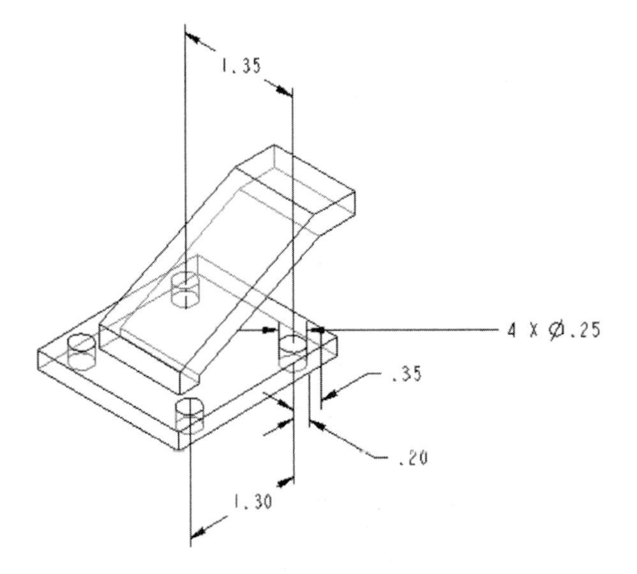


FIGURE 4.29 *Figure for exercise 4.10.*

4.11 For the model of the double side support from exercise 3.11, add the two holes. The geometry of the holes is given in Figure 4.30.

FIGURE 4.30 *Figure for exercise 4.11.*

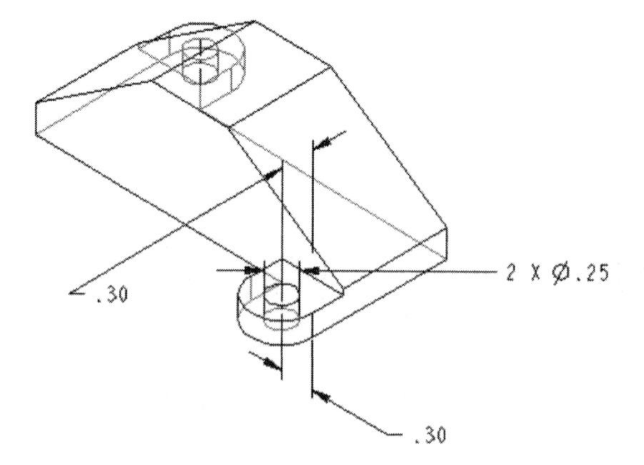


4.12 Add the holes to the model of the slotted support from exercise 3.12. Use the geometry shown in Figure 4.31.

FIGURE 4.31 *Figure for exercise 4.12.*

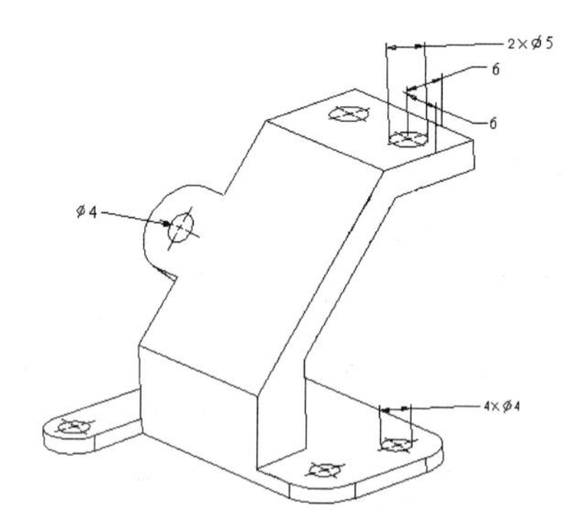


4.13 For the model of the clamp base from exercise 3.13, add the holes. Use Figure 4.32 for the geometry of the holes.

FIGURE 4.32 *Figure for exercise 4.13.*

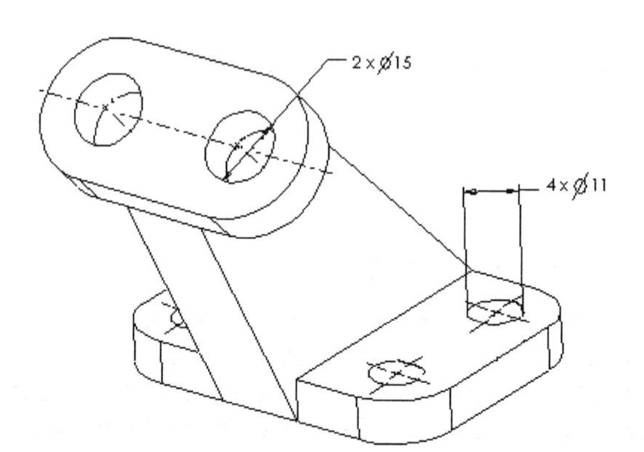

# CHAPTER 5

# Options That Remove Material: Cut, Slot, Neck, and Shell

## Introduction and Objectives

In this chapter, we consider options from the *SOLID* menu that remove material from a base feature. The first two of these options are *Cut* and *Slot.* From the procedural standpoint, both the *Cut* and *Slot* are created in the same manner. However, from the manufacturing point of view, the two are different. Simply put, a cut is created using a bulk material removal process such as punching or sawing. Using a milling process, a slot is created when material is removed. Examples of slots are keyways, key seats, and kerfs.

Pro/E has another feature for the removal of material, the *Neck* option. A neck is a slot that is revolved around an axis of revolution. Another option called *Shell,* removes a surface and then removes material below the surface.

The objective of this chapter is as follows:

- Acquaint the reader with the material removal options *Cut, Slot, Neck,* and *Shell.*

## 5.1 Adding a Cutout or Slot

A *Cut* or *Slot* may be placed on a base feature using the same procedure. Both these options may be either *Solid* or *Thin* features, in a manner analogous to a protrusion. However, in a *Thin* cut/slot, the thickness of the feature is the thickness of the material removed from the model.

### 5.1.1 TUTORIAL # 21: USING CUT TO REMOVE MATERIAL FROM A U BRACKET

One model upon which we have chosen to illustrate the *Cut* option is the U bracket shown in Figure 5.1. In this tutorial, we will construct the bracket by using the *Thin* option to create the base feature and then rounding off the top of the bracket with the *Cut* option.

Let us begin by quickly constructing the base feature with the *Thin* option. The overall dimensions of the feature are 3 units wide, 5 units high, and 3 units deep. We see from Figure 5.1 that the radius of the rounded top is 1.5 units.

Select *Feature, Create, Solid, Protrusion,* and *Thin.* Pro/E will open the *Sketcher.* Sketch the U-shaped geometry shown in Figure 5.2 and dimension as shown. Upon successful regeneration, enter a thickness of 0.25. Make sure that the arrow defining the direction for the thickness points inward. *Extrude* to a *Blind* depth of 3.00. Your part should appear similar to the one shown in Figure 5.3.

We are now ready to round off the top of the bracket. In order to accomplish this task, we will construct a cut with a circular boundary and remove all the material outside of the cut. The *Thru All* option will be used to extrude the cut through the entire feature. Thus, material will be removed from both arms of the base feature.

Choose *Create* and *Cut.* Notice that the cut can be either *Solid* or *Thin.* In this case, select *Solid.* Choose *One Side.* For the sketching plane, click on either side of the base feature. Choose the top of the U bracket to orient the base feature in the *Sketcher.* In the *Sketcher,* you should see a rectangular shape.

FIGURE 5.1 *The U bracket.*

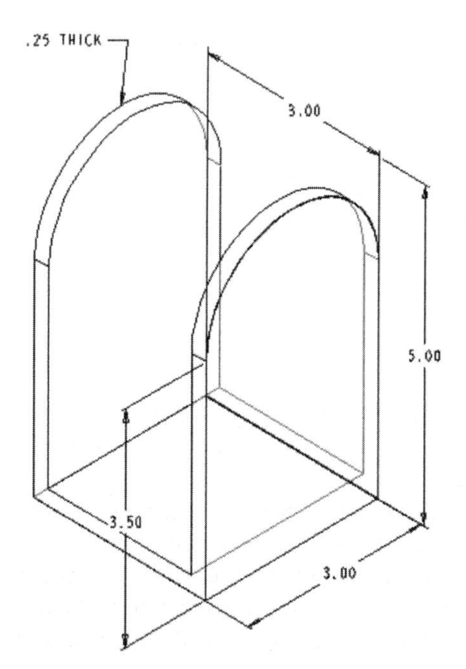

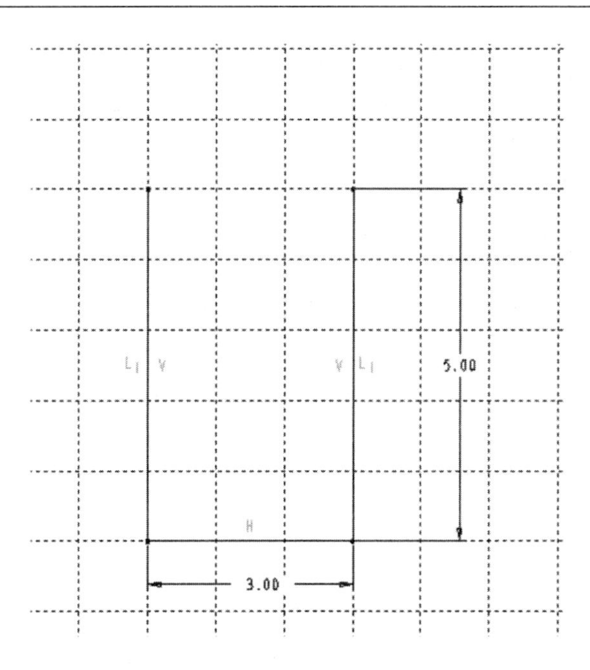


FIGURE 5.2 *The sketch used in generating the base feature for the U bracket.*

Sketch vertical and horizontal lines along the edge of the U bracket. Then add an arc at the top of the sketch, using your right mouse button. In Figure 5.4, we have reproduced the U bracket as it would appear in the *Sketcher* with the added lines and arc. The inside of the cut is defined by the region enclosed by the arc and the vertical and horizontal lines.

Align the vertical and horizontal lines to the edge of the base feature. Dimension the location of the center of the arc to the bottom of the base feature (the value is 3.50; see Figure 5.4). If drawn correctly, the ends of the vertical lines and the center of the arc should line up.

After successfully regenerating, select *Done.* We wish to remove the material outside of the cut. By placing the vertical and horizontal lines in alignment with the edges of the bracket, the only material removed from the base feature will be in the area outside of the arc. Make sure that the arrow is pointing away from the inside of the cut. If it is not, you may need to flip the arrow.

The cut must proceed “into” the screen. Choose *Thru All* for the depth and *Done.* The bracket should appear as in Figure 5.1. Save the part.

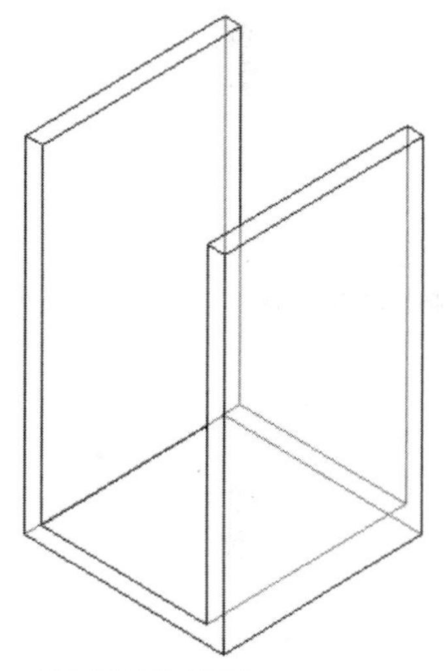

FIGURE 5.3 *The base feature for the U bracket created with the* Thin *option.*

## 5.1.2 Tutorial # 22: A Washer with a Cut

Consider the washer shown in Figure 5.5. In order to create the cut in the washer, we need to choose *Cut* as the option and then sketch the profile of the cut in the *Sketcher.*

First, create the base feature by constructing a protrusion of a solid disk. The disk has a diameter of 0.438 and a thickness of 0.063. Notice that because the disk has only one planar surface, you will need to create this part using default datum planes. If you do not use the datum planes, it will be impossible to orient the base feature in the *Sketcher* when it comes time to sketch the cut.

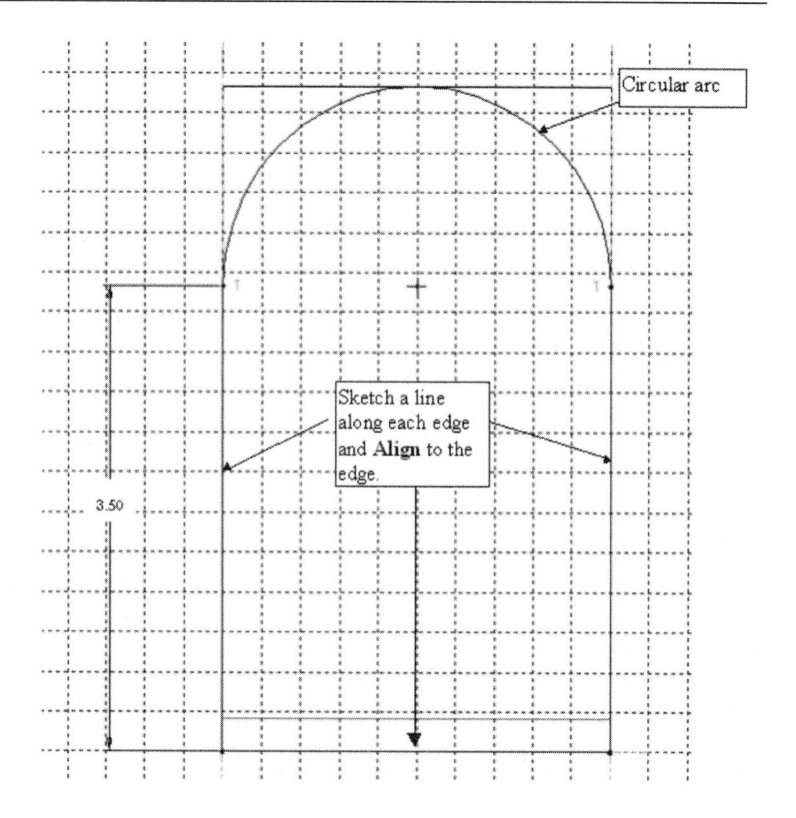


**FIGURE 5.4** *The U bracket in the* Sketcher *with sketched vertical and horizontal lines along the edges of the base feature.*

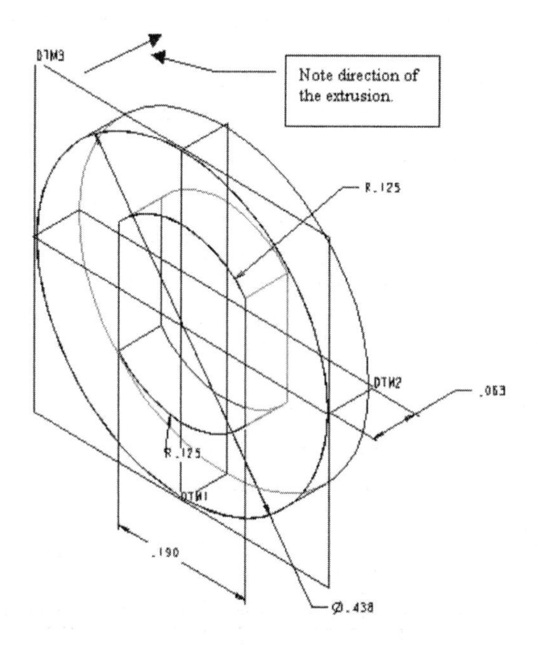


**FIGURE 5.5** *The dimensions of the part are shown in this drawing of the complete model of the washer.*

In order to construct the base feature:

1. Create a part called "Washer."
2. Add the default datum planes.
3. Select *Feature, Create, Protrusion, Extrude, Solid,* and *Done.*
4. Use the *One Side* option.
5. Select DTM3 as the sketching plane and DTM2 as the "*Top.*"
6. Sketch a circle whose center is aligned with the intersection of the DTM1 and DTM2 edges.
7. *Dimension* the circle and *Modify* to a value of 0.438.
8. Choose *Done.* Extrude *Blind* to a depth of 0.063.

Once the disk has been constructed, choose *Create, Cut,* and *Solid.* Choose the *One Side* option. Select DTM3 as the sketching plane and DTM2 as the "*Top.*"

In the *Sketcher,* add a circle and two vertical lines, as shown in Figure 5.6. Using *Geom Tools,* select the intersections of the lines with the circle and *Delete* the unwanted parts of the lines and circle.

Even after deleting the unwanted parts of the circle, the center for both arcs will be the center for the original circle. Align this center to

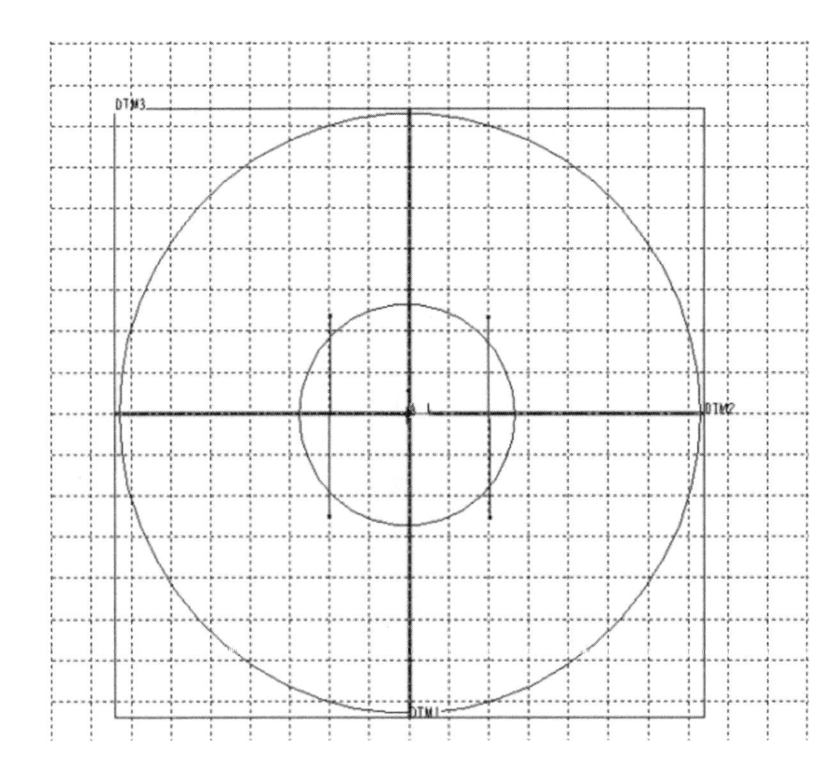


FIGURE 5.6 *The geometry needed to construct the cut in the washer.*

the intersection formed by the datum plane edges. Then, dimension the geometry as illustrated in Figure 5.7. Notice that both arcs must be dimensioned. In this case, it is easier to give radial dimensions. *Regenerate* and *Modify* to the proper dimensions. *Regenerate* again. Your sketch should now look like the one in Figure 5.7.

Choose *Done* and *Thru All* for the depth of the cut. Then select *OK*. Your final version of the washer should look like the one in Figure 5.5.

### 5.1.3 TUTORIAL # 23: AN AXLE WITH A SLOT

As an example of constructing a slot on a base feature, we consider the slow speed rear axle from the NDSU Mini Baja car. The base feature of the axle is a circular cylinder. A keyway is located near the end of one axle that is used to secure the single rear brake. The hubs at each end of the axle that connect the wheels to the axle will not be modeled at this time. We can create the keyseat by using a user-defined datum plane and the *Slot* option.

Begin the construction of the model by creating a file called "Axle." After creating the file, add the default datum planes.

The base feature will be constructed using DTM3 as the sketching plane and DTM2 to orient the model in the *Sketcher.* The sequence of steps to be followed in creating the base feature is:

1. Select *Feature, Create, Protrusion, Solid,* and *Done.*
2. Choose *Both Sides.*
3. Select DTM3 as the sketching plane. Define DTM2 as the "*Top.*"

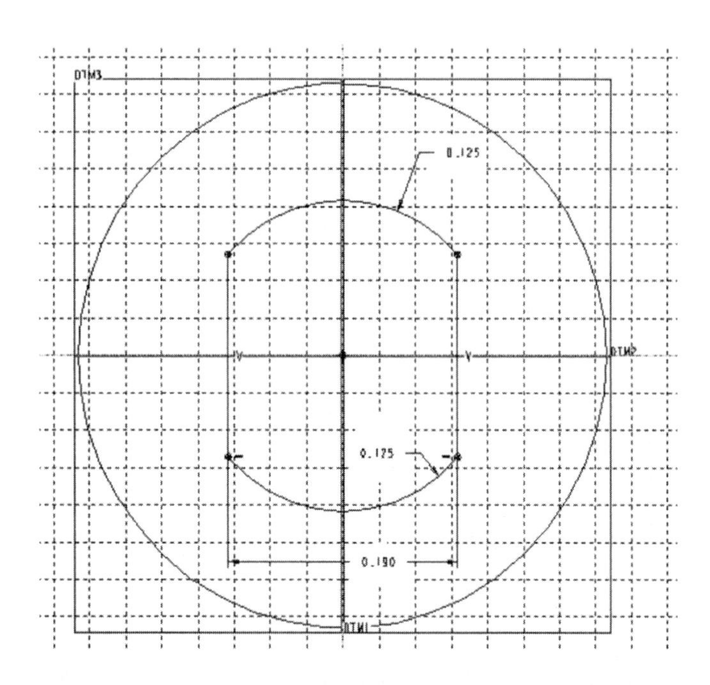


FIGURE 5.7 *Washer with dimensions defining the size of the cut.*

4. In the *Sketcher,* draw a circle whose center is aligned with the edges of DTM1 and DTM2.
5. Dimension the diameter of the circle to the value of two inches.
6. Extrude using the *Blind* option to a depth of 43 inches.

The axle after the extrusion is shown in Figure 5.8.

In order to create the slot, we will need a user-defined datum plane tangent to the cylinder. We can construct this datum using the *Tangent* option or simply by using the *Offset* option with a value of 1 inch from DTM2. We will use the approach using the *Offset* option.

Choose *Feature, Create,* and *Plane.* Select *Offset* from the *DATUM PLANE* menu (Figure 3.10c). Pick DTM2 and then choose *Done Sel.* Select *Enter* and provide the value of 1.00. The arrow should point "up." Finally, choose *Done.* Pro/E will place the datum (DTM4), which will become the sketching plane for the slot. The model with the added datum is shown in Figure 5.9.

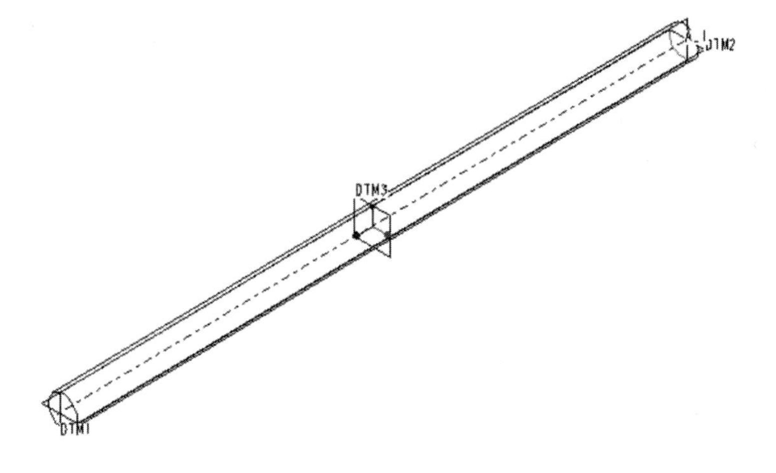


**FIGURE 5.8** *The base feature for the axle: a simple circular cylinder.*

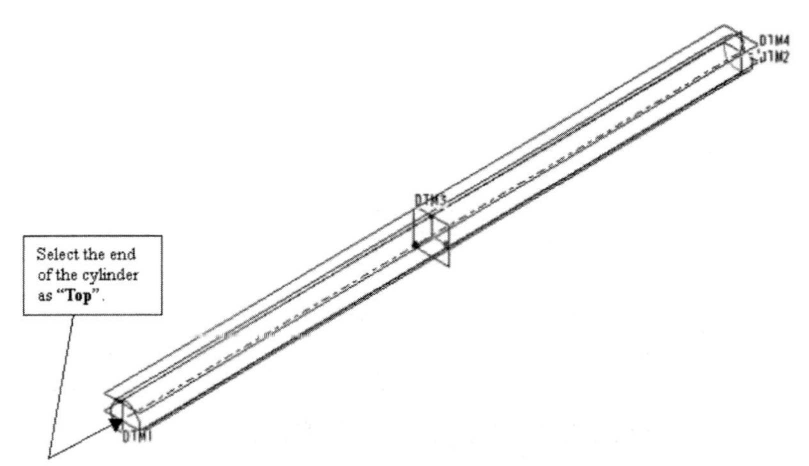


**FIGURE 5.9** *The model of the axle with an added user-defined datum plane (DTM4).*

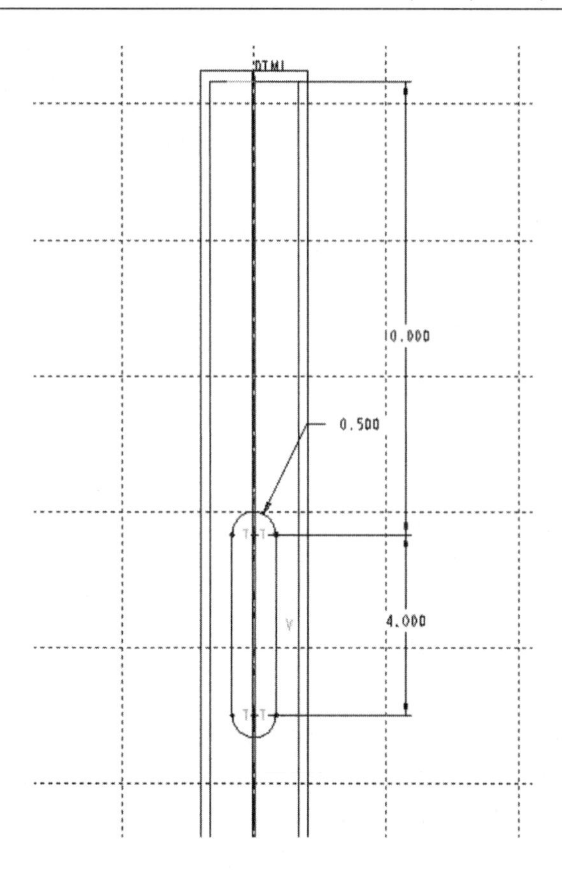


FIGURE 5.10 *A "blown up" view of the slot geometry.*

Now let us add the slot to the axle. Select *Feature, Create, Slot, Solid,* and *Done.* Use the *One Side* option and choose DTM4 as the sketching plane. In order to orient the model in the *Sketcher,* select the end of the cylinder as shown in Figure 5.9. Make this end the "*Top.*"

In the *Sketcher,* draw the geometry shown in Figure 5.10. *Dimension* as shown and *Modify* to the proper values. Note in the figure, that in order to show the geometry of the slot, we zoomed in on the part. The depth of the slot is 0.25. *Extrude* the slot into the part. The model with the slot is shown in Figure 5.11. Save the part.

## 5.2 ADDING A NECK

A *Neck* is a revolved slot; the angle of rotation is defined by the user. Rotation of the slot is performed around a centerline that the user must add to the sketch when creating the geometry of the slot. All neck features are sketched with *open* sections.

### 5.2.1 TUTORIAL # 24: AN ADJUSTMENT SCREW

Let us create a neck. First, we will need to make a base feature. Our part is an adjustment screw. We can create the base feature for this screw, which is

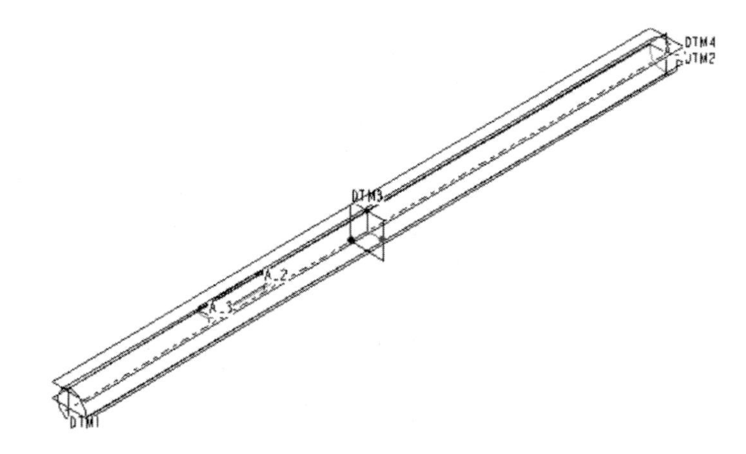


FIGURE 5.11 *The model of the axle with the slot.*

formed by a pair of dissimilar radius cylinders joined end to end. The base feature is shown in Figure 5.12. Construct the base feature. You will need to use datum default planes. Use the following approach:

1. Create a part called "AdjustmentScrew."
2. Add the default datum planes.
3. Select *Feature, Create, Protrusion, Extrude, Solid,* and *Done.*
4. Choose the *One Side* option.
5. Pick DTM3 as the sketching plane and DTM2 as the "*Top.*"
6. In the *Sketcher,* draw a circle aligned with the DTM1 and DTM2 edges.
7. *Dimension* and *Modify* the circle to a diameter of 0.500.
8. *Regenerate* and choose *Done.*

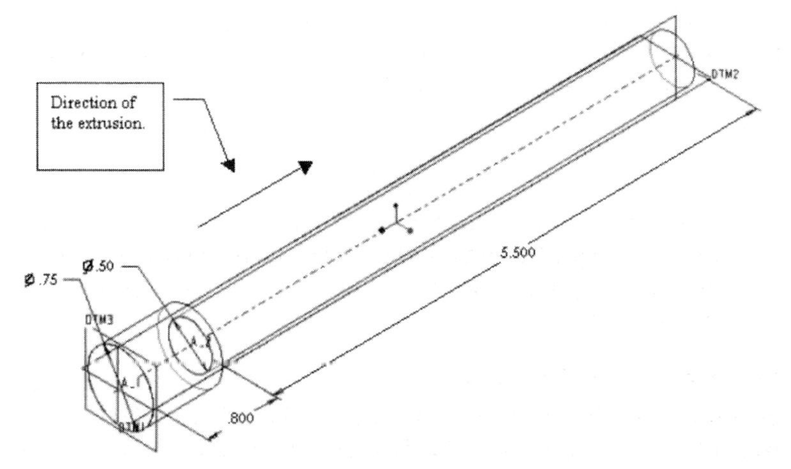


FIGURE 5.12 *The base feature for the adjustment screw part.*

9. Extrude the model in the direction shown in Figure 5.12 to a depth of 6.300 inches.
10. Again, select *Feature, Create, Protrusion, Extrude, Solid,* and *Done.*
11. Choose the *One Side* option.
12. Once again, use DTM3 as the sketching plane and DTM2 as the "*Top.*"
13. Draw a second circle aligned with the DTM1 and DTM2 edges.
14. *Dimension* and *Modify* the circle to a diameter of 0.750.
15. *Regenerate* and choose *Done.*
16. *Extrude* the model in the direction shown in Figure 5.12 to a depth of 0.800 inches.

Your model should be similar to the one shown in Figure 5.12. We are now ready to add the neck. The neck is shown in Figure 5.13. The adjustment screw is used in a vise, and the neck forms a circular slot for a set screw. As the adjustment screw is tightened or loosened, the set screw revolves in the neck and moves the jaw of the vise.

You may want to look briefly at Figure 5.13, which illustrates the base feature with the added neck. Select *Create* and *Neck.* Choose a rotation of 360° from the *ANGLE* menu (Figure 5.14). Select *Done.*

The program will ask for the sketching plane. What should the sketching plane be? In this case, the surface of the cylinder is curvilinear, meaning it has no plane that can be chosen as the sketching plane. Luckily, the cylinders were created symmetrically about datum DTM1. We can use this datum as the sketching plane.

Select DTM2 as "*Top*" to orient the part as shown in Figure 5.15. Then sketch the geometry of the neck. Align the endpoints of the vertical lines with the side of the cylinder. Change your line type to centerline. Place a cen-

FIGURE 5.13 *The adjustment screw with neck.*

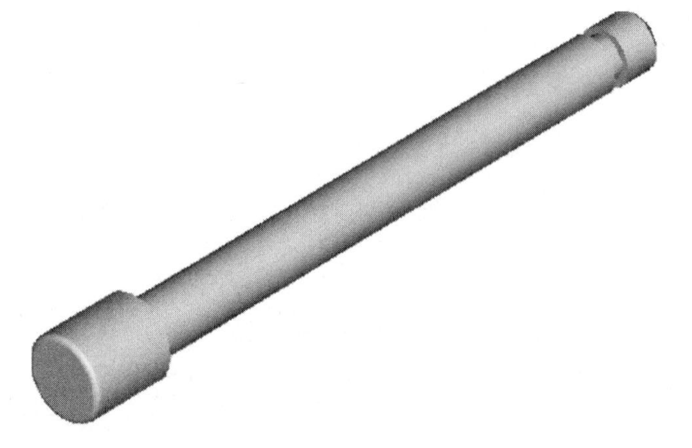

terline along the center of the cylinder. Align this centerline with the dotted axis line.

*Dimension* the sketch as illustrated in Figure 5.15. *Regenerate* and add the proper dimensional values. Upon successful regeneration, select *Done.*

The adjustment screw should appear as in Figure 5.13. However, you may need to change the orientation. Save the part. In a later tutorial, we will add threads to the screw.

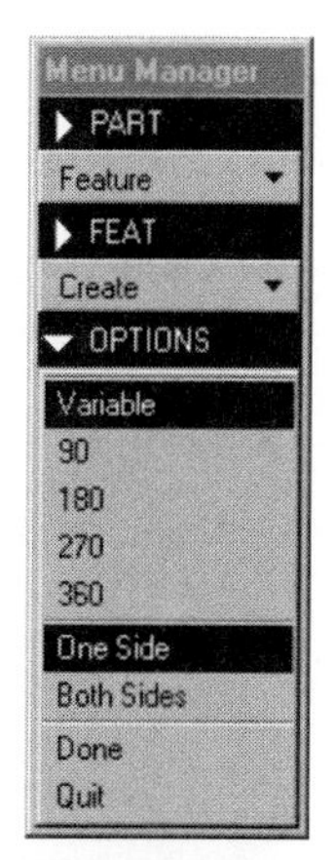


FIGURE 5.14 *The* ANGLE *menu is used to define the angle of rotation.*

## 5.3 USING THE SHELL OPTION

The *Shell* option is another built in Pro/E menu command for removing material from a base feature. This option removes a given surface and the material below it to create a hollow shell of specified thickness. It is important to remember that the *Shell* option removes material from all the features that exist on the model. Thus, if you do not want a certain feature to be shelled, you need to add that feature after shelling the rest of the model. If the feature cannot be added after shelling, the order in which it appears in the model tree must be changed with the *Reorder* command. We will examine this command in chapter 13.

### 5.3.1 TUTORIAL # 25: AN EMERGENCY LIGHT COVER

Let us first create a base feature to use as an example for the *Shell* option. Call this part "EmergencyLightBase." The model, a support base for an emergency light, is shown in Figure 5.16. The model is constructed by making a solid protrusion, adding holes to the model and using the shell option. After shelling, the holes will become bosses for screws to be inserted.

Start the construction of this model by adding default datum planes to the part. In order to create the base feature, select *Feature, Create, Protrusion,*

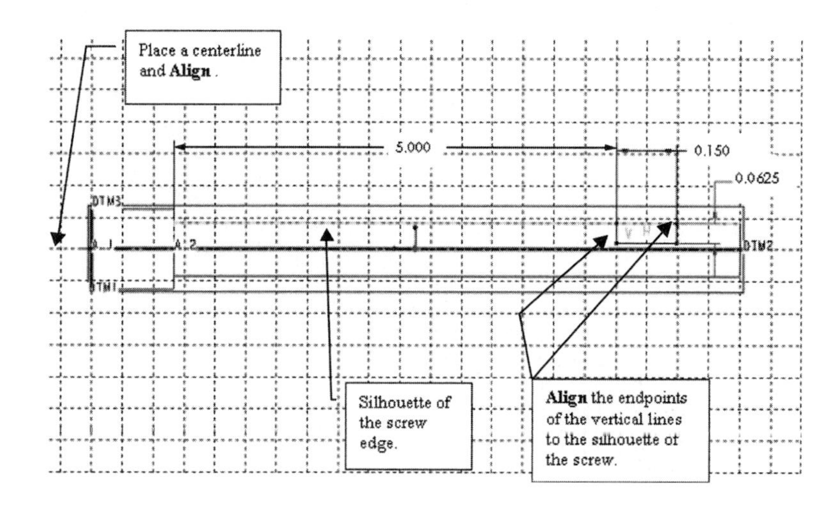


FIGURE 5.15 *The adjustment screw in the* Sketcher *with sketched geometry for the neck.*

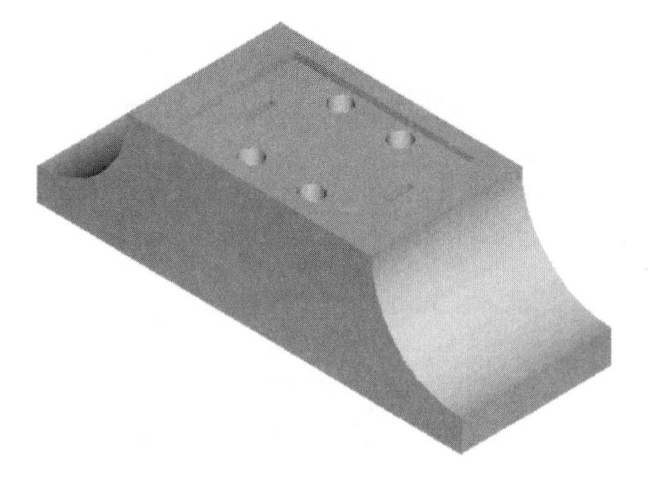

**FIGURE 5.16** *This shaded model of the emergency light base shows the recessed surface containing the bosses.*

*Solid,* and *Done.* Use the *Both Sides* option and select DTM3 as the sketching plane. Define DTM2 as the "*Top.*"

In the *Sketcher,* draw the geometry shown in Figure 5.17. *Dimension* to the given values and *Regenerate.* Extrude *Blind* to a depth of 5.00 inches. The model after extrusion is shown in Figure 5.18.

A recessed surface needs to be created on the model. We will construct this surface by using a cut to remove some material.

Choose *Feature, Create, Cut, Solid,* and *Done.* The sketching plane is shown in Figure 5.19. Select the *One Side* option. Use the surface shown in the same figure as the "*Top.*" The arrow should point "into" the model. If it does not, use *Flip* to change its direction.

The model in the *Sketcher* is shown in Figure 5.19. The geometry of the cut is a simple rectangle. Select the *Rectangle* option from the *SKETCHER*

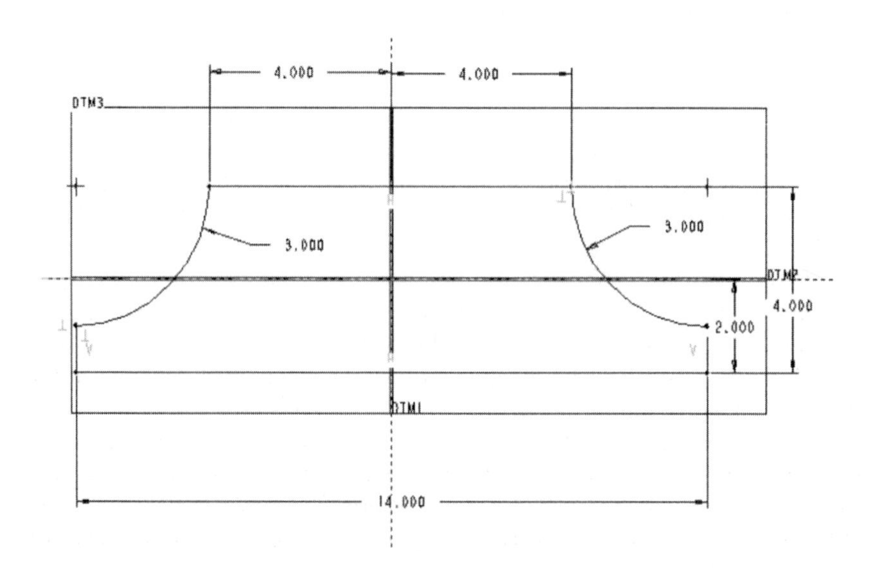


**FIGURE 5.17** *The geometry needed to construct the emergency light base.*

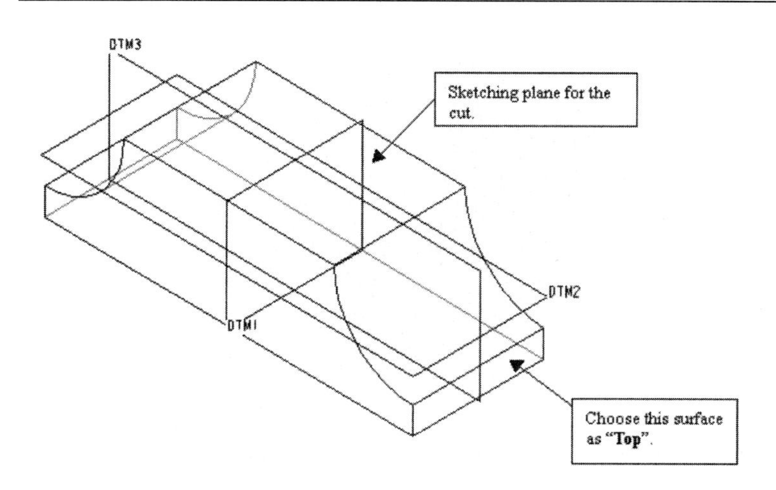


FIGURE 5.18 *The model of the emergency light base after the first protrusion.*

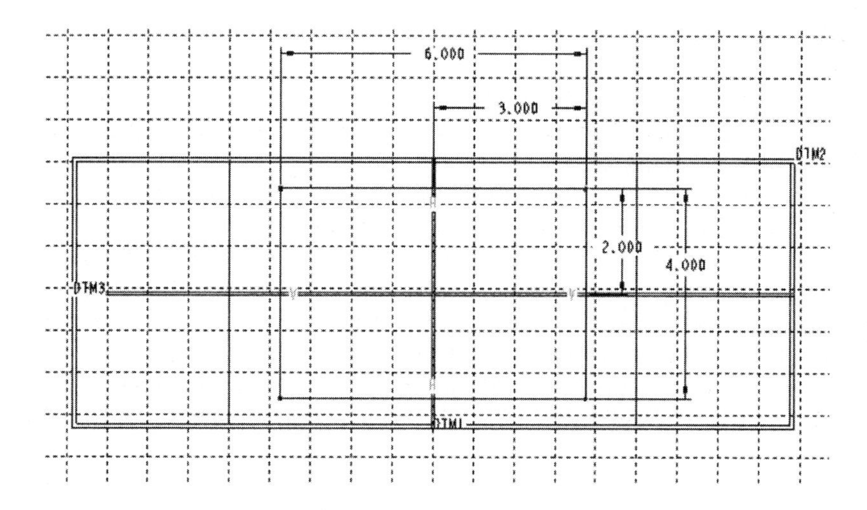


FIGURE 5.19 *The geometry of the cut is a rectangle.*

menu (Figure 2.10) and sketch the section. The dimensions of the cut are given in Figure 5.19. *Dimension, Regenerate,* and *Modify* the dimensions to the correct values. Choose *Done.* The depth of the cut is 0.25 inches. Use the *Blind* option in extruding the cut. The model after extruding the cut is shown in Figure 5.20.

After the recessed surface is created, we can add holes using the surface as the placement plane. After the model is shelled, each of these holes will become a boss. Upon assembly, a washer and hexagonal nut will be placed within these bosses. The combination of the washer and nut will be used to secure a bracket to the base.

We can add the holes by using the placement method outlined in section 4.1. Select *Feature, Create, Hole, Straight,* and *Done.* Use the *Linear* option and *Done.* Select the recessed surface to place the hole. Figure 5.21a shows the first hole placed on the model. Also given in the figure are the references

**FIGURE 5.20** *Using the* Cut *option to remove material yields a rectangular recessed surface.*

for locating the first hole. Place the holes using the dimensions. The diameter of the hole is 0.75 inches.

The model with all four holes placed is shown in Figure 5.21b. Use the same method in placing the remaining holes.

We are now in a position to shell the model. The shell option will remove material below the defined surface and generate four bosses because of the holes. After shelling, we will place a hole through each boss. These holes will accept studs from the bracket upon assembly.

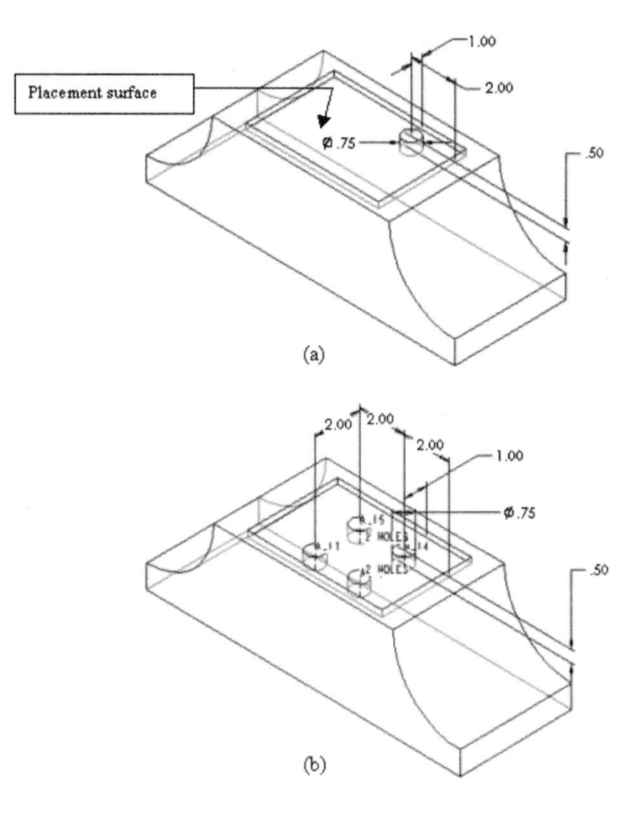


**FIGURE 5.21** *(a) The model is shown with the first hole placed. (b) The remaining holes have been placed. In both cases, the* Linear *option was used.*

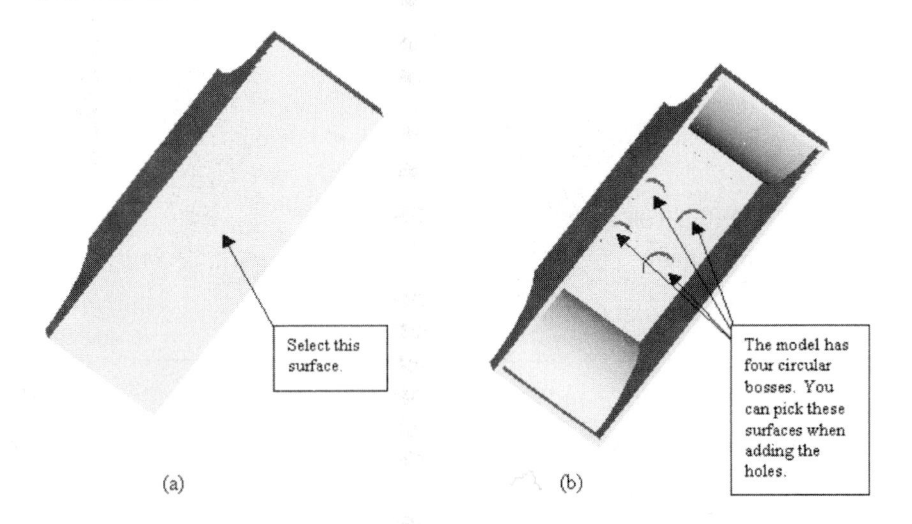


FIGURE 5.22 *Both (a) and (b) show the base. In (a), the model has yet to be shelled. After shelling material is removed from the model, the bosses are generated.*

In order to shell the model, the proper surface must be chosen. In order to make it easier to select the surface, you may wish to reorient your model as shown in Figure 5.22a. Otherwise, use *Query Sel* in choosing the proper surface.

Select *Create* and *Shell.* If you have not reoriented the model, then use *Query Sel.* Click on the surface shown in Figure 5.21a. Select *Done Sel,* then *Done Refs.* Enter a thickness of 0.2.

Let us add holes through each boss. Each cylindrical boss has an axis running through its center. We can use the axis and the *Coaxial* option to place the hole. The placement surface for each hole is the surface of the boss as shown in Figure 5.22b. The steps for adding each hole are: *Feature, Create, Hole, Straight, Done, Coaxial,* and *Done.* Select the placement plane and then the axis. The holes are 0.25 inch in diameter. Save the part.

In the *Shell* option, the order in which the features are created is critical. Suppose that the holes had not been placed on the base before shelling. Then the bosses would not exist. In order to create the bosses after shelling, protrusions would have to be made from the inside of the model.

We will examine the effect of changing the order in which features are created in a model in a subsequent chapter.

## 5.4 SUMMARY

In this chapter, we discussed the options *Cut, Slot, Neck,* and *Shell.* These options are used to remove material from a model.

*Cut* and *Slot* are identical from the procedural point of view. A *Cut* is material removed from a given side of a base feature. A *Slot* is material removed from within a base feature.

A *Neck* is a revolved slot. The geometry of the neck is sketched using an *open* section. Then, the section is revolved about an axis of revolution.

The *Shell* option is used to remove material below a given surface. Because the *Shell* option removes material from all the features that exist on the model, this option may be used to generate a boss. In order to generate a boss, a hole must be placed on the model before it is shelled.

### 5.4.1 STEPS FOR USING THE CUT OR SLOT OPTIONS

The following sequence of steps may be used to create a *Cut* or *Slot:*

1. Select *Feature* and *Create.*
2. Choose either *Cut* or *Slot.*
3. Select *Thin* or *Solid* and *Done.*
4. Pick either *One Side* or *Both Sides.*
5. Pick the sketching plane and an orientation plane.
6. In the *Sketcher,* sketch the section.
7. *Dimension, Regenerate,* and *Modify* to the proper dimensional values.
8. Select *Done* and for the *Cut* option, pick the direction of the material to be removed.
9. Enter the depth of the cut or slot.
10. Select *OK.*

### 5.4.2 STEPS FOR ADDING A NECK TO A BASE FEATURE

In general, a *Neck* may be created on a base feature by using the following steps:

1. Select *Feature, Create,* and *Neck.*
2. Choose the angle of rotation from the *ANGLE* menu (Figure 5.14).
3. Select *Done.*
4. Pick the sketching plane and an orientation plane.
5. In the *Sketcher,* sketch the section using an open section.
6. *Dimension* and *Modify* to the appropriate values.
7. Select *Done.*

### 5.4.3 STEPS FOR USING THE SHELL OPTION

The general sequence of steps for using the *Shell* option are:

1. Select *Create* and *Shell.*
2. Pick the shell surface.
3. Enter the thickness of the shell.

## 5.5 Additional Exercises

The problems chosen for this section require the use of the material removal options discussed in this chapter. The base features will require the use of options from previous chapters in this book. Construct the models as assigned. Save your work unless instructed not to do so.

5.1 Construct the guide plate. Use Figure 5.23 for reference.

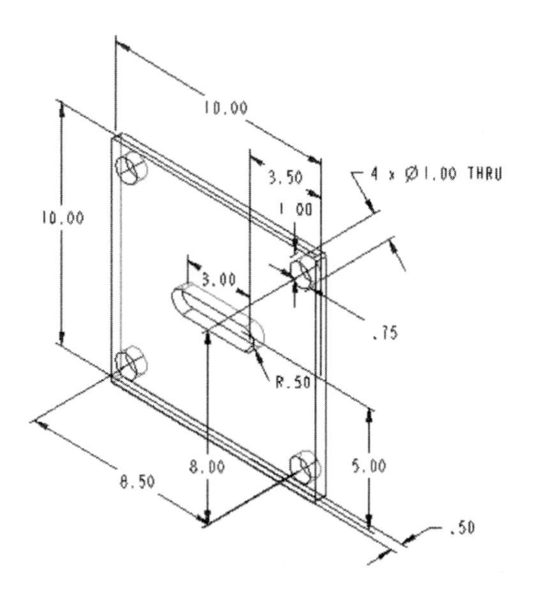


FIGURE 5.23 *Figure for exercise 5.1.*

5.2 Add the cut to the side guide (see also exercises 3.8 and 4.9). Use Figure 5.24 for the geometry of the cut.

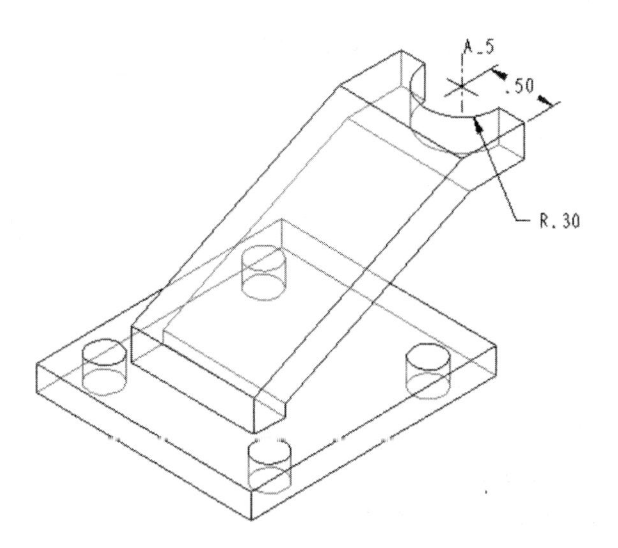


FIGURE 5.24 *Figure for exercise 5.2.*

5.3 Construct the slotted hold down. The geometry of this model is shown in Figure 5.25.

**FIGURE 5.25** *Figure for exercise 5.3.*

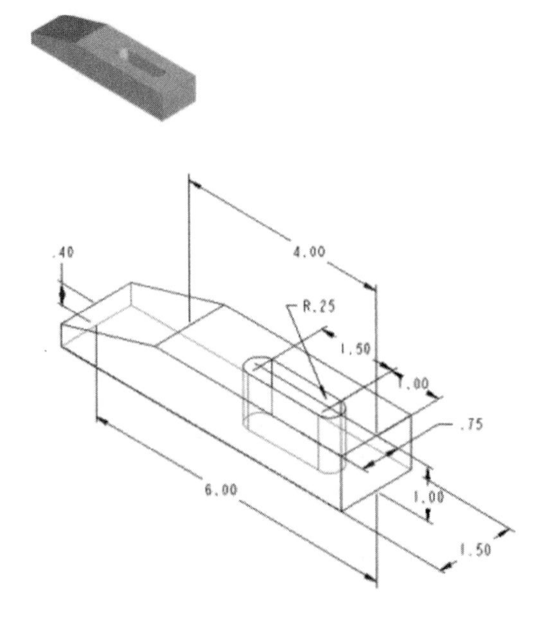


5.4 Create the model of the slide base illustrated in Figure 5.26.

**FIGURE 5.26** *Figure for exercise 5.4.*

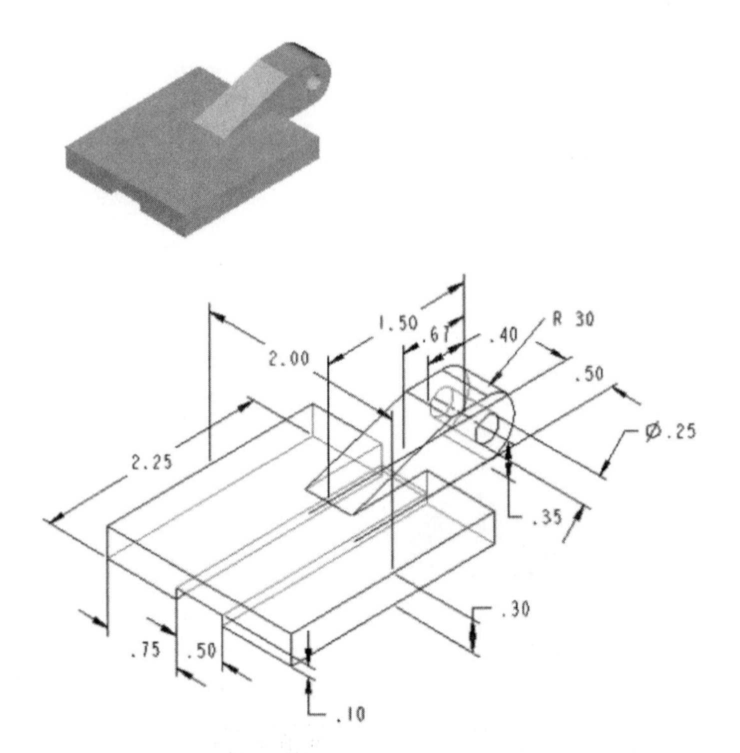

5.5 Use the *Neck* option to construct the groove on the V-pin shown in Figure 5.27.

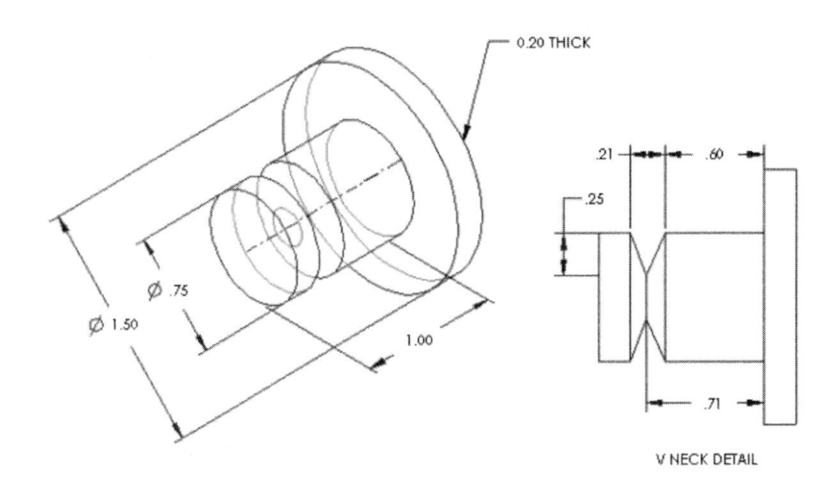


FIGURE 5.27 *Figure for exercise 5.5.*

5.6 Construct the model of the coupling. Use Figure 5.28 for reference.

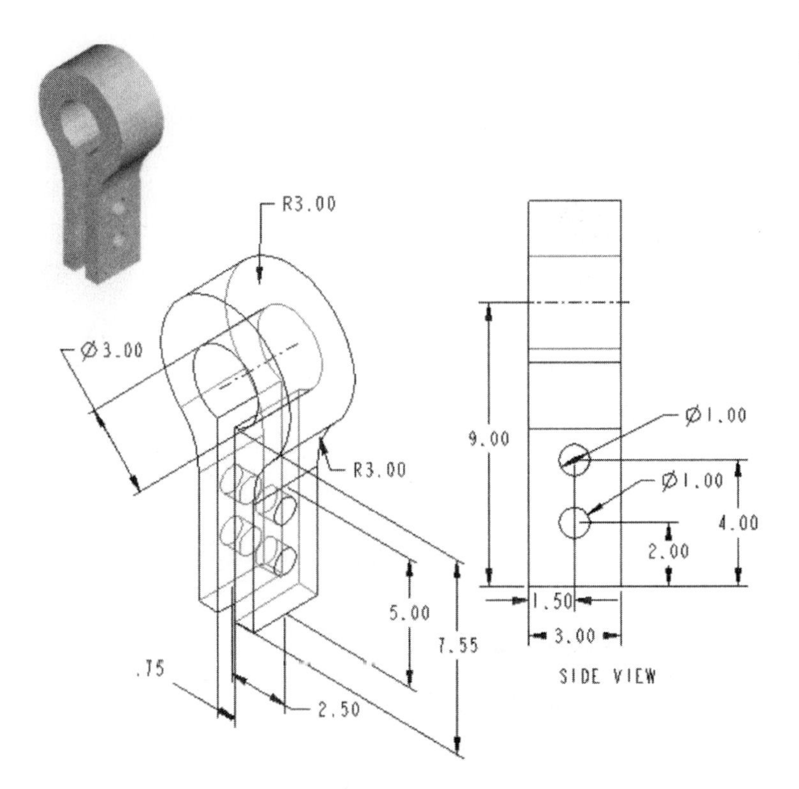


FIGURE 5.28 *Figure for exercise 5.6.*

5.7 Create the model of the pipe clamp top component. The geometry is given in Figure 5.29.

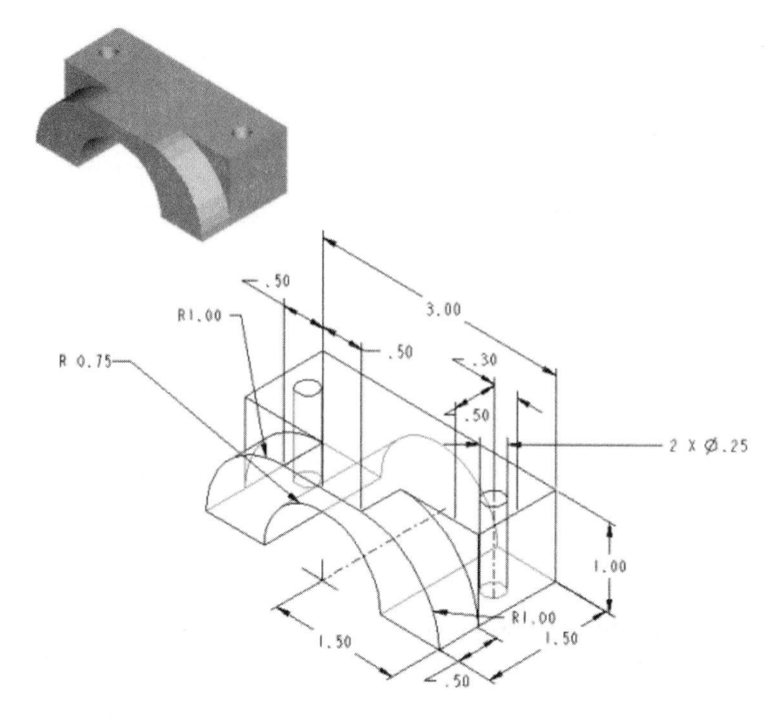


FIGURE 5.29 *Figure for exercise 5.7.*

5.8 Construct the model of the pipe clamp bottom component. Use the geometry given in Figure 5.30.

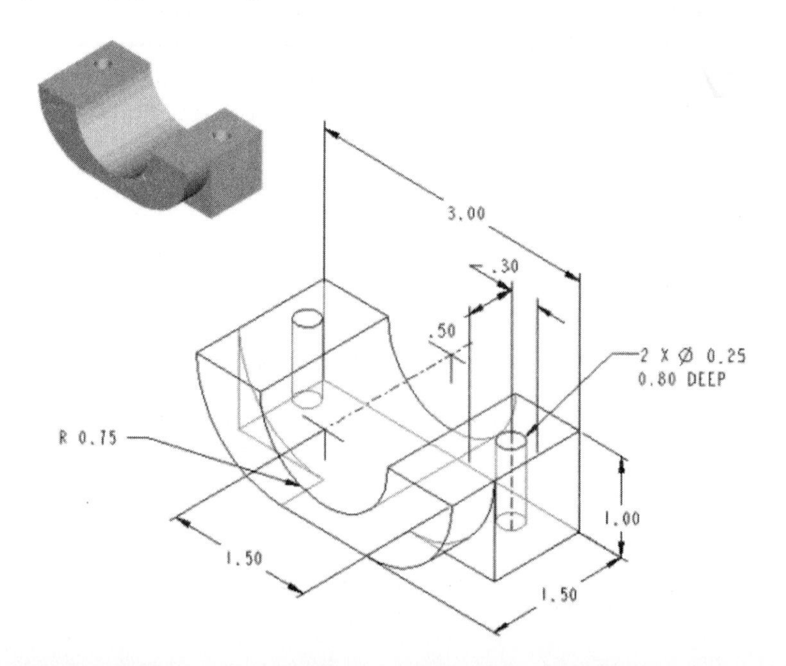


FIGURE 5.30 *Figure for exercise 5.8.*

5.9 Use the *Protrusion* and *Shell* options to create the model of the robot arm housing. The geometry of the model is given in Figure 5.31.

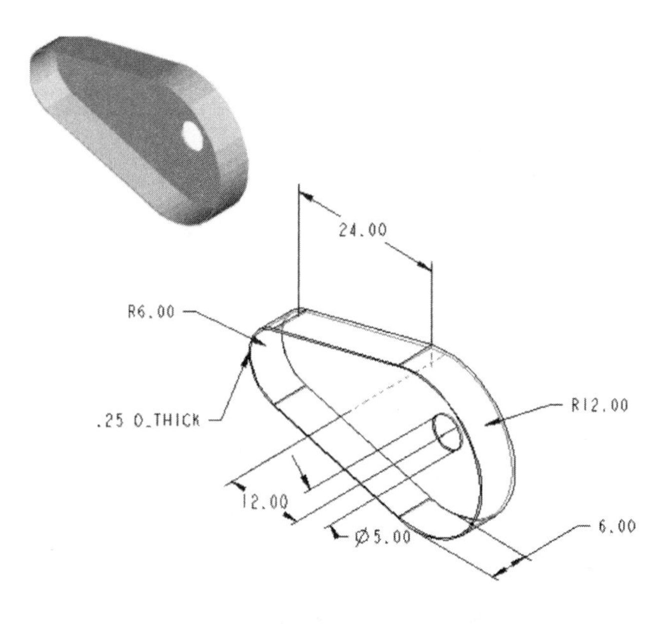


**FIGURE 5.31** *Figure for exercise 5.9.*

5.10 Use the *Shell* option to construct the bosses on the model of the electric motor cover. The geometry of the model is shown in Figure 5.32.

**FIGURE 5.32** *Figure for exercise 5.10.*

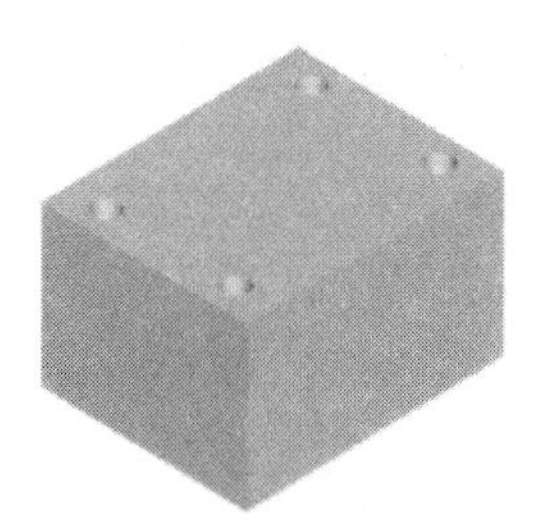

A shaded model of motor cover

The model rotated to show boss detail

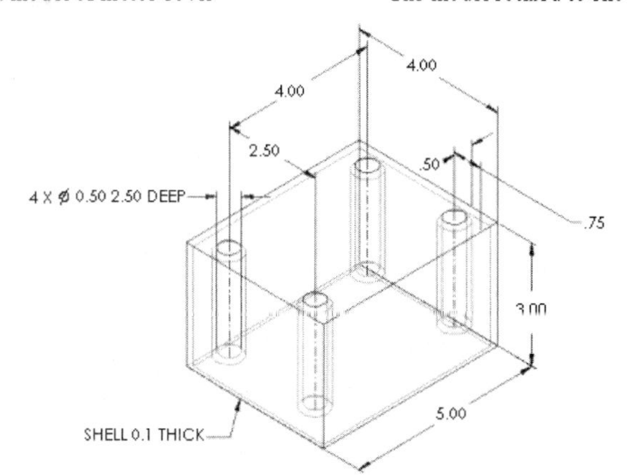

5.11 Add the slot shown in Figure 5.33 to the slotted support (see also exercise 4.12).

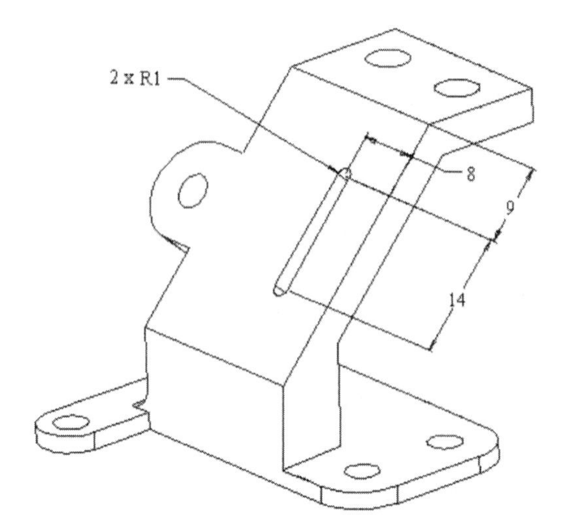


FIGURE 5.33 *Figure for exercise 5.11.*

# CHAPTER 6

# OPTIONS THAT ADD MATERIAL: FLANGE, RIB, AND SHAFT

## INTRODUCTION AND OBJECTIVES

This chapter deals with options that may be used to add a feature to a base feature. The options discussed in this chapter are *Flange, Rib,* and *Shaft.* All three of these options are accessible through the *SOLID* menu.

A flange is a revolved protrusion. Flanges are often used in joints as load-bearing surfaces. Ribs are thin sections that act as structural elements. Shafts are revolved sections that are often used to align parts or as threaded studs.

The objectives of this chapter are:

1. The reader will attain an ability to construct a flange.
2. Using the *Shaft* option, the Pro/E user will be able to place a shaft with varying cross section on a base feature.
3. The reader will utilize the *Rib* option to place a thin section between two base features.

## 6.1 FLANGES

A *Flange* is the opposite of a neck. In creating a flange, material is added by revolving a sketched section around a centerline. Consider again the adjustment screw in Figure 5.12. The larger cylinder could have been constructed by creating a flange around the smaller diameter cylinder. Of course, you would have had to construct the smaller cylinder longer. That is, the length of the cylinder would be $5.438 + 0.812 = 6.25$. Then the flange could be created along one end.

The procedure for creating a flange is the same as that for creating a neck. Flanges, like all necks, are sketched with an *open* section. A *Through Plane* must be created as the sketching plane. Furthermore, you must place a centerline, which becomes the axis of revolution.

### 6.1.1 TUTORIAL # 26: A LAMP HOLDER

As a further example in illustrating the construction of a flange, consider the lamp holder shown in Figure 6.1. The holder is used in an emergency light. The holder fits into the base of a parabolic reflector and is locked into place by twisting the holder to engage the alignment tabs. The flange keeps the holder in place.

The construction of the flange begins with the creation of the base feature, in this case, a right circular cylinder. Begin by creating a part called "EmergencyLightHolder." Add the default datum planes.

In order to construct the base feature, select *Feature, Create, Protrusion, Solid,* and *Done.* Use DTM1 as the sketching plane and DTM2 as the "*Top*" to reorient the part. Create the base feature shown in Figure 6.2 by drawing a circle in the *Sketcher* that is 0.718 inches in diameter. Align the center of the circle to the intersection of DTM3 and DTM2. Extrude to a *Blind* depth of 1.50 inches.

With the base feature constructed, we can now proceed to place the rest of the features. We will use solid cuts to create the tapers along the stem of the holder.

Select *Feature, Create, Cut, Solid,* and *Done.* Choose the *Both Sides* option from the *ATTRIBUTES* menu. Pick DTM2 as the sketching plane and DTM1 as "*Top.*" In the *Sketcher,* draw the geometry shown in Figure 6.3. You need to align the lines and endpoints as shown in the figure. *Dimension, Regenerate,* and *Modify* the drawing to the proper values. Upon successful regeneration, select *Done.* The extrusion arrow should point outward from the

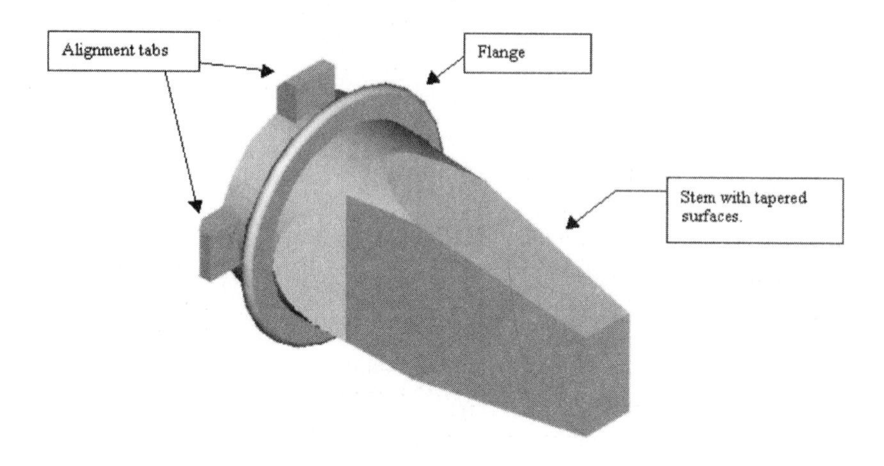


FIGURE 6.1 *The lamp holder to be constructed.*

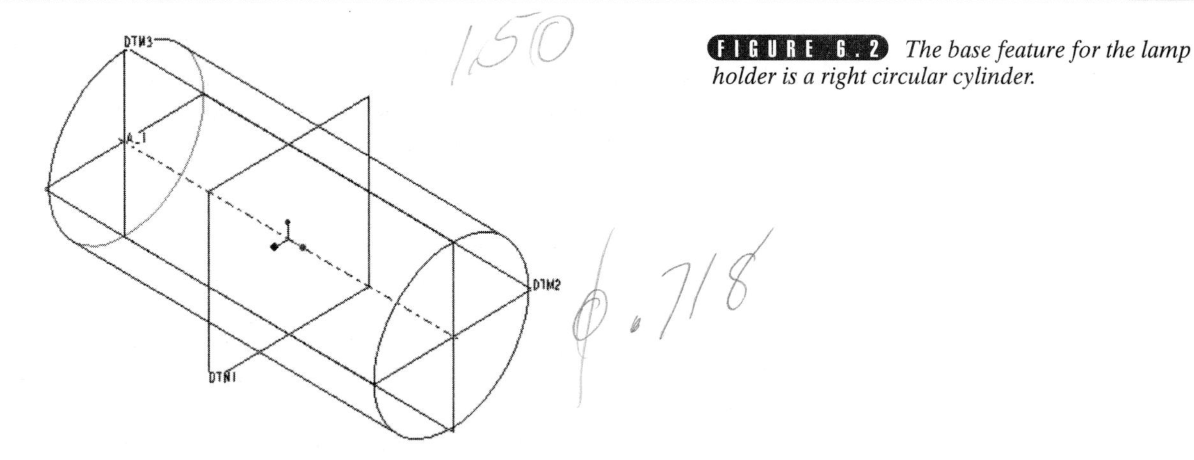


FIGURE 6.2 *The base feature for the lamp holder is a right circular cylinder.*

model. Use the *Thru All* option. The model after the first cut is shown in Figure 6.4.

In order to complete the stem, a second cut is needed. Again, select *Feature, Create, Cut, Solid,* and *Done.* Use the *Both Sides* option and DTM3 as the sketching plane. Select DTM2 as "*Top*" to orient the model in the *Sketcher* as shown in Figure 6.5.

Sketch the section shown in the figure and dimension to the given values. Be sure to add and align the lines as shown in the figure. The model, after the cut feature has been added, is shown in Figure 6.6.

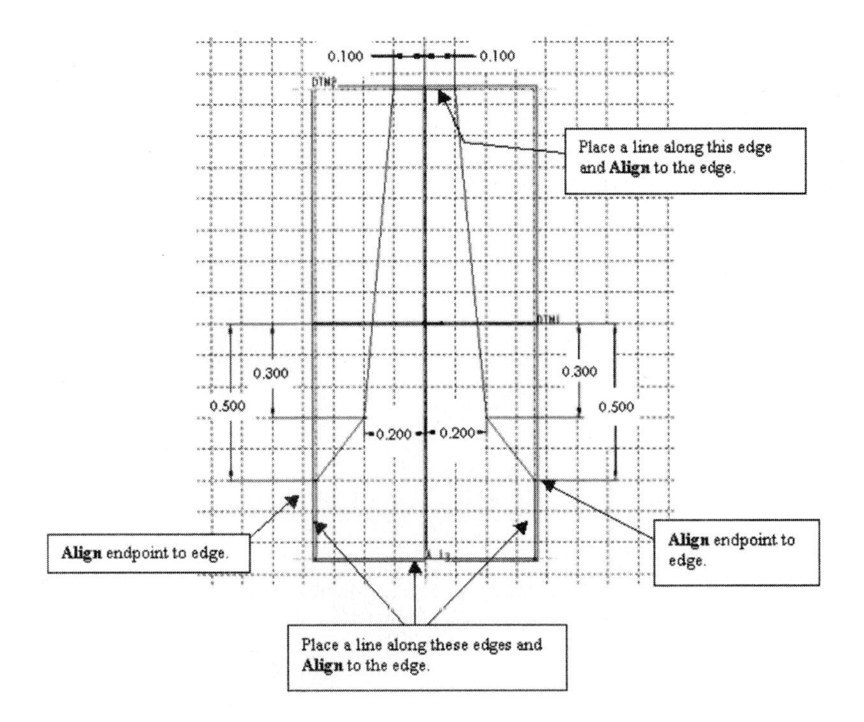


FIGURE 6.3 *The sketch required for constructing the first cut.*

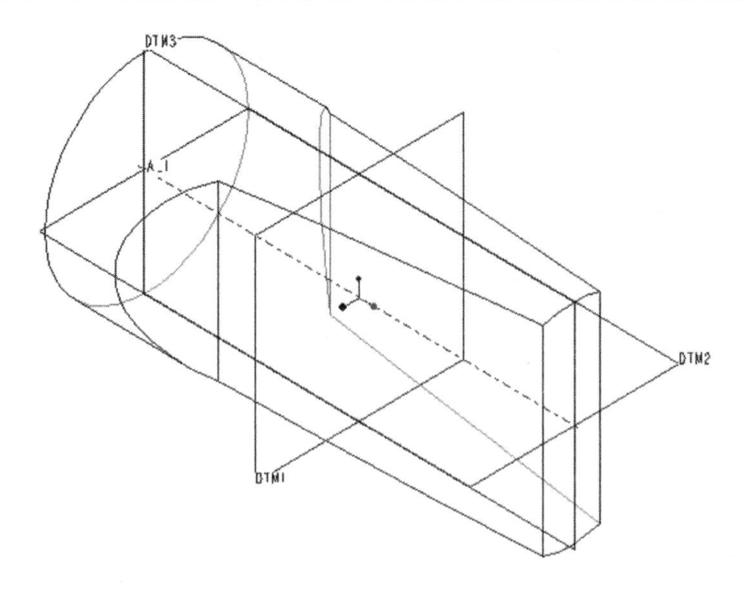


FIGURE 6.4 *The model after the first cut has a taper along the stem.*

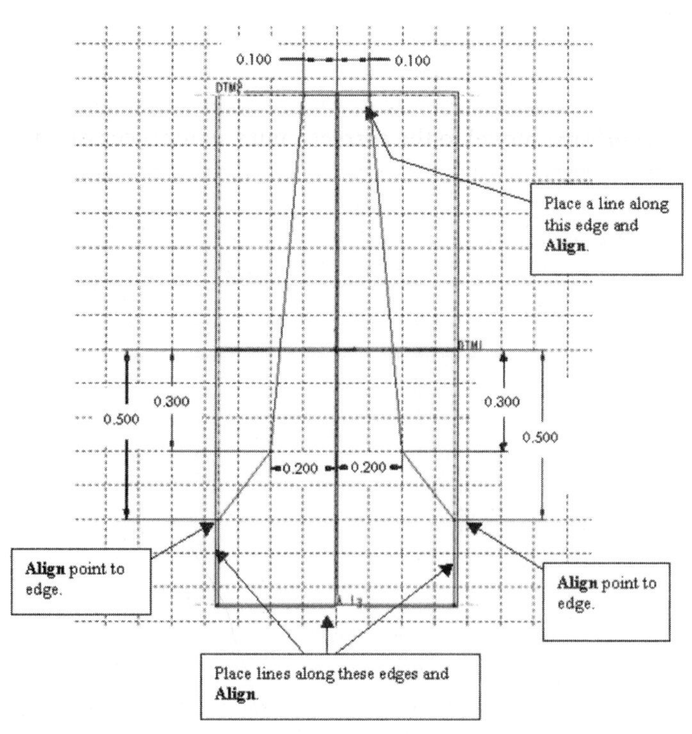


FIGURE 6.5 *The geometry for the second cut is given in this figure.*

We are now in a position to add the alignment tabs to the model. The two tabs will be added separately, using solid protrusions. Choose *Feature, Create, Protrusion, Solid,* and *Done.* Select *Both Sides* from the *ATTRIBUTES* menu. Then choose DTM3 as the sketching plane. Select DTM2 as the "*Top.*"

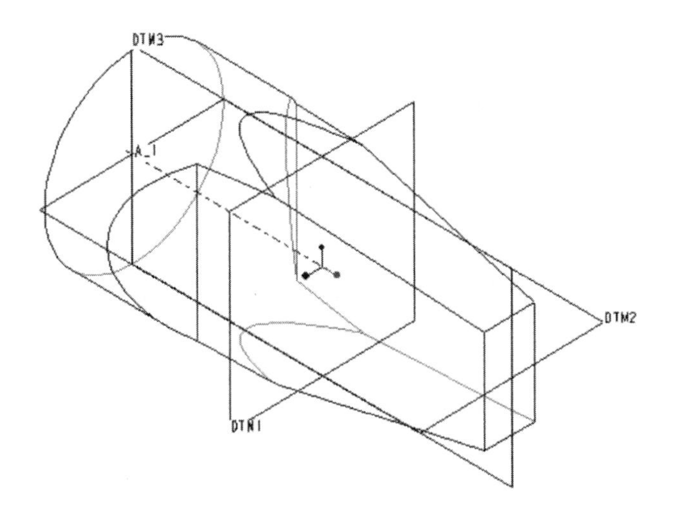


FIGURE 6.6 *After using the geometry to cut the model a second time, the model takes on this shape.*

Pro/E will deposit the model in the *Sketcher.* Sketch the geometry shown in Figure 6.7. Note that the tab extends into the model. We extend the tab so that when the extrusion process is performed, the tab is blended into the curvilinear surface of the model.

Be sure to align the tab as shown in Figure 6.7. *Dimension, Regenerate,* and *Modify* the model to the correct values. After successfully regenerating the model, extrude the tab to a *Blind* depth of 0.185. The model with the first tab added is shown in Figure 6.8.

The second alignment tab can be constructed in a similar fashion. Again, select *Feature, Create, Protrusion, Solid,* and *Done.* Choose the *Both Sides* option. Then select DTM2 as the sketching plane. Use DTM3 as the "*Top.*"

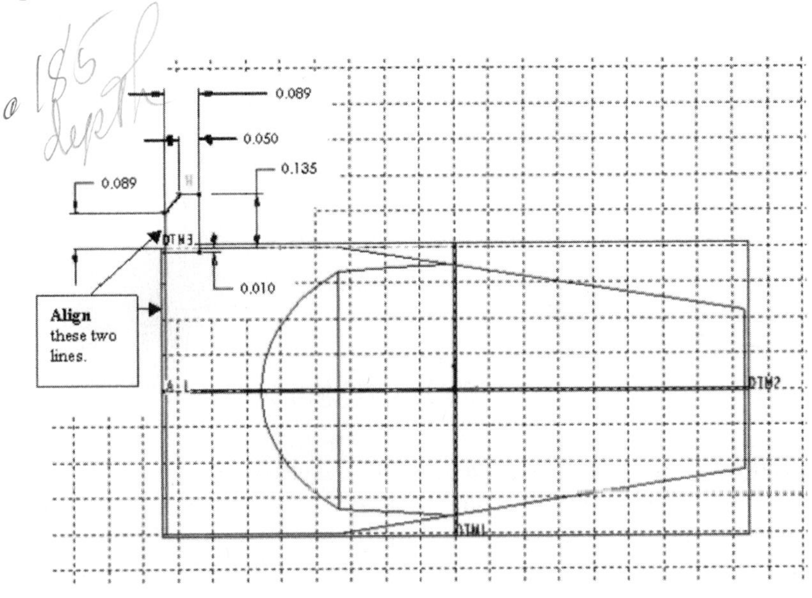


FIGURE 6.7 *The geometry of the first alignment tab with the appropriate dimensions. Note the two lines that need to be aligned.*

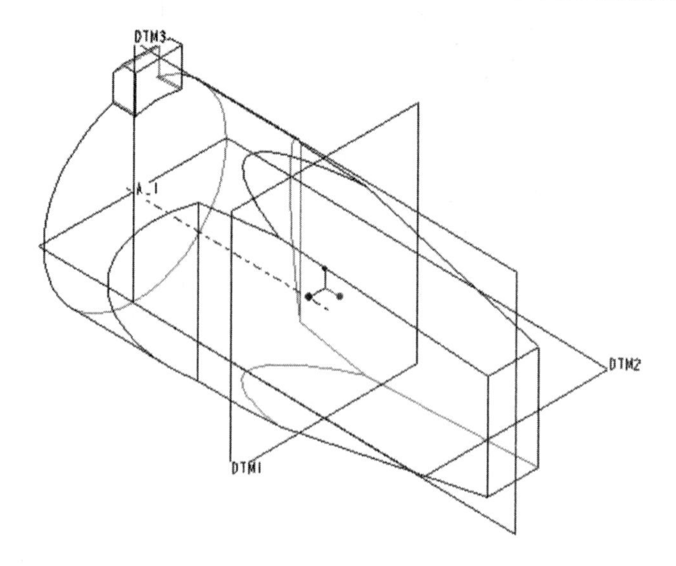


FIGURE 6.8 *The model after the first alignment tab has been added.*

The geometry for the second tab is given in Figure 6.9. Sketch the geometry and *Modify* to the dimensional values shown. Again, notice the two lines that need to be aligned. After regenerating, extrude to a *Blind* depth of 0.185. The model with the second tab is shown in Figure 6.10.

We are now in a position to add the flange. The flange will be created by revolving the geometry 360° about an axis of rotation. We will use DTM3 as the sketching plane.

Choose *Feature, Create,* and *Flange.* From the *OPTIONS* menu, select *360°* and *One Side.* Use DTM3 as the sketching plane with DTM2 as "*Top*" to reorient the model.

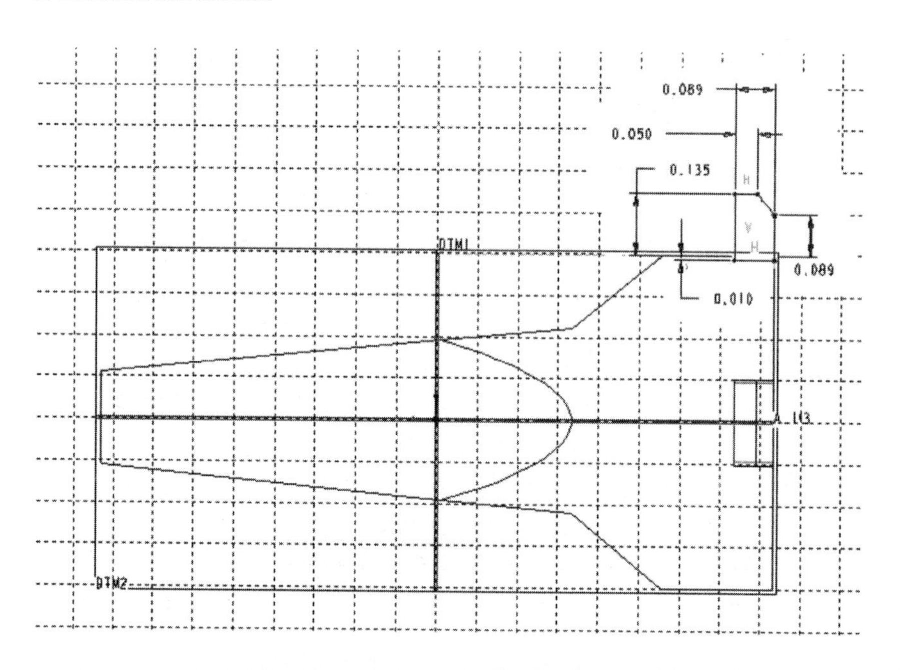


FIGURE 6.9 *The geometry of the second alignment tab. Again, notice the two lines that need to be aligned.*

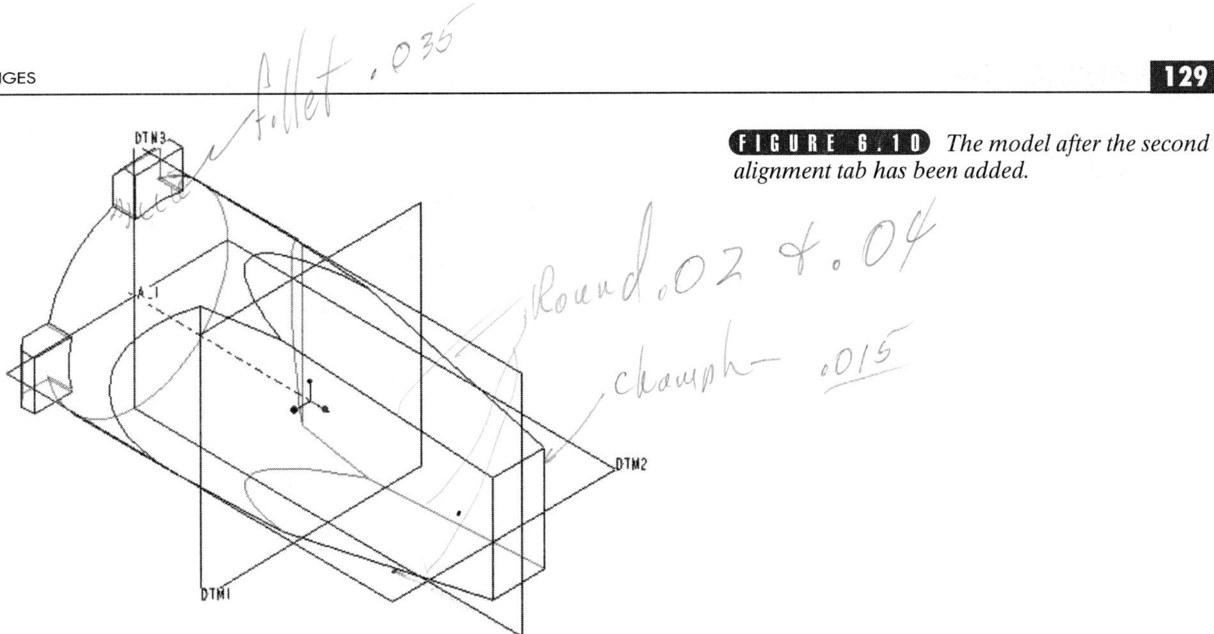


**FIGURE 6.10** *The model after the second alignment tab has been added.*

Add a centerline to the sketch by changing the line type to *CenterLine.* Place the line along the DTM2 edge and align to the edge. The geometry of the flange is shown in Figure 6.11, along with the appropriate dimensional values. Sketch the geometry and *Modify* to the appropriate dimensions. The flange is drawn with an open section. Note that the endpoints of the vertical lines defining the geometry of the flange need to be aligned to the silhouette of the model. The model with the flange added is given as Figure 6.1. Save the part.

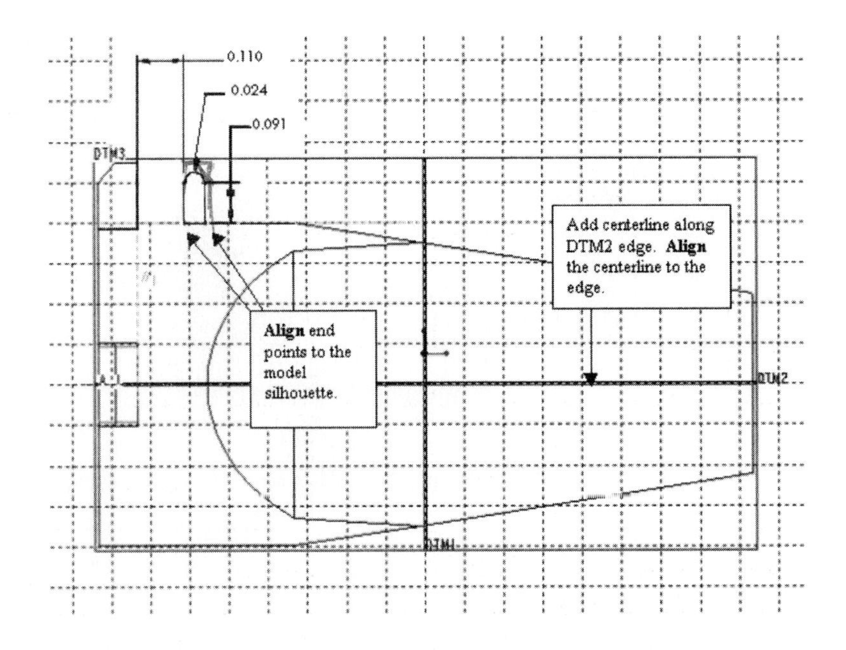


**FIGURE 6.11** *The geometry of the flange consists of two vertical lines and an arc between them. Notice the location of the centerline.*

## 6.2 RIBS

A rib or web is a thin protrusion attached to a base feature. A classic example of a part with a rib is the connecting arm shown in Figure 6.12. The base feature for the arm is constructed by making two cylinders of unequal diameter.

### 6.2.1 TUTORIAL # 27: A CONNECTING ARM

Create the base feature for this part by making a protrusion of the two cylinders equally spaced from a datum plane. You will want to extrude *Both Sides* from the sketching plane. The reason for this, is to set up a sketching plane for the rib. Then the rib can be created by extruding symmetrically from the sketching plane. Note that you will need to create each cylinder separately.

Sketch the section for the base feature as shown in Figure 6.13. The holes through the cylinders as well as the keyway will be added later. Use plane DTM3 as the sketching plane and define DTM2 as the "*Top.*" Use the *Both Sides* option for the protrusion, which will create the protrusion symmetrically with respect to the DTM3 datum plane. Later, we will use this plane as the sketching plane for the rib.

Evenly space the circles with respect to the DTM1 plane. Then *Dimension* as shown. Extrude *Blind* to a depth of 0.625.

The rib can be added with the *Rib* feature. Select *Create, Solid,* and *Rib.* Choose the DTM3 default datum plane as the sketching plane. Then select DTM2 as the "*Top*" to orient the part so that the section can be drawn.

All sections for ribs are drawn as *open* sections. In order to sketch the section for this particular rib, draw two horizontal lines equally spaced above and below the DTM1 datum plane edge (see Figure 6.14). Then add a vertical line. Place the line from the endpoints of each line through the center of

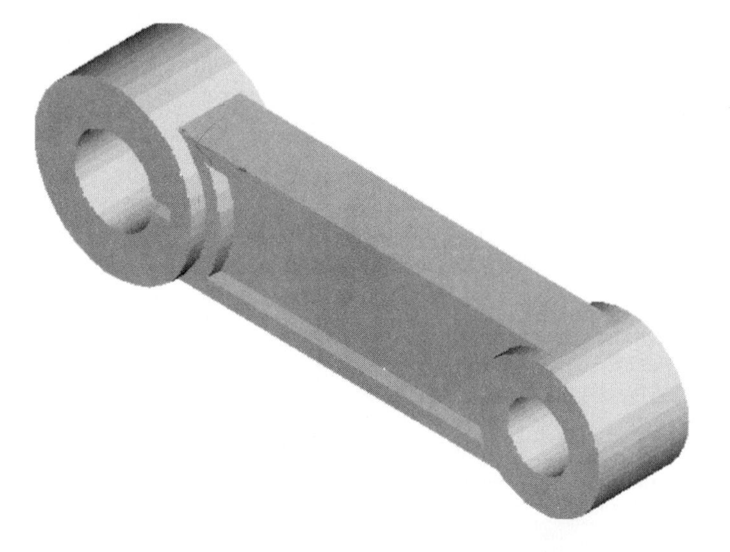

FIGURE 6.12 *After regenerating, this shaded model of the connecting arm can be created. The rib is the slender blended rectangular feature between the two cylinders.*

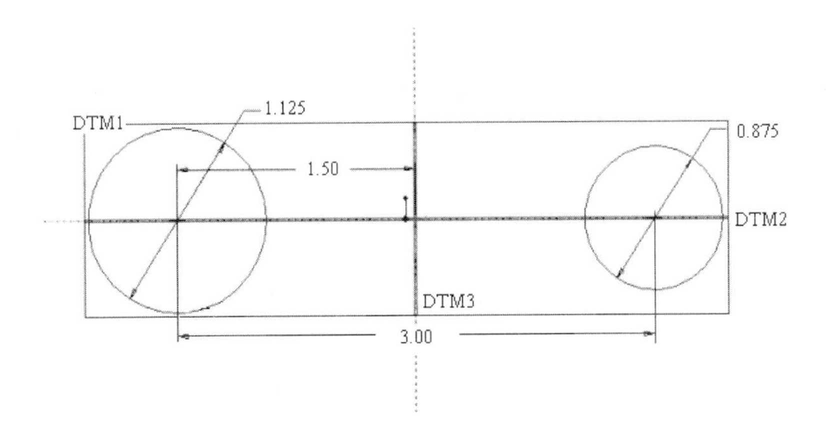


FIGURE 6.13 *Use this sketch of the two circles in the* Sketcher.

the smaller cylinder. Although this section is an open section, one end of the section must be closed. Otherwise, two loops would result. If you try to regenerate a section with two loops, the program will display an error indicating more than one loop. When extruded, the vertical line will not add to the part, because material already exists where the line is defined. Pro/E tolerates this positive interference quite well.

You will need to align the endpoints of the horizontal lines to the edges of the cylinders by clicking on the end of each line and then on the edge of the cylinder. If the vertical line and the horizontal lines have been placed properly, then alignment of the ends of the horizontal lines with the smaller cylinder will automatically dimension the width of the rib to the diameter of the smaller cylinder. After aligning the endpoints, select *Regenerate.*

*Fill* the rib "inward" and select *OK*. The thickness of the rib is 0.438. Once you have created the rib, your part should appear as in Figure 6.15.

Add the coaxial holes for each cylinder. The dimensions are given in Figure 6.16. Then add the 0.077 x 0.126 keyway. Create the keyway by using a cut. Sketch a rectangle as shown in Figure 6.16. Notice that adding the vertical line encloses the section (which is required), and it does not remove

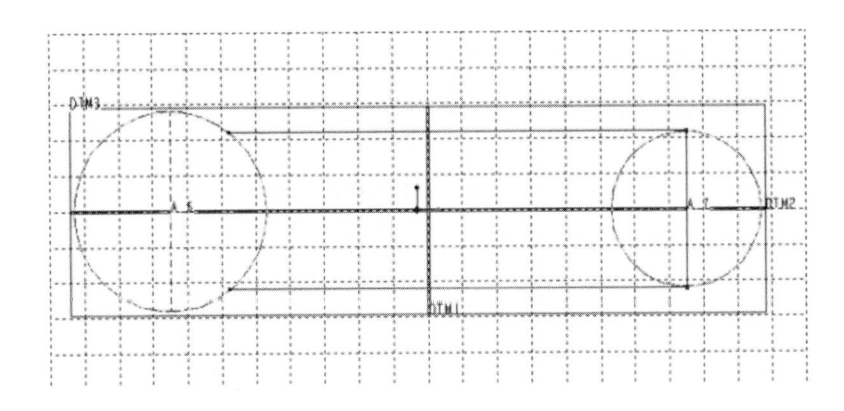

FIGURE 6.14 *The proper positioning of the lines to create the rib.*

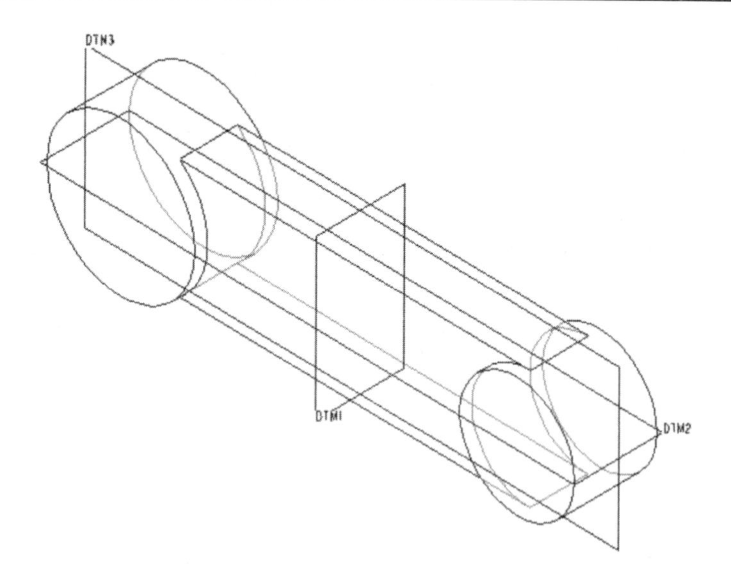


**FIGURE 6.15** *Connecting arm with added rib.*

material from the part because it goes through the hole (another example of positive interference). Be sure to align the ends of the horizontal line to the curved edge of the cylinder. *Dimension* and *Regenerate* the model. The depth of the keyway is *Thru All.*

The rib is not complete. We need to remove some material from the rib. In order to remove this material, we will use the *Cut* option twice, once from each side of the rib.

Enter *Feature* mode and choose *Cut.* Use one side of the rib as the sketching plane and enter the *Sketcher.* Sketch two horizontal lines equidistant from the DTM2 default datum plane. The lines should appear similar to the ones shown in Figure 6.17.

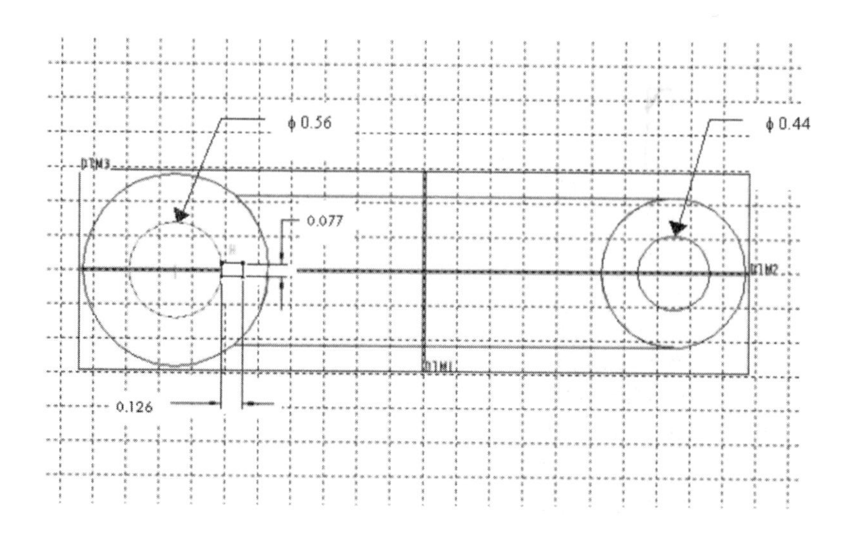


**FIGURE 6.16** *The 0.077 × 0.126 keyway may be added to the connecting arm with the geometry shown.*

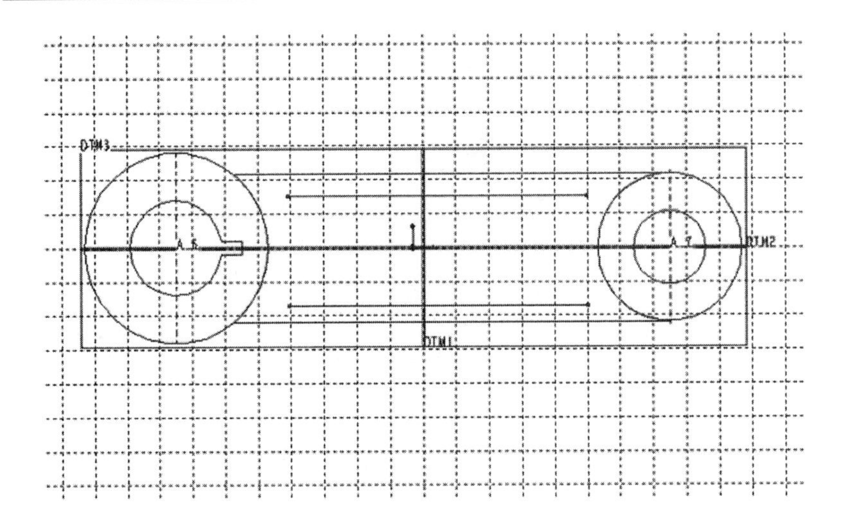


FIGURE 6.17 *The connecting arm in the* Sketcher *with the horizontal lines located.*

Create circular arcs that pass through the endpoints of these lines and are concentric with the circular ends of the arm. In order to create these arcs, use the *Arc* option from the *GEOMETRY* menu (Figure 2.20). From the *ARC TYPE* menu shown in Figure 6.18, choose *Concentric.* Then click on the circle shown in Figure 6.19. In order to define the length of the arc, click on the endpoint of the top horizontal line and move the mouse pointer downward. When the arc passes through the next endpoint, click the left mouse button. The arc should be passing through or near the endpoints of the lines. Your sketch should look similar to the one in Figure 6.19. Repeat for the other side.

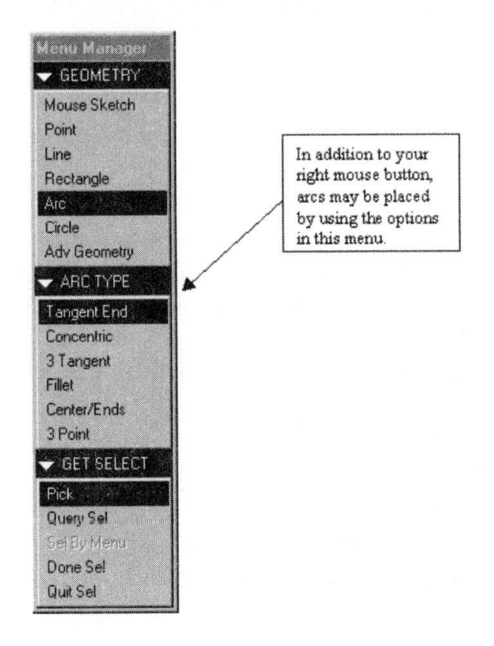


FIGURE 6.18 *Concentric circles may be created using the option from the* ARC TYPE *menu.*

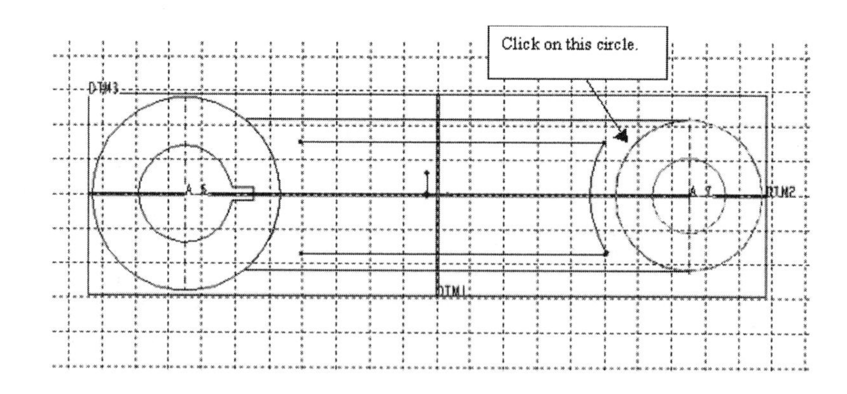


FIGURE 6.19 *The connecting arm with one arc located.*

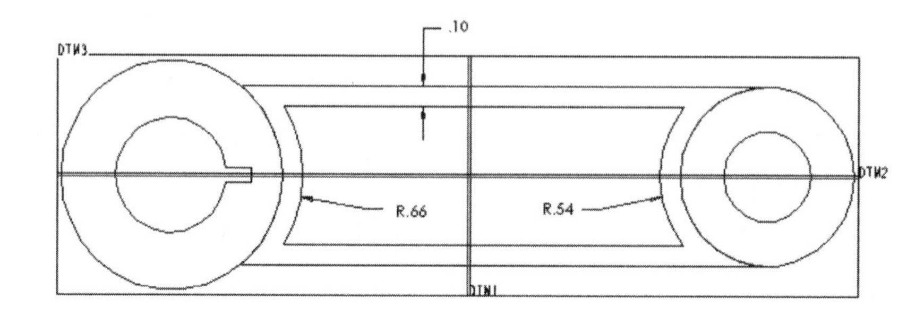


FIGURE 6.20 *Use these dimensions to properly generate the cut.*

*Dimension* the cut-out sketch. The horizontal lines are 0.1 unit from the edge of the arm. The radii of the arcs are 0.6625 and 0.5375. These dimensions are shown in Figure 6.20. The depth of the cut is 0.1.

*Regenerate* the sketch and then choose *Done.* Create another cut on the opposite side of the arm using the other side of the arm as the sketching plane. The final version of the connecting arm should appear as in Figure 6.12.

Save this part to use for a future tutorial on rounds and revolved sections.

## 6.3 SHAFTS

A shaft is a surface of revolution that extends from a placement surface. Shafts are created by sketching a cross section. This section must be closed. Then, the section is revolved about a centerline, which the Pro/E user must place in the sketch.

The shaft is located on the base feature using a method analogous to the placement of holes. That is, the shaft is located using the *Linear, Coaxial, On Point,* and *Radial* options discussed in chapter 4.

### 6.3.1 TUTORIAL # 28: A SHAFT FOR THE U BRACKET

As our example of a shaft, we take the U bracket from chapter 5 (Figure 5.1) and place a shaft at the base of the bracket. Retrieve the U bracket.

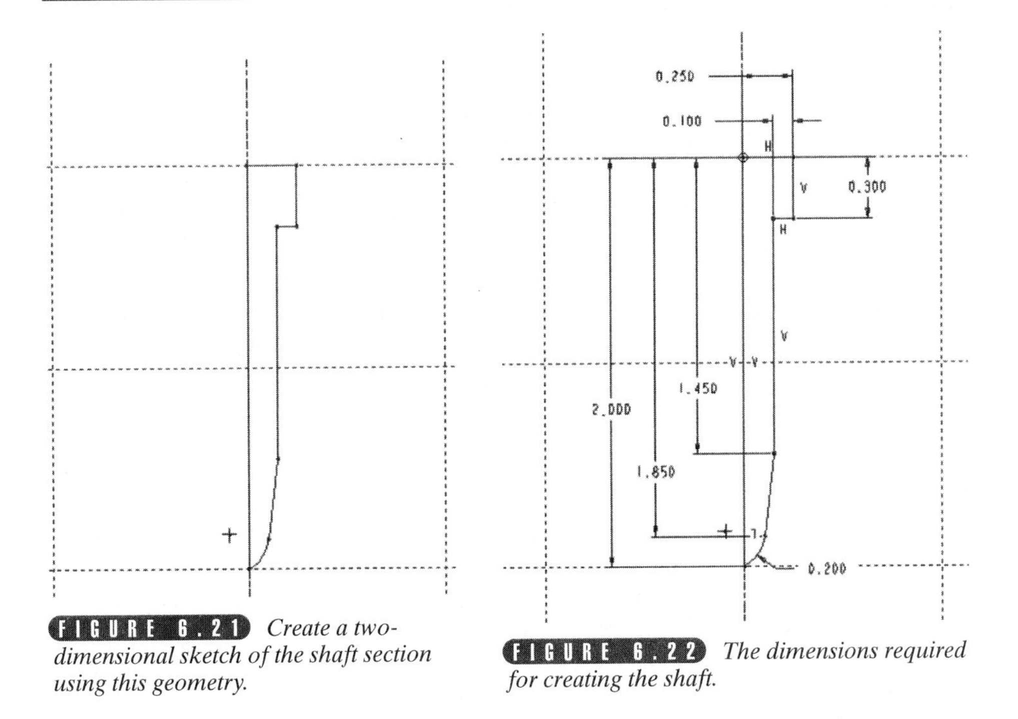


**FIGURE 6.21** *Create a two-dimensional sketch of the shaft section using this geometry.*

**FIGURE 6.22** *The dimensions required for creating the shaft.*

Select *Feature, Create, Solid,* and *Shaft.* The shaft will be located on the part using the same options as those for locating holes. Thus, you need to pick a placement plane. With your mouse, select the bottom of the base as the placement plane. Choose *Linear* and *Done* as your placement options. We will locate the shaft at the center of the bottom of the bracket.

At this point, Pro/E will open and display a second window. The second window is to be used to sketch the profile of the shaft. In Figure 6.21, we have reproduced the profile of the shaft as it would appear in the *Sketcher.* Notice that a centerline has been placed. Also, notice that the sketch is closed.

Sketch the profile, as shown in Figure 6.21. Then, place the dimensions as shown in Figure 6.22. *Regenerate* and *Modify* to the proper values.

Place the shaft by selecting two orthogonal sides of the part. Because the shaft is to be centered on the bottom of the part, and the bottom is 3.00 units square, any two sides will do. Enter a value of 1.5 to locate the shaft. Select *OK.* Your bracket should appear as shown in Figure 6.23.

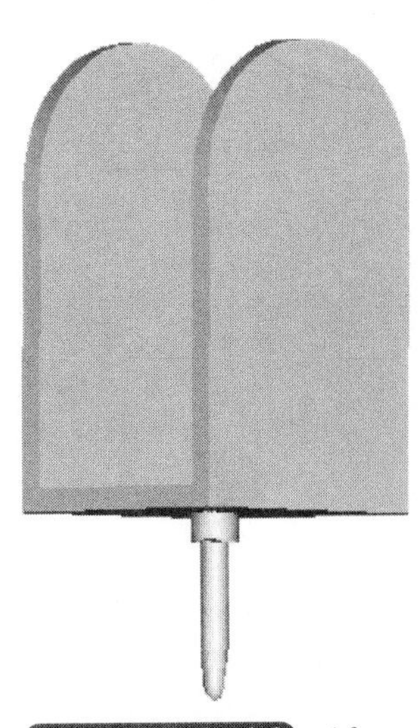

**FIGURE 6.23** *After adding the shaft, a shaded model of the U bracket with the shaft may be created.*

## 6.4 SUMMARY

In this chapter, we considered three options that are used to add material to a part. These options are *Flange, Rib,* and *Shaft.* A *Flange* is created by adding a revolved section around a centerline. Both the cross section and the centerline must be provided by the user. A *Rib* is a thin protrusion attached to one or more features. Ribs are constructed by sketching the section of the rib on a sketching plane and extruding the section to a given depth. A *Shaft* is a closed revolved section that extends from a surface. Shafts are located by the method used to locate holes. The user must sketch the profile of the shaft, as well as locate an axis of revolution.

### 6.4.1 STEPS FOR ADDING A FLANGE

A flange must be added to a base feature. In general, the sequence of steps for adding a flange is as follows:

1. Choose *Feature, Create,* and *Flange.*
2. Select the angle of revolution.
3. Choose the sketching plane and the orientation plane. Pro/E will load the model in the *Sketcher* using your selections.
4. Sketch the geometry of the section. The section must be *open.*
5. Place a centerline. This centerline will be used as the axis of revolution.
6. *Dimension* and *Modify* the section.
7. Select *Done.* The software will revolve the section.

### 6.4.2 STEPS FOR ADDING A RIB

After the base feature has been constructed, a rib may be added to the model. The general steps for adding the rib are:

1. Select *Feature, Create, Solid,* and *Rib.*
2. Select the sketching plane.
3. Choose the orientation plane.
4. After Pro/E has reoriented and displayed the model in the *Sketcher,* sketch the section.
5. *Dimension* and *Modify* the section. *Regenerate.*
6. Choose *Done.* Then, select the fill direction.
7. Choose or enter the depth of the extrusion.
8. Select *OK.*

### 6.4.3 STEPS FOR ADDING A SHAFT

Like ribs and flanges, a shaft is added to a base feature. In general, the following sequence of steps may be followed in order to add a shaft to a model:

1. Select *Feature, Create, Solid,* and *Shaft.*
2. Pick the placement plane with your mouse.

3. Choose the placement option from the *PLACEMENT* menu (Figure 4.1b).
4. Pro/E will open a new window. Sketch the section of the shaft. Also, be sure to place a centerline.
5. *Dimension* and *Modify* the section. *Regenerate.*
6. Choose *Done.*
7. Place the shaft by entering or selecting the appropriate features as required by your selection in step 3.
8. Select *OK.*

## 6.5 ADDITIONAL EXERCISES

The exercises in this section can be constructed using the options discussed in this chapter. Create models of the various parts. For some parts, additional options from previous tutorials are necessary to complete the part. Save your work unless instructed not to do so.

6.1 Construct the two-sided bushing, using the *Protrusion* and *Flange* options. Use Figure 6.24 for the geometry of the model.

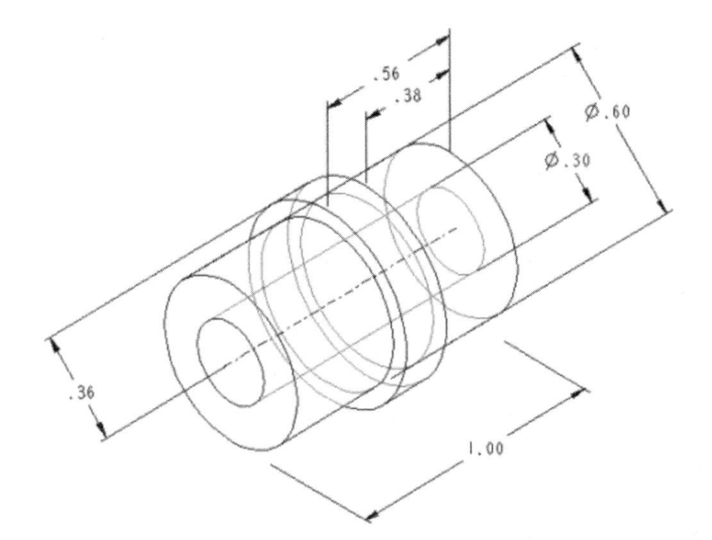


FIGURE 6.24 *Figure for exercise 6.1.*

6.2 Use the *Flange* option and the geometry in Figure 6.25 to construct the model of the lock plunger.

FIGURE 6.25 *Figure for exercise 6.2.*

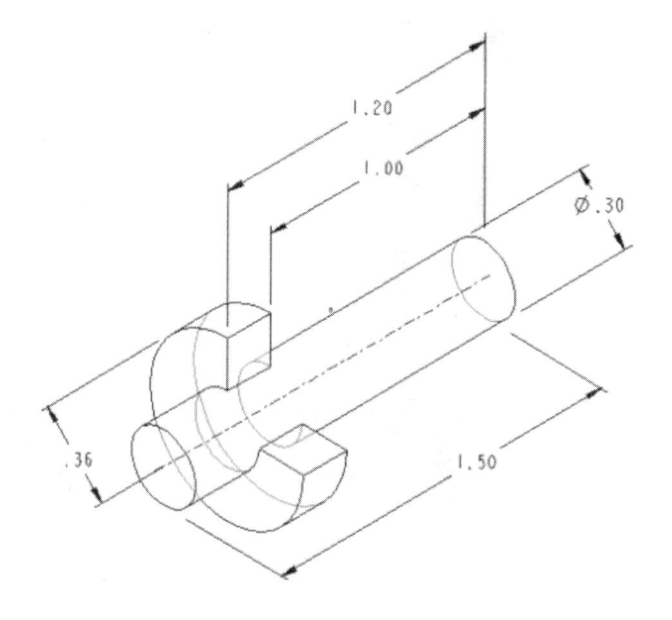


6.3 Construct the model of the lifter as shown in Figure 6.26.

FIGURE 6.26 *Figure for exercise 6.3.*

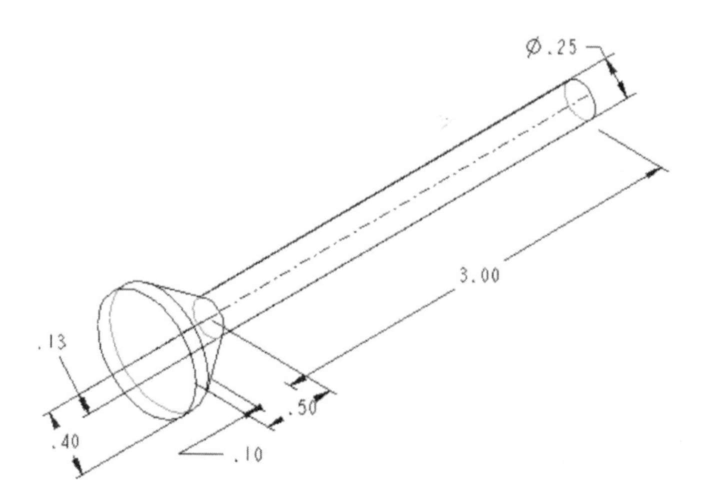

6.4 Create the model of the pin. Use the geometry in Figure 6.27.

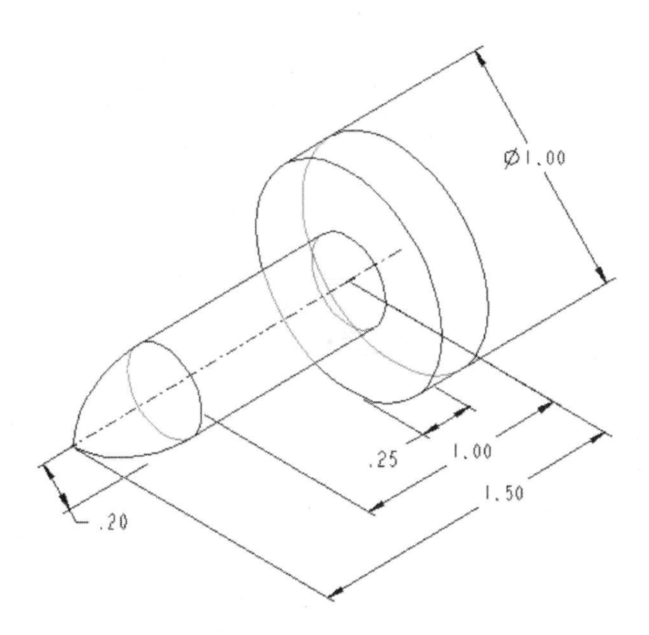


FIGURE 6.27 *Figure for exercise 6.4.*

6.5 Use the geometry shown in Figure 6.28 to create the model of the cone head rivet.

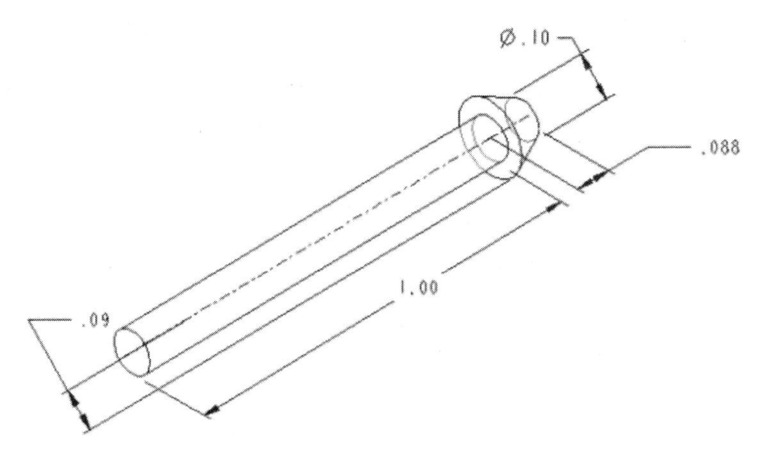


FIGURE 6.28 *Figure for exercise 6.5.*

6.6 Using the geometry shown in Figure 6.29, create the model of the flat top countersunk rivet.

FIGURE 6.29 *Figure for exercise 6.6.*

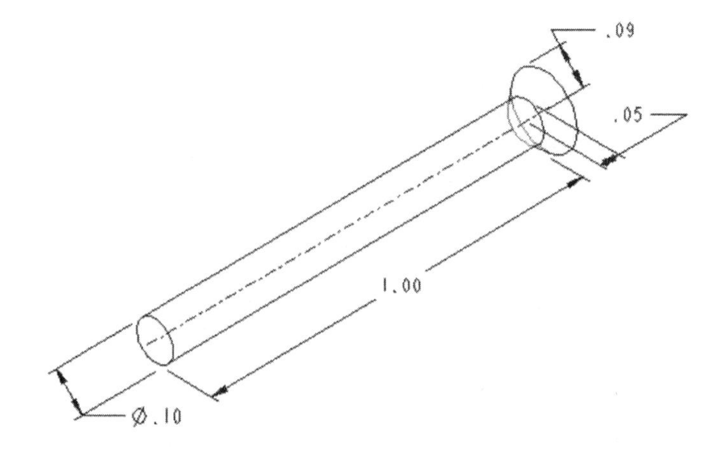


6.7 Construct the model of the valve poppet. Use the model shown in Figure 6.30.

FIGURE 6.30 *Figure for exercise 6.7.*

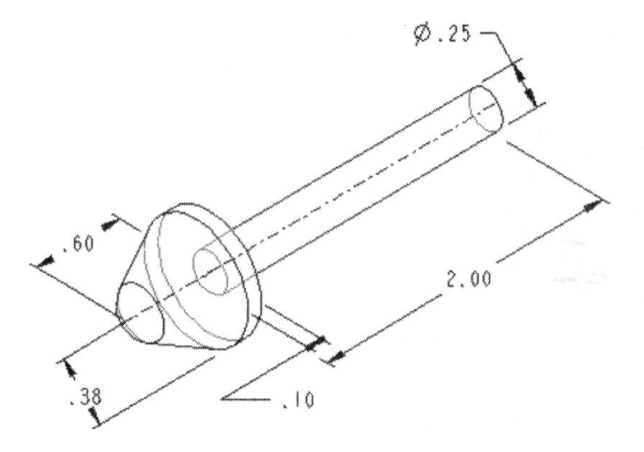

6.8 Use Figure 6.31 to create the model of the detent pin.

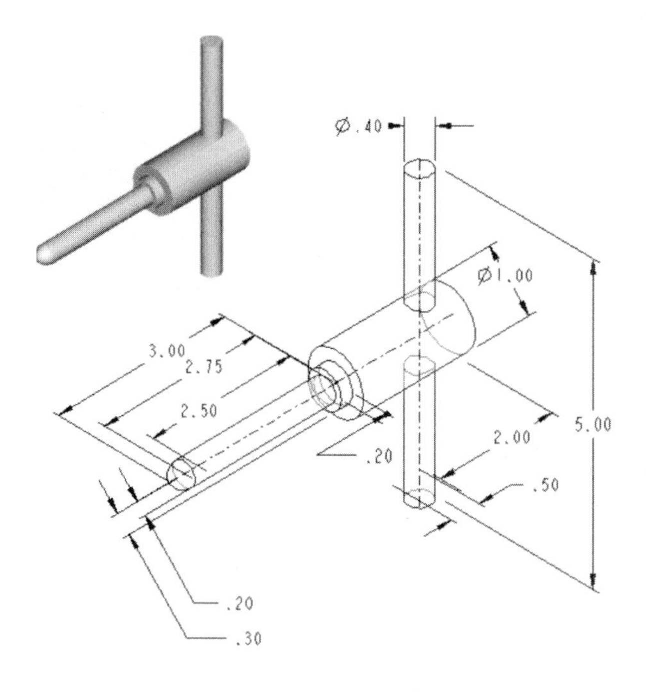


FIGURE 6.31 *Figure for exercise 6.8.*

6.9 Construct the model of the coupler. Use Figure 6.32 as a reference.

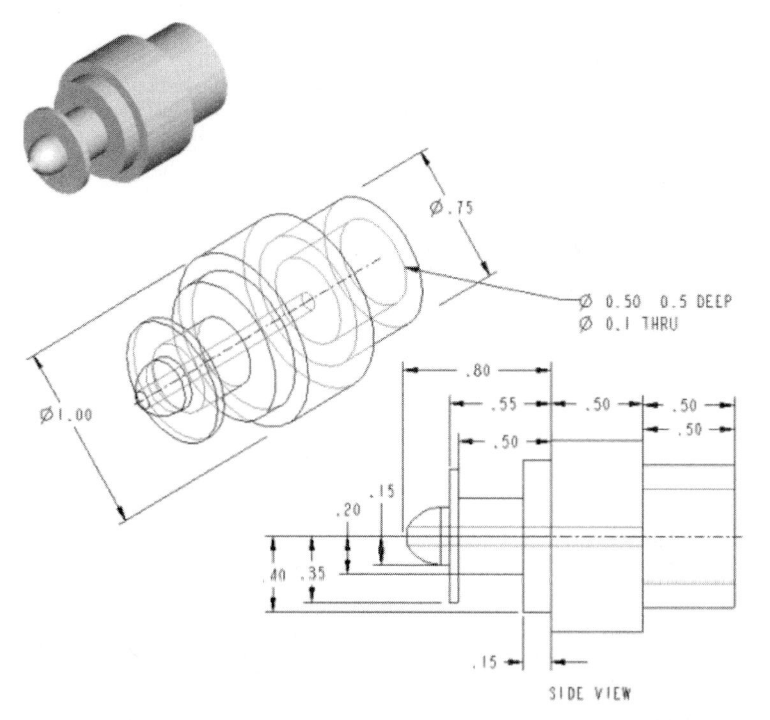


FIGURE 6.32 *Figure for exercise 6.9.*

6.10 Create the model of the lift lever shown in Figure 6.33.

**FIGURE 6.33** *Figure for exercise 6.10.*

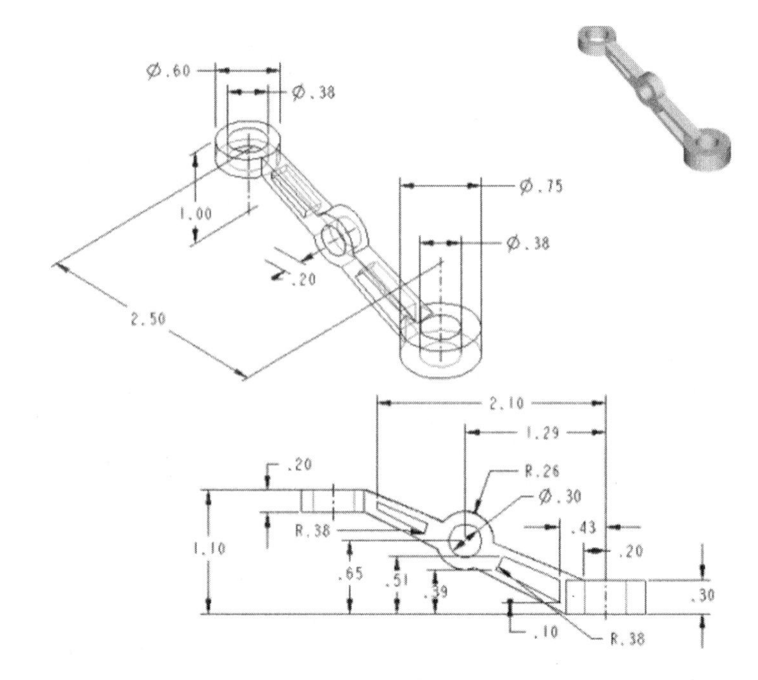


6.11 Construct the model of the angle bracket with side reinforcements. Use the geometry shown in Figure 6.34.

**FIGURE 6.34** *Figure for exercise 6.11.*

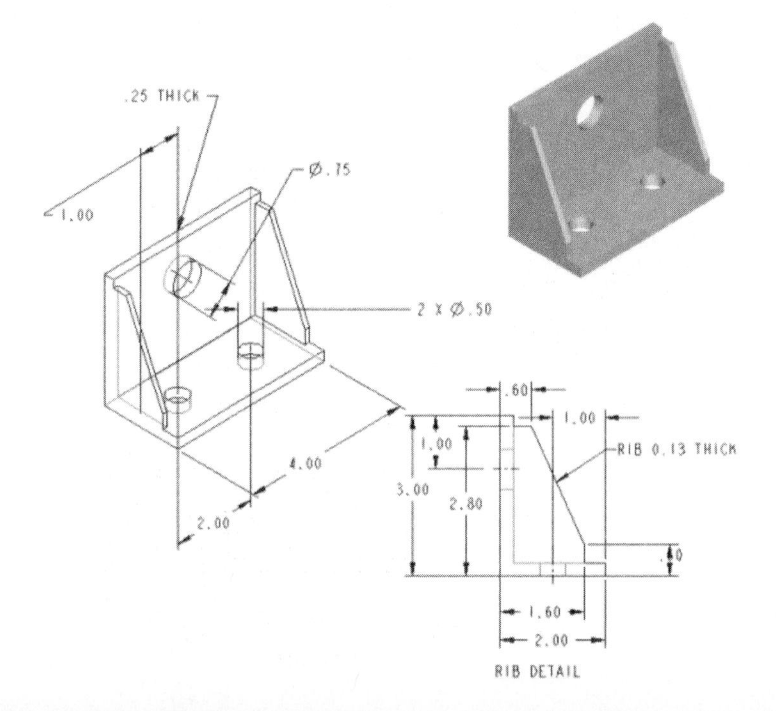

6.12 Use the geometry in Figure 6.35 to create the model of the single bearing bracket.

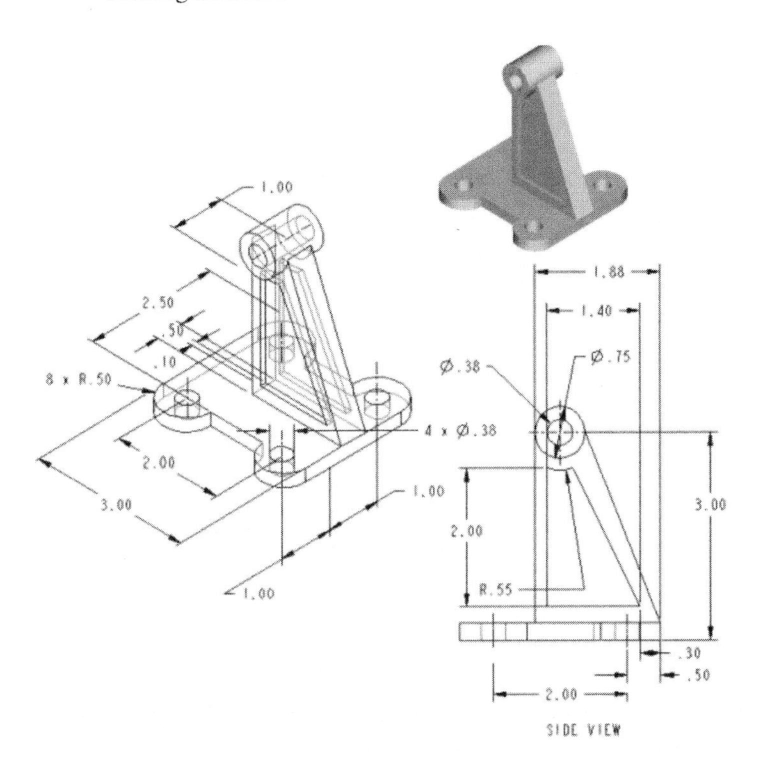


**FIGURE 6.35** *Figure for exercise 6.12.*

6.13 Using Figure 6.36, construct the model of the double bearing bracket.

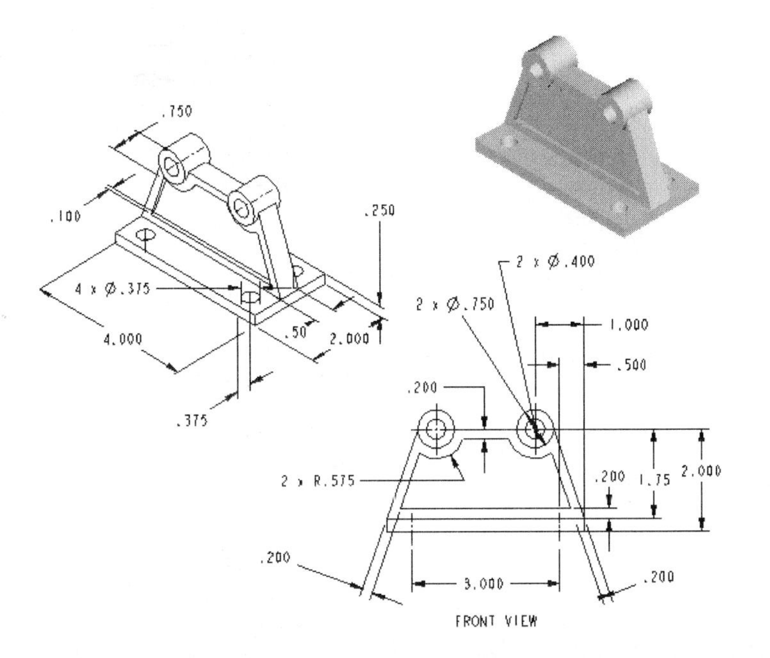


**FIGURE 6.36** *Figure for exercise 6.13.*

# Options That Speed Up Model Construction: Pattern, Copy, Group, Mirror Geom, and UDF

## Introduction and Objectives

Many of the models that we have constructed contain multiple copies of the same feature. So far, you have been asked to place all the copies individually, using the fundamental methods outlined for construction of a feature. You may have wondered whether an easier way to place these features is available.

Furthermore, some models that we have examined have features that are symmetric about a plane. One may ask whether the symmetry can be used to place the feature.

You may have also noticed that some parts have features that are contained in a separate part. You may have wondered whether the feature can be taken from one part and incorporated into another part.

The answer to all these queries is, of course, yes. Multiple copies of a feature may be generated by using the *Copy* or *Pattern* options. Several features may be collected into a single entity using the *Group* option and then copied. Symmetry about a plane may be exploited to complete a part by using the *Mirror Geom* option. Finally, a feature from a given part may be stored as a *UDF* (user-defined function) and retrieved into a new part.

In this chapter, we will examine how the *Pattern, Copy, Group, Mirror Geom,* and *UDF Library* options can be used to increase model construction efficiency. These options are accessed via the *FEAT* menu as shown in Figure 7.1.

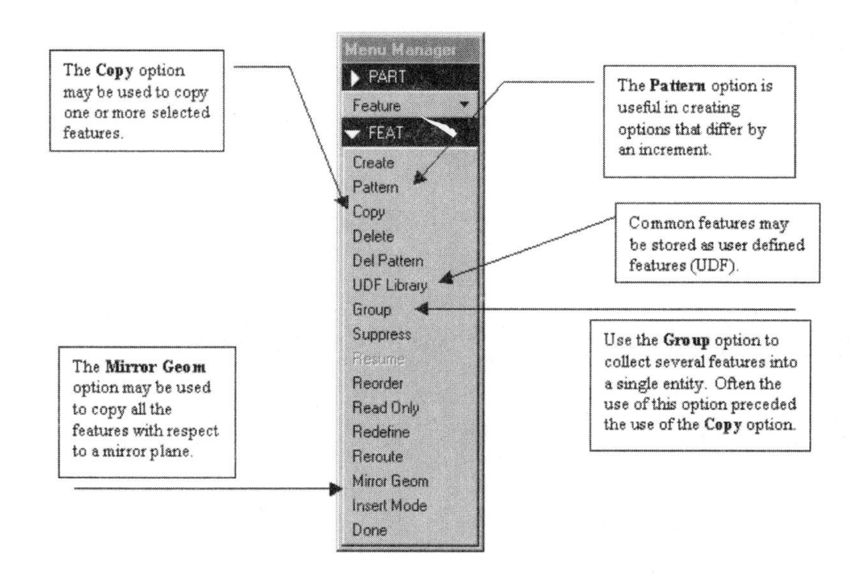


FIGURE 7.1 *The* FEAT *menu and the options* Pattern, Copy, UDF Library, Group, *and* Mirror Geom *that are used to increase model construction efficiency.*

By the end of the chapter, the reader will:

1. Be able to construct copies of an original feature by using the *Copy* or the *Pattern* option.
2. Understand the difference between the *Mirror Geom* and *Copy Mirror* options.
3. Construct rotational patterns of sketched features by creating datums on the fly.
4. Place additional features on patterned features by using the *Ref Pattern* option.
5. Collect multiple features into a single-named feature using the *Group* option and copy the group.
6. Create user-defined features (*UDF*) using the *UDF Library* options and load a UDF into a new part.

## 7.1 USING THE PATTERN OPTION

One option that can be used to place features more easily is the *Pattern* option. In chapter 4, we were faced with placing multiple holes that were identical except for an increment among their centers. The holes were placed individually, but they could have been placed by simply making a copy of the original hole and locating the new hole by incrementing from the original hole, which is how the *Pattern* option works.

A pattern can be one of three types as shown in Figure 7.2. In creating *Identical* patterns, Pro/E makes some assumptions. These assumptions are:

1. All the pattern instances are of the same size.
2. The patterns are to be placed on the same surface as the original.
3. The only intersection between the patterns is with the placement surface.

The last assumption implies that the pattern(s) cannot intersect another feature or pattern. Because of these assumptions, *Identical* patterns are generated faster than all the other types of patterns. If the first two assumptions are relaxed, then a *Varying Pattern* is generated. A *General Pattern* has no assumptions, so it becomes the slowest pattern to generate.

In the sections that follow, we will illustrate how to use the *Pattern* option.

### 7.1.1 TUTORIAL # 29: A PATTERN WITH A LINEAR INCREMENT

Our first example concerning the use of the pattern option involves placing the remaining sketched holes in the plate from chapter 4. Retrieve the plate containing the single hole. The final model with all four holes is shown in Figure 7.3. Notice that these holes are identical except for the location of their centers.

Note that the additional holes can be created by incrementing the original hole in two directions. The first direction requires an increment of 10.50 and the second increment a value of 4.00.

We wish to make identical holes several times by incrementing the dimensions defining the original hole. Click on *Feature* and then *Pattern*. The *SELECTION* menu will appear (Figure 1.16b). Click on the original hole with your mouse pointer. The *PAT OPTIONS* menu (Figure 7.2) will become available. Select *Identical* from the *PAT OPTIONS* menu.

The dimensions of the hole will become visible. In general, we can define two directions for the pattern. These directions are represented by the current dimensions 0.5 and 0.75. It is immaterial which direction is incremented

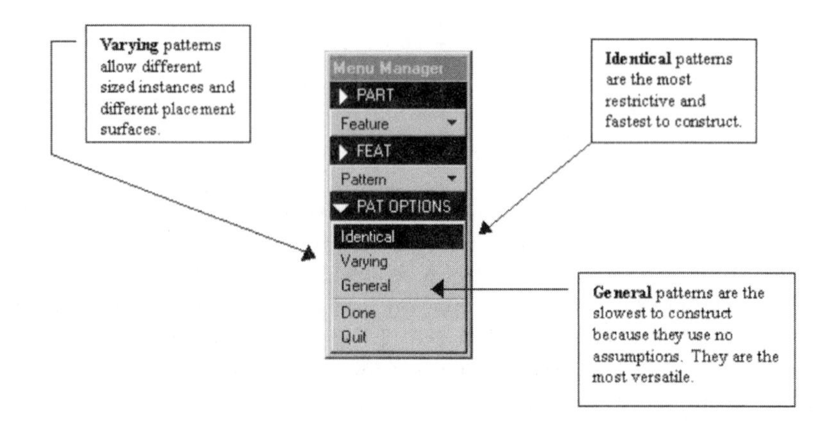


FIGURE 7.2 *Use the* PAT OPTIONS *menu to activate options for creating a* Pattern.

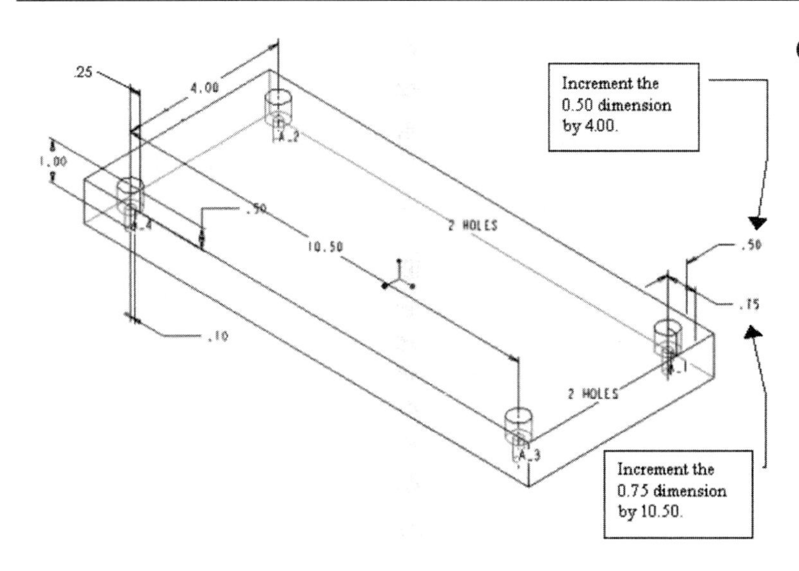


FIGURE 7.3 *The plate from chapter 4 with all four counterbored holes placed.*

first. Our first pattern will be constructed by incrementing the 0.5 dimension by four, while incrementing the 0.75 dimension to 10.5. Click on the 0.75 label with your mouse pointer. Then enter 10.5 as your increment.

Because the pattern has no other *First* direction to increment, click on *Done.* We need to let Pro/E know what is the total number of instances along the *First* direction, which includes the original. Here we want to make one copy of the original so enter two as the number of instances. Now, click on 0.5 for the *Second* direction. Enter a value of four for the increment. With no other *Second* direction to increment, click on *Done.* Again, two instances along the second direction include the original, so enter two as the number of instances. Then click on *Done.* The plate should appear as shown in Figure 7.3. Save the part.

## 7.1.2 TUTORIAL # 30: A PATTERN WITH A RADIAL INCREMENT

The pipe flange from chapter 4 is incomplete at this point. What remains is to place three more holes along a bolt circle. As mentioned in the chapter, the additional holes can be placed individually. However, it is more convenient to make a pattern of the original hole and increment the pattern, thereby constructing the additional holes.

Retrieve the pipe flange part file. Recall that in locating the original hole, the *Radial* option was used with a value of 0° from datum plane DTM3. This angular value will serve as our increment. In general, if *N* holes are to be located on a bolt circle, then the holes are 360/*N* degrees apart. In this case, which requires four holes, the holes will be 90 degrees apart.

Select *Feature, Pattern,* and click on the hole. Use the *Identical* option. The increment is the 0° dimension. Click on this dimension and enter a value of 90 for the increment. The pattern has no other *First* direction, so select *Done.* Enter a value of four for the total number of instances.

No other direction is required, so choose *Done* in answer to the *Second* direction. Pro/E will add the instances. Your part should appear as in Figure 7.4.

### 7.1.3 TUTORIAL # 31: A NUT WITH A ROTATIONAL PATTERN

The part considered in the previous section had a built-in radial dimension (the 0° from datum plane DTM3), which will not always be the case. For features that do not have built-in radial dimensions, the *Make Datum* option must be used during construction. Patterns that fall in this group are called "rotational patterns" by Pro/E.

It must be noted that when using the *Make Datum* option, the datum, which has been created, appears only during the construction of the feature. The datum will not appear on the model *after Regeneration.*

Let us illustrate the construction of a rotational pattern. We have chosen the nut shown in Figure 7.5 as our example. All but one of the cuts on the sides of the nut is to be constructed by using the pattern option.

The nut is a new part, so the base feature must be constructed before proceeding to the cuts. In order to create the base feature, follow these steps:

1. Create a new part file called "Nut" and load the default datum planes.
2. Select *Feature, Create, Protrusion, Extrude, Solid,* and *Done.* Use DTM3 as the sketching plane and DTM2 as the "*Top.*"
3. Sketch the 0.5 diameter disk, which is the base feature.
4. Extrude the disk to a depth of 3/8 inch.
5. Select *Feature, Create Slot, Extrude, Solid,* and *Done.* Place a 0.30 inch diameter coaxial hole.

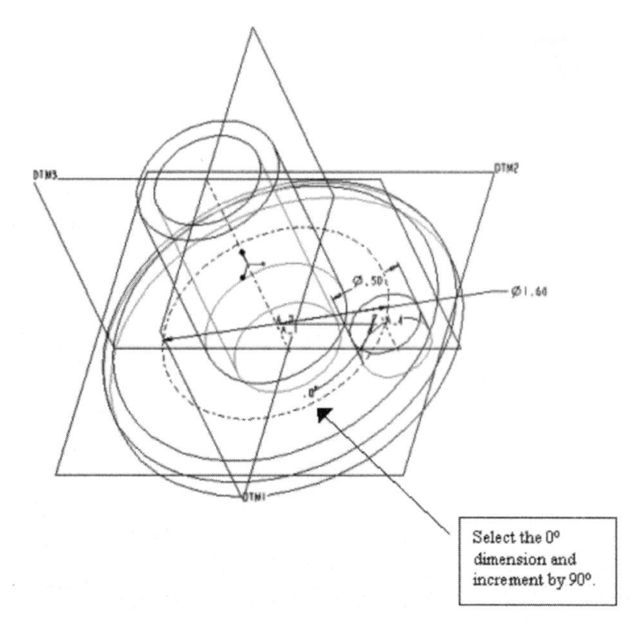


FIGURE 7.4 *The pipe flange with all the holes added.*

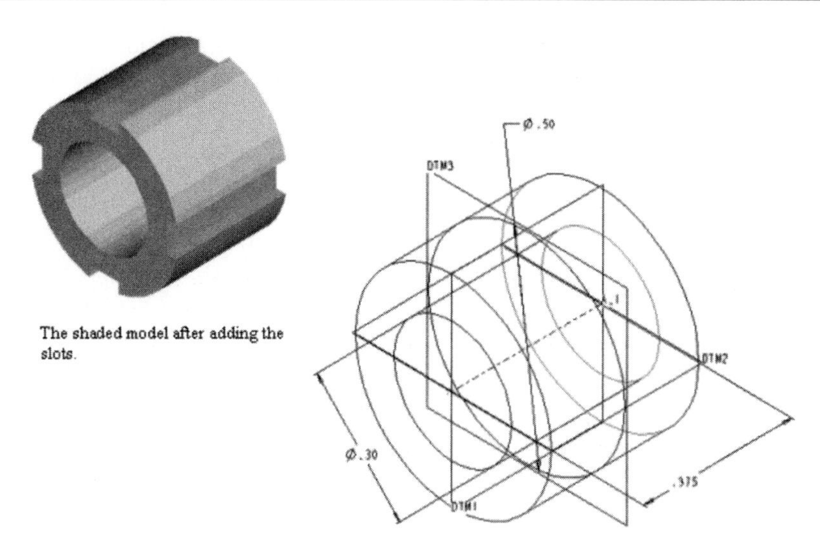


FIGURE 7.5 *A nut constructed from a cylindrical base feature and a patterned cut.*

Let us proceed to add the first cut. Select *Feature, Create,* and *Slot.* Choose *Solid* and *Done.* Select *One Side* because we will be using the front of the cylinder as the placement plane. Select *Make Datum* from the *SETUP PLANE* menu shown in Figure 7.6. Use the *Through* option and click on axis A_1 shown in Figure 7.6. Then, select *Angle.* Choose datum DTM2 and *Done.* Select *Enter Value* and enter 0.0.

Pro/E will create datum DTM4 aligned with DTM2. The angular measurement DTM4 will be used to construct the rotational pattern.

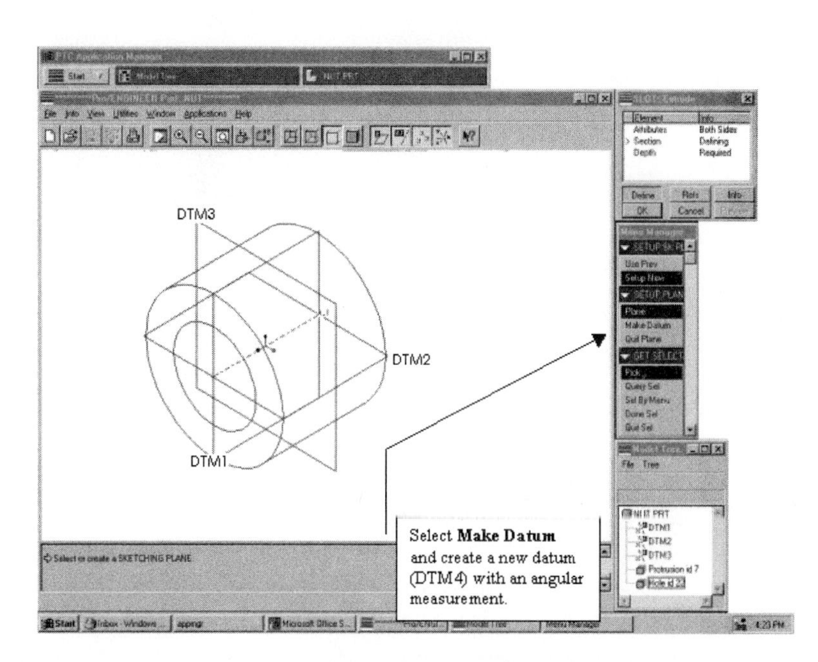


FIGURE 7.6 *Use the* Make Datum *option to create an additional datum "on the fly."*

A red arrow will appear on the newly constructed datum as shown in Figure 7.7. You may accept this direction as the direction of the slot. With the direction of the arrow as shown in Figure 7.6, choose DTM3 as the "*Bottom*" to reorient the model in the *Sketcher* as shown in Figure 7.8.

Now we can proceed to sketch the slot. Use the *Rectangle* option from the *GEOMETRY* menu and sketch the slot. Be sure to align the edges of the rectangle as shown in Figure 7.8. *Regenerate* the slot and *Modify* using the dimensions shown in the figure. Select *Done.* Then extrude *Blind* with a depth of 0.100. The nut should appear as shown in Figure 7.9. Notice that after regenerating that datum plane, DTM4 is no longer visible.

The rest of the slots can be created using the *Pattern* option and the dimension to DTM4 as the incrementing direction. In order to construct the remaining slots, select *Feature* and *Pattern.*

With your mouse, choose the slot as the feature to be patterned. Select the *Varying* option from the *PATTERN* menu. Notice that DTM4 becomes visible again. Click on the angular measurement (0°) and enter a value of 360/4 = 90°. The pattern has no other *First* direction, so select *Done.* The total number of instances is four. The pattern has no other directions, so choose *Done* once again. The nut should appear as in Figure 7.5. Save the part. It will be revisited in chapter 8.

## 7.1.4 TUTORIAL # 32: A PROPELLER

In the previous tutorial, a part was considered wherein a rotational pattern was useful in constructing the feature. In order to create the rotational pat-

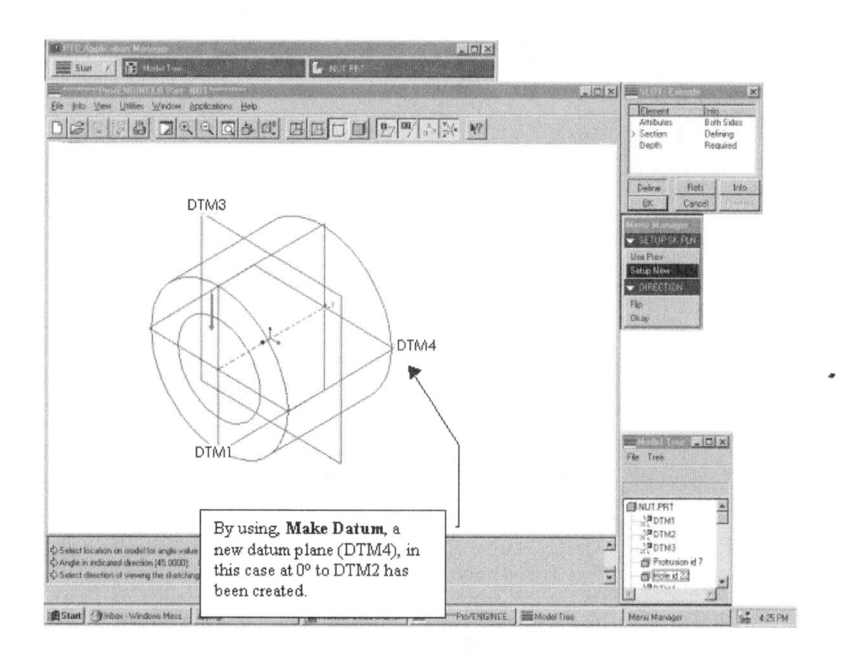


FIGURE 7.7 *Notice the plane DTM4 that has been created by using the* Make Datum *option and located from the datum DTM2.*

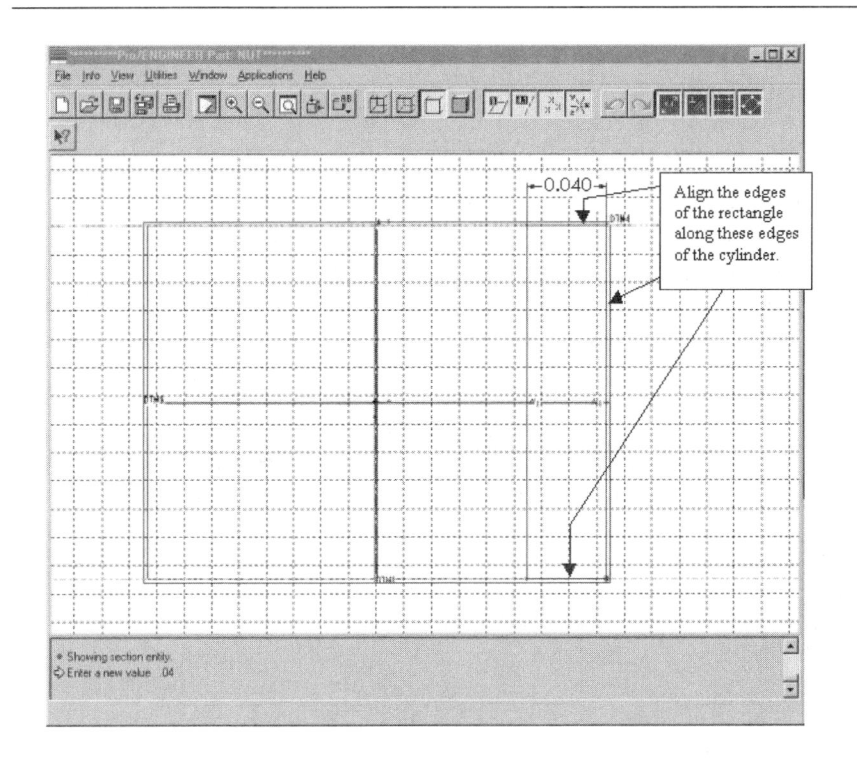


**FIGURE 7.8** *The nut with the dimensioned slot in the* Sketcher.

tern, a datum was added on the fly in order to create a reference with an angular dimension. This angular dimension was then used to construct the rotational pattern.

The additional datum was used as the sketching plane. There was no need to create any additional datum planes. For parts that contain patterned protrusions, such as the propeller in Figure 7.10, an additional datum is

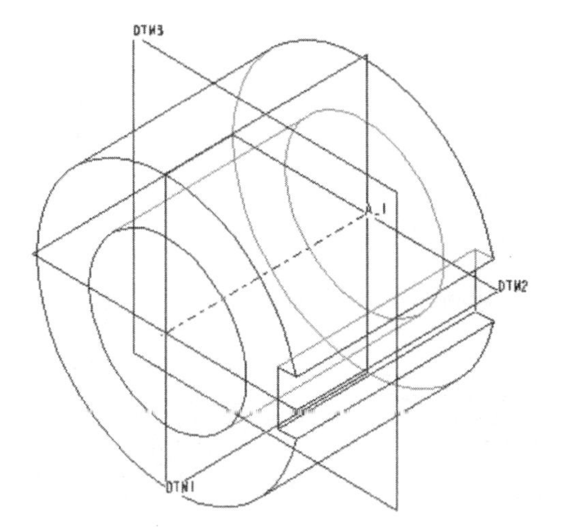


**FIGURE 7.9** *The nut with one slot.*

required, because the sketching plane must be constructed offset from the reference datum.

The base feature for the blades of the propeller is a rectangular protrusion. The *Pattern* option will only pattern this feature. The cuts that are used to round the edges of the blade may be attached to the pattern by using the *Ref Pattern* option. We will see how this option works as we proceed through the tutorial.

As always, begin by creating a part. Call this part "Propeller." The propeller has metric units, so you will need to change the units to mm. In order to do this, select *Set Up* from the *PART* menu (Figure 1.7), then *Units, Millimeter, Done,* and *Done.*

After changing the units, add the default datum planes. The base features will be constructed by using several protrusions shown in Figure 7.10. The first protrusion is a circular cylinder with a diameter of 90 mm.

Select DTM3 as the sketching plane. Use the *One Side* option with the protrusion direction into the screen as shown in Figure 7.11. Select DTM2 as the "*Top.*" Sketch a circle. Align the center of this circle to the edges of datum planes DTM1 and DTM2. *Dimension, Regenerate,* and *Modify* to 90. Extrude the section a *Blind* depth of 60.

Add the second protrusion by selecting the back of the first protrusion as the sketching plane as shown in Figure 7.11. Again, select DTM2 as the "*Top.*" Sketch the circular section and align its center to the DTM1 and DTM2 datum plane edges. *Dimension* the diameter and *Modify* to a value of 65 mm. Extrude a *Blind* depth of 90 mm.

We need to construct a circular cavity for the placement of the radial bolt holes. We can construct this cavity by using a cut. Select *Feature, Create,* and

**FIGURE 7.10** *The blades of the propeller are created by constructing a rotational pattern. In this case, two additional datums must be created using the* Make Datum *option. The first datum contains the angular dimension, while the second is the sketching plane.*

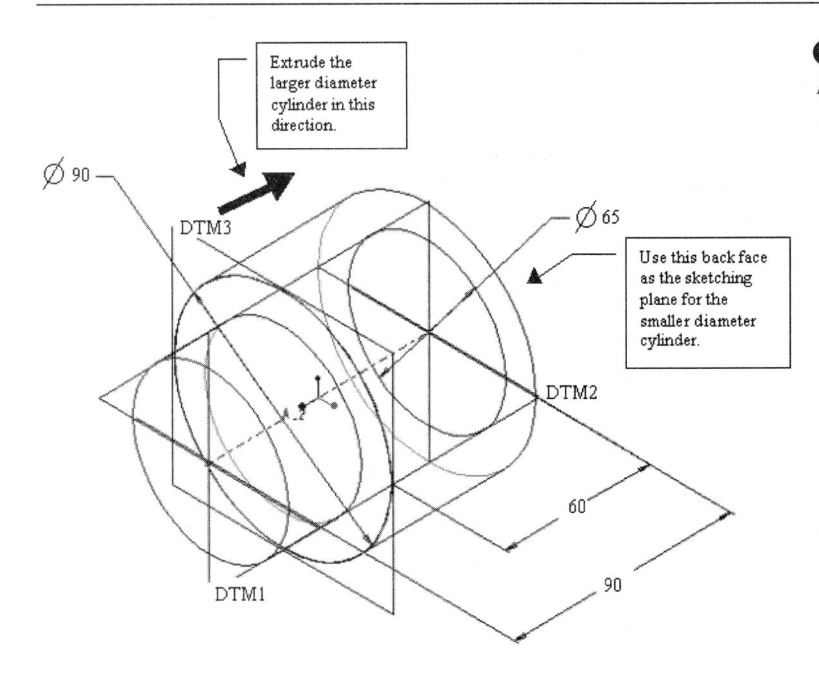


FIGURE 7.11 *The base features for the propeller are two concentric circular cylinders.*

*Cut.* Use the *Thin* and *One Side* options. Then select Surface 2 in Figure 7.10. Define DTM2 as "*Top.*"

Sketch a circle whose center is aligned with the DTM1 and DTM2 datum planes. Dimension the diameter of this circle to 30 mm. *Regenerate.* Grow the *Thin* section outward to a thickness of 15. *Extrude* the cut a blind depth of five. The propeller hub with the cut is shown in Figure 7.12.

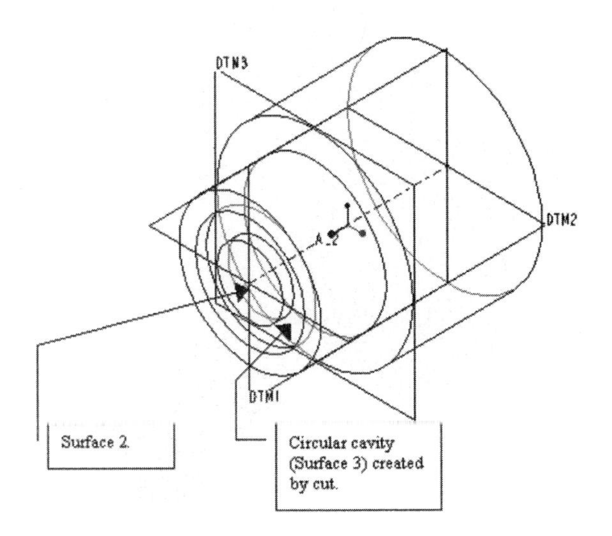


FIGURE 7.12 *The propeller hub after using the* Cut *option to remove some material. This action creates a circular cavity wherein holes for bolts can be placed.*

Using axis A_2 in Figure 7.12, place a coaxial hole. For the placement plane, select Surface 2 in Figure 7.12. Project the hole with the *One Side* option and *Thru All.* The size of the hole is 15 mm in diameter.

The holes for the bolts can be placed using the method outlined in chapter 4. Create the first hole by selecting *Feature, Create, Hole, Straight, Done, Radial,* and *Done.*

A bolt circle with a diameter of 45 mm has eight holes. Select the cavity (Surface 3) as the surface to place the holes. Pick A_2, shown in Figure 7.12, as the axis using *Query Sel.* Choose datum DTM2 as the reference plane. Enter a value of 0°.

Choose the *Diameter* option and enter a value of 45. Use the *One Side* option and *Thru All.* The diameter of the hole is 5 mm.

Add the remaining holes by using the *Pattern* option as outlined in Tutorial # 30. Select *Feature* and *Pattern.* Pick on the radial hole and then select *Done.* Then click on the 0° measurement and enter 360/8. Choose *Done.* Enter eight for the total number of instances. Select *Done* one last time. The model, at this point in the construction, is shown in Figure 7.13.

We will add the first propeller blade to the model using a single protrusion. The remaining blades will be added using a rotational pattern. Because of the orientation of the blades, and the need to create a rotational pattern, two additional datum planes will be constructed on the fly using the *Make Datum* option. The first datum is an orientation plane with an angular dimension. This datum will be used to construct the rotational pattern. The second datum is the actual sketching plane. In this case, this plane is offset from the first datum by 140 mm.

Select *Feature, Create, Protrusion, Thin,* and *Done.* Use the *One Side* Option and *Done.* From the *SETUP PLANE* menu (Figure 7.6), select *Make Datum.*

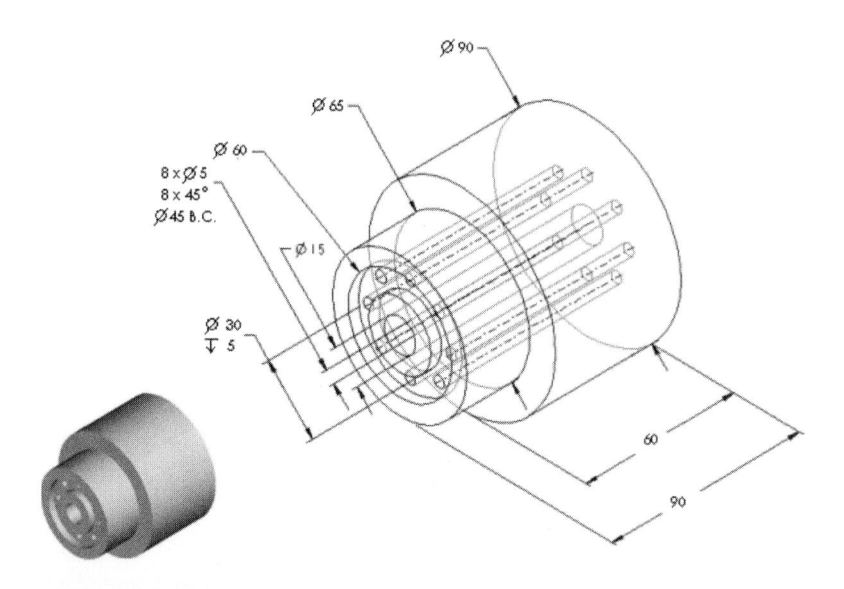


FIGURE 7.13 *The propeller hub after adding the holes for the bolts. The propeller blades can now be added.*

Choose *Through* and click on A_2. Then select *Angle* and datum DTM2. Pick the *Done* option and then *Enter Value.* Enter 360/3. Pro/E will add the new datum (DTM4).

Because DTM4 is not the sketching plane, select *Setup New* from the *SETUP SK PLN* menu. Then choose *Make Datum* and the *Offset* option. Click on DTM4. Enter a value of 140. Make sure that the protrusion direction is properly chosen. The appropriate direction is shown in Figure 7.14.

At this point, we can go ahead and select the orientation and reorient the model in the *Sketcher.* Datum plane DTM3 is orthogonal to DTM5, so select "*Bottom*" and then click on DTM3.

The software will load the model in the *Sketcher* as shown in Figure 7.15. Sketch a diagonal line and then *Dimension* as shown. By selecting *Relation* and then *Add,* add the following relations to the model:

$$sd1 = 120^* \text{Cos}\,(sd0)$$

$$sd2 = sd1/2$$

The first relation defines the horizontal length of the blade using geometry and the 120–mm slanted length of the blade. The second relation ensures that the blade is symmetrically placed about DTM2.

Now, *Regenerate* the model. Because of the relations, only sd0 needs to be modified. *Modify* sd0 to 15.00°. Upon successful regeneration, select *Done* and then fill the *Thin* protrusion in the direction shown in Figure 7.16.

The thickness of the blade is 10 mm. Extrude the protrusion *Blind* to a depth of 140-65/2. This value will blend the blade base feature with the rest of the model. The model with this base feature is shown in Figure 7.17.

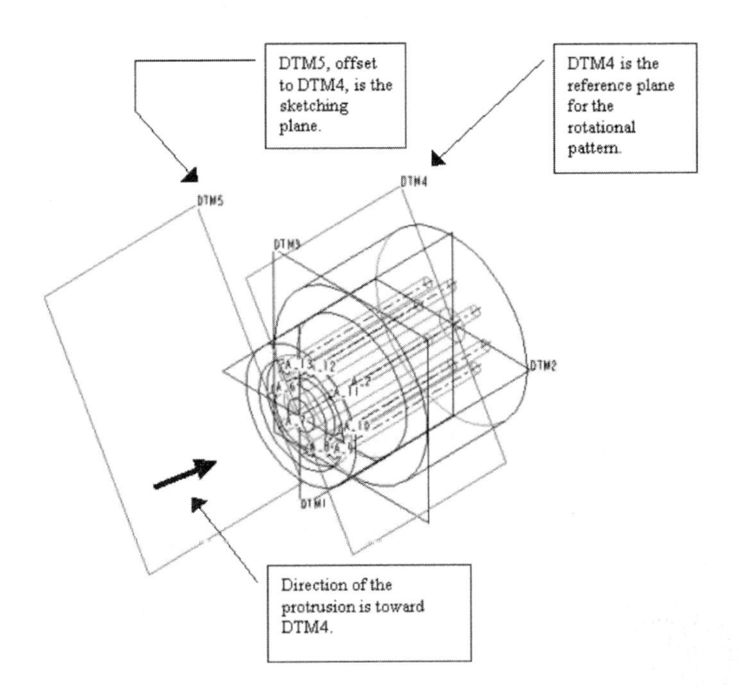


**FIGURE 7.14** *Two additional datums must be added to the model in order to add the propeller blades.*

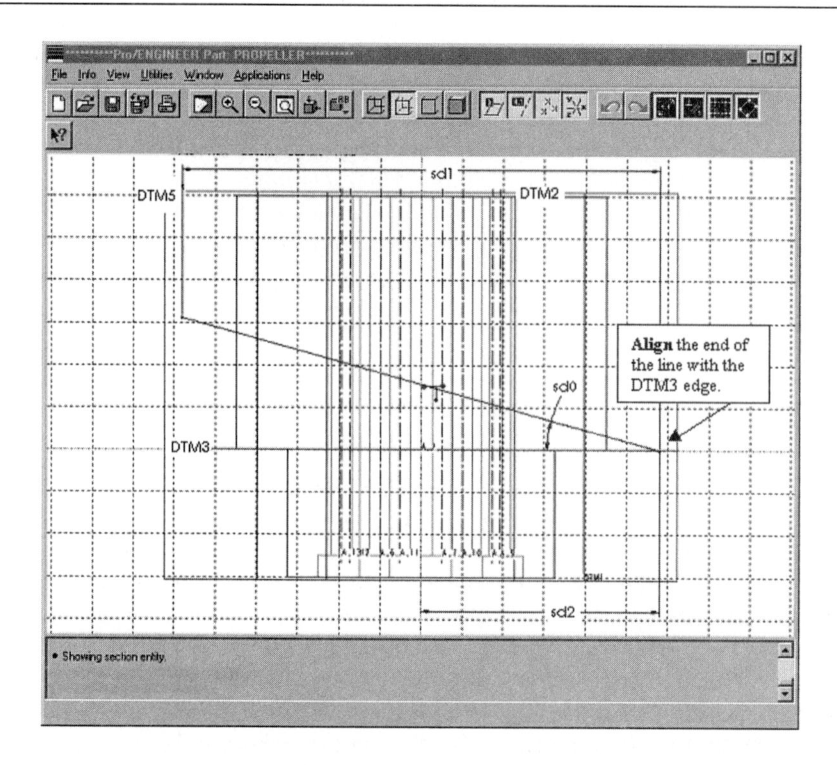


FIGURE 7.15 *The model is shown in the* Sketcher *with the geometry of the first blade.*

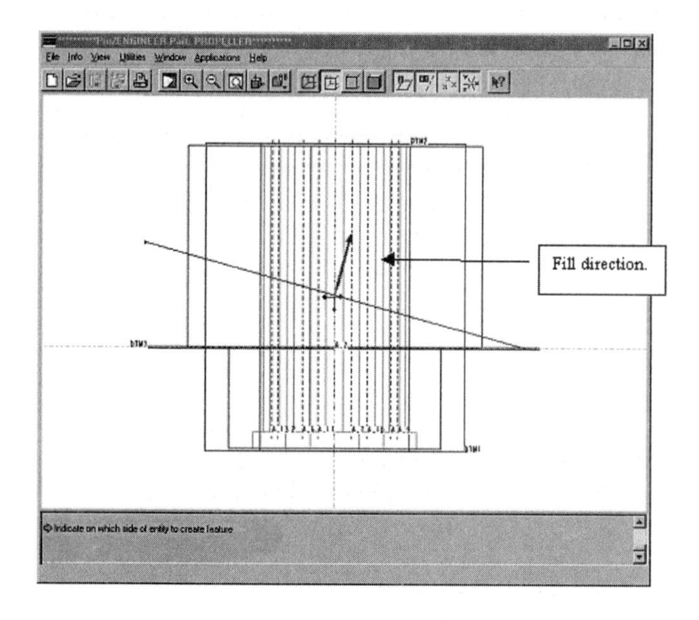


FIGURE 7.16 *Because the protrusion is constructed using a* Thin, *a fill direction must be prescribed.*

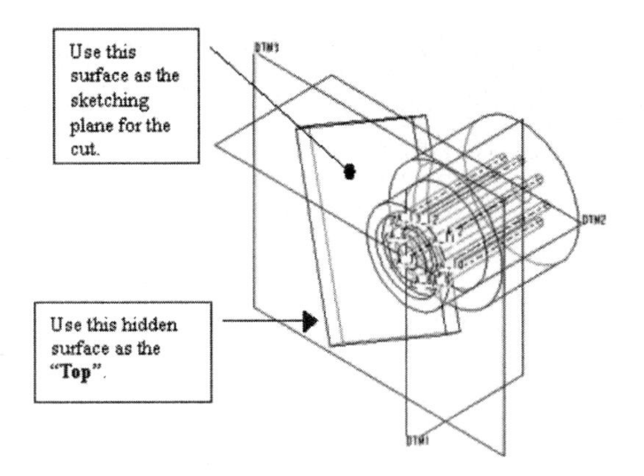


FIGURE 7.17 *After protruding the section shown in Figure 7.16, a rectangular protrusion is created. Note that datums DTM4 and DTM5, which were constructed using the* Make Datum *option, are no longer visible.*

The blade base feature can now be patterned. The cuts that round the edges of the blade will be added after the pattern has been constructed.

Select *Pattern* from the *FEAT* menu. Because the feature intersects surfaces on the model, a *General* pattern must be used. Select *General.* Select the blade base feature.

Notice that DTM4 and DTM5 have become visible again. Click on the angular measurement (120°) and enter an increment value of 360/3 = 120°. No *First* direction means you can select *Done.* The total number of instances is three. With no other directions for this pattern, choose *Done* once again. Pro/E will now construct the pattern.

Additional features can be placed on a pattern provided they are associated and dimensioned to the original feature in the pattern. The *Ref Pattern* option can be used to add the additional feature to the rest of the pattern.

Select *Feature, Create, Cut, Solid,* and *Done.* Use the *One Side* option. The sketching plane for this cut is shown in Figure 7.17. Select the surface shown in Figure 7.17 as the "*Top*" to reorient the model in the *Sketcher.*

The *Cut* geometry is shown in Figure 7.18. We have zoomed in to the model in order to show the cut detail. *Sketch* the geometry, *Regenerate,* and *Modify.* Be sure to align the lines as shown. Extrude the cut a *Blind* depth of 10.

We can now add the cut to the rest of the pattern. The model with the added cut is shown in Figure 7.19. Select *Pattern, Ref Pattern,* and *Done.* Click on the cut. You may need to use *Query Sel.*

The software will add the additional feature to the rest of the blades. Repeat the process and place the cut on the opposite side of the blade. The part with all the cuts added is shown in Figure 7.20.

At this point, save the model. We will complete the propeller using the *Round* option in chapter 8.

## 7.2 USING THE GROUP OPTION

The *Group* option is a useful utility feature from the *FEAT* menu that allows the user to gather several features into a single group. A name is given to the group. The group can be treated as a feature.

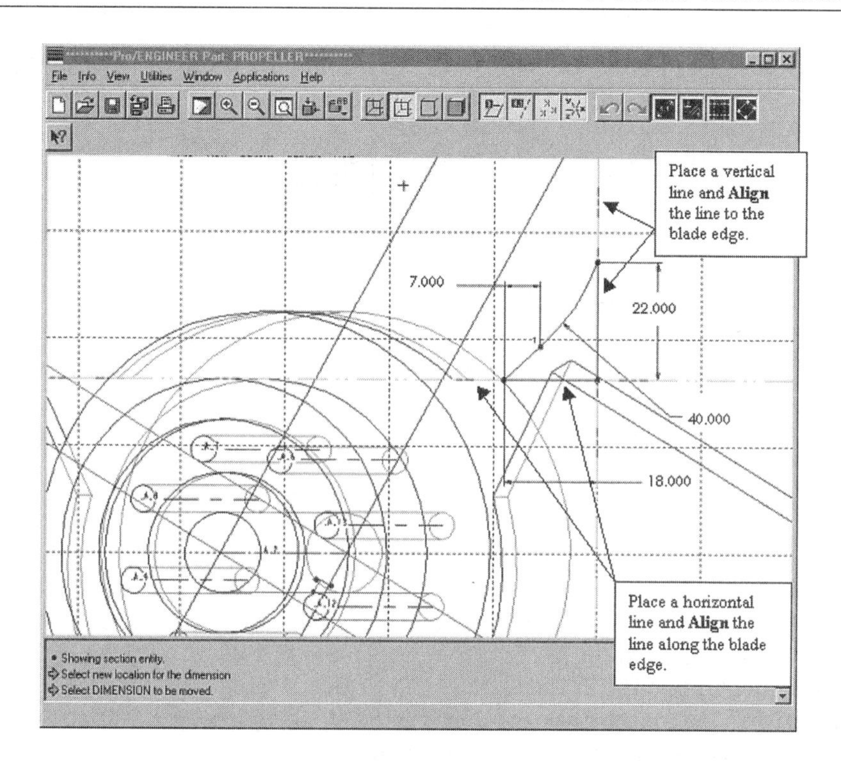


**FIGURE 7.18** *The geometry of the cut is composed of three lines and an arc.*

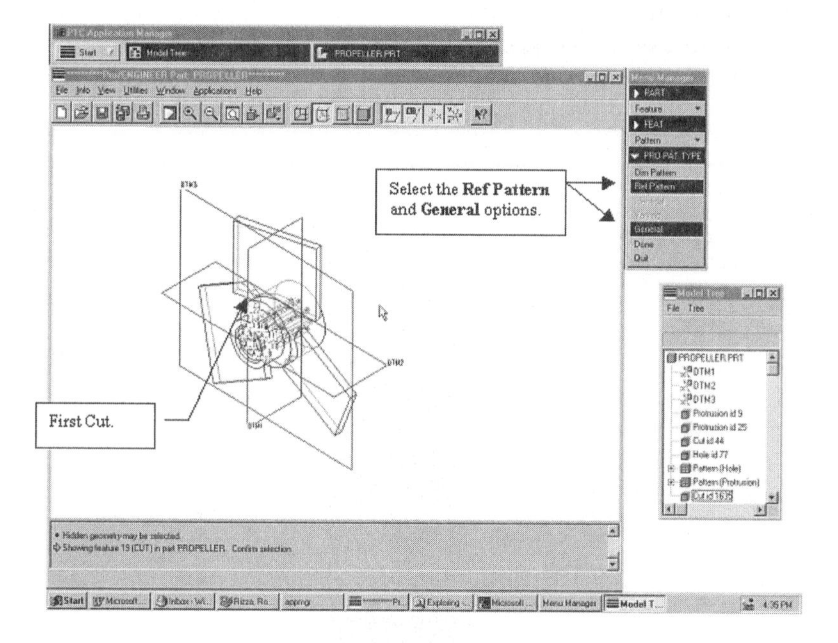


**FIGURE 7.19** *The original blade after adding the cut.*

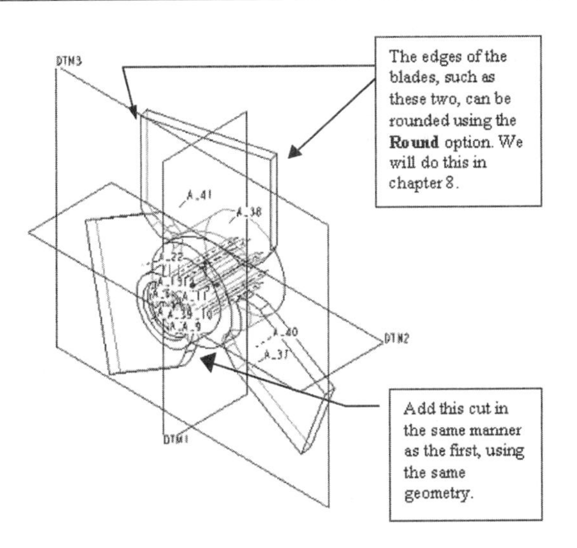


FIGURE 7.20 *The model after adding the cuts on the lower edges of the blades.*

The *Group* option is helpful when copying multiple features because the features can be collected into a single-named group and then copied as a single entity. The group option is accessed via the path *Feature* and *Group*. The *GROUP* menu, shown in Figure 7.21, contains the options for constructing and, to some extent, manipulating the group.

Note the *Pattern* option within this menu. Grouped features cannot be patterned with the *Pattern* option discussed in section 7.1. Instead, the creators of Pro/E have provided a *Pattern* option within the *GROUP* menu that can be used for the same purpose.

The *Ungroup* option is located in this menu. This option can be used to ungroup the elements contained within a group.

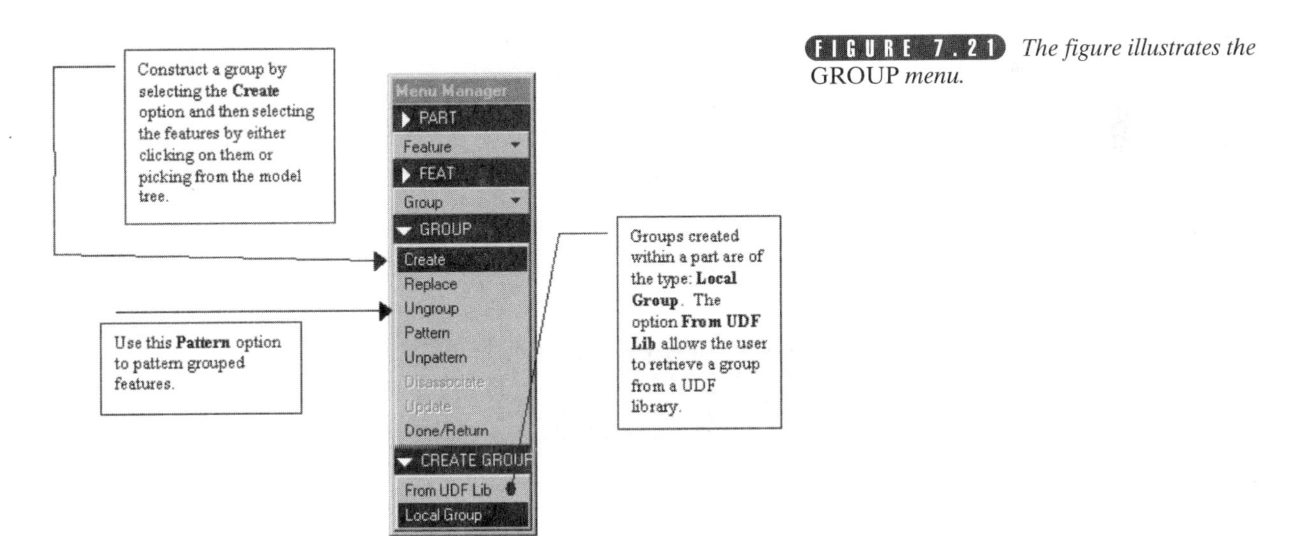


FIGURE 7.21 *The figure illustrates the* GROUP *menu.*

A group of features constructed by the user within the part is called a *Local Group.* Select *Local Group* from the *GROUP* menu.

The option *From UDF Lib* allows the user to retrieve features that have been grouped and stored in a library. It allows the user to place the common features in different parts.

### 7.2.1 TUTORIAL # 33: A VIBRATION ISOLATOR PAD

Let us construct a simple model to illustrate the utility of using the *Group* option. The part, a vibration isolator pad, is shown in Figure 7.22. Let us begin by quickly constructing the base feature. Create a part called "VibrationPad" and load the default datum planes.

Select *Feature, Create, Protrusion,* and *Solid* and construct the triangular base of the support part. The dimensions for constructing the base are given in Figure 7.23. Use DTM2 as the sketching plane. The plate is extruded *Blind* to a depth of 0.20.

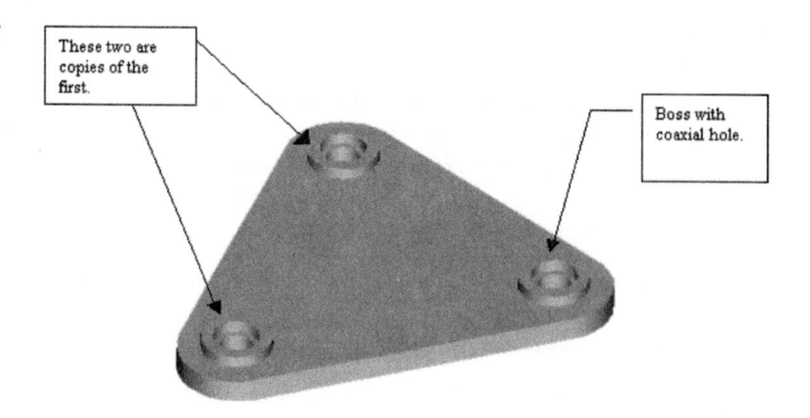


**FIGURE 7.22** *The vibration isolator pad is constructed with two protrusions and a hole. The first protrusion is used to construct the triangular base feature. The second protrusion forms the boss. After a coaxial hole is added to the boss, the two can be grouped and copied.*

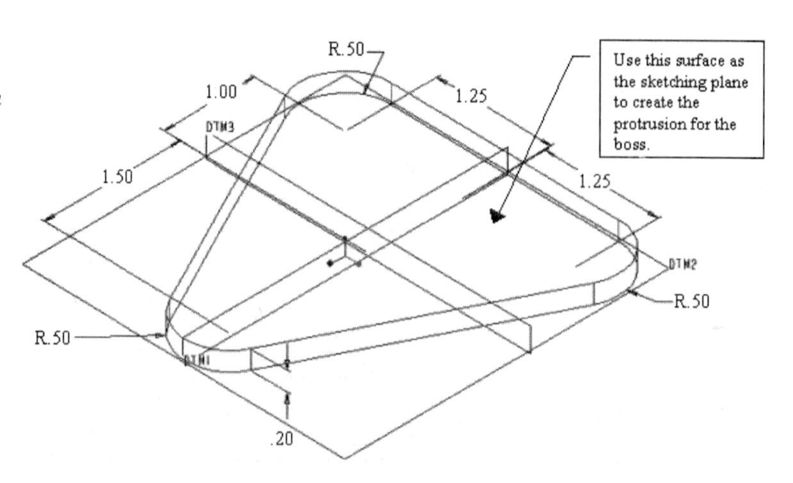


**FIGURE 7.23** *Construct the triangular base feature by selecting DTM2 as the sketching plane and DTM3 as* "Bottom." *Sketch the section shown and extrude* Blind *to a depth of 0.20.*

The pad contains bosses that allow the pad to properly align against a cover plate when the bosses are seated within holes. The bosses contain coaxial holes. The bosses are constructed with one-sided protrusions. The protrusions and the holes form a group.

Of the three bosses shown in Figure 7.22, the original is constructed by using a *Protrusion* and a *Hole.* The second boss is a copy that may be created by using the *Copy Mirror* option after the two features have been grouped. The last instance is added by using the *Copy* option. In this tutorial, we will construct the original boss and hole features. Then, we will group the two into a single-named entity. In a later tutorial in this chapter, we will add the remaining copies.

Let us construct the first boss-hole combination. The procedure to follow is:

1. Select *Feature, Create, Protrusion,* and *Thin.*
2. Use the top of the base, as shown in Figure 7.23, as the sketching plane and reorient the base so that it appears as in Figure 7.24.
3. Sketch a circle. *Dimension* this circle as shown in Figure 7.24.
4. *Modify* to the values given in the figure.
5. Upon successful regeneration, select *Done* and grow the part inward with a thickness of 0.100.
6. Extrude the protrusion to a depth of 0.100.

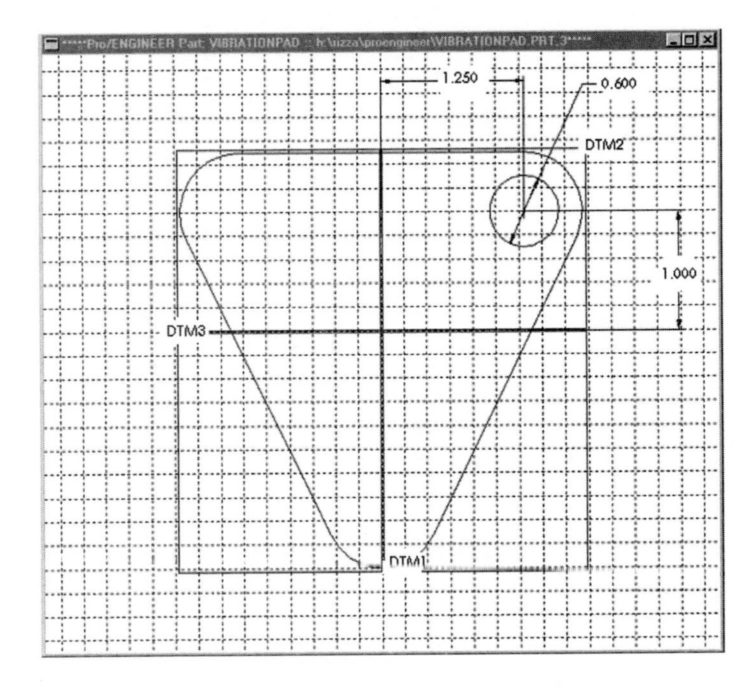


FIGURE 7.24 *The second protrusion is constructed using the* Thin *option and this sketch of the geometry.*

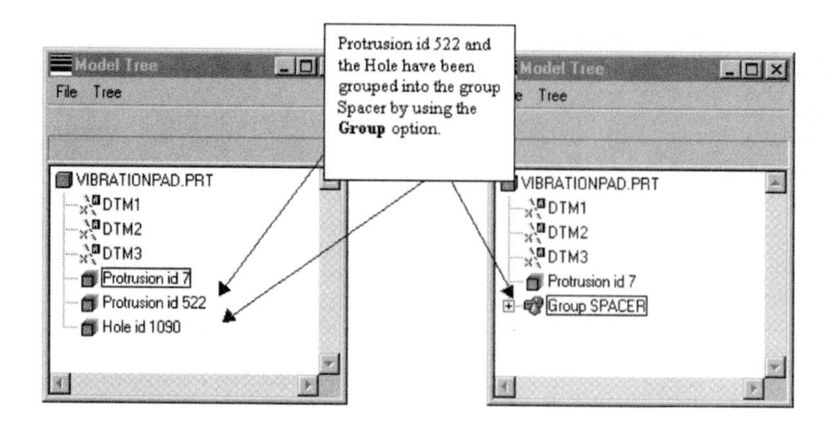


FIGURE 7.25 *By using the* Group *option, the boss and coaxial hole may be grouped.*

Now use the axis of the circular protrusion to place a 3/8″ diameter hole. Use the surface shown in Figure 7.23 as the placement surface. The depth of the hole is *Thru All.*

We are now in a position to group the protrusion and hole. Group the two features by selecting *Feature* and *Group.* Pro/E will load the *GROUP* menu. From this menu select *Local Group.* Then enter the name "Spacer" for the group.

Using your mouse, select the protrusion and then the hole. Choose *Done Sel, Done,* and *Done/Return.* Pro/E will indicate whether the group was created successfully. Note in the *Model Tree* for the part that the protrusion and the hole have been replaced with a single feature called "Spacer" as shown in Figure 7.25. Save this model for the next tutorial.

## 7.3 THE COPY OPTION

The *Copy* option (Figure 7.1) is another useful option from the *FEAT* menu that may be used to increase model construction efficiency. The *Copy* option allows the user to make additional copies of one or more features by varying the dimensions and/or references describing the location of the feature. Features may be copied between the same or different models.

Features may be copied from a different part or assembly into the current part by using new references. When doing so, you must use the new references (*New Refs* option). Pro/E will prompt the user to provide the proper information for the new references.

The options available when using the *Copy* command are shown in Figure 7.26. The *Mirror* option in the menu is different from the *Mirror Geom* option to be discussed in the next section. The *Mirror* option copies one or more features by using a given plane as a mirror. The *Mirror Geom* option, as we will see in the next section, copies *all* geometry by performing a mirroring operation. Do not confuse the two.

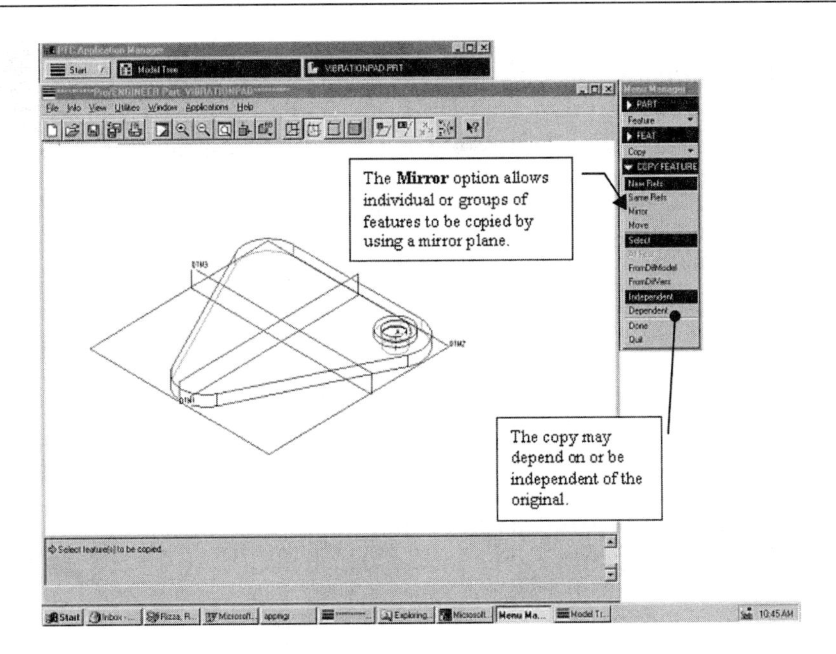


FIGURE 7.26 *The* Copy Feature *menu contains the options available for copying a desired feature.*

Notice the options *FromDifModel* and *FromDifVers* in Figure 7.26. These two options allow the user to copy features from a different model or different version of the same model.

The *Independent* and *Dependent* options allow the user to prescribe whether the copy should depend on the original feature. Use the *Dependent* option if you want the copy of the feature to be automatically updated when the original feature is updated.

The *Move* option provides the user with the ability to make a copy of a given feature by performing a simple translation or rotation. Translation of the feature must be along a desired direction and an offset distance. When using the *Rotation* option, the user must define the rotation direction as well as the rotation angle. The translation or rotation is defined by selecting a *Plane, Curve, Edge, Axis,* or coordinate system (*Csys*).

### 7.3.1 TUTORIAL # 34: USING COPY AND COPY MIRROR

We are now ready to copy the group created in Tutorial # 33. After retrieving the part, select *Feature* and *Copy.* Choose *Mirror* from the *Copy* options. Then select *Dependent* and *Done.*

Click on the group "Spacer" in the *Model Tree.* Then choose *Done Sel* and *Done.* Pick the DTM1 datum plane as the mirror plane. Pro/E will place the copy of the group. Notice in the *Model Tree* that a new feature exists called Group "Copied_Group." Note that this copy of the original group is a group itself, named "Copied_Group."

The third group may be added by using the copy feature again. In this case, we can use the same references as the original group (*Same Refs* option, Figure 7.24) and simply change the dimensions locating the group. This approach is valid because the original and the copy are identical except for their location.

Select *Feature, Copy, Same Refs,* and *Done.* Select the original group "Spacer" (not the first copy). Then choose *Done Sel* and *Done.*

The software will load the *Group Elements* box and *GP* (Group) *Var* (Vary) *DIMS* (Dimensions) menu. Both the box and menu are shown in Figure 7.27. The box functions in the same manner as the *Feature Creation* box during feature construction. It lists all the elements required in creating the copy. The *GP VAR DIMS* menu allows the user to scroll through the various dimensions that define the feature and select the ones that need to be changed.

Only three dimensions need to be changed. The dimensions on your model may not be numbered in the same order as those in Figure 7.27, which may occur if you defined your dimension in a different order.

When you scroll through the listed dimensions, either on the screen or in the menu, the dimensions will be highlighted on the model. Select the 1.250 dimension and click. Pro/E will place a check mark in the box next to the label for this dimension. Next, click on the 1.000 dimension. Then select *Done.*

The software will now prompt for the updated values to the dimensions just chosen. For the 1.25 dimension, enter an updated value of 0.00. Enter −1.5 for the 1.00, (the negative is used because of the direction from DTM3). Select *OK.*

Pro/E will now place the copy using the values that you just entered. Save the part.

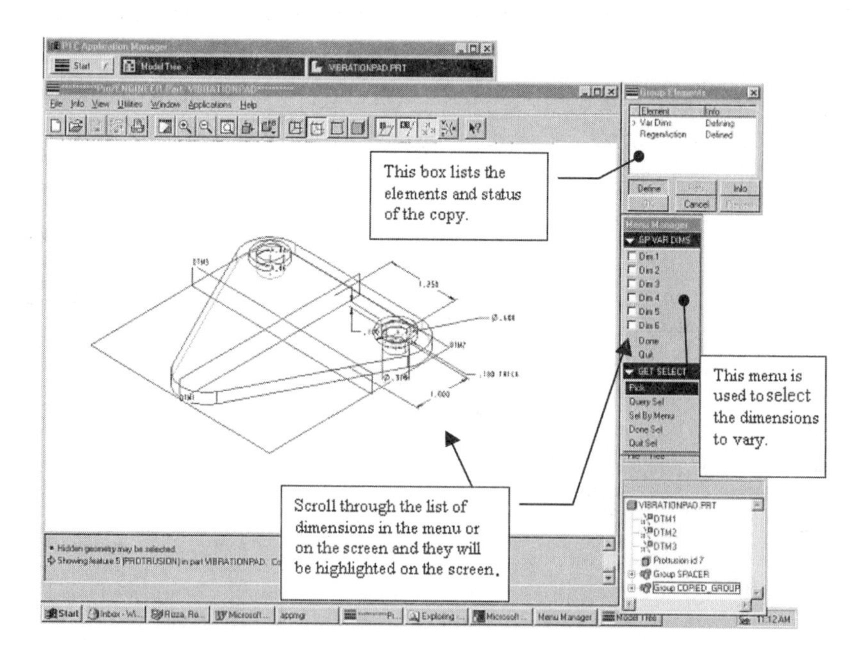


FIGURE 7.27 *The second copy may be constructed by using the same references as the original and changing the dimensional values.*

## 7.4 THE MIRROR GEOM OPTION

The *Mirror Geom* option from the *FEAT* menu may be used to create a copy of the existing feature(s) by reflecting the feature(s) from a mirror plane. The mirrored feature(s) will be constructed as merged features; that is, they will be connected to the original features.

The *Mirror Geom* option is helpful in constructing features with symmetry. For example, consider the plate in Figure 7.2. The model could have been created by constructing one quarter of the model and then using the *Mirror Geom* option to complete the rest of the model.

### 7.4.1 TUTORIAL # 35: COMPLETING THE HINGE WITH MIRROR GEOM

Let us gain some experience in using this option. We will use the half-completed model of the hinge from chapter 2. As you may recall, half of the model was constructed using a simple protrusion. The model at this stage is shown in Figure 7.28.

Now let us construct the rest of the model by using the *Mirror Geom* option. Retrieve the model and choose *Feature.* From the *FEAT* menu, select *Mirror Geom.*

In order to copy the features using this option, a mirror plane must be defined. If the model contains datum planes, then one of the datum planes may be used as a mirror plane. This model does not contain any datum planes, but it does have a plane that is a plane of symmetry. This plane is shown in Figure 7.19.

Select this plane as the mirror plane. Pro/E will mirror all the features behind the plane and merge the reflected features to the original model. After applying *Mirror Geom* once, the model takes on the appearance shown in Figure 7.29. Save the model.

Notice in the *Model Tree* (Figure 7.30) the order of the features. Because of the order of creation, the copy (merge id 69) depends on the original protrusion (First Feature id 1). If a change is made to the original protrusion, then the merged feature will be automatically updated.

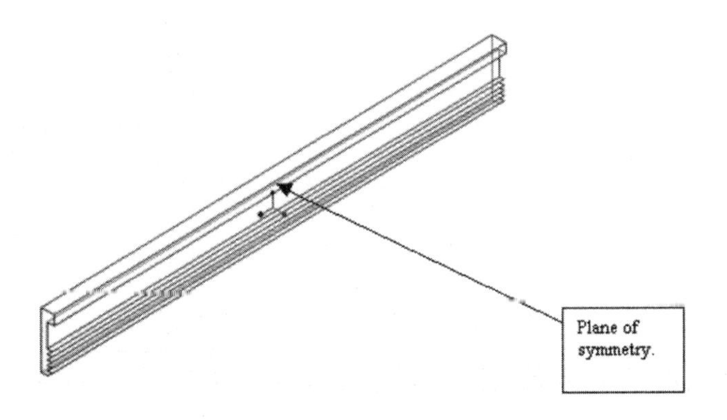


FIGURE 7.28 *The part to be used as an example for the* Mirror Geom *option. This part was constructed in chapter 2.*

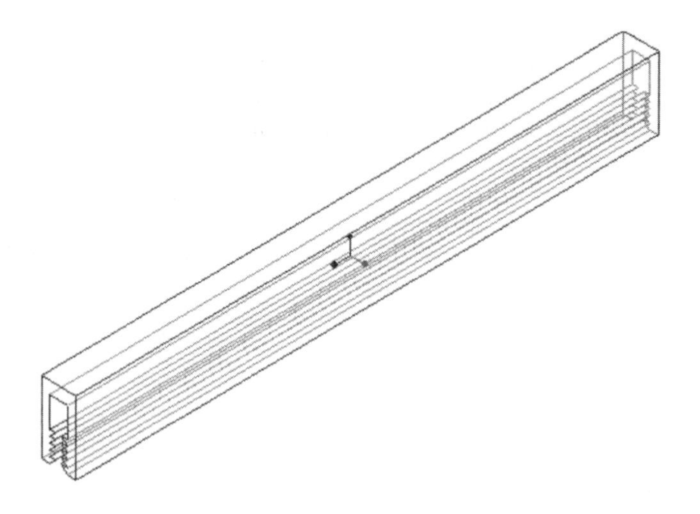

FIGURE 7.29 *The hinge after applying the* Mirror Geom *option.*

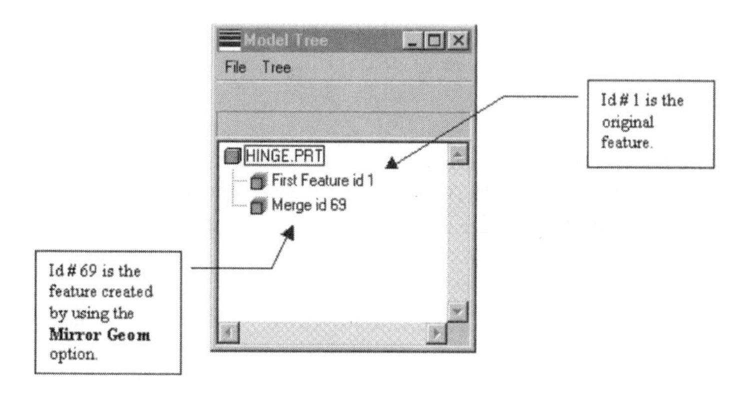


FIGURE 7.30 *The* Model Tree *for the hinge part after using the* Mirror Geom *option.*

In addition, because the merged feature depends on the original protrusion, it cannot be redefined without changing the original protrusion. The merged feature is said to be a *child* of the protrusion. The protrusion is the *parent.*

The relationship between a child and a parent can be important in feature construction and modification. This relationship will be discussed further in chapter 13. A method for changing the order of feature construction, as well as the consequences of such an action, will also be examined in that chapter.

Notice that in using the *Mirror Geom* option, all of the features reflected from the mirror plane were used. If individual features are to be copied by mirroring, then use the *Copy Mirror* option discussed in the previous section.

## 7.5 USER-DEFINED FEATURES

For features that occur on many parts, Pro/E allows the user to save the feature or features as a user-defined feature (*UDF*). A library containing user-defined features is known as a *UDF Library.* The software will place a UDF in the working directory unless instructed otherwise. Most organizations will set up a special folder to be used as a UDF library. You will need to check with your system administrator to see if a UDF library has been created.

When creating a UDF, the feature may be saved with or without the reference part on which it was originally constructed. Saving the UDF with its reference part is useful in illustrating how the UDF was originally created. Of course, storing the reference part with the UDF will increase the memory required for storage of the UDF.

When adding a UDF to a new model, the group may be one of four types. These types are as follows:

1. *Standalone.* All the information describing the UDF is associated and copied with the UDF when the UDF is loaded into a new part. This capability requires more storage space. Furthermore, any changes made to the reference part will not show up in the UDF.
2. *Subordinate.* A subroutine UDF retrieves its values from the original reference part during execution. Therefore, the reference part must also be present. Use of the subordinate option will require smaller memory allocations than the *Standalone* option. Furthermore, any changes made to the reference part will automatically be updated in the UDF.
3. *Independent Groups.* The use of this option allows the user to construct UDF groups that are independent to any changes in the UDF.
4. *UDF Driven Groups.* In some cases, you may wish to have changes in the reference part automatically updated in the UDF group. In such cases, use the *UDF Driven* instead of the *Independent* option.

User-defined features are created by selecting the appropriate features. If a pattern is chosen, then only one member of the pattern need be selected. In defining the UDF, the dimensions and placement reference must also be selected. The dimensions and placement reference may be defined to be variable, allowing the user to change the dimensions and/or placement reference when loading the UDF into a new part. In order to create variable dimensions and/or placement references, select the *Var Dims* option when creating the UDF.

In some cases, you may wish to create a UDF that refers to an entire family of features. For example, you may wish to create a UDF of a cap screw. You may construct a family of cap screws wherein different dimensions such as the thread pitch or screw diameter vary by a given amount.

In order to create a family, select the *Family Table* option during the creation of the UDF and select the dimensions to add to the table. Pro/E requires that a symbol be attached to each dimension.

When loading a UDF, the software allows the user to display the dimensions in three different ways: *Normal, Read Only,* and *Blank*. These options are characterized as:

1. *Normal.* The dimensions are defined in the manner traditional to Pro/E. That is, the user may modify the dimensions.
2. *Read Only.* The software will display the dimensions, but they cannot be modified.
3. *Blank.* Dimensions will not be displayed, nor can they be modified.

## 7.5.1 TUTORIAL # 36: CREATING A UDF

Many plates and surfaces contain a rectangular array of four holes. Thus, it is advantageous to create a UDF containing such features.

The plate with the sketched holes from chapter 4 requires such holes. In this tutorial, we will add the holes to this plate and then proceed to create a UDF for subsequent use.

Retrieve the plate. The plate should appear as in Figure 7.3. We wish to complete the part by adding the rectangular array of holes shown in Figure 7.31.

The holes are placed by using a single call to the *Hole* option and the *Pattern* option. In order to place the first hole, follow the sequence of steps:

1. Select *Feature, Create,* and *Hole.*
2. Select the *Linear* option.
3. For the placement plane, use the surface shown in Figure 7.31.

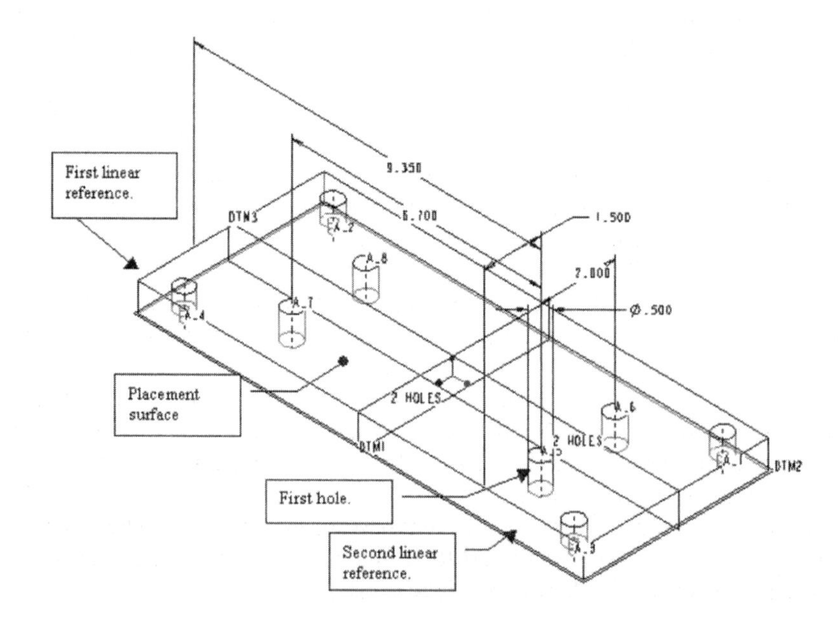


FIGURE 7.31 *The plate is shown with the added rectangular array of holes.*

4. Use the linear values (9.350 and 1.500) to place the hole.
5. The diameter of the hole is 0.500.

Create the remaining holes by using the *Pattern* option. Use the method outlined in Section 7.1. The sequence of steps is as follows:

1. Select *Feature* and *Pattern.*
2. Choose *Identical.*
3. Select *Done.*
4. Choose the first hole.
5. Select 9.350 as the first direction.
6. Enter 6.700.
7. Select *Done.*
8. Enter 2 for the total number of instances along the *First* direction.
9. Select 1.500 for the second direction.
10. Enter 2.00 for the increment.
11. Select *Done.*
12. Enter 2 for the total number of instances in the *Second* direction.

We are now in a position to create a UDF of the four holes. You may want to save the part before proceeding.

Choose *Feature, Create,* and *UDF Library.* All UDFs must have a name. When prompted to do so, enter "Four_Holes" as the name of this UDF. Then select *Standalone* and *Done.*

The software will ask whether to include the reference part with the UDF. In this case, the reference part (the plate) is simple and illustrative, so answer *Yes.*

The software will load the *UDF FEATS* menu as shown in Figure 7.32. This menu allows the user to select the features to be added to the UDF. In this case, because the holes are a pattern, only one hole needs to be selected. Using *Query Sel,* choose the first hole with your mouse. After accepting the proper feature, choose *Done Sel, Done,* and then *Done/Return.*

The software will load the *PROMPTS* menu as shown in Figure 7.33. It will also highlight, in turn, all the references relating to the hole. Because these references are used by the sketched holes on the plate, you will need to choose *Single* from the *Prompts* menu.

One nice aspect of a UDF created in Pro/E is that the user may enter a descriptive text for each of the references. Then when the UDF is loaded into a new part, the user is presented with the text. This text makes identifying the features in the UDF easier.

In this case, three references define the location of the hole. The first reference is the placement plane, and the remaining two are the surfaces used as the linear references (consider Figure 7.31).

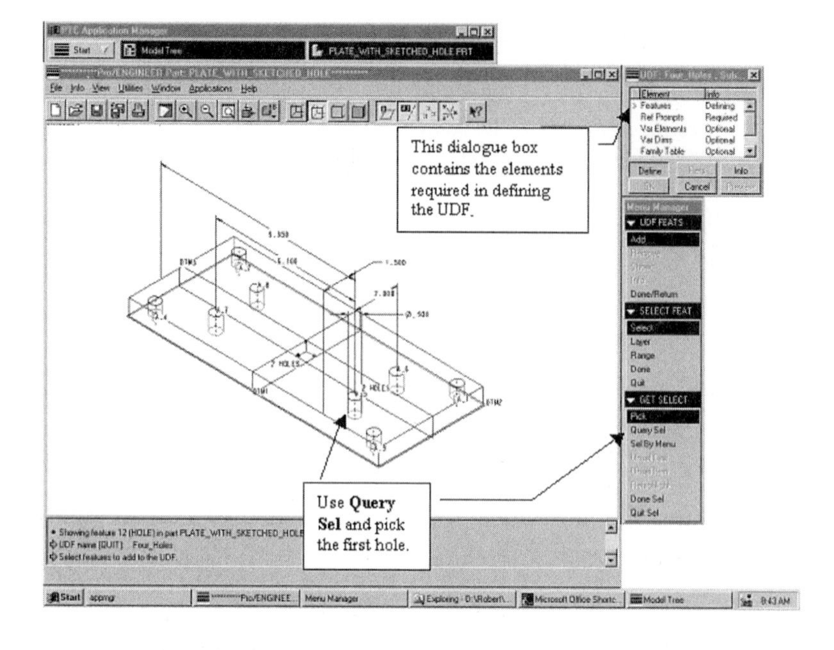


FIGURE 7.32 *The* UDF FEATS *menu allows the user to choose the features to be included in the UDF.*

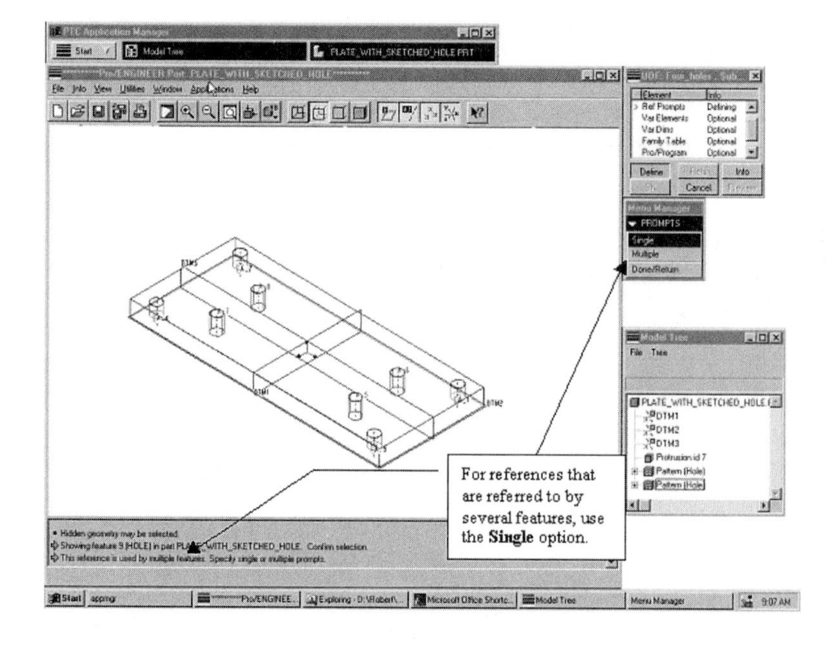


FIGURE 7.33 *The software allows the user to enter descriptive text called* Ref Prompts. *These* Ref Prompts *are used when loading the UDF into a new part.*

Using Figure 7.31 and the highlighted reference on your screen, enter the prompts in the following manner: select *Single, Done,* and then the prompt from the following:

1. Placement _Plane for the placement plane.
2. First_Reference for the first reference.
3. Second_Reference for the second reference.

The prompts may be changed by using the options in the *MODIFY PROMPT* menu shown in Figure 7.34. Every time you choose *Next,* a different reference will be highlighted. Change the prompt by using the *Enter Prompt* option. If you are satisfied with the prompts, select *Done/Return.*

We may set some or all of the dimensions defining the UDF as variable. We may also define a *Family Table.* This example has little need to define a family table. However, it is advantageous to make the dimensions locating the original hole and increments for the pattern variable. Thus, when the UDF is retrieved into a new part, the user will have the opportunity to fine-tune the location of the holes and pattern.

From the UDF creation box in Figure 7.34, choose the *Var Dims* option. The software will load all the dimensions for the holes as shown in Figure 7.35. The dimensions that we wish to make variable are the diameter of the hole (0.500), the location of the original hole (9.350 and 1.500), and the pattern increments (6.700 and 2.000). After choosing the *Var Dims* option, select the previously named dimensions. After you have chosen all the dimensions, select *Done Sel, Done/Return,* and *Done/Return.*

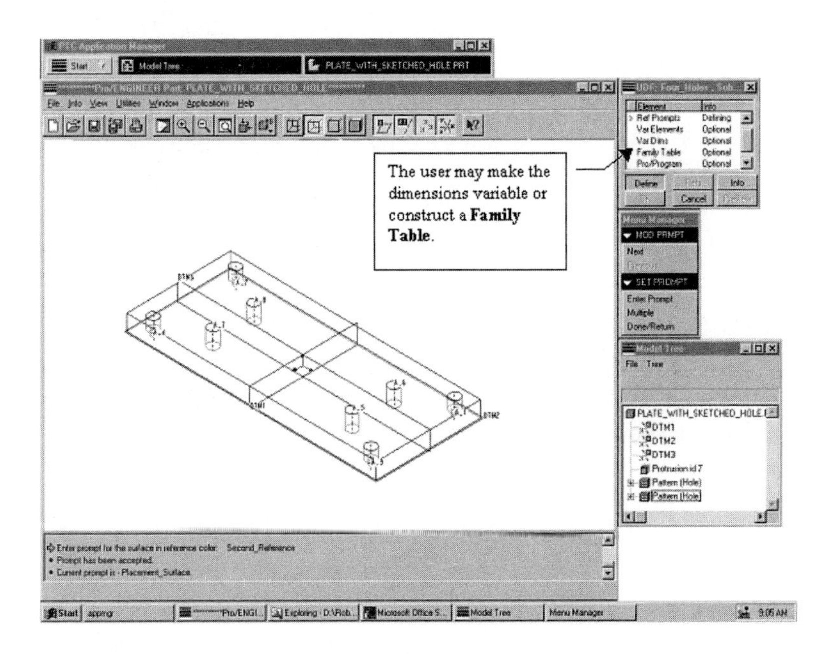


**FIGURE 7.34** *After all the prompts have been defined, the user may cycle through the prompts by choosing* Next *and* Previous *from the menu.*

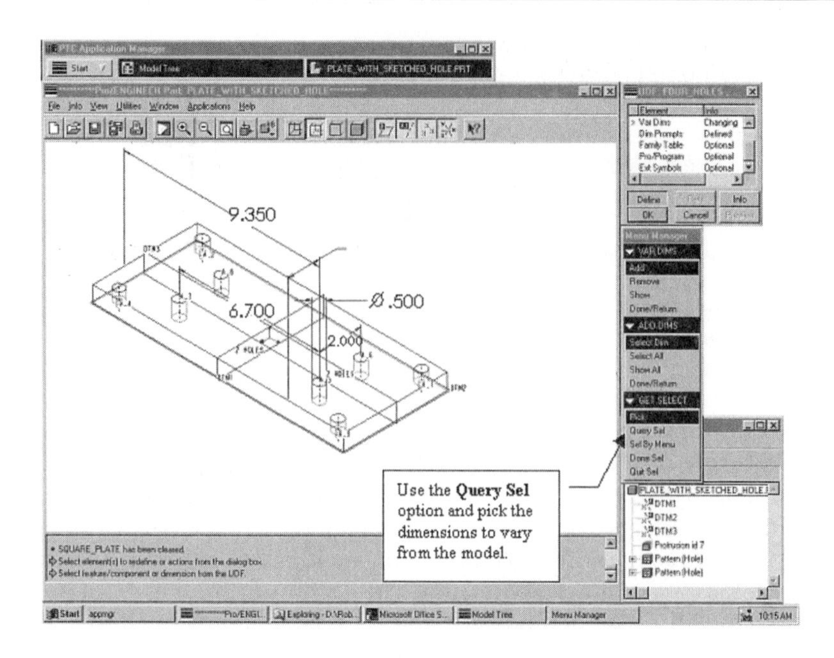


FIGURE 7.35 *The user may select dimensions from the model that may vary in the UDF.*

As in the case of the references, the software allows the user to define prompts for the dimensions. The software will highlight in turn each of the chosen dimensions. Enter the following prompts:

1. For 0.500, type: Enter the diameter of the hole.
2. For 9.35, type: Enter distance from first reference.
3. For 1.500, type: Enter the distance from the second reference.
4. For 6.700, type: Enter the increment along the first reference.
5. For 2.000, type: Enter the increment along the second reference.

Choose *OK*. The software will save the UDF in the working directory. You may confirm the directory's existence by using the *List* option shown in Figure 7.36. Additional options for the management of the UDF are located in the *DBMS* option. Select *Done/Return.* If you have not saved the model, please do so now.

## 7.5.2 TUTORIAL # 37: ADDING A UDF TO A PART

After a UDF has been placed in a UDF library, it may be retrieved into a new part. In this tutorial, we will illustrate the placement of a UDF by using the UDF created in tutorial # 36.

First, we must create a base feature. This base feature is a 10-inch by 10-inch plate that is 0.25 inches in thickness.

Create a new part called "Square_Plate" and load the default datum planes. Construct the plate by using these steps:

1. Select *Feature, Create, Protrusion, Extrude,* and *Solid.*
2. Select DTM2 as the sketching plane and DTM3 as the "*Bottom.*"

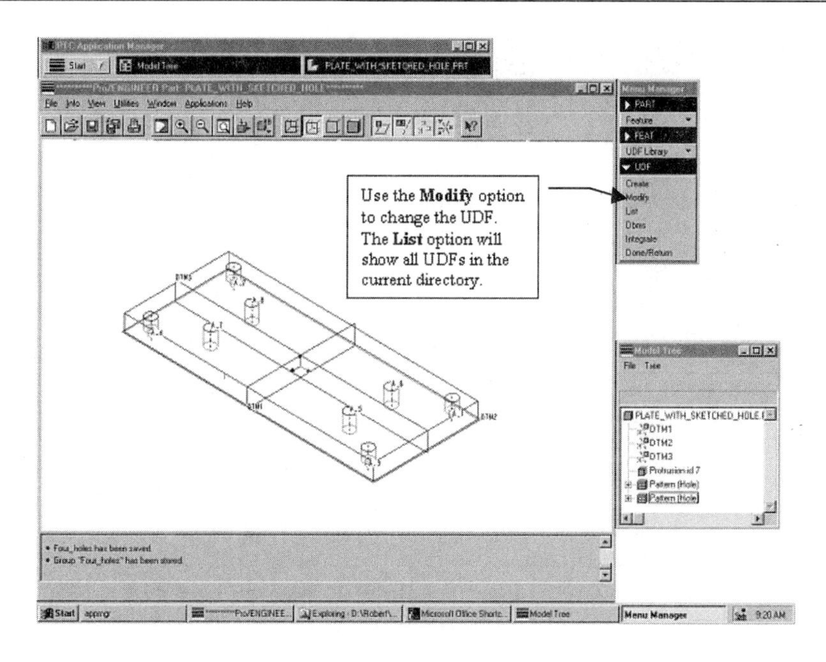


FIGURE 7.36 *A UDF may be defined or modified by using the options from the* UDF LIBRARY *menu.*

3. In the *Sketcher,* sketch a rectangle, centered with respect to the DTM1 and DTM3 datum plane edges.
4. *Dimension, Regenerate,* and *Modify* the plate.
5. Extrude *Blind* to a depth of 0.25.

The plate, including the holes, is shown in Figure 7.37. Note the references. These references will be used in adding the UDF.

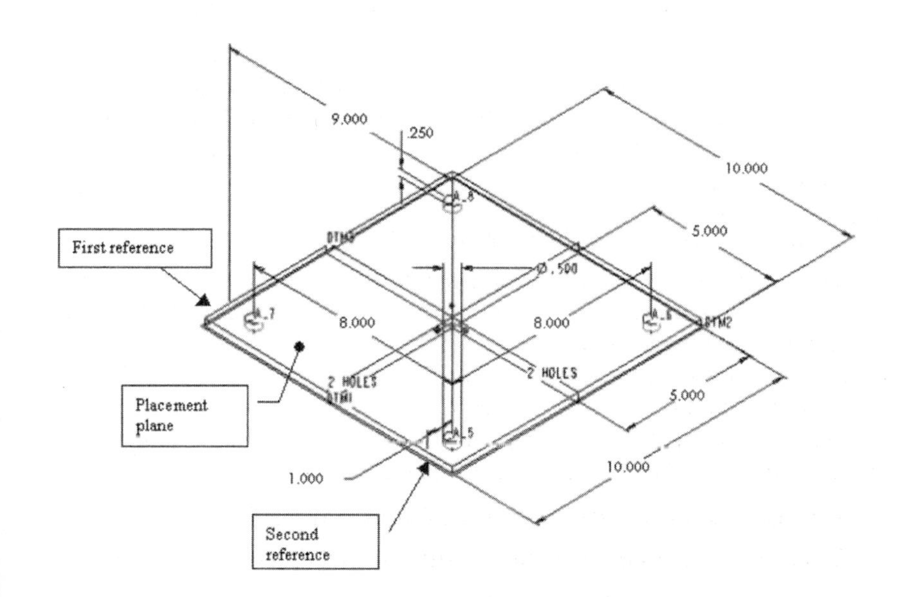


FIGURE 7.37 *The square plate is shown with the added holes.*

Now let us place the UDF as a group, which may be done in two different ways. The first approach is to use *Create* and the *User Defined* options. The second approach is to use *Group, Create,* and the *From UDF Lib.* This second approach was discussed in section 7.2.

In this tutorial, select *Create* and the *User Defined* options. The software will open a window listing all the user-defined features. You should see a UDF named "Four_holes." Select this UDF by double clicking on the name or clicking on the *Open* button.

After you have selected the UDF, the software will open a second window containing the UDF and its reference part. This UDF is shown in Figure 7.38.

Choose *Independent* and *Done.* At this point, the user is presented with the option to scale the UDF. This option is useful if the UDF has different units than the model. Because both the UDF and the model have the same units, select *Same Dims* and *Done.*

Pro/E will now cycle through the prompts. For each prompt shown in Table 7.1, enter the given value. The dimensions are taken from Figure 7.38.

Select *Normal* for the dimension display option and then *Done.* The software will now cycle through the reference prompts. When prompted, use *Query Sel* to pick the references shown in Figure 7.39.

In Figure 7.40, the plate is shown with the added UDF. After the UDF has been added to the model, the user may use *Redefine* to change the UDF. The UDF should be the same as the one shown in the figure. If so, select *Done.* If not, select the *Redefine* option and modify the UDF as necessary. When you are finished, save the Plate.

**FIGURE 7.38** *The software has opened a second window containing the UDF and its reference part.*

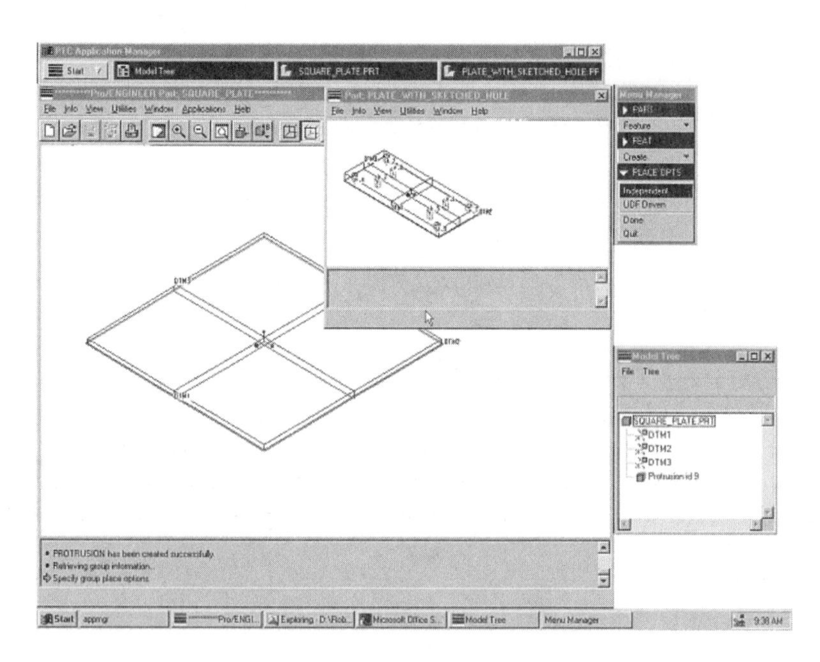

TABLE 7.1 VALUES FOR THE PROMPTS

| Prompt | Value |
|---|---|
| Enter the diameter of the hole. | 0.500 |
| Enter distance from first reference. | 9.000 |
| Enter the distance from the second reference. | 1.000 |
| Enter the increment along the first reference. | 8.000 |
| Enter the increment along the second reference. | 8.000 |

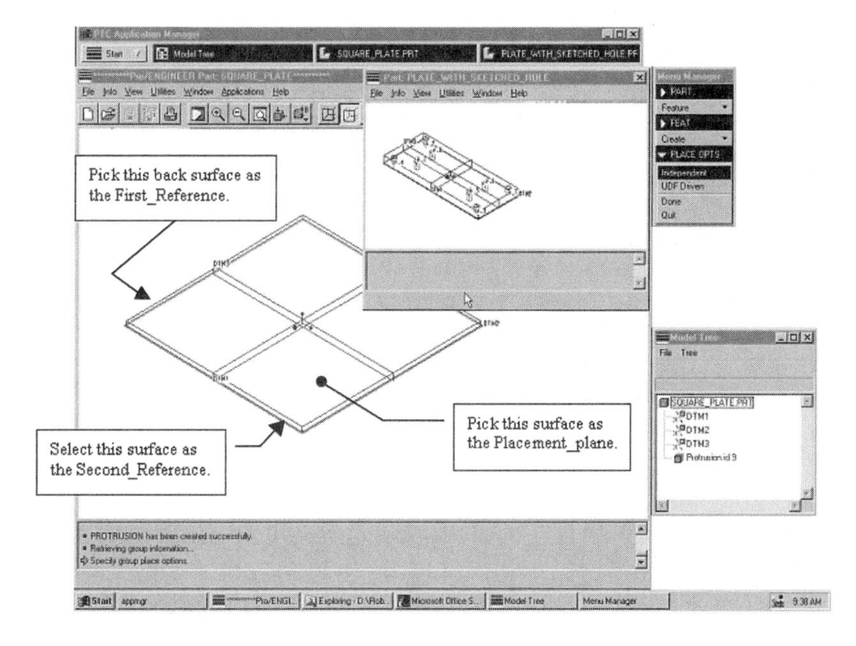


FIGURE 7.39 *Pro/E will cycle through the reference prompts. Select the given surfaces as prompted.*

## 7.6 SUMMARY

In this chapter, we considered options that increase model construction efficiency. These options are the *Pattern, Group, Copy, Mirror Geom,* and *UDF Library.*

### 7.6.1 REVIEW AND STEPS FOR THE PATTERN OPTION

The *Pattern* option allows the user to create additional features from an original if an increment separates the instances. The original is called the pattern leader. The *Pattern* option will only pattern a single feature. After a pattern has been created, the *Ref Pattern* option may be used to place additional features on the pattern.

Three types of patterns are available. The first type, the *Identical* pattern, has stringent requirements. If a pattern leader intersects another feature or

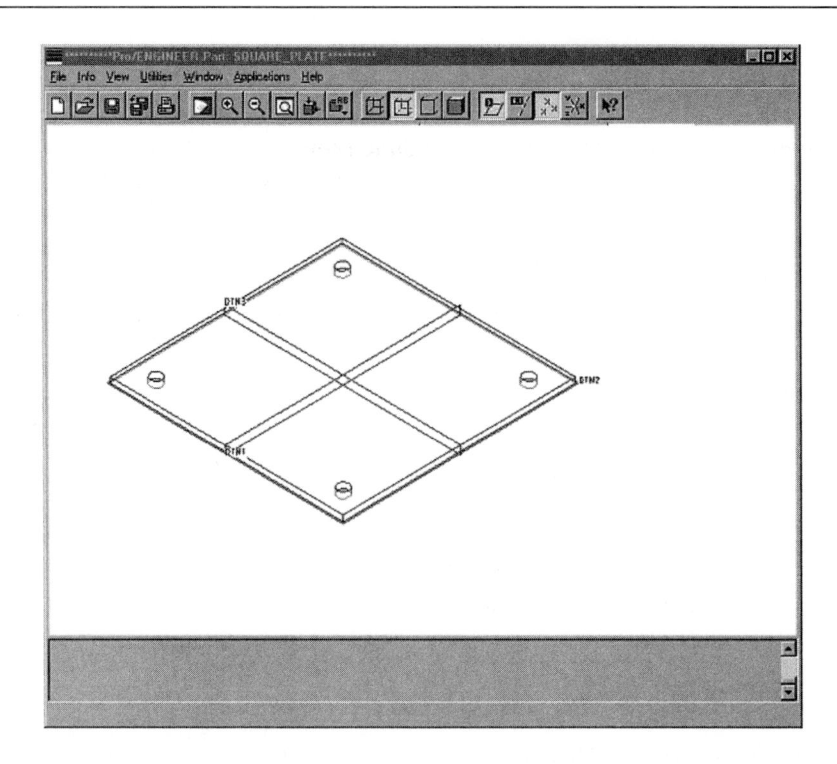

FIGURE 7.40 *The square plate with the UDF added.*

another pattern, it cannot be patterned using the *Identical* option. The *Identical* pattern is the fastest to generate.

If the pattern is to vary in size or is to be placed on a different placement surface from the leader, then it may be constructed by using the *Varying* option. The *Varying Pattern* is the second type of pattern and the second fastest to generate.

The *General* pattern is the slowest to create, but uses no assumptions. It allows features to be patterned that intersect another feature or pattern.

The general steps for constructing a pattern are the following:

1. Select *Feature* and *Pattern.*
2. Choose the pattern type: *Identical, Varying,* or *General.*
3. Select *Done.*
4. Choose the pattern leader.
5. Select the first direction to increment.
6. Enter the increment to the *First* direction.
7. Select another *First* direction, otherwise select *Done.*
8. Enter the total number of instances along the *First* direction.
9. Select a dimension along the *Second* direction.
10. Enter the increment for the *Second* direction.

11. Select another *Second* direction, otherwise choose *Done.*
12. Enter the total number of instances in the *Second* direction.

For a *Rotational Pattern,* an angular dimension must exist prior to the construction of the pattern. This angular dimension may be added to the model during the construction of the feature by using the *Make Datum* option.

### 7.6.2 REVIEW AND STEPS FOR THE GROUP OPTION

The *Group* option is used to collect multiple features into a single-named entity. After a group is created, it may be treated as a single feature. If the group is to be patterned, then the *Group Pattern* option must be used. Grouped features cannot be patterned using the *Pattern* option summarized in section 7.5.1.

The general steps for creating a group are:

1. Select *Feature* and *Group.*
2. Choose *Local Group* and *Done.*
3. Enter the name of the group.
4. Select the features to be placed in the group using your mouse.
5. Choose *Done Sel, Done,* and *Done/Return.*

### 7.6.3 REVIEW AND STEPS FOR THE COPY OPTION

The *Copy* option may be used to copy features. The *Copy Mirror* option allows the user to copy individual features or a group from a mirror plane.

In addition, features may be copied by changing one or more of the dimensions of the original feature. In such cases, either the same references are used (*Same Refs* option) or new references are determined (*New Refs* option).

The steps for creating a copy of a feature or group using the *Copy Mirror* option are:

1. Select *Feature* and *Copy.*
2. Choose *Mirror* and *Done.*
3. Select the feature or group to be copied.
4. Choose the mirror plane.

The sequence of steps for constructing a copy using the *Same Refs* or *New Refs* option is:

1. Select *Feature* and *Copy.*
2. Choose either *Same Refs* or *New Refs.*
3. Select *Done.*
4. Choose the feature or group to be copied.
5. Scroll through the dimensions on the screen or in the *GP VAR DIMS* menu and check the dimensions to be changed.

6. Select *Done.*
7. Enter the new values for the dimensions to be changed.
8. If the *New Refs* option was chosen, select the new references for the dimensions.
9. Select *OK.*

### 7.6.4 REVIEW AND STEPS FOR THE MIRROR GEOM OPTION

The *Mirror Geom* option may be used to copy *all* features relative to a mirror plane. If a single feature or group is to be copied by mirroring, then use the *Copy Mirror* option.

In general, the following steps may be used to construct a copy by using the *Mirror Geom* option:

1. Select *Feature* and *Mirror Geom.*
2. Choose the mirror plane.

### 7.6.5 REVIEW AND STEPS FOR CREATING AND PLACING A UDF

A user-defined feature (UDF) may be created and placed in a UDF library. The benefit in creating a UDF is that common features occurring on multiple parts need be created only once.

User-defined features are created and placed in a UDF library using the *UDF Library* option. They may be retrieved and placed on a new part with either the *User Defined* option or the *From UDF Lib* option in the *CREATE* menu.

The general steps for creating a UDF are:

1. Select *Feature, Create,* and *UDF Library.*
2. Enter a name for the UDF.
3. Choose *Standalone* and *Done.*
4. Decide whether to save the reference part with the UDF.
5. Select all the features to place in the UDF.
6. Enter the prompts for the placement references.
7. Choose *Done/Return.*
8. If desired, use the *Var Dims* option to create the variable dimensions. Select the variable dimensions. After all the variable dimensions have been selected, enter the corresponding prompts.
9. If so inclined, choose *Family Table* to add a family table to the UDF. Choose *Dimension* and add the dimensions to the family table. Enter a symbol name for the dimension. Choose *Done/Return.*

A UDF may be retrieved from the UDF library and placed on a new part as a group. Of course, you must have permission to access the folder. In general, a UDF may be retrieved as follows:

1. Choose *Create* and *User Defined* or *Group, Create,* and *From UDF Library.*
2. Select *Independent* or *UDF Driven.*
3. Choose the placement scale: *Same Size, Same Dims,* or *User Scale.*
4. The software will prompt for input needed for any variable dimensions. Respond with the appropriate values.
5. Select the *Display* option: *Normal, Read Only,* or *Blank.*
6. Select the placement references when prompted to do so.
7. The software will place the UDF. Use *Redefine* to modify the UDF, otherwise choose *Done.*

## 7.7 ADDITIONAL EXERCISES

7.1 Use Figure 7.41 for the model of the pipe clamp top (see also exercise 5.7). Complete the model using *Mirror Geom.*

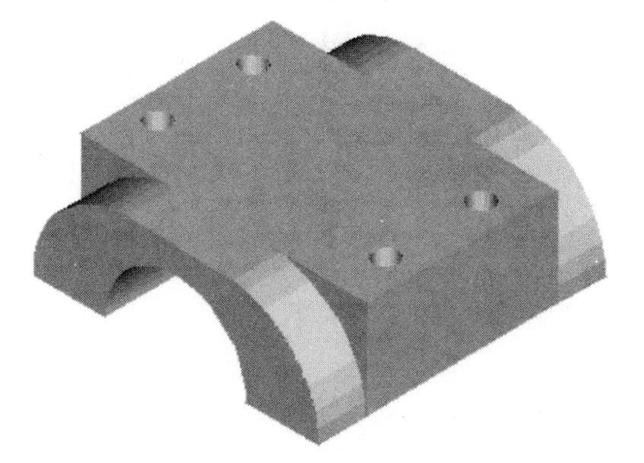

FIGURE 7.41 *Figure for exercise 7.1.*

7.2 Using Figure 7.42, complete the model of the pipe clamp bottom (see also exercise 5.8), using *Mirror Geom.*

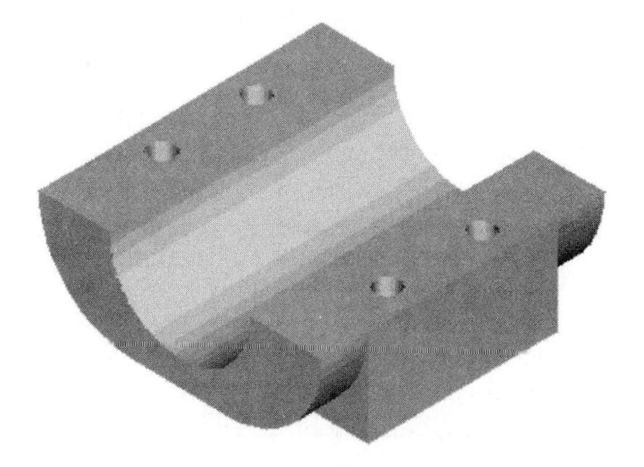

FIGURE 7.42 *Figure for exercise 7.2.*

7.3 Add the holes on the part of the side support (see also exercises 3.10 and 4.6), using the *Pattern* option and Figure 7.43.

FIGURE 7.43 *Figure for exercise 7.3.*

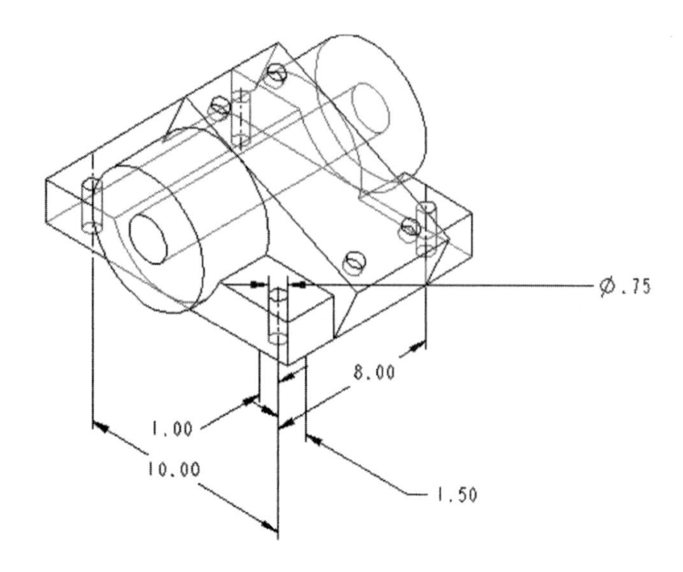


7.4 Add the holes to the two inclined planes on the double side support (see also exercises 3.11 and 4.10). Use Figure 7.44 for reference.

FIGURE 7.44 *Figure for exercise 7.4.*

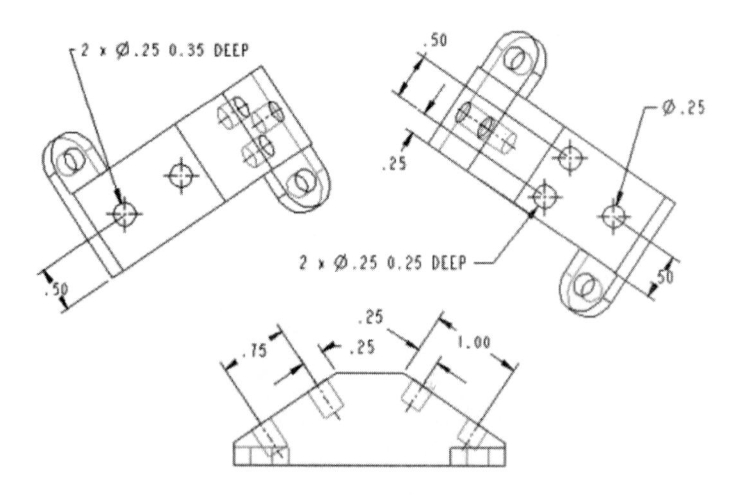

7.5 Construct the gasket by using a single protrusion and the *Mirror Geom* option. Use Figure 7.45 as a reference.

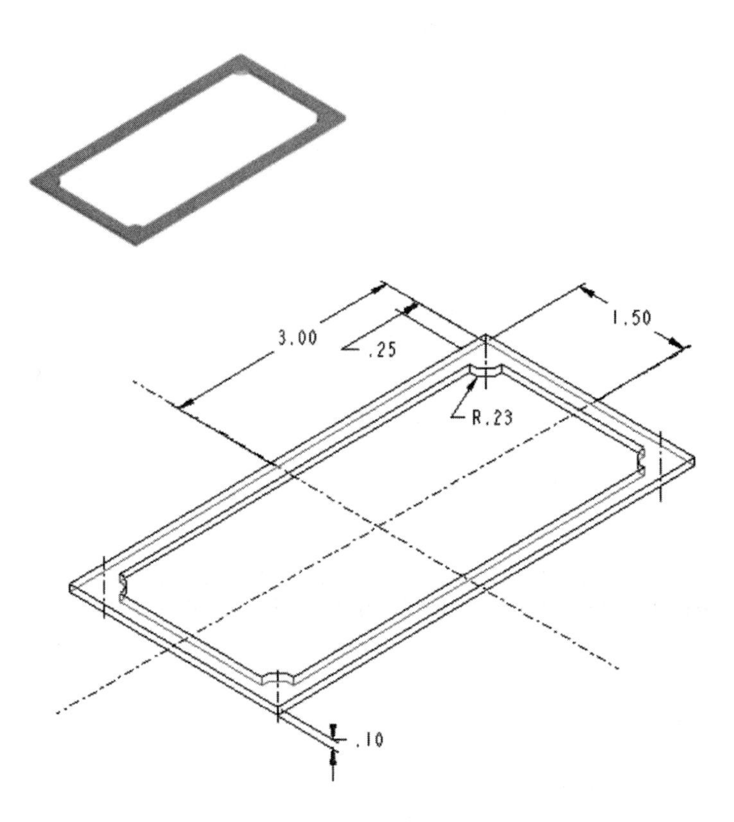


FIGURE 7.45 *Figure for exercise 7.5.*

7.6 Construct the wing plate using a single protrusion and the *Mirror Geom* option. Consult Figure 7.46 for the geometry of the model.

FIGURE 7.46 *Figure for exercise 7.6.*

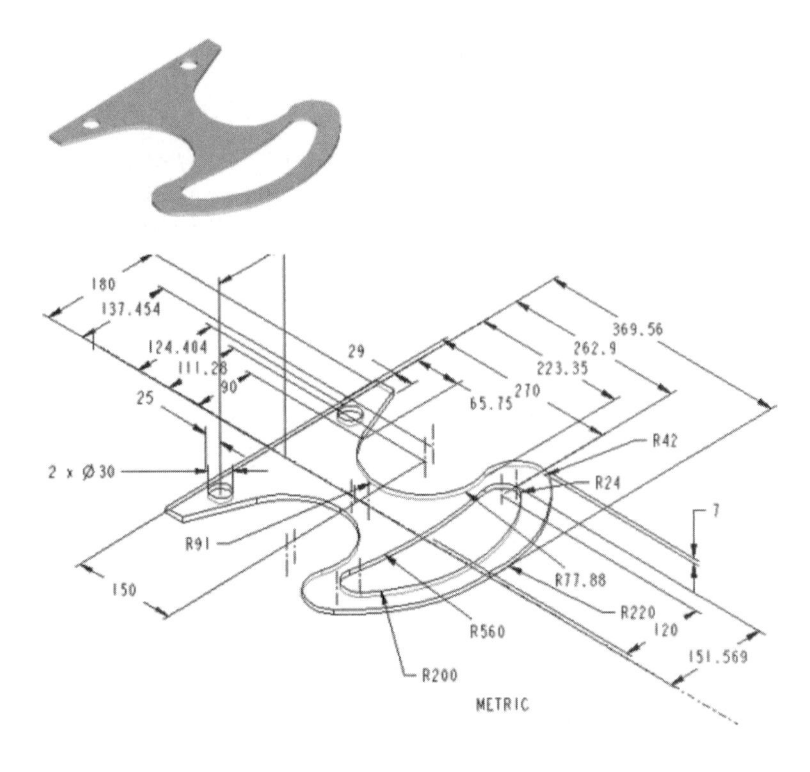


7.7 Create the model of the skew plate. Use the *Pattern* and the *Copy* options to place the holes. Notice that it is a metric part. The geometry of the part is given in Figure 7.47.

FIGURE 7.47 *Figure for exercise 7.7.*

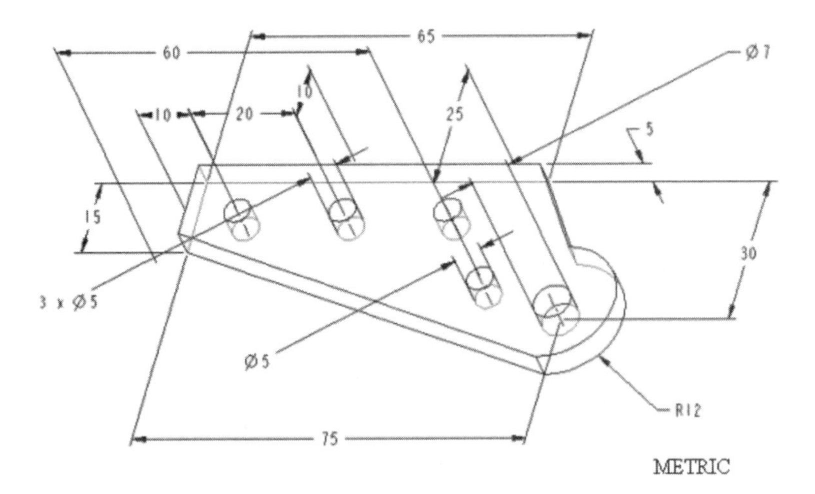

7.8 Construct the ratchet, using the *Pattern* option to place the teeth. The geometry of the part is given in Figure 7.48.

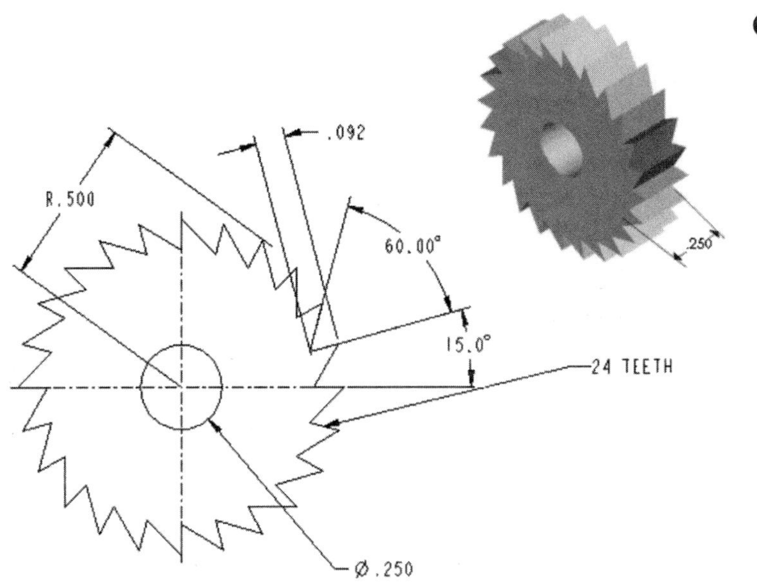


FIGURE 7.48 *Figure for exercise 7.8.*

7.9 Use the *Pattern* option to add the teeth to the index. The geometry is given in Figure 7.49.

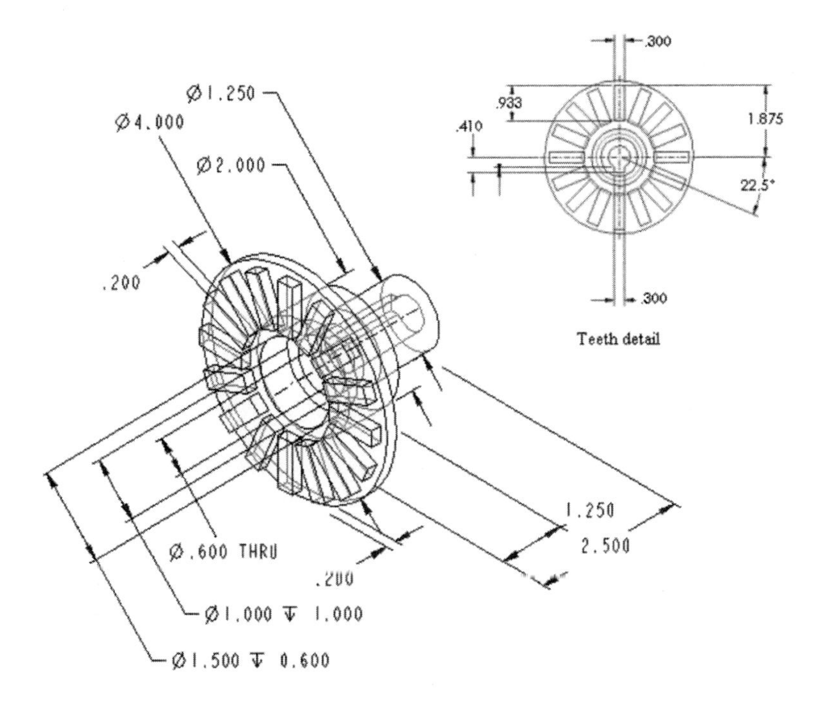


FIGURE 7.49 *Figure for exercise 7.9.*

7.10 Construct the post hinge. Hint: the *Group* option is helpful in completing this part. Use Figure 7.50 as a reference.

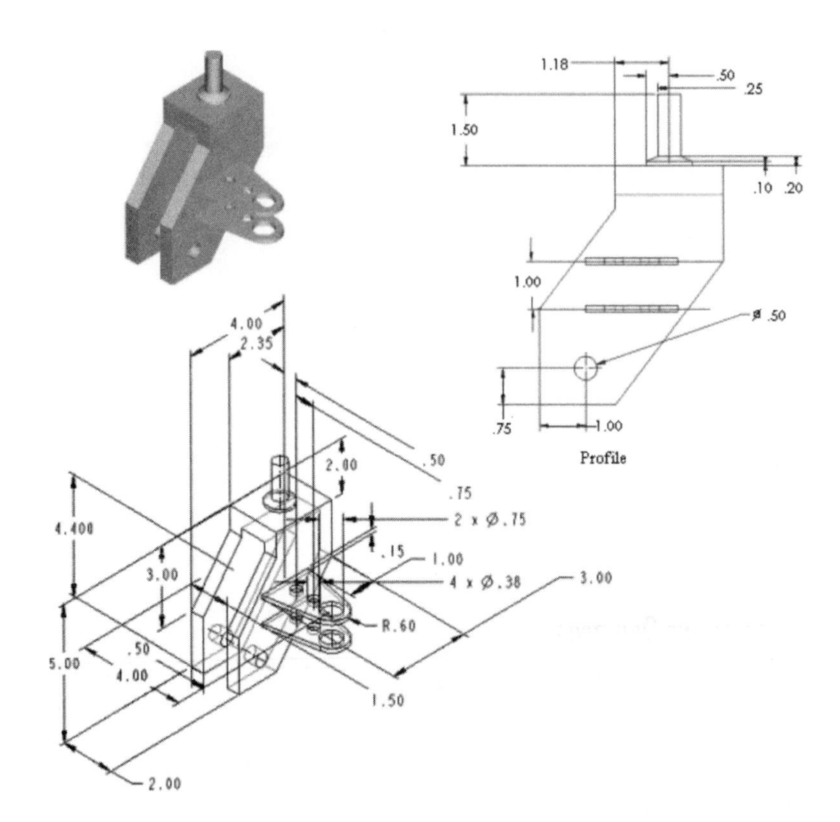


FIGURE 7.50 *Figure for exercise 7.10.*

# CHAPTER 8

# FILLETS, ROUNDS, AND CHAMFERS

## INTRODUCTION AND OBJECTIVES

In this section, we will examine the features *Round, Fillet,* and *Chamfer* that can be used to remove a sharp edge. These features remove material from a model.

*Round* and *Fillet* are essentially identical options. Fillet removes material from an inside corner, while round removes material from an outside corner. The *Chamfer* option is used to bevel an edge.

In engineering practice, rounds are placed to remove sharp edges or at corners where the geometry leads to stress concentrations. For molded parts, the rounding or beveling of sharp corners allows the part to be removed from the mold easier.

The typical value for the radii of the round is 1/8 inch (3 mm). Larger radii rounds yield parts with better strength characteristics. However, it is not always possible to regenerate a part with larger radii rounds due to possible interference between the round/fillet and other features. In general, it is best to place rounds/fillets on a model after all other features have been added.

If a part with a round/fillet fails to regenerate, the *Resolve* option may be used to investigate the failure. This option is discussed at some length in chapter 13. Often the problem can be fixed by simply reducing the radius of the round/fillet or changing the order in which the rounds and/or fillet are placed.

At the conclusion of this chapter, the reader will be able to:

1. Place rounds/fillets at an edge or between two surfaces.
2. Place an edge chamfer.

## 8.1 THE USE OF ROUNDS AND FILLETS

A *Round* is an *exterior* corner that has been given a curvature. The curvature is indicated by a small arc with a given radius. A *Fillet* is an *interior* corner that has been given a curvature. Like rounds, the size of the fillet is dimensioned with a radius.

In Pro/E, rounds and fillets may have a constant radius along their length (the *Constant* option), or the radius may vary (the *Variable* option), to produce a tapered round. These options are shown in Figure 8.1 and are selected from the *ROUND SET ATTRIBUTE* menu (*RND SET ATTR*) during the definition of the round geometry.

A *Full Round* is placed on planar surfaces bounded by two parallel edges as shown in Figure 8.1.

The *RND SET ATTR* menu also contains options for selecting the edges or surfaces to round. As can be seen in Figure 8.1, the software allows the user to place rounds on one or more edges (*Edge Chain* option), between two surfaces (*Surf-Surf*), or between an edge and a surface (*Edge-Surf* option).

Figure 8.2 illustrates the *CHAIN* menu by Pro/E. The options contained in this menu are as follows:

1. *One By One.* Using this option, the user may select individual edges and curves one at a time. The software will highlight the features after they are chosen. The default color is blue.

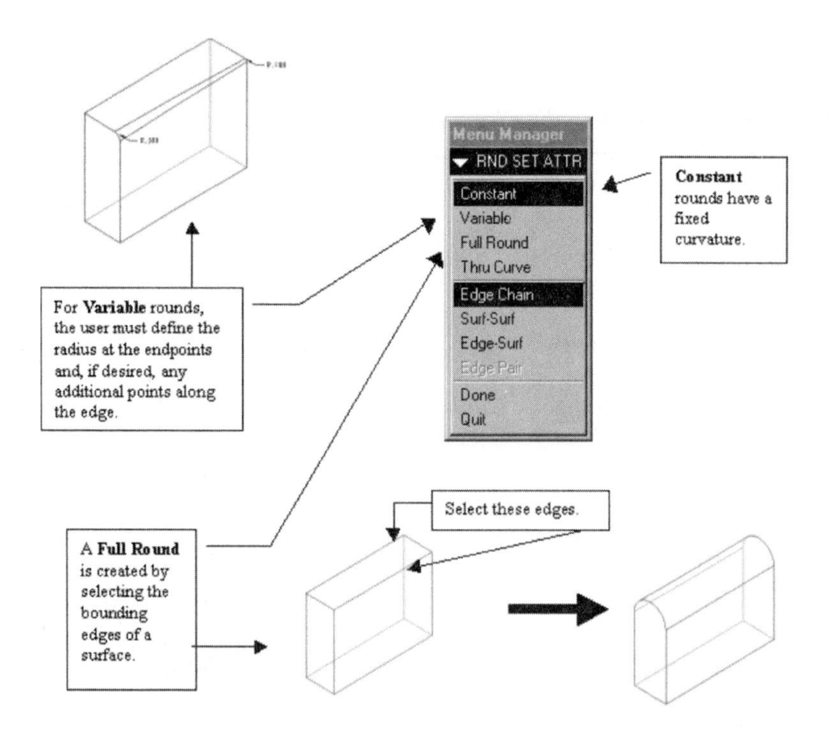


**FIGURE 8.1** *For constructing a* Round *or* Fillet, *you may use the options found in this menu.*

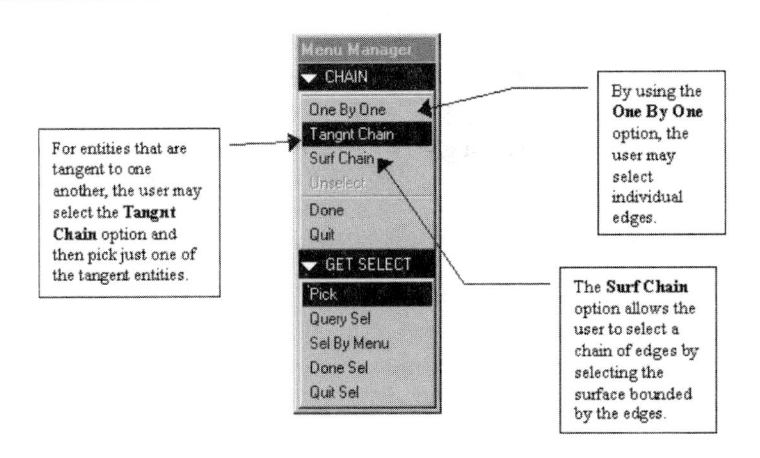


FIGURE 8.2 *The* CHAIN *menu contains options for selecting multiple edges in creating rounds or fillets.*

2. *Tangnt Chain.* In cases where the edges are all tangent to one another, the *Tangnt* (tangent) option is more convenient than the *One By One* option. This option allows the user to select all edges, which are tangent to a chosen edge, by picking just one of the edges.
3. *Surf Chain.* The *Surf Chain* option defines a chain of edges by selecting a surface. The edges form the boundary of the surface. When using this option, Pro/E allows the user to accept all or some of the edges.

### 8.1.1 TUTORIAL # 38: A ROUND AND FILLET FOR THE ANGLE BRACKET

Let us make a few rounds and fillets. Retrieve the angle bracket last seen in chapter 4. We wish to add a 0.125 radial fillet and a 0.625 radial round to the part. After we are done adding the fillet and round, the part will look like the one shown in Figure 8.3.

Now, let us create the round. Select *Create, Round, Simple,* and then *Done.* You can make several different types of rounds/fillets depending on their placement. The different types are shown in Figure 8.1. The menu appears after choosing the *Simple* option.

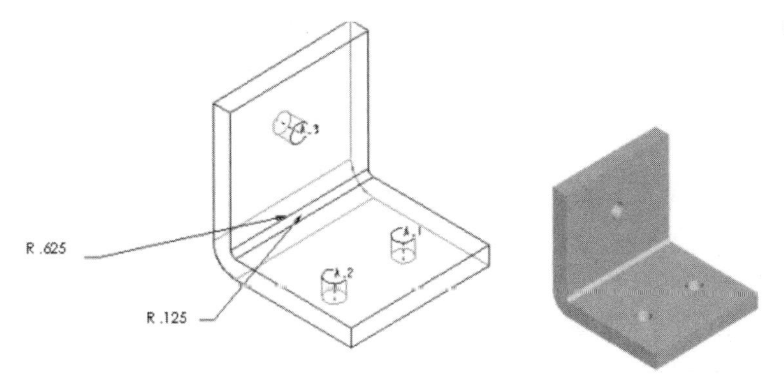


FIGURE 8.3 *Wireframe and shaded models of the angle bracket with added round and fillet.*

The rounds/fillets can vary along their length or be constant. This variation can be obtained with the *Variable* option in Figure 8.1. The round and fillet in this case are of constant radius along their length.

Select *Edge Chain.* We will place the 0.625 round first. Use your mouse and select the edge. You may need to use *Query Sel* to pick the correct edge. This case needs no other edges rounded with this radius, so choose *Done.* Enter the value 0.625 for the radius of the round. Then choose *OK.*

Now let us create the *Fillet.* Choose *Feature, Create, Round,* and *Simple.* Again, choose *Edge Chain* and pick the appropriate edge with your mouse pointer. Select *Done* and then enter the value 0.125. Finally, click on *OK.* The angle bracket should now have both the round and fillet and should appear as in Figure 8.3. Save this version of the bracket for use again in chapter 16.

### 8.1.2 TUTORIAL # 39: EDGE ROUNDS FOR THE CONNECTING ARM

Many parts contain multiple edges that must be rounded. If these edges require a round with the same radii, then the edges can all be rounded at the same time by creating a chain. As an example, recall the connecting arm from chapter 6.

The outer edges of the rib require a 0.1 radius round. Because the radius for each round is the same, we can create them all at the same time.

Select *Feature, Create,* and *Round.* Then choose *Simple.* Accept the *Constant* and *Edge-Chain* options by clicking on *Done.* From the *CHAIN* menu, reproduced in Figure 8.3, select *One By One* to pick each edge one at a time. We will use the *Query Sel* command to make sure that the proper feature is chosen.

Use *Query Sel* and click on each outer edge of the rib. Pro/E will highlight each of the four edges in blue (the default color). If you accept the selection, then the edge will remain highlighted in blue. Rotate the model if you desire. After you have selected all the edges, choose *Done Sel.* Then select *Done.* Enter a value of 0.1 for the rounds. A shaded model of the connecting arm with the added rounds is shown in Figure 8.4.

### 8.1.3 TUTORIAL # 40: A ROUND ON A CIRCULAR EDGE

Often, parts with circular features have edges that must be rounded. Adding a round to a circular edge is as simple as adding a rounding to a linear edge. Because the circular edge is continuous around the part, it need be selected only once.

A part with a circular base feature is the pipe flange from chapters 4 and 7. Retrieve this part. Select *Feature, Create, Round,* and *Simple.* Choose *Constant* and *Edge-Chain.* Using *Query Sel,* click anywhere on the circular edge of the base. Enter a value of 0.1 for the radius of the round. A shaded model of the part is shown in Figure 8.5.

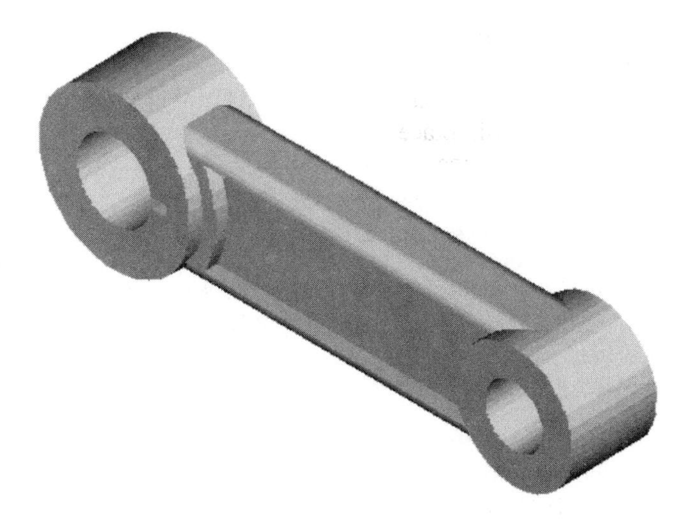

FIGURE 8.4 *The connecting arm with added rounds (compare with Figure 6.12).*

### 8.1.4 Tutorial # 41: Edge Rounds for the Propeller Blades

We can round off the remaining edges of the blades using the *Round* option. Select *Feature, Create, Round, Single,* and *Done.* The rounds are *Constant* type. Select the *One By One* option and pick the edges, two of which are shown in Figure 8.6. Use *Query Sel* to pick the hidden edges. The radius of the rounds is 30 mm. Save the completed model.

### 8.1.5 Tutorial # 42: A Part with Surface-to-Surface Rounds

Some parts require a round (or fillet) between two surfaces. Such is the case in the pipe flange from chapter 4. A 0.05-inch round is required between the circular cylinder and the top of the base.

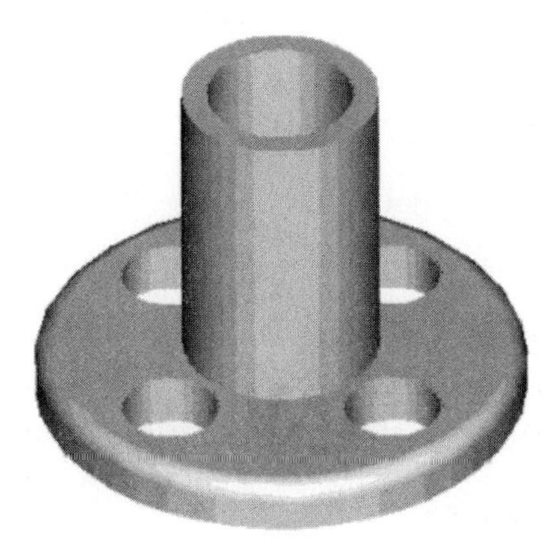

FIGURE 8.5 *The pipe flange with a 0.1 radius round.*

FIGURE 8.6 *The completed model of the propeller is shown after using* Round.

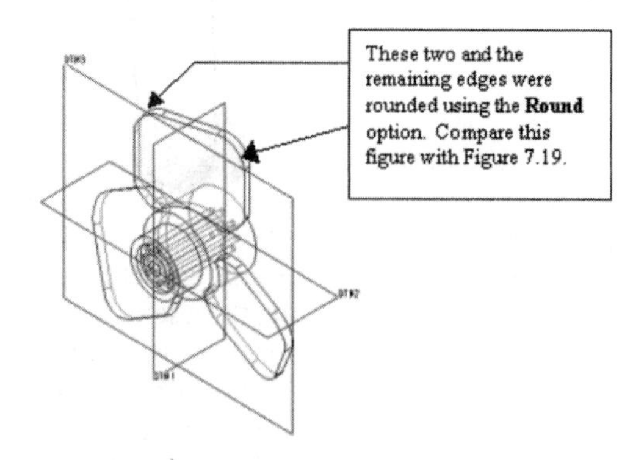


Let us begin by retrieving this part. We can readily place the edge round by using the method outlined earlier. Select *Feature, Create, Round,* and *Simple.* Choose *Constant* and *Edge-Chain.*

Click on the edge of the base and enter a value of 0.05, as shown in Figure 8.7. Now add the surface-to-surface round. Select *Feature, Create,* and *Simple.* The round has a constant cross section, so select *Constant.* Choose the *Surf-Surf* option from the *RND SET ATTR* menu (Figure 8.1).

Pro/E will now ask you to pick the surfaces. The order in which the surfaces are chosen is immaterial. Use *Query Sel* and pick the cylindrical surface. Accept when the proper surface is highlighted. Then use *Query Sel* again and pick the top of the base. Enter a value of 0.1. Choose *OK.* Pro/E will place the round.

FIGURE 8.7 *The pipe flange with the added surface-to-surface round.*

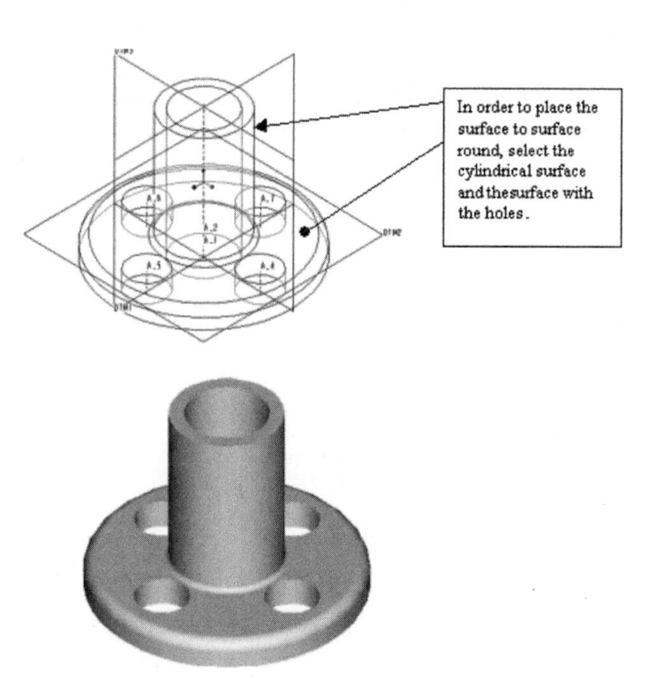

## 8.2 CHAMFERS

Sharp corners can also be removed by beveling the corner; such a bevel is called a *Chamfer.* Chamfers are used at the ends of cylinders as well as on an edge. A chamfer that occurs on a corner is called a corner chamfer. Chamfers are specified by giving a linear and an angular dimension.

In Pro/E, for corners that are formed by surfaces at 90°, a feature called *45 × d* is available. For such a feature, the user need only give the linear dimension (the value *d*). For edges that are not at 90°, several options are available. The first of these options is the *d × d* chamfer. For this option, a chamfer is created by specifying the distance, d, from the edge along each surface. If the distance along each surface needs to be different, then use the *d1 × d2* option. Of course, you will need to give the program the values d1 and d2. Finally, the chamfer can be defined with an angle and linear distance using the *ang × d* option. The options for dimensioning a chamfer are shown in Figure 8.8.

### 8.2.1 TUTORIAL # 43: ADDING CHAMFERS TO THE ADJUSTMENT SCREW AND THE NUT

Let us now add a chamfer and a round to the Adjustment Screw from chapter 5. Retrieve this part. We wish to add a 45° × .05 chamfer at one end of the screw.

Select *Feature, Create, Chamfer,* and *Edge.* At this point, we can construct either an edge or a corner chamfer. In the case of this part, we want an edge chamfer. Select *Edge.* Because the angle desired for the chamfer is 45°, we can choose the *45° × d* option. If the angle had been something besides 45°, the option of choice would have been the *ang × d* option.

After choosing the *45° × d* option, enter 0.05 for the value of d. With your mouse, click on the circular edge shown in Figure 8.9. Then select *Done Sel* and *Done Refs.* Finally, click on *OK.* Pro/E will add the chamfer to the screw.

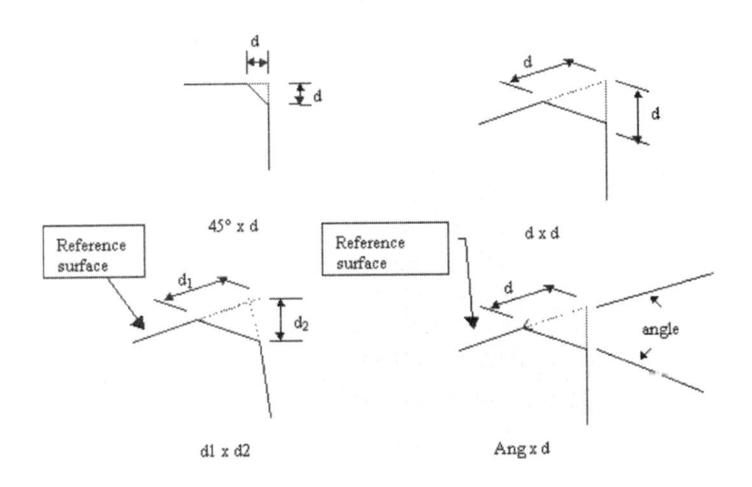


FIGURE 8.8 *The different ways to define a chamfer in Pro/E.*

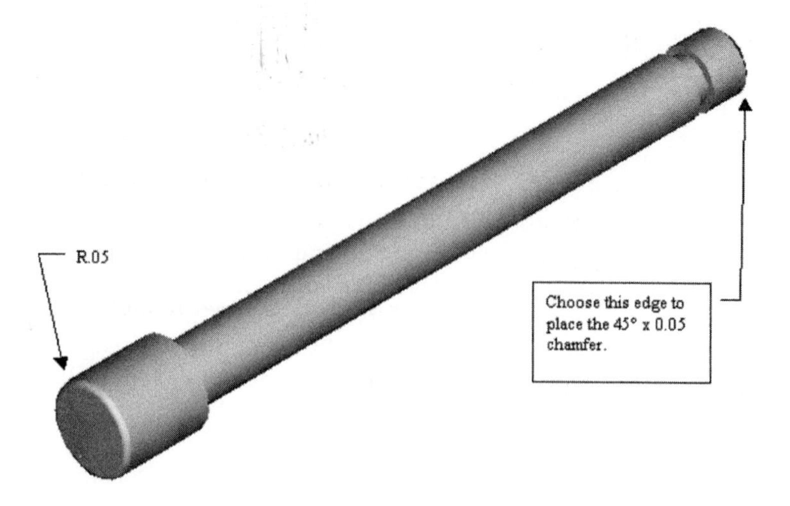


**FIGURE 8.9** *Adjustment screw with one end chamfered.*

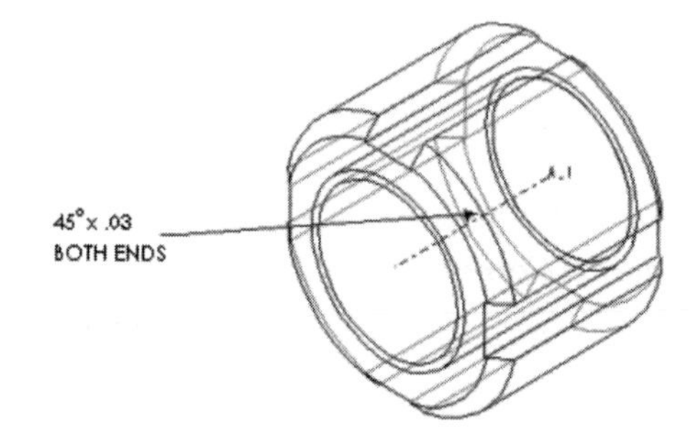


**FIGURE 8.10** *The nut from chapter 7 with chamfers added at both ends.*

Add the edge round by selecting *Feature, Create,* and *Round.* After selecting the edge, enter a value of 0.05 for the radius of the round. Your version of the part should appear as in Figure 8.9. Save the adjustment screw.

Retrieve the part "Nut" from chapter 7 and add a 45° × 0.03 chamfer on the ends of the nut. The nut with the added chamfer is shown in Figure 8.10.

## 8.3 SUMMARY

In engineering practice, sharp edges and corners must be rounded or beveled to reduce stress concentrations. The *Round, Fillet,* and *Chamfer* are the Pro/E options for removing material at model corners and edges.

Because the placement of these features leads to problems in regeneration, it is best to place rounds, fillets, and chamfers after all other features have been located on the model. If the software is unable to regenerate the model, often the problem can be remedied by simply reducing the radius of the round/fillet or changing the order in which the rounds and/or fillets are placed.

### 8.3.1 STEPS FOR ADDING A ROUND OR FILLET

The general steps for placing a round or fillet are identical. The steps are:

1. Select *Feature, Create, Round* (or *Fillet*), *Simple,* and *Done.*
2. Choose the round attribute *Constant, Variable,* or *Full Round.*
3. Select *Edge Chain, Surf-Surf,* or *Edge-Surf.*
4. Choose *Done.*
5. From the *CHAIN* menu, select *One By One, Tangnt Chain,* or *Surf Chain.*
6. Pick the edges or surfaces on the model.
7. Select *Done Sel* when you are finished picking all the entities.
8. Select *Done.*
9. Choose *OK.*

### 8.3.2 STEPS FOR ADDING A CHAMFER

In general, the procedure for adding a chamfer to a model is as follows:

1. Select *Feature, Create,* and *Chamfer.*
2. Choose *Edge* or *Corner.*
3. Pick the option for dimensioning the chamfer. For example, you may select $45° \times d$.
4. Enter the increment or increments values required by the option chosen in step 3.
5. Add additional chamfers by choosing the *Add* option. Select *Done Sel.*
6. Choose *Done Refs.*
7. Select *OK.*

## 8.4 ADDITIONAL EXERCISES

8.1 Complete the model of the cement trowel (also see exercise 3.5) using the geometry in Figure 8.11.

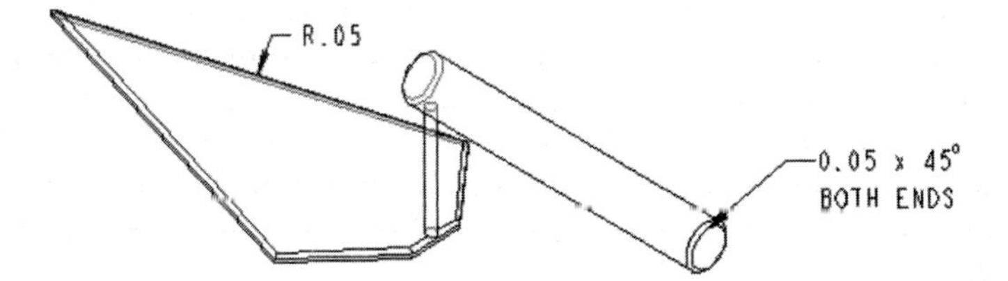


FIGURE 8.11 *Figure for exercise 8.1.*

8.2 Add the round to the model of the side guide (also see exercises 3.8, 4.9, and 5.2). Use Figure 8.12 as a reference.

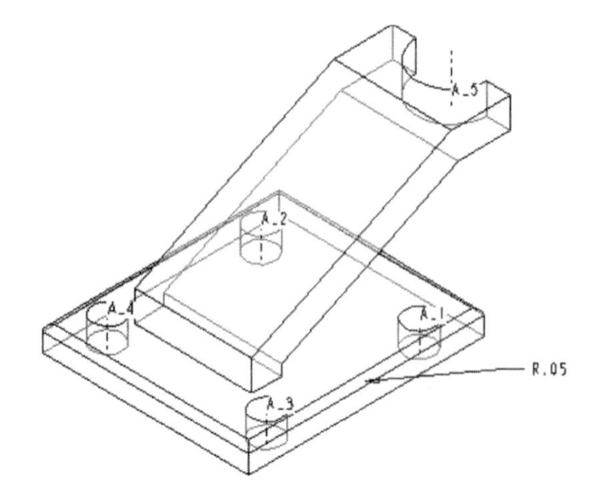


FIGURE 8.12 *Figure for exercise 8.2.*

8.3 Add the chamfer to the model of the control handle (also see exercises 3.6 and 4.7). The geometry of the chamfer is given in Figure 8.13.

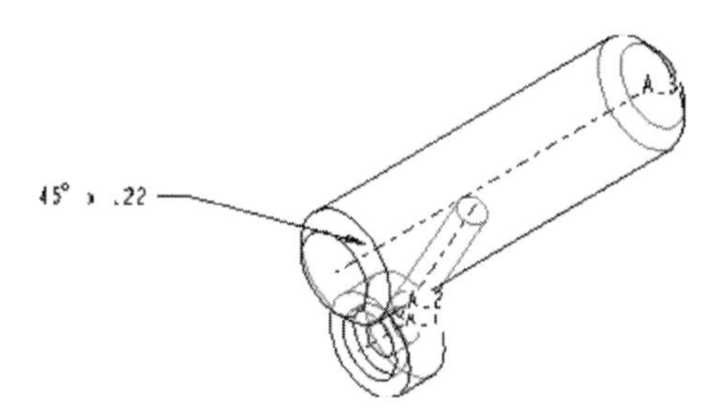

FIGURE 8.13 *Figure for exercise 8.3.*

8.4 As shown in Figure 8.14, add the chamfer to the bushing (also see exercises 3.2 and 4.3).

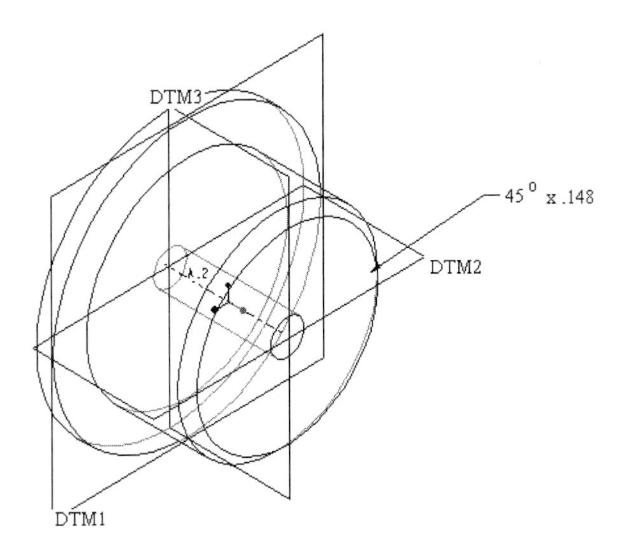


FIGURE 8.14 *Figure for exercise 8.4.*

8.5 Add the chamfer to the model of the pipe clamp top (also see exercises 5.7 and 7.1). Use Figure 8.15 for reference.

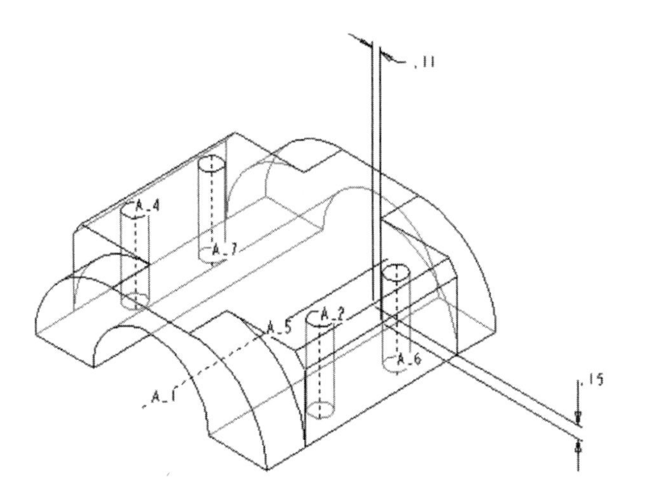


FIGURE 8.15 *Figure for exercise 8.5.*

8.6 Add the chamfer to the model of the pipe clamp bottom (also see exercises 5.8 and 7.2). Use Figure 8.16 for reference.

**FIGURE 8.16** *Figure for exercise 8.6.*

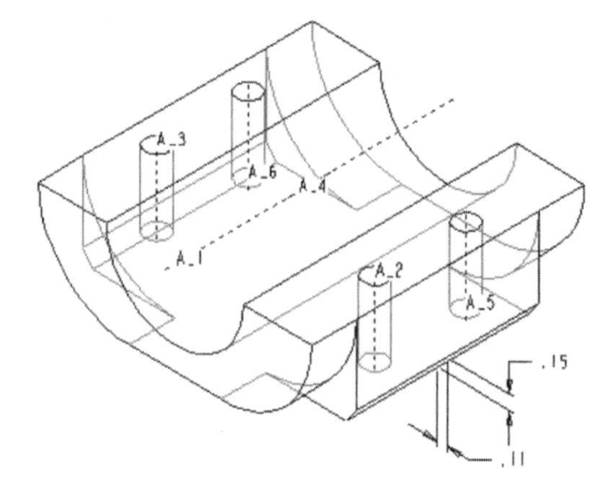


8.7 Add the round and chamfer to the model of the side support (also see exercises 3.10, 4.6, and 7.3). The geometry of the round and chamfer are given in Figure 8.17.

**FIGURE 8.17** *Figure for exercise 8.7.*

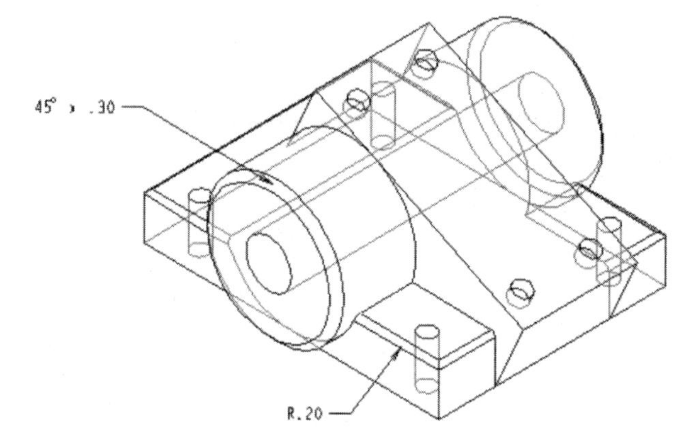

8.8 Construct the guide plate. Use Figure 8.18 for reference.

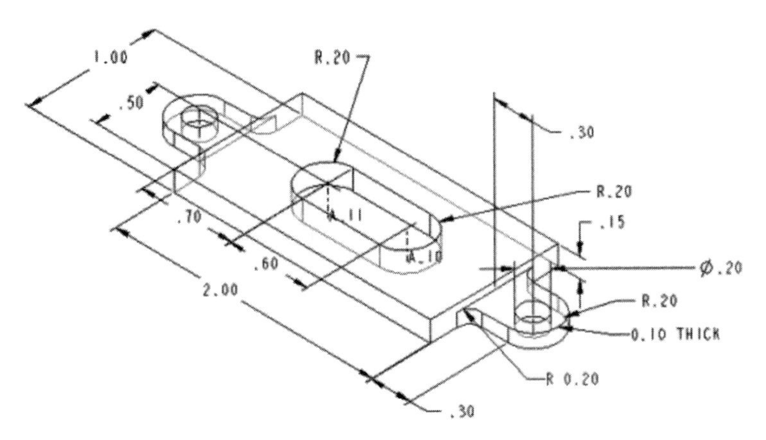


FIGURE 8.18 *Figure for exercise 8.8.*

8.9 For the single bearing bracket (also see exercise 6.12) in Figure 8.19, add a round to base and edges of the rib of radius 0.05. Also, add a round between the base and rib of radius 0.10.

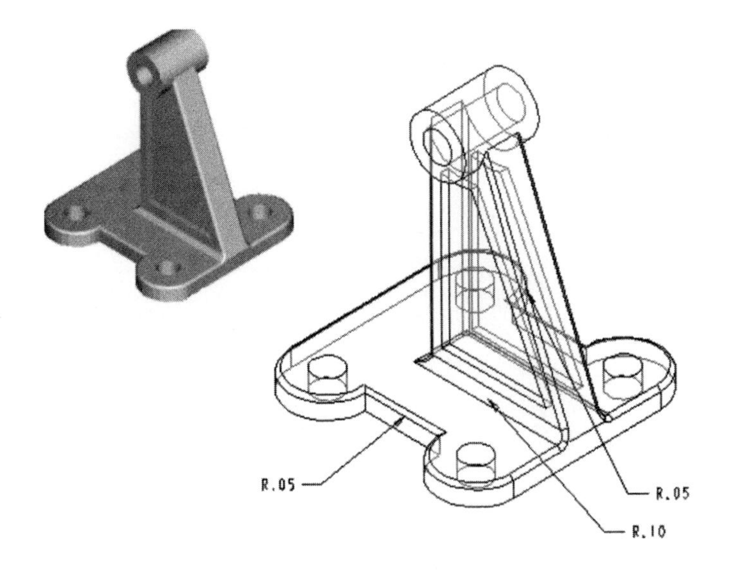


FIGURE 8.19 *Figure for exercise 8.9.*

8.10 For the model of the double bearing bracket (also see exercise 6.13) shown in Figure 8.20, add rounds to the rib, base edge, and the surface between the rib and the base. The value of the rounds is R 0.05.

FIGURE 8.20 *Figure for exercise 8.10.*

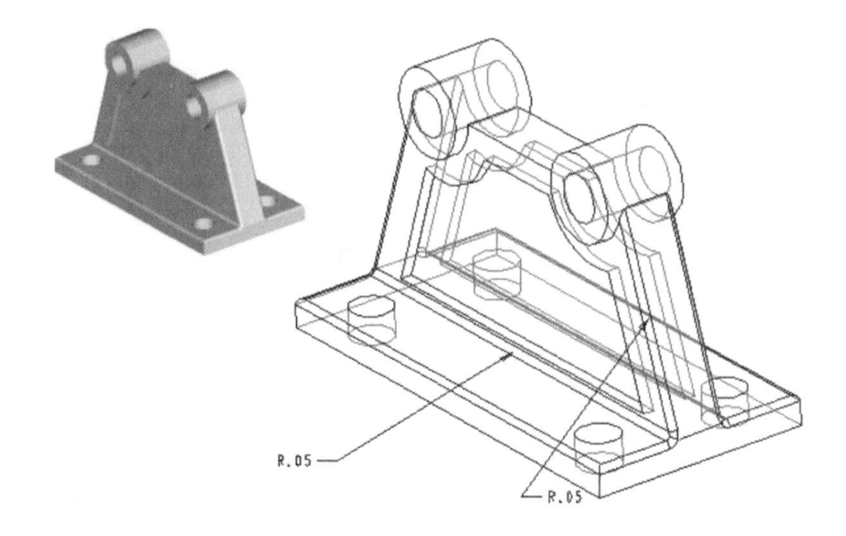


8.11 Add the rounds and chamfer to the model of the post hinge (also see exercise 7.10) as shown in Figure 8.21.

FIGURE 8.21 *Figure for exercise 8.11.*

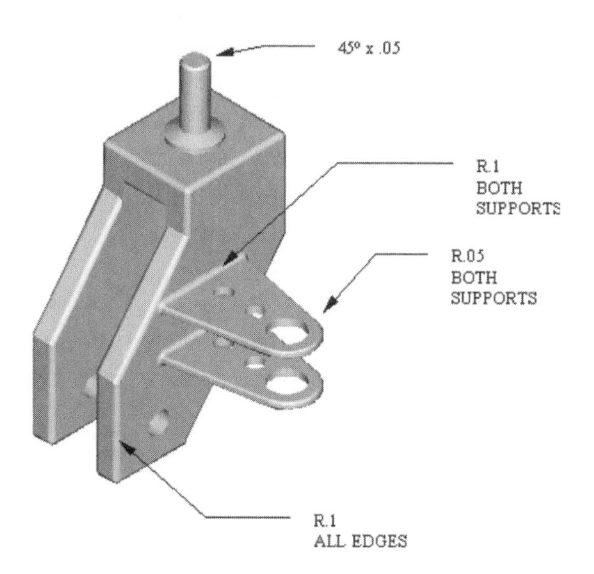

8.12 Round the edge of the clamp base (exercise 4.13). Use Figure 8.22 as a reference.

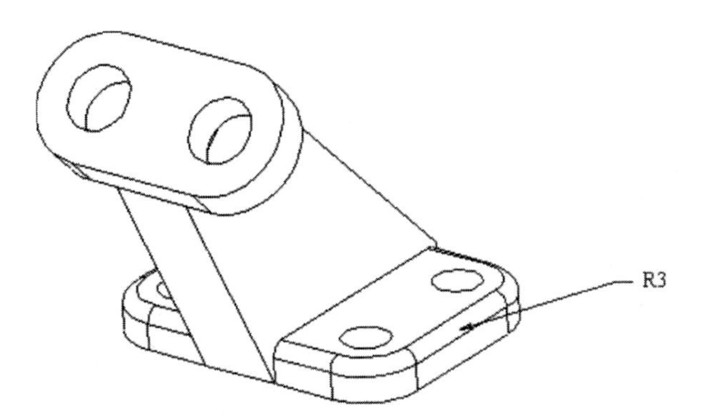


**FIGURE 8.22** *Figure for exercise 8.12.*

8.13 Add the surface-to-surface round to the slotted support (exercise 4.12). The geometry of the round is shown in Figure 8.23.

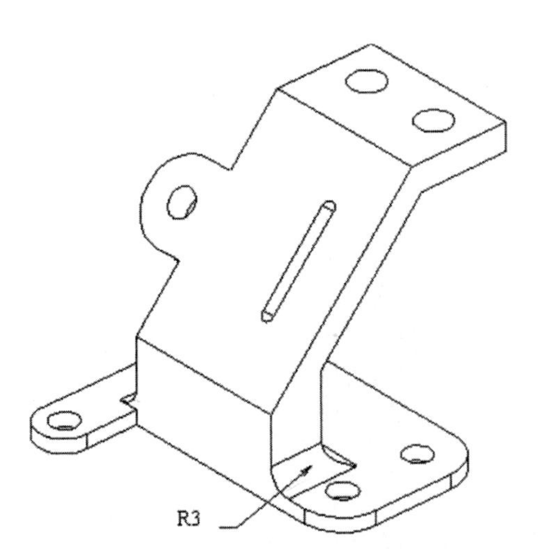


**FIGURE 8.23** *Figure for exercise 8.13.*

# CHAPTER 9

# DATUM POINTS, AXES, CURVES, AND COORDINATE SYSTEMS

## INTRODUCTION AND OBJECTIVES

In chapter 2, we considered datum planes that can be used in construction of base features. These planes were either of the default type or user-created. In addition to datum planes, Pro/E allows the user to define datum axes, points, curves, and coordinate systems (Figure 9.1).

Datum axes can be used in creating or placing a feature by using the axes as a reference. Datum points are used to specify target points for placement of features or to specify load application points in mesh generation. The points may be referenced by a datum coordinate system. Datum curves, which can be constructed through a series of datum points or from model geometry, are used to create features and for definition of sweep trajectory (sweeps are discussed in chapter 11).

The objectives of this chapter are in two parts. By the end of this chapter, the reader will be able to:

1. Construct a datum axis, datum coordinate system, datum points, and a datum curve.
2. In addition, he/she will be able to use these datum features to place additional features on a model.

## 9.1 CONSTRUCTING A DATUM AXIS

Because a datum axis may be used to reference a feature, the creation of an axis is similar to a datum plane. The option for constructing a datum axis is obtained via the *DATUM* menu (Figure 9.1). A datum axis is given a name by Pro/E of the form A_#, where # is the number of the datum axis. As with datum planes, datum axes are selected by choosing the name label of the axis.

The options for constructing a datum axis are shown in Figure 9.2. These options are:

FIGURE 9.1 *In addition to datum planes, Pro/E allows the user to construct datum, axes, curves, points, and coordinate systems.*

1. *Thru Edge.* By selecting an edge, a datum axis can be created, which passes through the edge.
2. *Normal Pln.* A datum axis is created normal to a chosen plane or surface.
3. *Pnt Norm Pln.* A point and a plane are used to construct a datum axis that is normal to the plane and passes through the point.
4. *Thru Cyl.* A datum axis may be created through the axis of a surface of revolution.
5. *Two Planes.* The intersection of two planes (or surfaces) is used to construct the datum axis.
6. *Two Pnt/Vtx.* By selecting two datum points or vertices, a datum axis may be created between the datum points or vertices.
7. *Pnt on Surf.* For a datum point located on a surface, a datum axis is constructed, which passes through the datum point. The axis is oriented so that it is normal to the surface at the point.
8. *Tan Curv.* A datum axis is created, which is tangent to a curve or edge at the endpoint of the curve or edge.

### 9.1.1 TUTORIAL # 44: A PART REQUIRING A DATUM AXIS

A part, which could use the advantage of a datum axis, is the collar shown in Figure 9.3. The hole on the lateral surface of the cylinder can be created in several ways. The approach, which we believe is the easiest and is used in this tutorial, is to construct a datum axis. The axis can then be used to locate the hole by using the coaxial option.

We begin by creating a part and placing the default datum planes. Select *Feature, Create, Protrusion,* and *Solid.* Use DTM3 as the sketching plane and DTM2 as the "*Top*" and enter the *Sketcher.* Sketch a circle with a diameter of 0.65. This diameter and all dimensions for the collar are given in Figure 9.4. Use the *Both Sides* option in creating the protrusion.

We need to construct an additional datum plane that is tangent to the surface of the cylinder. The plane, DTM4, is shown in Figure 9.5. This plane is used as the placement surface for the hole. In order to create this plane, select *Feature, Create, Datum,* and *Plane.* From the *DATUM PLANE* menu (Figure 3.8) select the *Offset* option. Pick DTM1 and *Done.* Enter a value of 0.65/2. The direction of the arrow really doesn't matter. However, in order to create the datum, which is shown in Figure 9.5, you will need to make sure that the arrow points to the right.

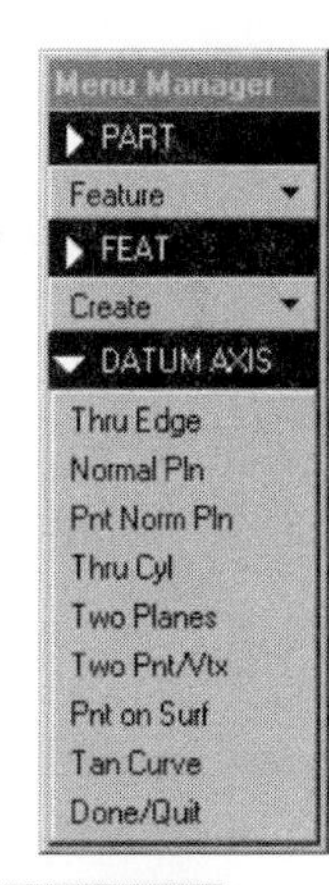


FIGURE 9.2 *The* DATUM AXIS *menu that contains the options for creating datum axes.*

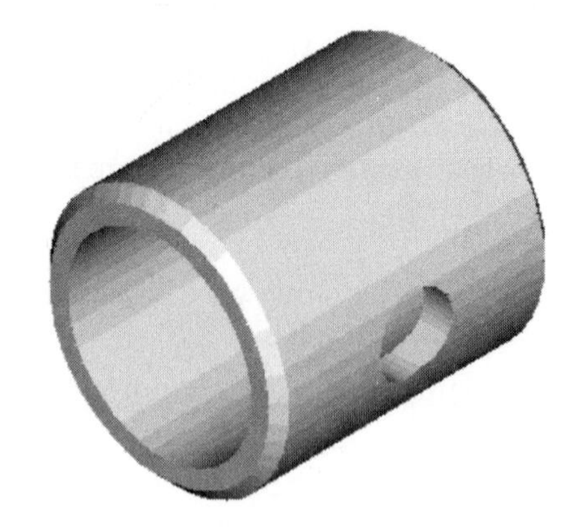

**FIGURE 9.3** *The hole in the side of the collar is located by constructing a datum axis.*

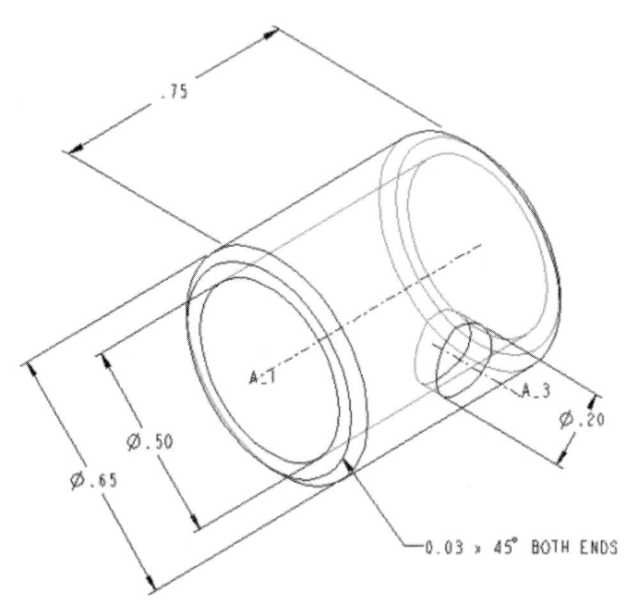


**FIGURE 9.4** *A solid model of the collar, defining appropriate dimensions.*

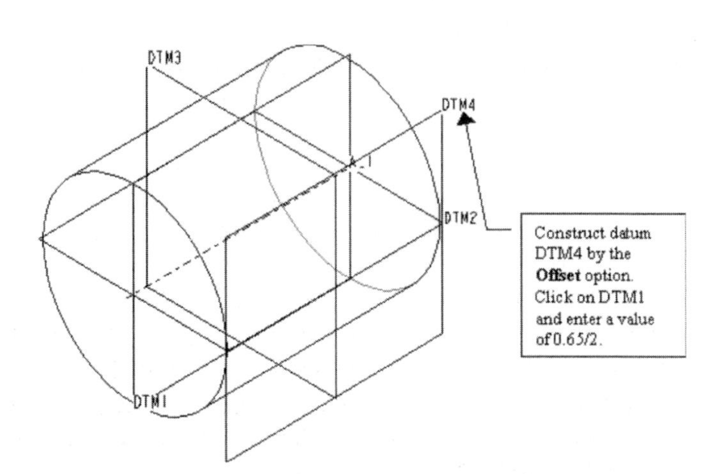


**FIGURE 9.5** *The base feature for the collar, along with the additional datum plane DTM4.*

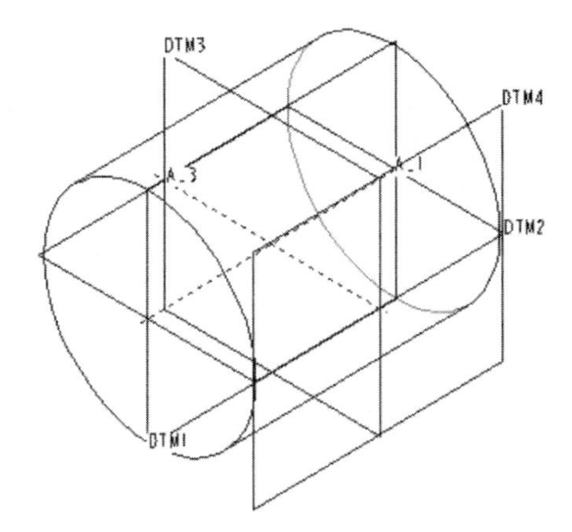


FIGURE 9.6 *The model of the collar after creating the datum axis (axis A_3).*

In order to construct the datum axis, select *Feature, Create, Datum,* and *Axis.* Select the *Two Planes* option from the *DATUM AXIS* menu (Figure 9.2). Then, click on DTM1 and DTM2. The software will construct the datum axis as shown in Figure 9.6.

At this point, the hole in the side of the collar may be placed, using DTM4 as the placement plane and the datum axis. Use *Create, Hole, Straight,* and the *Coaxial* option. Pick the datum axis when asked to do so. Pick DTM4 as the placement plane. Use the *One Side* option and *Thru All.* Make sure that the arrow points toward the cylinder. Enter a diameter of 0.20 for the hole.

Using the *Coaxial* option, place a 0.50 diameter hole *Thru All* along the A_1 axis. Finally, construct a 0.03 × 45° chamfer on each end to complete the model. Save the part.

## 9.2 DATUM COORDINATE SYSTEMS

Datum coordinate systems can be created to reference datum points and other features. Such coordinate systems are constructed by using the options found in the *COORDINATE SYSTEM OPTIONS* menu reproduced in Figure 9.7. Overall, nine options can be used in defining a coordinate system. These options are defined in the following manner:

1. *3 Planes.* If the default datum planes have been used, this option can be easily used to construct a coordinate system. In using the three planes option, Pro/E constructs the coordinate system at the intersection of the three planes.
2. *Pnt + 2Axes.* The point is used as the origin of the coordinate system. A plane is constructed through the one point and one of the axes. The second axis defines the orientation of the plane.
3. *2Axes.* The origin is located at the intersection of two axes. As in the *Pnt + 2Axes* option, a plane is defined as it passes through the point and is parallel to the first axis.

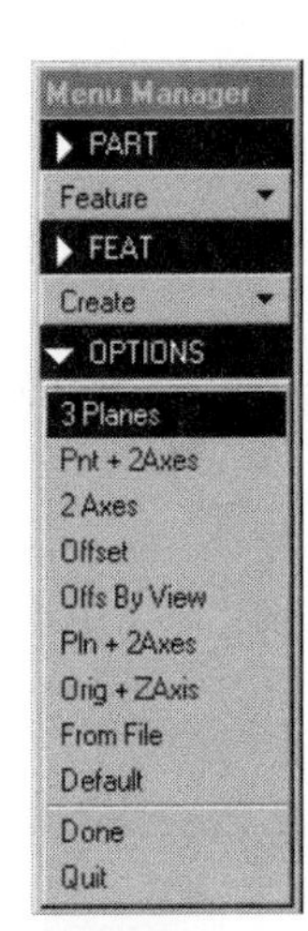


FIGURE 9.7 *Coordinate systems may be created using several options. The options are located in this* OPTIONS *menu.*

4. *Offset.* The offset option can be used to construct a coordinate system offset from a preexisting one. The new coordinate system is located using the *MOVE* menu, which allows translation and rotation.
5. *Offs By View.* This option allows the user to construct a new coordinate system that is offset from a preexisting system and parallel to the screen.
6. *Pln + 2Axes.* The intersection of the plane and the first axis chosen is used to locate the origin of the coordinate system. The second axis is used to define a plane that passes through the origin and is parallel to the first axis.
7. *Orig + Zaxis.* The origin is defined by selecting a point. The direction of the X and Y are chosen next. Then the software uses the right-hand rule to obtain the direction of the Y-axis.
8. *From File.* This option may be used to construct a new coordinate system when an old one already exists. A file containing the elements of a transformation matrix is entered. The file must have an extension ".trf." The format of the file is:

| | | | |
|---|---|---|---|
| X1 | X2 | X3 | T1 |
| Y1 | Y2 | Y3 | T2 |
| Z1 | Z2 | Z3 | T3 |

   The columns define vectors. The first column in the matrix determines the X axis direction while the second column determines the Y direction. Any values can be used for the third column, because the Z direction is determined by using the right hand rule. The last column contains the elements that define the translation of the new origin.
9. *Default.* A coordinate system may be created at a default location using this option. For base features, this location provides the anchor point of the section for the feature. The positive X direction is defined so that it points in the direction to the right, while the Y-axis is taken so that it points in the vertical direction.

### 9.2.1 TUTORIAL # 45: A COORDINATE SYSTEM FOR A TWO-DIMENSIONAL FRAME

In this section, we begin the construction of a two-dimensional frame that will eventually be completed using datum points, datum curves, and a sweep. The sweep will be added in chapter 11. In order to construct the datum points, we need to first place a coordinate system.

Let us retrieve the metric start part created in chapter 3 called "StartPartMetric." Because it is often advantageous to have a metric start with a coordinate system, we will save the part after adding the datum coordinate system. Then, we will rename the part to allow the start part to be used again.

We will construct the coordinate system so that its origin is located at the intersection of the three planes. In this endeavor, we will use the *3 Planes* option.

Select *Feature, Create, Datum* and *Coord System*. Choose the *3 Planes* option and select DTM1, DTM2, and DTM3, in turn. The software will place a coordinate system at the intersection of the three planes (which is a point), as shown in Figure 9.8.

The X, Y, and Z axes need to be defined. Pro/E will display *THE COORDINATE SYSTEM* menu shown in Figure 9.8 and will highlight each of the arrows in red. We want the X-axis to point to the right and the Y-axis to point in the upward direction. Use the *Reverse* option and make sure that the axes point in the appropriate direction. After the directions are defined, click on the X, Y, or Z option in the menu. After all the axes have been defined, you should have a coordinate system similar to the one shown in Figure 9.9. Note that the third axis will be determined by using the right-hand rule. Thus, all you have to define are the first two.

The datum planes, with the properly defined and oriented coordinate system, is shown in Figure 9.9.

We can save the start part. Select *File* and *Save*. Hit *Return* to create a new version of the start part with the added coordinate system.

Now, we can save the part under a new name. Select *File* and *Save As*. Hit *Return*. Then, enter "2Dframe." This series of actions will create a part called "2Dframe."

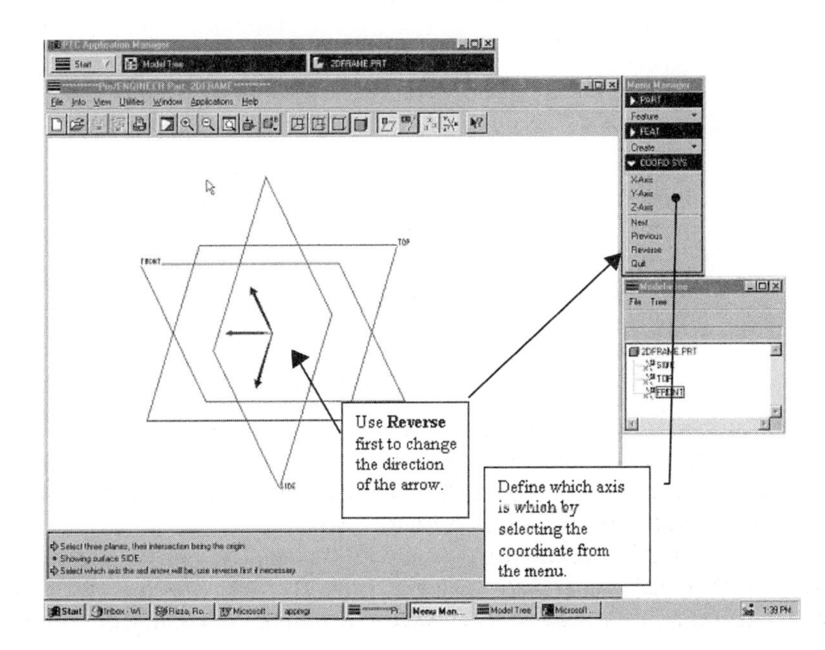


**FIGURE 9.8** *A coordinate system has been placed at the intersection of the three default datum planes. The arrows are used to indicate the positive direction of the coordinate axes.*

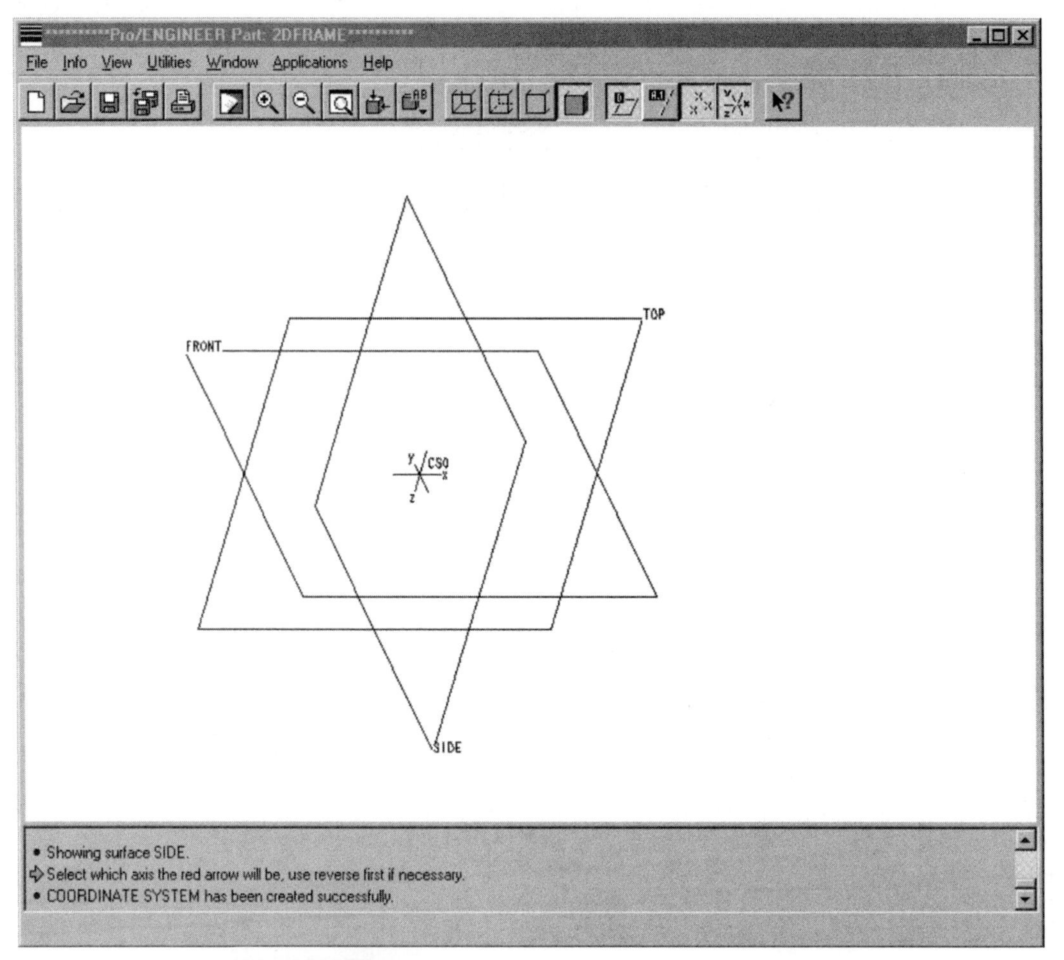


FIGURE 9.9 *The coordinate system "CS0" has been created at the intersection of the three planes.*

In section 9.3, we will place datum planes using the coordinate system placed in this section.

## 9.3 DATUM POINTS

Datum points are displayed with an "X" and given a name in the form *Pnt N,* where N is the number of the point. An array containing multiple points can be created. This array is a single feature, which makes modifying or referencing the points easy.

Datum points are specified by three coordinates. The coordinates of the points may be *Cartesian* (x,y,z), *Cylindrical* (r,$\theta$,z), or *Spherical* (r,$\theta$,$\phi$) as shown in Figure 9.10a. The coordinates of the datum points may be entered

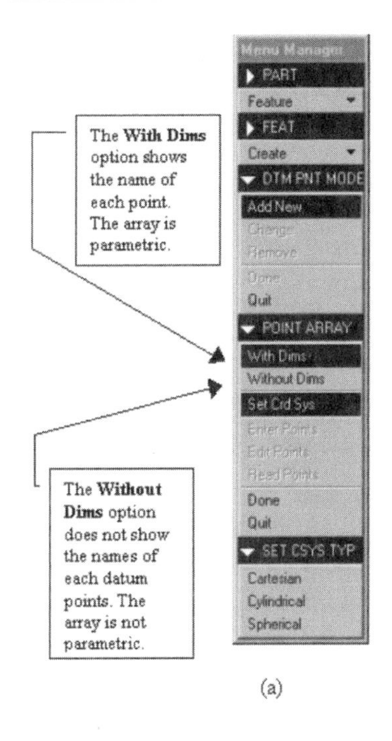


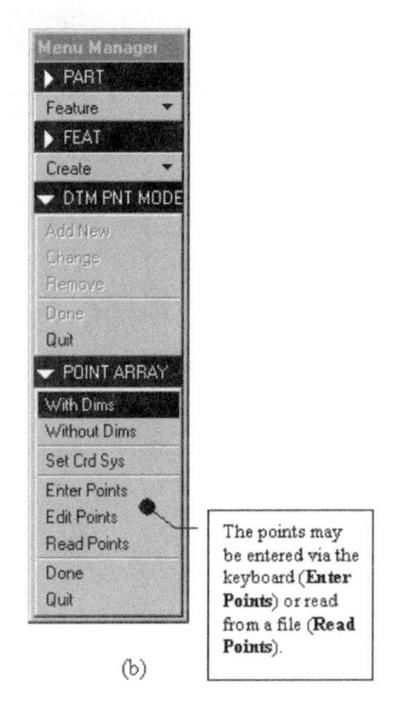


FIGURE 9.10 *(a) The type of coordinate system may be defined. (b) The method of entering the coordinates of a datum point may be prescribed.*

via the keyboard (*Enter Points* option) or by file (*Read Points* option) by using the *POINT ARRAY* menu shown in Figure 9.10b.

If the coordinates of the datum points are entered by file, the file, which must have an extension ".pts," must be an ASCII file containing rows of numbers. If each row contains three numbers, Pro/E assumes that the three numbers are the coordinates of the point. If the rows have more than three numbers, then the software will read the second, third, and fourth numbers as the coordinates of the point. It allows the user to use the first number in each row as a point number.

Pro/E allows the user to treat the points as parametric or nonparametric features. If it is desired that the points should be parametric, then the *With Dims* option should be selected from the *POINT ARRAY* menu in Figure 9.10b. If instead, the *Without Dims* option is chosen, then the datum point array will not be parametric and no names will be given to the individual points in the point array. This definition can be changed by using the *Redefine* option, but the number of points contained in a datum array cannot.

A datum point may be constructed using the options given in the *DATUM POINT* menu. This menu is reproduced in Figure 9.11.

As can be seen from Figure 9.11, 10 options can be used to locate a datum point. Some of these options involve surfaces and curves. If the coordinates of the point are known, relative to some coordinate system, then the *Offset Csys* option is the method of choice.

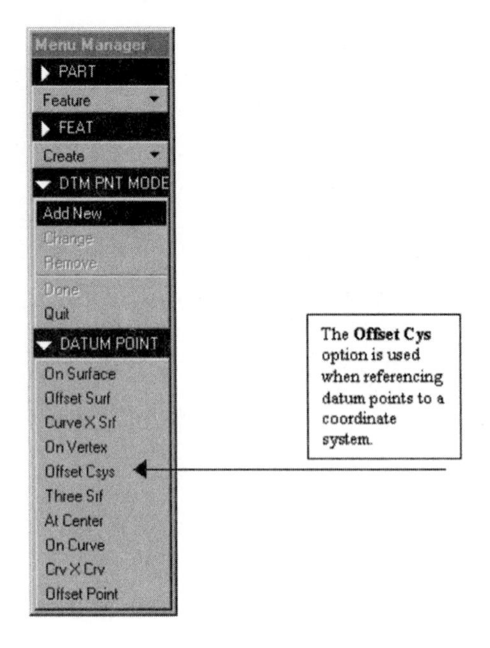


FIGURE 9.11 *The* DATUM POINT *menu, which contains options for locating a datum point.*

The options work in the following manner:

1. *On Surface.* The datum point is placed on a surface. Its location is prescribed by dimensioning from two planes or edges.
2. *Offset Surf.* A datum point is created on a surface that is offset along a chosen direction from two planes or edges.
3. *Curve X Srf.* This option allows the user to construct a datum point at the intersection of a curve and a surface.
4. *On Vertex.* If a vertex exists, a datum point can be constructed on the vertex.
5. *Offset Csys.* Using a coordinate system and coordinates relative to that coordinate system, a datum point (or array of points) is constructed. The coordinates may be entered via file or from the keyboard.
6. *Three Srf.* For parts containing three surfaces that intersect, a datum point may be constructed at the intersection.
7. *At Center.* The center of an arc or circle can be used to place a datum point.
8. *On Curve.* The datum point is placed on a prescribed curve or edge. The location of the point along the curve must be given as a function of the distance from one endpoint of the curve.

9. *Crv X Crv.* For two curves that do not necessarily intersect, a datum point may be constructed, which lies at the location that is the minimum distance between the two curves.
10. *Offset Point.* A point or array of points is constructed, which is offset from another point or vertex.

### 9.3.1 TUTORIAL # 46: ADDING DATUM POINTS TO THE TWO-DIMENSIONAL FRAME

Let us add datum points to the 2Dframe model from section 9.2.1. We will use the *Enter Points* option and enter the coordinates directly from the keyboard for each of the five points that need to be entered.

When you use the *Enter Points* option, Pro/E will ask for the X, Y, and Z coordinates of each point in succession. Begin by selecting *Feature, Create, Datum,* and *Point.* The points will be referenced to the coordinate system created in 9.2.1. So, choose the *Offset Csys* option from the *DATUM POINT* menu (Figure 9.11). Select the *With Dims* option and pick on the coordinate system label (CS0 in Figure 9.9). Since CS0 is a cartesian coordinate system, select *Cartesian.* Finally, choose *Enter Points* from the *POINT ARRAY* menu.

The coordinates of the points are given in Table 9.1. In section 9.4.1, we will add a datum curve that goes through each of the datum points. This datum curve will start at the first point, loop through the points and end at the second point. This succession will properly close the curve for later use in constructing a sweep. It is the reason why the second point (coordinates 0,1,0) is repeated.

After all the coordinates have been entered, hit the *Return* key on your keyboard. Pro/E will not automatically save the points to disk. If the points were to be saved, they would be written to the file "CS0.pts." We will not save the points.

If you wish to edit the coordinates of the points, you must use the *Edit Points* option. Click on *Done.* The part should appear as in Figure 9.12.

TABLE 9.1 THE DATUM POINTS COORDINATE FOR THE "2DFRAME."

| POINT # | X | Y | Z |
|---|---|---|---|
| 0 | 0.0 | 0.0 | 0.0 |
| 1 | 0.0 | 1.0 | 0.0 |
| 2 | 0.0 | 1.9 | 0.0 |
| 3 | 0.9 | 1.0 | 0.0 |
| 4 | 0.0 | 1.0 | 0.0 |

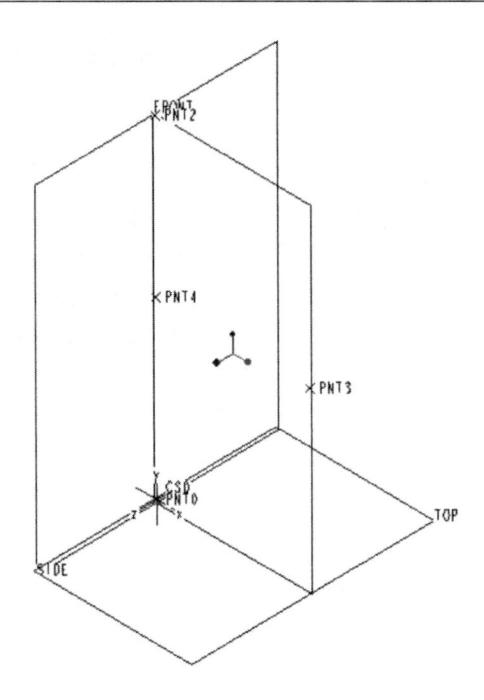


FIGURE 9.12 *The part "2Dframe" with the added datum points.*

## 9.4 DATUM CURVES

Datum curves can be used to construct surfaces or as trajectories in sweeps (the sweep option in examined in chapter 11). The options for constructing a datum curve are given in the *CRV* (Curve) *OPTIONS* menu. This menu is reproduced in Figure 9.13.

The *Thru Points* option may be used to place a curve through a series of datum points. The *Composite* option allows the user to construct a curve that is composed of other datum curves or edges. When selecting edges, the user must make sure that the edges are adjacent to each other. The order in which the curves and edges are chosen is immaterial. The selected edges may form open or closed loops.

The options *Intr. Surfs, OffsetFromSrf,* and *Offset In Srf* refer to construction of datum curves relative to surfaces. The *Intr. Surfs* option may be used to construct a curve at the intersection of two or more surfaces. The remaining options refer to curve construction offset from an existing curve and normal to a surface (*OffsetFromSrf*), or offset from an existing curve and along a surface (*Offset In Srf*).

Datum curves can be sketched using the *Sketch* option or entered through a file with the *From File* option. When importing a datum curve from a file, the file must have an extension ".ibl" or be imported as an IGES, SET, or VDA file.

The *Projected* option allows the user to construct a projection of a curve. The projection is created by sketching a curve and then projecting the curve onto one or more surfaces.

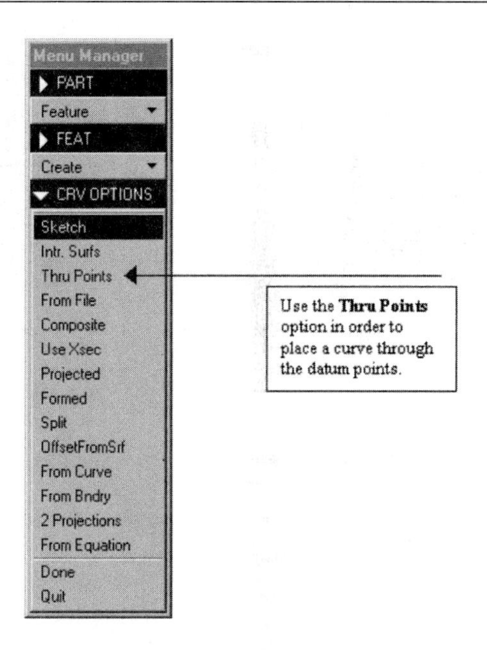


FIGURE 9.13 *The options in the* CRV *(Curve)* OPTIONS *menu are used to construct different types of curves.*

Datum curves constructed with the *Thru Points* option may be constructed with the additional options given in the *CONNECT TYPE* menu. This menu is reproduced in Figure 9.14. The *Spline* option allows the user to place a three-dimensional spline through the points. For a curve that twists and bends through a three-dimensional shape, a radius may be prescribed along the bends by using the *Single Rad* or *Multiple Rad* options. In cases where the same radius is used throughout the curve, use the option *Single Rad* otherwise use the *Multiple Rad* option.

The *Whole Array* option allows selection of datum points that were created as a single array. This option saves time, because each point does not have to be chosen. The option selects the points in sequential order.

Additional points may be added to a curve using the *Add Point* or *Insert Point* options. The *Add Point* option is used to add existing points to a curve, while the *Insert Point* option is used to add a new point between preselected endpoints. In order to remove a point from a curve, use the *Delete Point* option.

## 9.4.1 TUTORIAL # 47: ADDING A DATUM CURVE TO THE TWO-DIMENSIONAL FRAME

Let us add a datum curve to the two-dimensional frame. Because we have chosen to place the datum points in a single array, the *Whole Array* option may be used. In order for the datum curve to pass through all the points, it must bend in the X-Y plane; thus, a bend radius must be prescribed.

Select *Feature, Create, Datum,* and *Curve.* From the *CRV OPTIONS* menu (Figure 9.12), select *Thru Points* and *Done.* We will use a single radius of 0.03, so select *Single Rad* from the *CONNECT TYPE* menu. Select

FIGURE 9.14 *The* CONNECT TYPE *menu.*

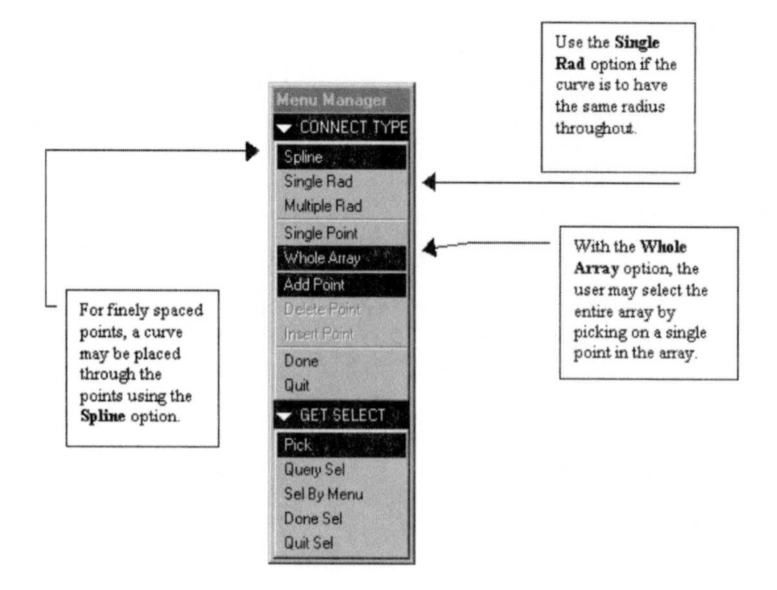


"Pnt0" and enter the bend radius of 0.03. Then choose *Done.* The two-dimensional frame with added datum curve is shown in Figure 9.15. Save the part. In chapter 11, we will use the sweep function to add the protrusion of the frame.

FIGURE 9.15 *The two-dimensional frame with added datum curve through the datum points.*

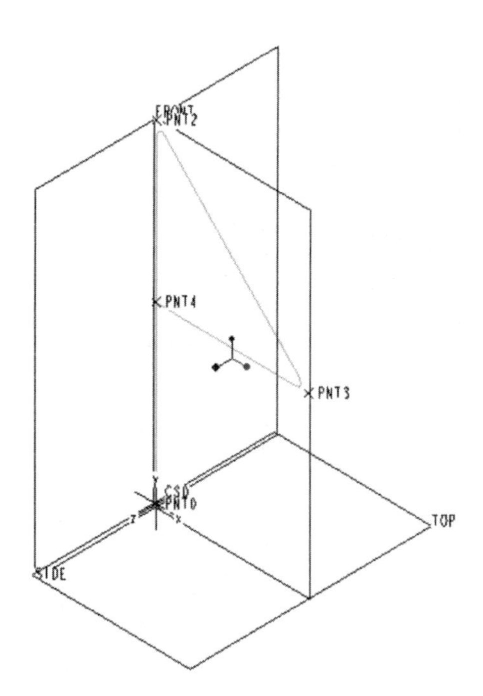

### 9.4.2 TUTORIAL # 48: A KING POST TRUSS

In the construction of the datum curve for the "2Dframe," a single datum curve was passed through all the points using a single bend radius. This method is sufficient, as in the case of the "2Dframe," when the frame is to be manufactured by bending a single piece of stock into the prescribed shape. However, many parts are composed of several structural elements. Datum curves may be generated for such parts by using datum points and the *Spline* option.

Consider the King Post Truss shown in Figure 9.16. The elements of the truss are composed of straight 2 × 4 dimensional lumber (actually 1.5 × 3.5). These elements may be created in Pro/E in three steps. The first step is to use datum points to define the ends of the elements. The second step is completed by creating a datum curve for each element using the *Spline* option. Finally, by sweeping a rectangular section along each datum curve, the truss is completed. Notice that because of the symmetry in the model, we may create half of the truss and use the *Mirror Geom* option to complete the model.

For this part, the units are in feet. Begin by creating a part called "KingPostTruss." Add default datum planes. Select *Set Up* and change the units to *Feet.* Create a coordinate system at the intersection of the three

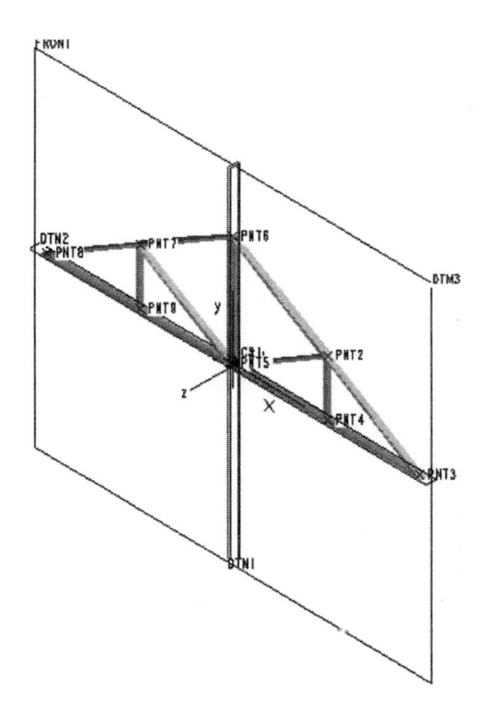


FIGURE 9.16 *The model of the king post truss.*

datum planes following the approach used in section 9.2. The sequence of steps adding the coordinate system is:

1. Pick *Feature, Create, Datum,* and *Coord System.*
2. Choose the *3 Planes* option and select DTM1, DTM2, and DTM3 one at a time.
3. Select the proper direction of the coordinate axes using Figure 9.16 as a guide.
4. Choose the names of the axes using Figure 9.16.

The coordinates of the datum points are given in Table 9.2. To create the datum points, use the following approach:

1. Select *Feature, Create, Datum,* and *Point.*
2. Choose the *Offset Csys* option.
3. Select the *With Dims* option and pick on the coordinate system label.
4. Pick *Cartesian* and *Enter Points.*
5. Enter the values given in Table 9.2.

After the datum points have been added to the part, the datum curves may be added. The procedure for adding the datum curves is as follows:

1. Select *Create, Datum Curve, Thru Points,* and *Done.*
2. Choose *Spline, Single Point,* and *Add Point* from the *CONNECT TYPE* menu (Figure 9.14).
3. Pick points "Pnt1," "Pnt2," and "Pnt3."
4. Select *Done Sel, Done,* and *OK.*

TABLE 9.2 COORDINATES FOR THE DATUM POINTS.

| POINT # | X | Y | Z |
|---|---|---|---|
| 0 | 0.0 | 0.0 | 0.0 |
| 1 | 0.0 | 6.93 | 0.0 |
| 2 | 6.0 | 3.5 | 0.0 |
| 3 | 12.0 | 0.0 | 0.0 |
| 4 | 6.0 | 0.0 | 0.0 |

The software will create the datum curve. The datum curve, constructed by using the preceding steps, is shown added to the model in Figure 9.17. You will need to repeat the steps four more times to place the rest of the datum curves. In doing so, follow this procedure:

1. Repeat steps 1 through 4 but replace step 3 with "Pnt0" and "Pnt1."
2. Repeat the steps a second time replacing step 3 with "Pnt0," "Pnt4," and "Pnt3."
3. Follow the procedure one more time replacing step 3 with "Pnt4" and "Pnt2."
4. For the last time, replace step 3 with "Pnt2" and "Pnt0."

The model, with all the curves added, is shown in Figure 9.18. Save the part, it will be completed in chapter 11.

### 9.4.3 TUTORIAL # 49: DATUM CURVES FOR AN AIRFOIL SECTION

In the construction of the datum curve for the part "2dframe," a datum curve was placed through all the points using the *Single Rad* option. Thus, a single radius was used to pass the curve through the prescribed points. However, in some cases, the geometry of the part is dictated by an equation. In such cases, datum points may be created by reading their locations into the part via a file. Then the *Spline* option may be used to place a datum curve through the points.

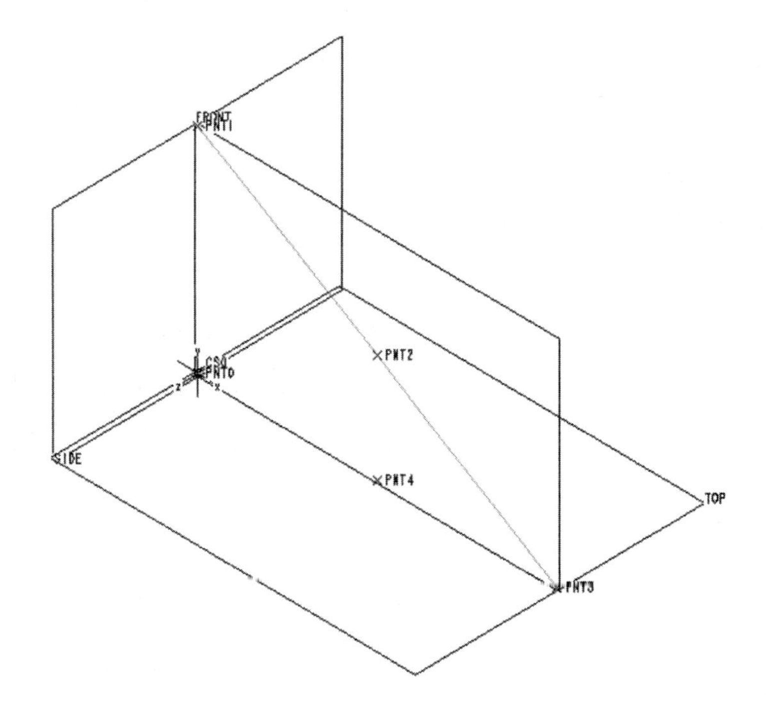


FIGURE 9.17 *The first curve is added to the model by passing through the points "Pnt1," "Pnt2," and "Pnt3."*

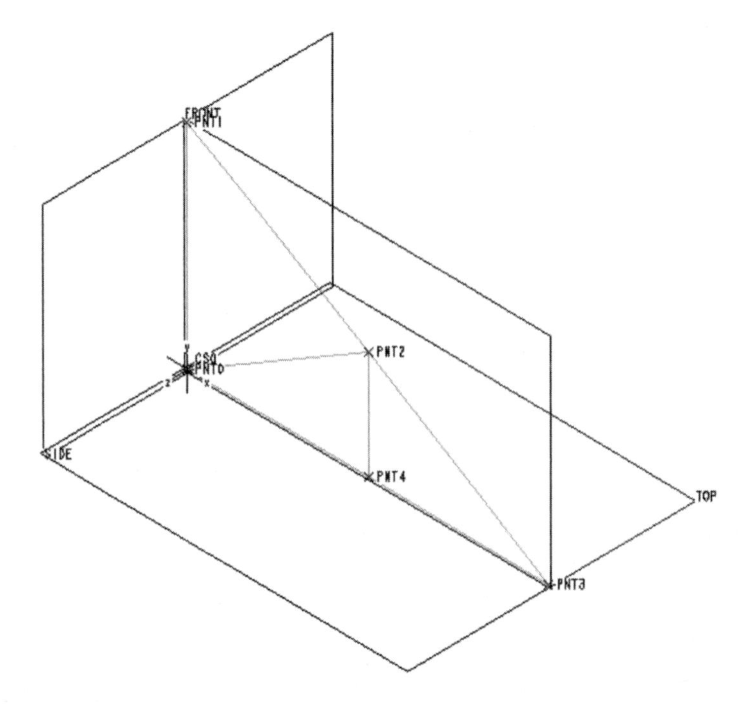


**FIGURE 9.18** *Using the appropriate procedure, the rest of the curves may be added.*

In this tutorial, we wish to use this option to construct two datum curves through several closely spaced points. The part under consideration is the section of an airfoil. We will construct the upper and lower ribs of the airfoil section. In chapter 11, the ribs will be completed using the *Sweep* option.

In order to speed up the construction of this part, we have bundled the start part "Wing" with this book. This start part contains an English version of the start part from section 9.1. You need to make sure that you have access to this file. Retrieve this part.

The airfoil under consideration is an NACA 4412 section[1] to be used for a model in a wind tunnel. Data for this and other airfoil sections are available in Abbott and Doenhoff (1959). The total number of datum points is 35, which describe an airfoil three inches in length. Because of the number of points in the list, we have bundled the file "Cs0.pts" that contains these datum points.

Add the datum points by selecting *Feature, Create, Datum, Point,* and *Offset Csys.* Select the *With Dims* option and pick on the coordinate system label. Select *Cartesian* and *Read Points.* Then enter the name "Cs0." The part, with the added datum points, is shown in Figure 9.19. In the part, points 1 through 18 run along the upper part of the airfoil from the leading to the

[1] See the Bibliography for this reference.

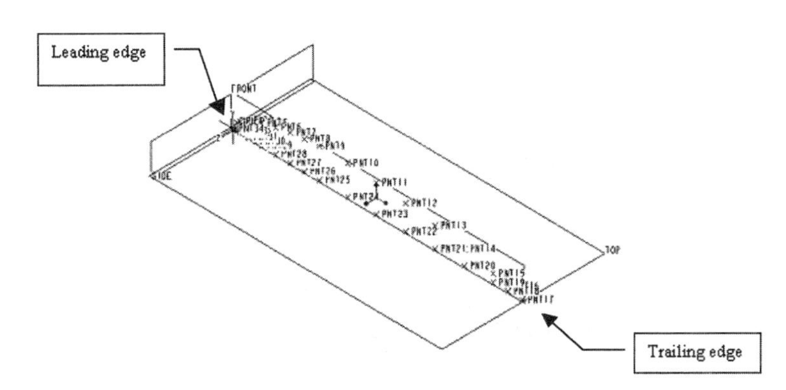


FIGURE 9.19 *The airfoil section begins to take shape after loading the datum points.*

trailing edge. Points 19 through 35 run along the lower part of the airfoil from trailing to leading edge.

Because of the geometry of the trailing and leading edges of the airfoil, we need to construct a separate trajectory for the top and bottom of the airfoil. Otherwise, the datum curve will not be able to regenerate because it will intersect at the trailing and leading edges.

Select *Feature, Create, Datum,* and *Curve.* Then choose *Thru Points* and *Done.* From the *CONNECT TYPE* menu (Figure 9.14), select *Spline* and *Single Point.* At this point, you will most likely need to *Zoom in* to the model. Place the curve through each of the points by selecting each point individually. Start with the first point (Pnt 0) and finish with the last point (Pnt 17). Repeat the process for the lower half. This time start with point 34 and end with point 17.

The airfoil section, with the two datum curves, is shown in Figure 9.20. In the figure, we have removed the display of the datum points for better visibility. Notice in the figure, that because of the finely distributed points, the shape of the airfoil section is accurately represented. Save the part. We will complete the ribs in chapter 11.

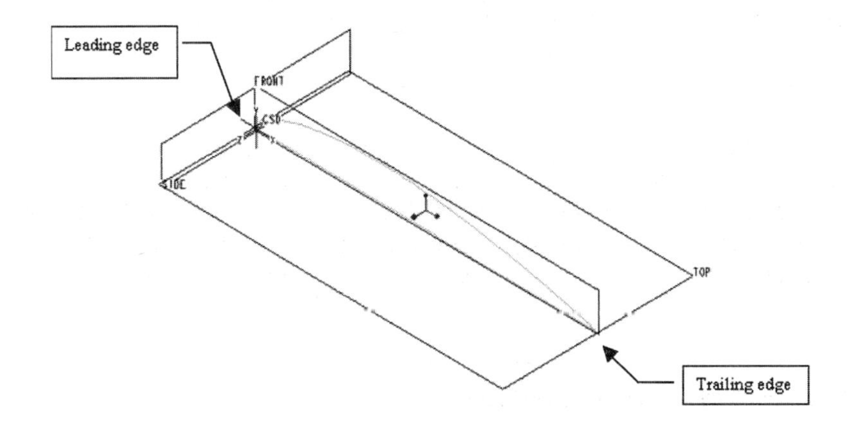


FIGURE 9.20 *The airfoil with the datum curves. The display of the datum points has been turned off.*

## 9.5 SUMMARY

In addition to datum planes, the user may construct datum coordinate systems, points, axes, and curves.

The coordinate system may be used as a reference for datum points. In addition, it may be used as a coordinate system for calculating material properties, such as moments of inertia, or in exporting the model, such as a reference for an IGES translation.

Datum points may be created as a parametric or nonparametric array. If the datum points are to be updated on an individual basis in the future, then they should be created as a parametric array.

Datum curves may be placed through datum points. These curves may be used as trajectories for sweeps. The use of datum points and datum curves allows the user to construct structural components such as frames and trusses.

Datum axes are often helpful in locating additional features in a model. The datum axis may be constructed by using options from the *DATUM AXIS* menu. If a datum may be constructed on a model, then the *Coaxial* option may be used to place a hole. This capability is beneficial in cases where the *Linear* option may not be practical.

### 9.5.1 STEPS FOR CREATING A DATUM AXIS

The options for the actual placement of a datum axis are found in the *DATUM AXIS* menu (Figure 9.2). The general steps for placing an axis are as follows:

1. Select *Feature, Create, Datum,* and *Axis.*
2. Choose the appropriate placement option from the *DATUM AXIS* menu.
3. Pick the appropriate references as required by the option selected from the *DATUM AXIS* menu. For example, if the *Two Planes* option was selected in step 3, then pick the two appropriate planes.

### 9.5.2 STEPS FOR CONSTRUCTING A DATUM COORDINATE SYSTEM

In general, the procedure for constructing a datum coordinate system is:

1. Select *Feature, Create, Datum,* and *Coord System.*
2. Choose the placement option from the *Options* menu (Figure 9.7).
3. Select the type of coordinate system from the choices: *Cartesian, Cylindrical,* and *Spherical.* Choose *Done.*

4. Select the appropriate features on the model, as required by the option chosen in step 2.
5. *Reverse* the directions of the arrows as required.
6. Scroll through the arrows and define each coordinate direction.

### 9.5.3 STEPS FOR ADDING DATUM POINTS TO A MODEL

The general steps for adding datum points is as follows:

1. Select *Feature, Create, Datum,* and *Points.*
2. Pick the placement option from the *DATUM POINT* menu shown in Figure 9.11.
3. Choose either the *With Dims* option or the *Without Dims* option.
4. If a coordinate system is being used to reference the points, select the type of coordinate system. Then select *Done.*
5. Enter the datum point coordinates either from the keyboard (*Enter Points* option) or from a file (*Read Points* option).
6. Edit the coordinate locations, if needed, by using the *Edit points* option.
7. Select *Done.*

### 9.5.4 STEPS FOR CREATING A DATUM CURVE

A datum curve may be placed through a series of points or through other features as defined by the *CRV* (Curve) *OPTIONS* menu (Figure 9.13). In general, the sequence of steps for creating a datum curve is:

1. Select *Feature, Create, Datum,* and *Curve.*
2. From the *CRV OPTIONS* menu, select the appropriate placement option.
3. Select *Spline, Single Rad,* or *Multiple Rad.*
4. Choose the appropriate feature or features as dictated by the option chosen in step 2. Choose *Done Sel.*
5. Choose *Done.*

## 9.6 ADDITIONAL EXERCISES

For the exercises that follow, construct the trajectories using datum coordinates, points, and curves. Use a start part if desired. The trajectories will be used to construct sweeps in the additional exercises to chapter 11.

9.1 Construct a datum curve through the points shown in Figure 9.21. The trajectory may be used to create a model of the bell crank.

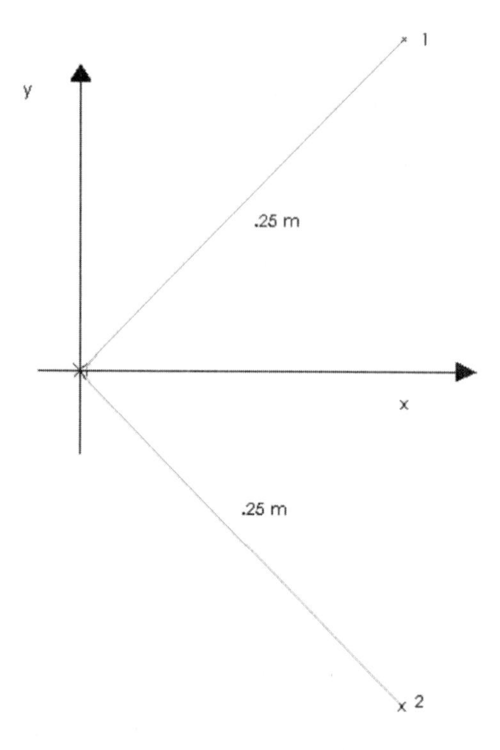


FIGURE 9.21 *Figure for exercise 9.1.*

9.2 Use the *Single Rad* option (0.75 radius) and the given geometry in Figure 9.22 to construct the datum curve through the points of a highway sign. Note the units of this model are in feet.

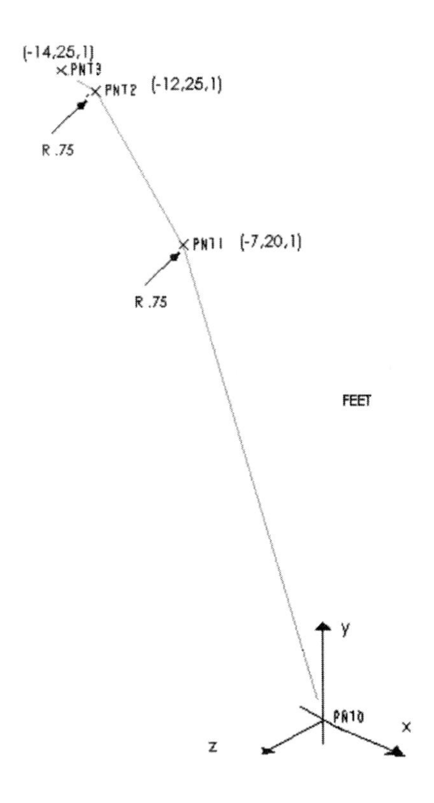


FIGURE 9.22 *Figure for exercise 9.2.*

9.3 The geometry of a flat roof truss is given by:

$$y = A \cos n\pi x.$$

In this equation, $x$ and $y$ are horizontal and vertical points determining the centerline of the truss in feet. $A$ is amplitude describing the maximum vertical height of the truss. The parameter $n$ describes the number of peaks in the interval. Generate a series of points for a 2–foot span of the truss similar to the one shown in Figure 9.23. Save these points in a text file. Construct a datum curve through these points.

**FIGURE 9.23** *Figure for exercise 9.3.*

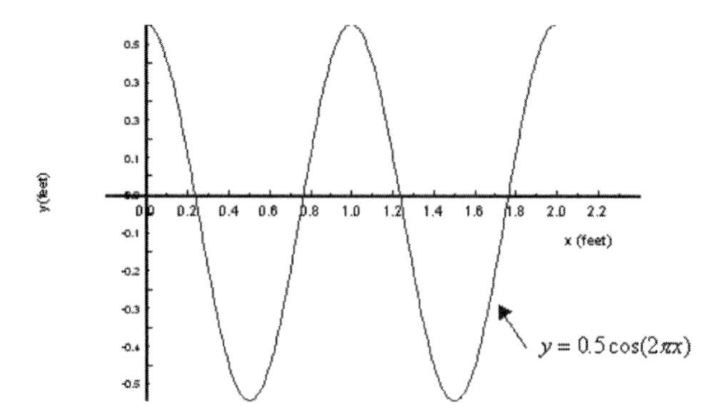


Use $A$ and $n$ from Table 9.3 as determined by your instructor.

**TABLE 9.3** VALUES OF $A$ AND $n$ FOR EXERCISE 9.3

| | $A$ | $n$ |
|---|---|---|
| A | 0.5 | 1 |
| B | 2.0 | 1 |
| C | 0.5 | 2 |
| D | 2.0 | 2 |

9.4. Use the *Through Edge* option and a datum axis to place the hole in the rest shown in Figure 9.24.

**FIGURE 9.24** *Figure for exercise 9.4.*

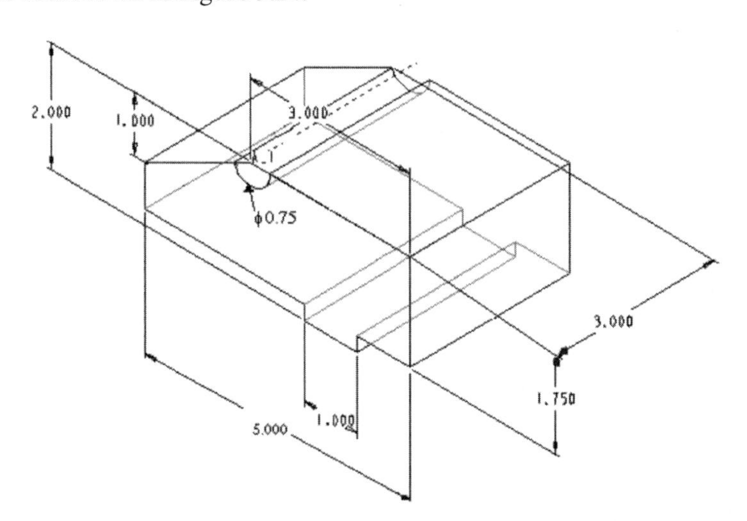

9.5 Construct the model of the key collar illustrated in Figure 9.25. Use a *Datum Axis* to locate the hole.

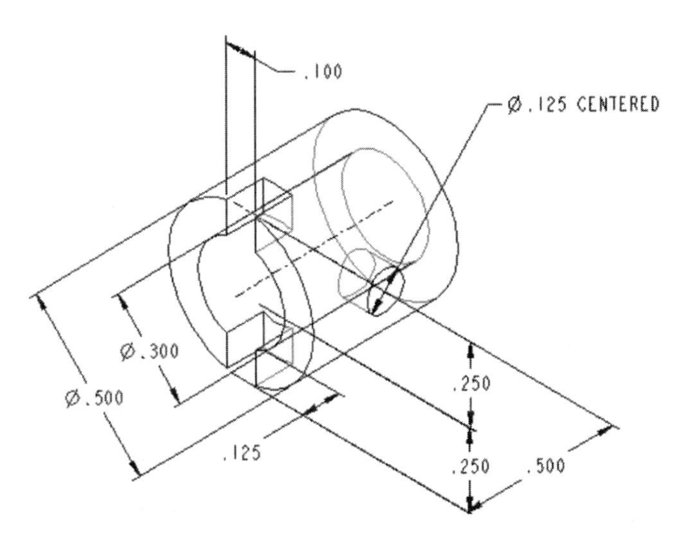


FIGURE 9.25 *Figure for exercise 9.5.*

9.6 Complete the model of the key bushing given in Figure 9.26. Use a *Datum Axis* to locate the hole.

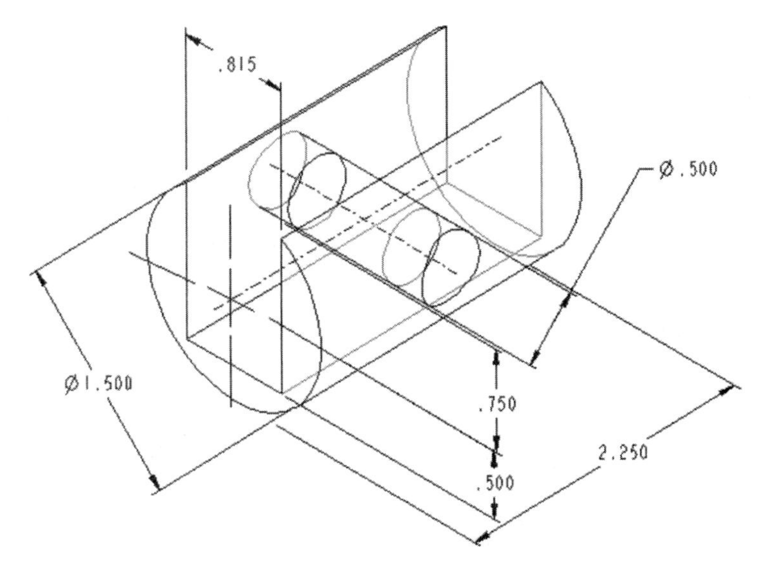


FIGURE 9.26 *Figure for exercise 9.6.*

9.7 Complete the model of the external lug by using a *Datum Axis* to locate the hole. Consult Figure 9.27 for the geometry and location of the hole.

FIGURE 9.27 *Figure for exercise 9.7.*

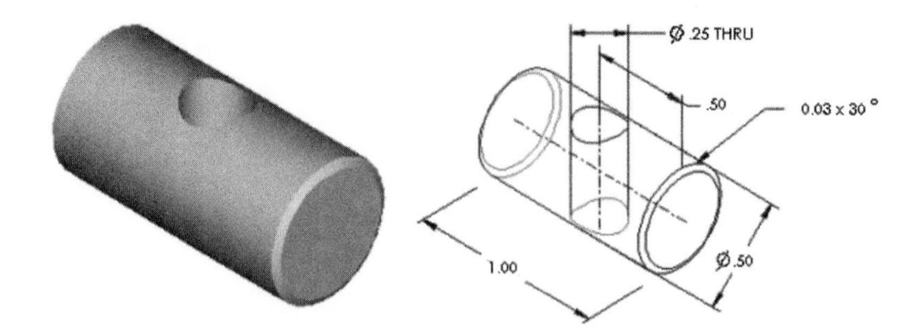


9.8 Using the geometry in Figure 9.28, create a model of the fink truss (half section).

FIGURE 9.28 *Figure for exercise 9.8.*

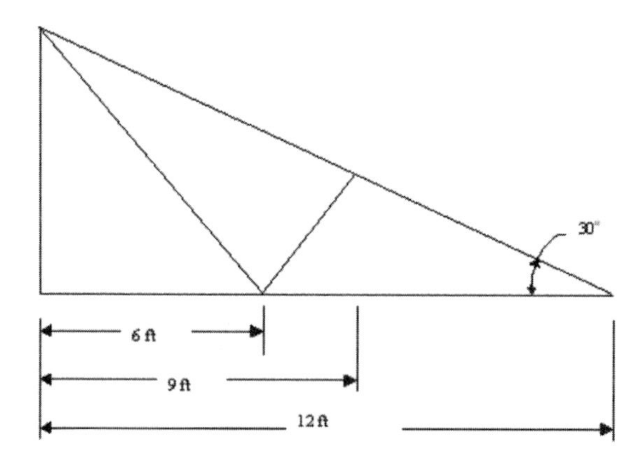


9.9 Create a model of the howe roof truss (half section) shown in Figure 9.29.

FIGURE 9.29 *Figure for exercise 9.9.*

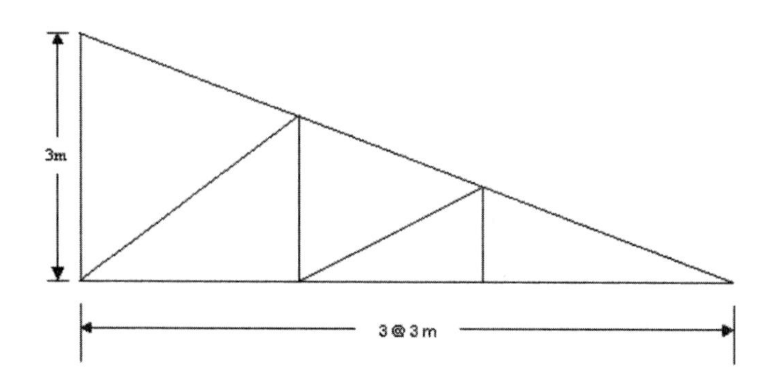

9.10 For the bridge truss (half section), construct a part with datum curves using the geometry shown in Figure 9.30.

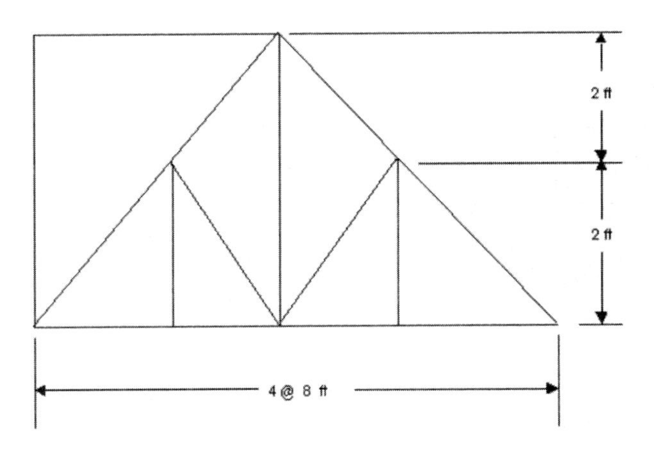


**FIGURE 9.30** *Figure for exercise 9.10.*

# CHAPTER 10

# The Revolve Option

## Introduction and Objectives

The *Revolve* option is used to construct a feature whose section may be revolved around a centerline. Thus, adding a revolved section to a base feature requires a sketch and a centerline. Actually, we have already constructed a revolved section. For example, both the neck and flanges in chapters 5 and 6, respectively, are sections that were created by revolving a sketch about a given centerline.

Revolved sections may be solid or thin in an analogous manner to solid and thin protrusions. A *Cut* may also be revolved. In this chapter, we examine three models. The first model is constructed using the *Solid* option, whereas the second is created using the *Thin* option. The third model is a thin revolved parabolic section.

Parabolic (conic) sections were discussed in some detail in chapter 2. The third part in this chapter serves to illustrate two points. The first is the construction of a conic section, and the second is the creation of a more advanced thin revolved section.

The objectives of this chapter are:

1. At the end of the chapter, the reader will be able to create solid and thin features using the *Revolve* option.
2. In addition, the reader will be able to construct a conic section using the *Adv Geometry* option and revolve the section.

## 10.1 REVOLVED SECTIONS

For a revolved section, Pro/E allows the user to choose the angle of rotation with the *Variable* option, for any angle less than 360°. When using the *Variable* option, the user needs to input the value of the angle of rotation. Options are provided for preset revolution angles; these preset values are 90°, 180°, 270°, and 360°. The amount of rotation can be defined by revolving to a known point, vertex, or plane. These revolutions can be accomplished with the *UpTo Pnt/Vts* and *Upto Plane* options. See Figure 10.1 for the menu containing these options.

A revolved section can be constructed either as a *Solid* or as a *Thin*. The option to select a *Solid* or *Thin* is given in the *SOLID OPTIONS* menu, where the option for a revolved section may also be found.

For the *Revolve* option, the section can only be sketched on one side of the centerline. If you sketch on both sides, then Pro/E will issue an error message.

### 10.1.1 TUTORIAL # 50: A PULLEY WHEEL

Often, part of an entire model can be constructed by revolving a section about a centerline. The pulley wheel shown in Figure 10.2 is such a model. The figure shows internal geometry of the model (the shaded area). In this particular case, by revolving this geometry, the entire model of the pulley can be constructed.

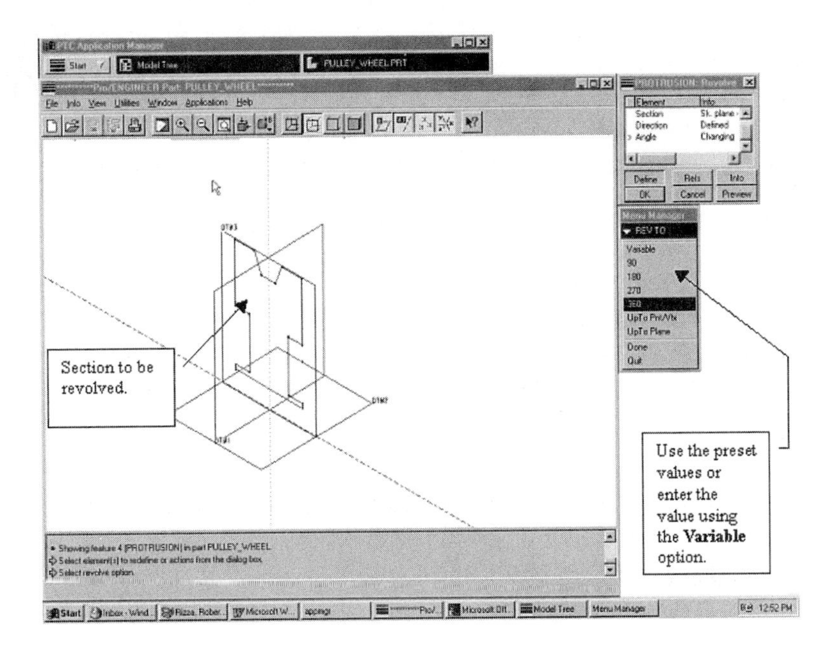


FIGURE 10.1 *The Pro/E interface with a section of pulley wheel about to be revolved. The Pro/E* REV TO *options menu is used to indicate the angular measurement of revolution.*

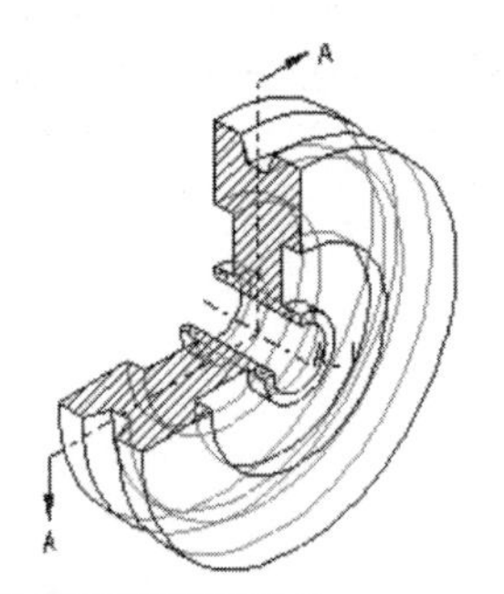


**FIGURE 10.2** *A pulley wheel with an offset cut section. This section shows the interior geometry of the wheel. This geometry can be used to generate the model.*

Let us construct the pulley wheel shown in Figure 10.2. Create a part called "PulleyWheel" and load the default datum planes. Select *Feature, Create,* and *Protrusion.* From the *SOLIDS OPTIONS* menu, choose *Revolve.* We will make a solid wheel, so click on *Solid.* Use the *One Side* option and click on DTM3 as the sketching plane. Make sure the red arrow points into the screen. Select DTM2 as the top. Pro/E will open the *Sketcher.*

You need to change the line type to centerline and place a horizontal centerline along DTM2. *Align* the line to DTM2. Actually, the location of the centerline is arbitrary, but it is often advantageous to place the datum plane, such that the symmetry of the model is exploited. By placing the centerline along DTM2, the datum planes are located at the center of the wheel.

Sketch the geometry shown in Figure 10.3. Dimension as shown in the figure. The locations of the dimensions have been chosen so that the section will regenerate symmetrically about the edge of the DTM1 plane.

Upon successful regeneration, choose *Done.* The software will reorient the model as shown in Figure 10.1. From the *REV TO* menu, select 360° and revolve the section 360°. The completed model is shown in Figure 10.4.

## 10.1.2 TUTORIAL # 51: CREATING A MUG USING A REVOLVED THIN

*Thin* revolves are constructed just as easily as solid sections. However, the user must give the thickness of the thin section. Let us construct a thin revolve. The part is a coffee mug. Create a part and call it "CoffeeMug." We will create the mug first and add the handle later in chapter 11.

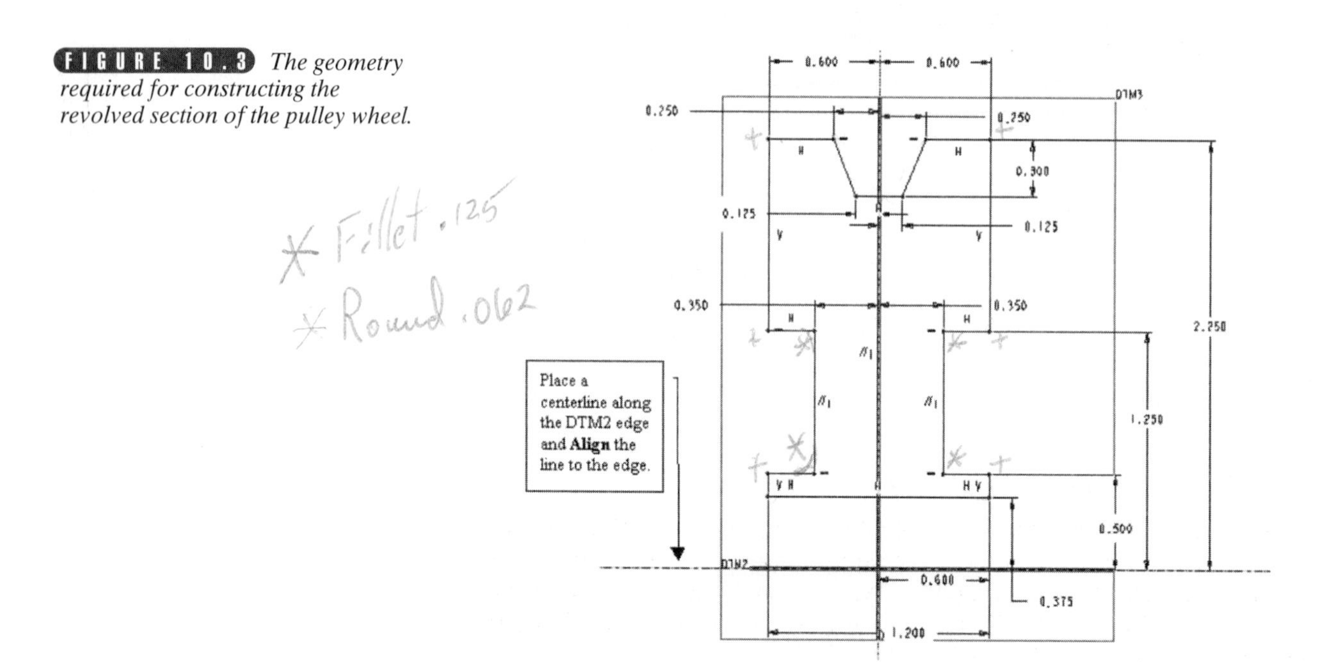


**FIGURE 10.3** *The geometry required for constructing the revolved section of the pulley wheel.*

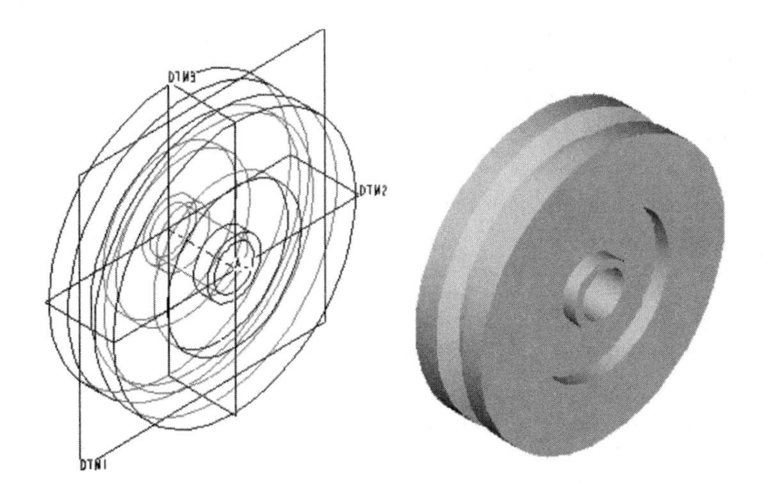


FIGURE 10.4 *An isometric and shaded model of the pulley wheel generated by revolving the section shown in Figure 10.3.*

In order to create the thin revolve, no datum planes are needed. However, when the time comes to add the handle, a sketching plane is required. Therefore, it is imperative that you create datum planes before creating this thin revolve.

After creating the default datum planes, select *Feature, Create, Protrusion,* and *Revolve.* Select *Thin* from the *SOLID OPTIONS* menu. Choose *One Side.* Select DTM3 as the sketching plane and DTM2 as the top. Pro/E will open the *Sketcher.*

Sketch the section as shown. The section is constructed from two lines. One line is horizontal and the other is vertical. Place the horizontal line along the DTM2 edge.

This line is hard to see in Figure 10.5. After revolving, it will become the bottom of the mug. Add a centerline along the DTM1 edge. Align the centerline along the DTM1 edge and align the horizontal line along the DTM2 edge. Dimension the section as shown in Figure 10.5. Upon successful regeneration, choose *Done.*

Define the thickness of the mug by growing the part inward. The arrow should point inward. The value of the thickness is 0.125. *Revolve* the part through 360°.

Use the *Round* option discussed in chapter 8 and place a 0.1 radius round on the top edge of the mug. The mug should appear as shown in Figure 10.6. A shaded model is used to represent the part.

### 10.1.3 TUTORIAL: # 52: A PARABOLIC REFLECTOR

Our last example for the use of the *Revolve* option in constructing surfaces of revolution is a parabolic reflector used in an emergency light. The cross section of the reflector is a parabola, a type of conic. The creation of conic sections in the *Sketcher* was discussed in section 2.9. The reader might want to refer to that section before continuing.

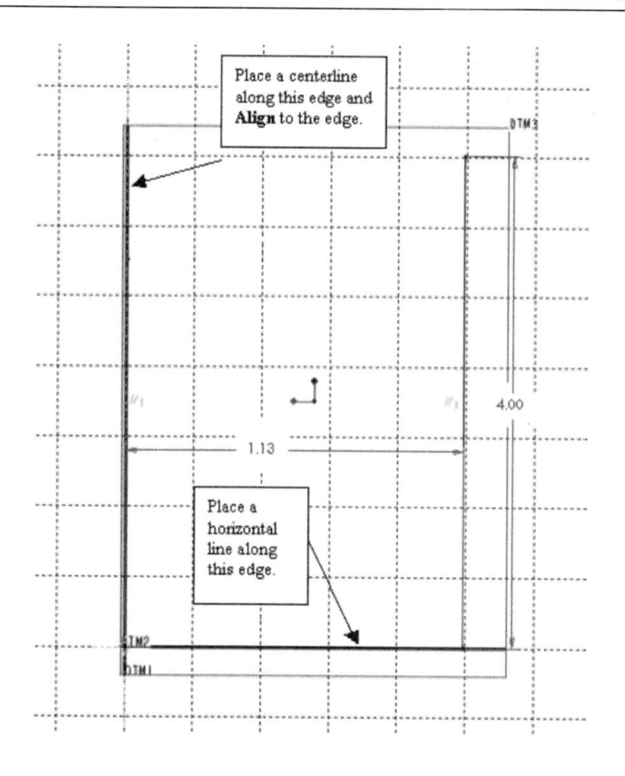


FIGURE 10.5 *The section needed to create the body of the mug.*

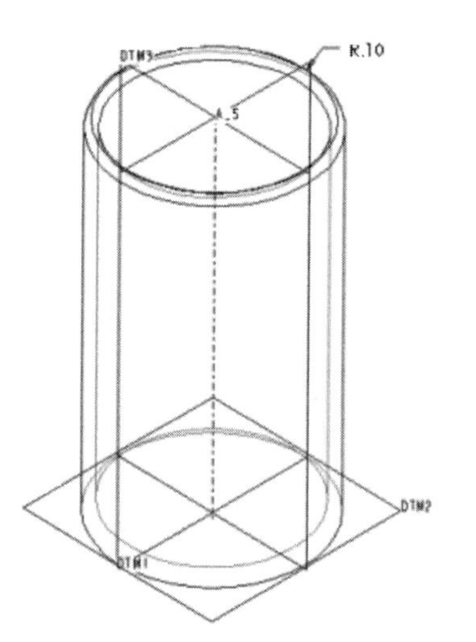

FIGURE 10.6 *After revolving the section given in Figure 10.5, the mug begins to take shape.*

Create a part called "EmergencyLightReflector" and add default datum planes. We will use DTM3 as the sketching plane and construct a parabola. The parabola will then be revolved through 360°.

Select *Feature, Create, Protrusion, Revolve, Thin,* and *Done.* Choose *One Side.* Make sure the arrow points into the screen. Select DTM3 as the sketching plane and DTM2 as "Top." Pro/E will open the *Sketcher.* Change the line type to centerline, place, a line along DTM1 and align. The centerline will become the axis of revolution.

Choose *Adv Geometry* from the *GEOMETRY* menu (Figure 2.20). From the *ADVANCED GEOMETRY* menu, select *Conic.* Because the section is going to be revolved, only half of the parabola needs to be drawn. We will sketch the conic section to the right of the centerline.

At this time, we can also sketch the support for the light bulb that fits into the reflector. The support is simply created by sketching a vertical and horizontal line. Sketch these two lines as shown in Figure 10.7. The tip of the vertical line will form the starting point for the conic. Align the tip of the vertical line to DTM2 and the end of the horizontal line to DTM1.

In Pro/E, a conic is drawn by locating the starting and endpoint of the conic. This will create a line. Then, some curvature is given to the line. Select the tip of the vertical line as the start point for the conic. This start point, as well as the end of the conic, is shown in Figure 10.7.

After locating the start and endpoints of the line, "Tweak" the conic; that is, add curvature to the line by dragging your mouse sideways. It does not matter how much curvature you add to the conic, because we will dimension the sketch shortly. After adding curvature, your sketch should look something like the one shown in Figure 10.7.

The reflector has a thin lip along the edge of the parabola. We can create this lip by adding a short horizontal line that intersects with the endpoint of the conic as shown in Figure 10.8.

In order to dimension the conic, we will use the "rho" method discussed in section 2.9. This method requires three dimensions for the conic. The dimensions are a radius defining the curvature, and two angles, one at each end of the conic, describing the tangency of the endpoints.

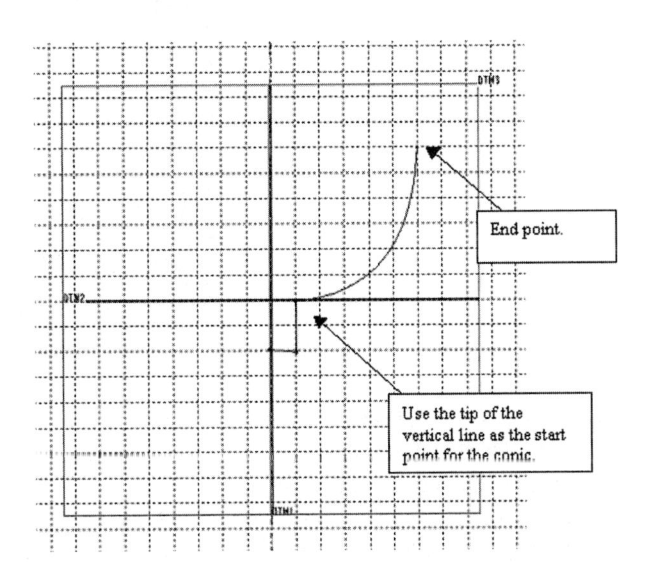


**FIGURE 10.7** *The section for the parabolic reflector may be constructed by using the* Conic *option.*

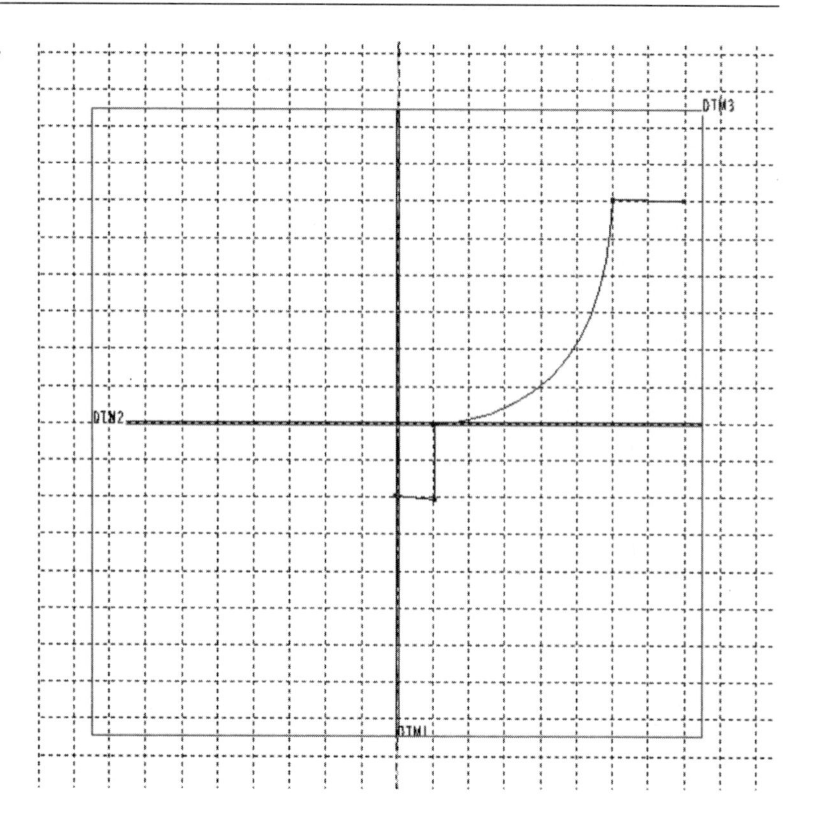


**FIGURE 10.8** *Add a horizontal line to the sketch to create a thin lip upon revolution.*

Now, let us dimension the sketch. Let us begin by dimensioning the tangency of the endpoints. In order to place the 90° dimensions, first select the conic, then the start point of the conic, and finally the vertical line. Place the dimension by clicking the middle button on your mouse.

The 270° dimension can be placed in a similar manner. Select the conic, then the endpoint of the conic, and finally the short line. Place the dimension.

Add the remaining dimensions in the normal manner. Regenerate and modify the dimensions to the values shown in Figure 10.9. Note that since the conic section is a parabola, "rho" has a value of 0.5.

Select *Done.* The thickness of the part is 0.1 inch. *Revolve* the section 360°. The reflector is shown in Figure 10.10 using a shaded solid model to represent the part.

The reflector can be completed by placing a hole through the support for the bulb. The support also needs two keyways, which are used in practice, to lock the bulb holder in place. Both the keyway and the hole may be created as a single feature using a cut, or as separate features, using a hole and two cuts. Obviously, the first approach is quicker.

We have rotated the model in Figure 10.11 in order to show the support more clearly. You may want to rotate the model as well. If you decide not to rotate, be sure to use the *Query Sel* option in order to select the correct sketching plane.

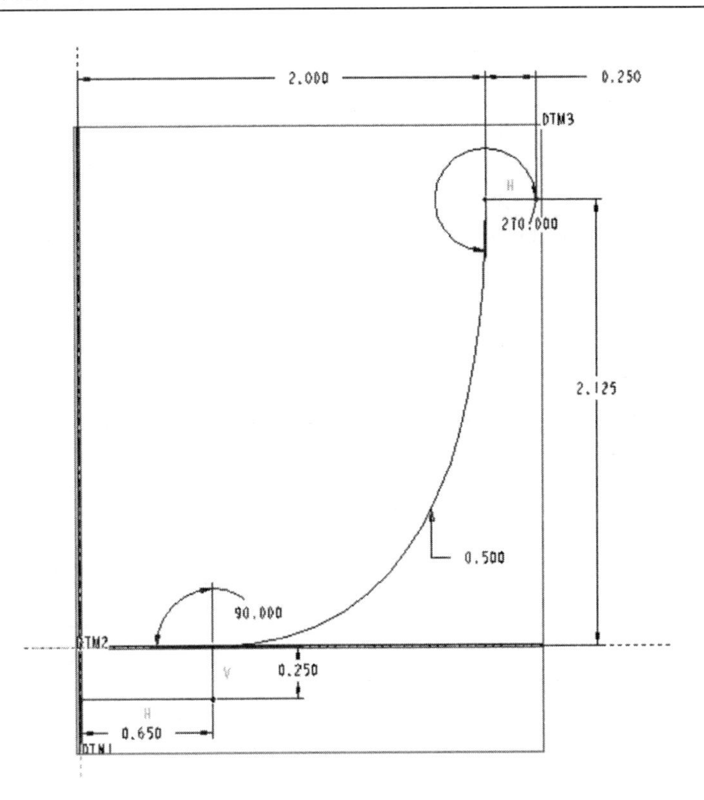


FIGURE 10.9 *The dimensions for the conic section.*

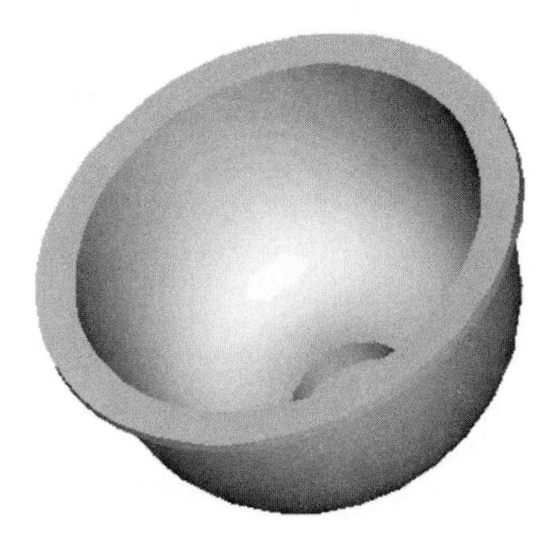

FIGURE 10.10 *The reflector after revolving the section in Figure 10.9 through 360°.*

Select *Feature, Create, Cut, Extrude, Solid,* and *Done.* Use the *One Side* option and select the surface shown in Figure 10.11 as the sketching plane. Make sure that the extrusion arrow is pointing toward the reflector and choose DTM3 as "*Top.*"

After Pro/E has loaded the model in the *Sketcher,* use Figure 10.12 and sketch the cut as shown. Begin by sketching a circle whose center is aligned with the edges of DTM3 and DTM1. Then sketch the keyway. Use *Geom Tools* to find the appropriate intersections and delete any unwanted part

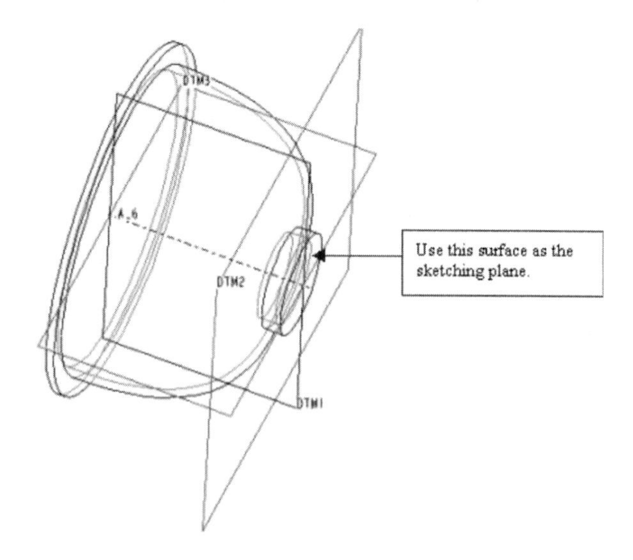


**FIGURE 10.11** *The parabolic reflector rotated to show the base.*

of the circle. Dimension the sketch using the values shown in Figure 10.13. In Figure 10.13, we have *Zoomed in* to show the detail and dimensions of the cut.

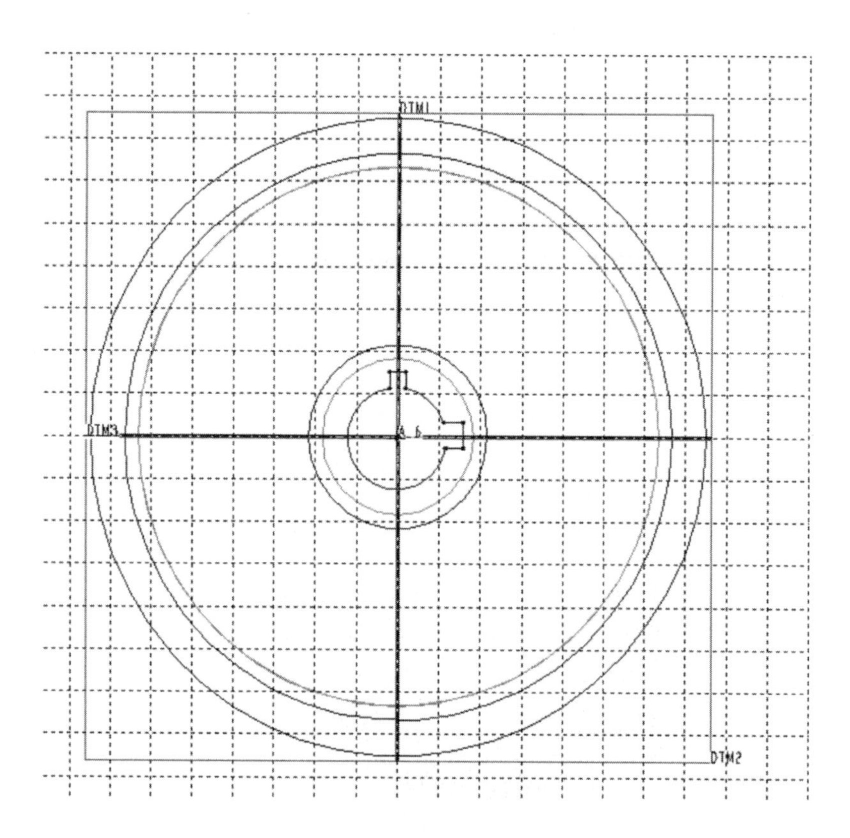


**FIGURE 10.12** *The cutout for the bulb holder.*

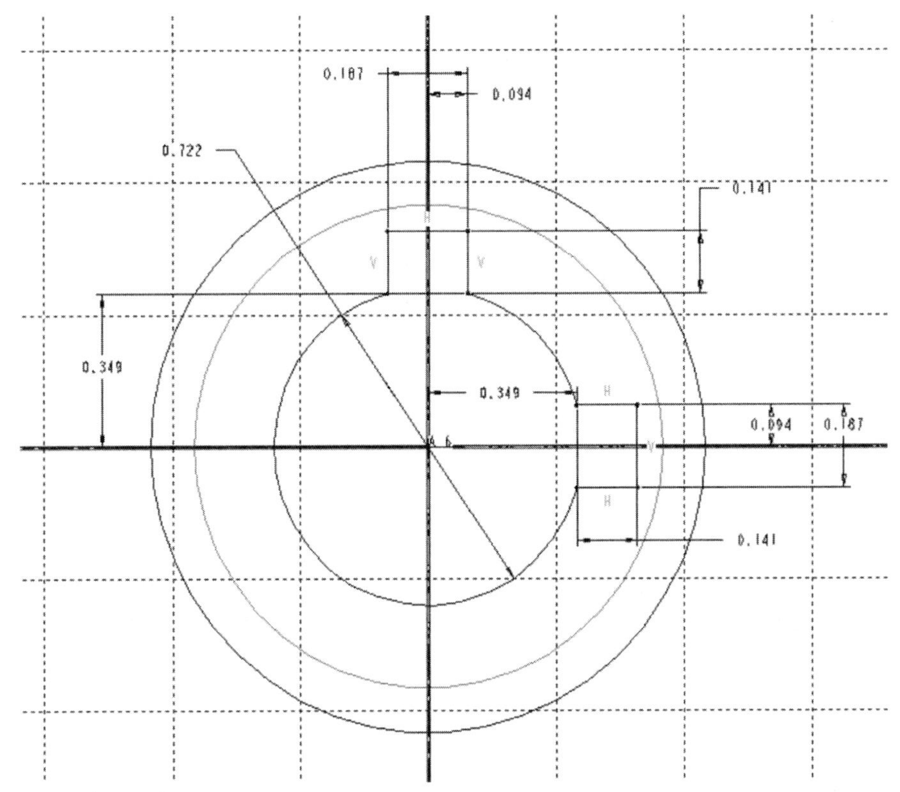


FIGURE 10.13 *The dimensions needed for regenerating the cutout.*

After successful regeneration, select *Done.* Cut the part inward. Use the *Thru All* option. The completed part is shown rotated in Figure 10.14. Save the part.

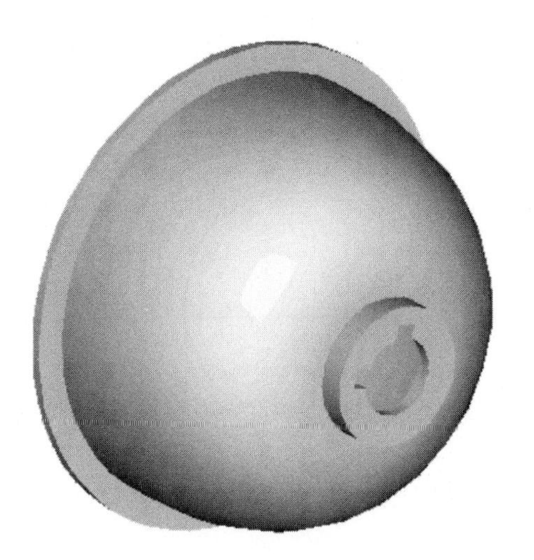

FIGURE 10.14 *The parabolic reflector after adding the cutout for the light bulb.*

## 10.2 SUMMARY AND STEPS FOR USING THE REVOLVE OPTION

Some models have features that may be constructed by revolving a section about an axis of revolution. In Pro/E, the option for carrying out such a task is called the *Revolve* option. In using this option, the user must define a sketching plane. Once a sketching plane is defined and the model reoriented in the *Sketcher,* the cross section is sketched. A centerline must be placed by the user. This centerline acts as the axis of revolution. The user must define the angular dimension through which the section is revolved by entering any value from 0° to 360° or by using the predefined values in the *REV TO* menu (Figure 10.1).

The general sequence of steps for constructing a revolved section is:

1. Select *Create* and *Feature.*
2. Choose *Protrusion* or *Cut.*
3. Choose either the *Solid* or *Thin* option.
4. Select *Revolve.*
5. Choose the sketching and orientation plane.
6. In the *Sketcher,* place a centerline and sketch the cross section.
7. *Dimension, Regenerate,* and *Modify.*
8. Select or enter the angular dimensions for the revolve using the *REV TO* menu (Figure 10.1).
9. Choose *OK* or *Preview* the model.

## 10.3 ADDITIONAL EXERCISES

10.1 Use the section shown in Figure 10.15 to create the model of the roller.

FIGURE 10.15 *Figure for exercise 10.1.*

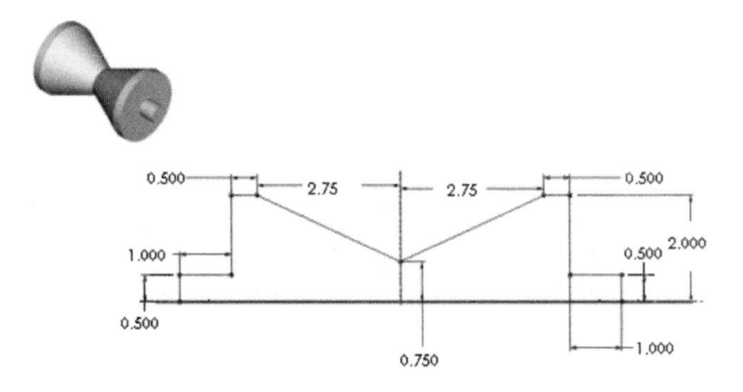

10.2 Use the section shown in Figure 10.16 to construct the model of the baseball bat.

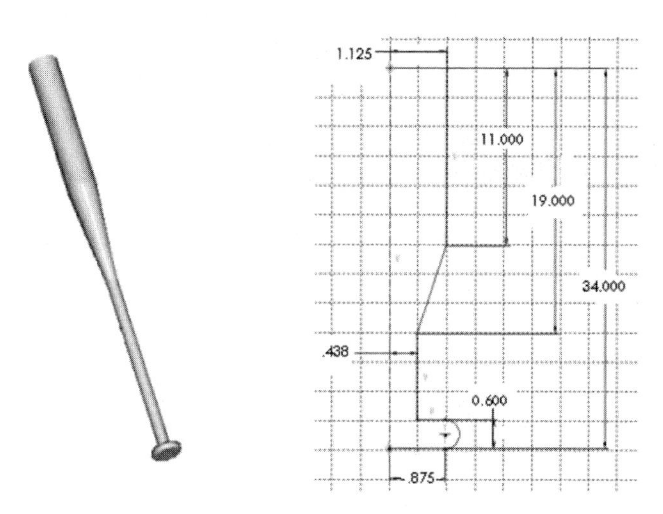


FIGURE 10.16 *Figure for exercise 10.2.*

10.3 Use the geometry in Figure 10.17 to create the model of the doorknob.

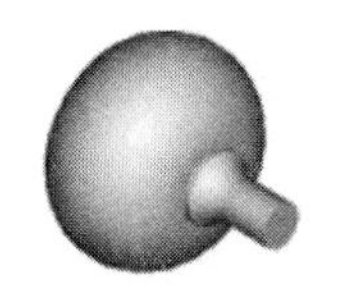


FIGURE 10.17 *Figure for exercise 10.3.*

10.4 Use the *Revolve* option and the given sketch in Figure 10.18 to create the model of the handle ball. The hole has a diameter of 0.20 inches and 0.25 inch deep.

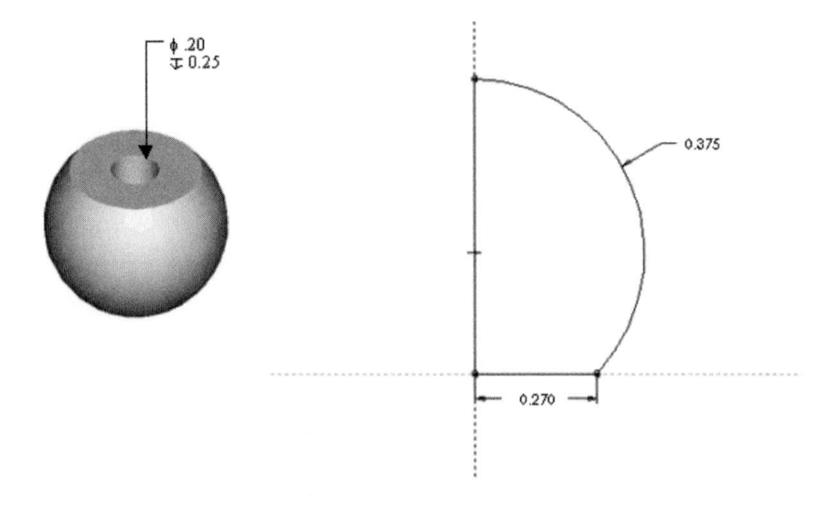


**FIGURE 10.18** *Figure for exercise 10.4.*

10.5 Construct the steady rest base, shown in Figure 10.19, using the dimensions provided.

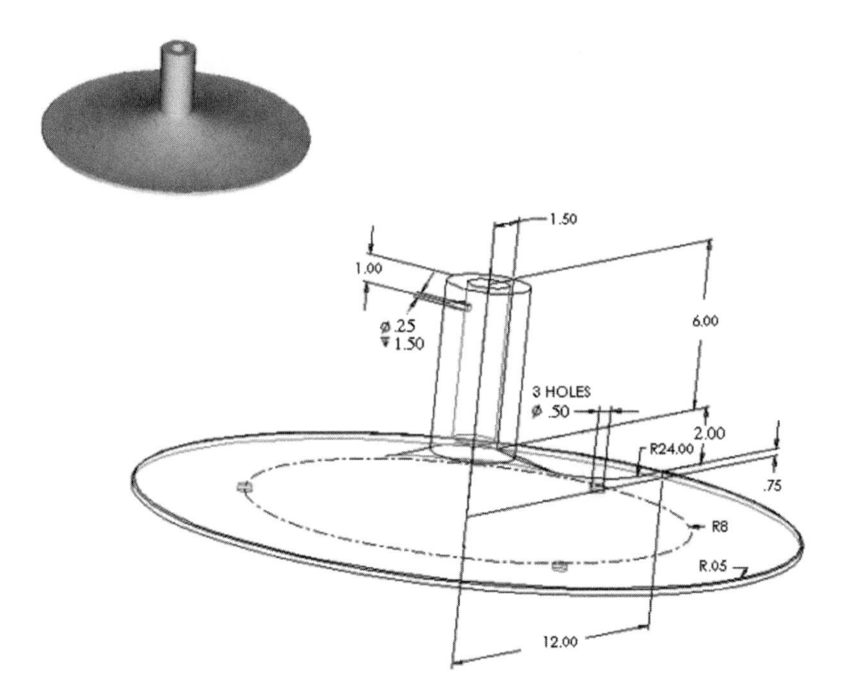


**FIGURE 10.19** *Figure for exercise 10.5.*

10.6 The ANSI standard Venturi nozzle shown in Figure 10.20 has an elliptic inlet (0.400 rho). Generate the geometry and revolve the section. The part is a thin with a thickness of 0.1 inches.

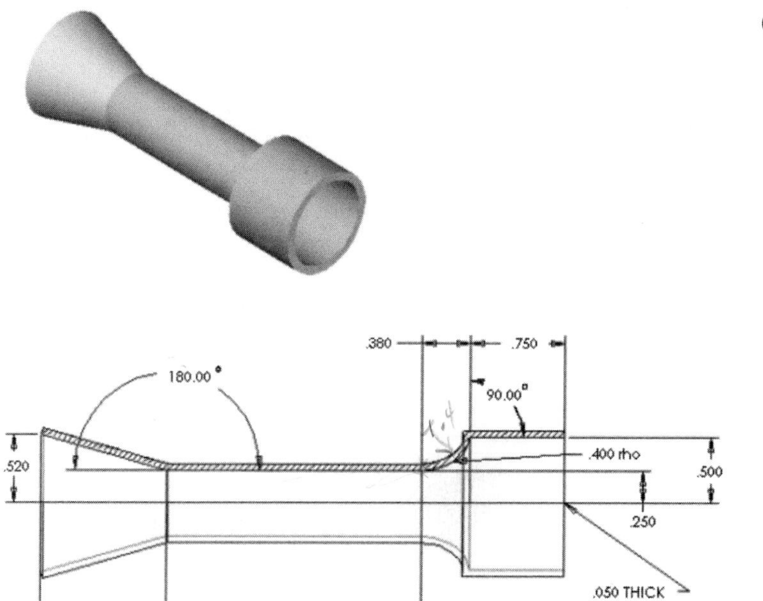


FIGURE 10.20 *Figure for exercise 10.6.*

10.7. Use a 360° revolution and the given section in Figure 10.21 to create the base feature for the fixed bearing cup. Add the three equally spaced holes and the round as shown for this metric part.

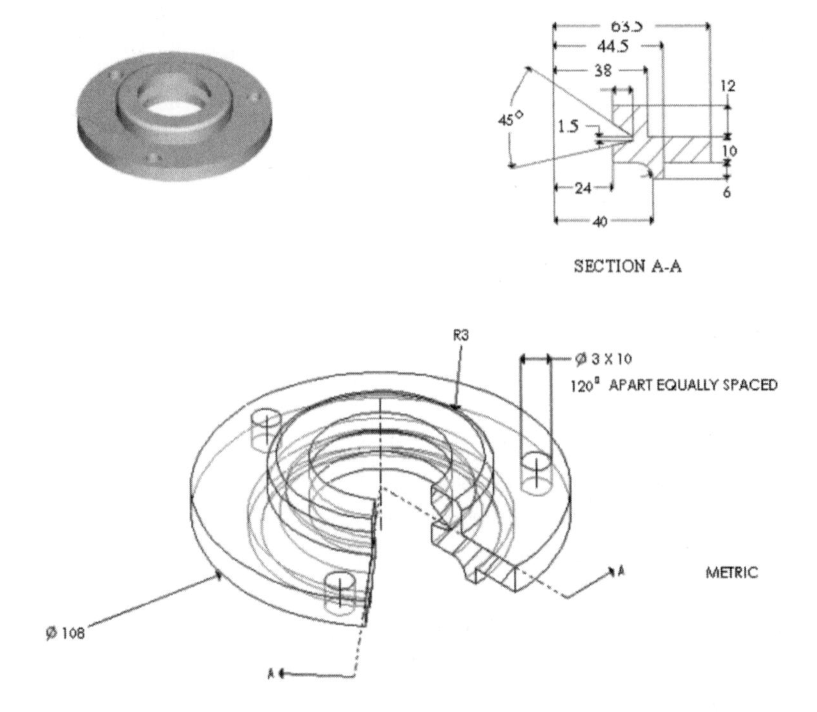


FIGURE 10.21 *Figure for exercise 10.7 (metric).*

10.8 For the round adapter, which is a metric part, use the *Revolve* option to construct the base feature. Add the slots and holes as required. The part is shown in Figure 10.22.

FIGURE 10.22 *Figure for exercise 10.8.*

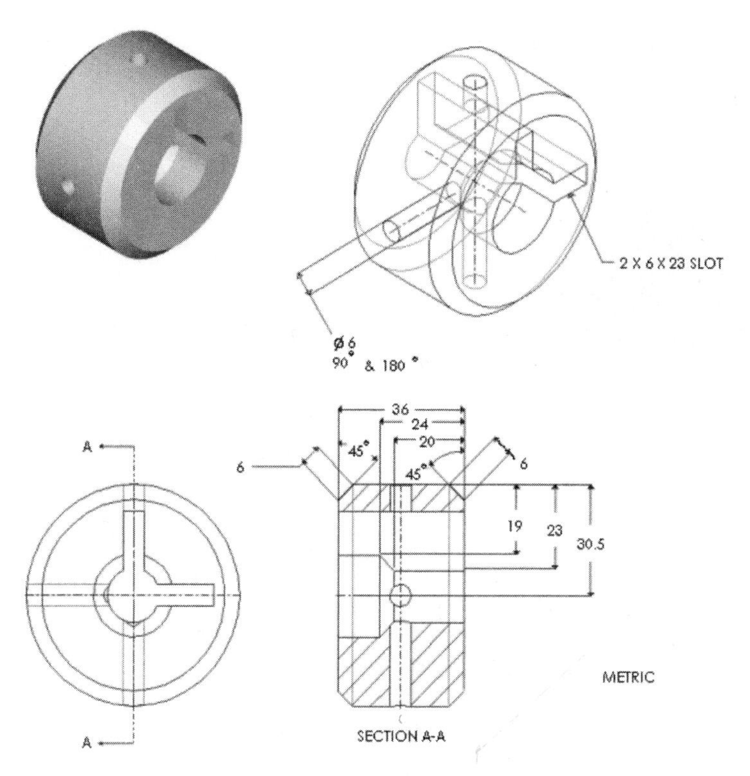


10.9 Construct the tapered portion of the taper collar using the section shown in Figure 10.23. Add the rest of the features as necessary.

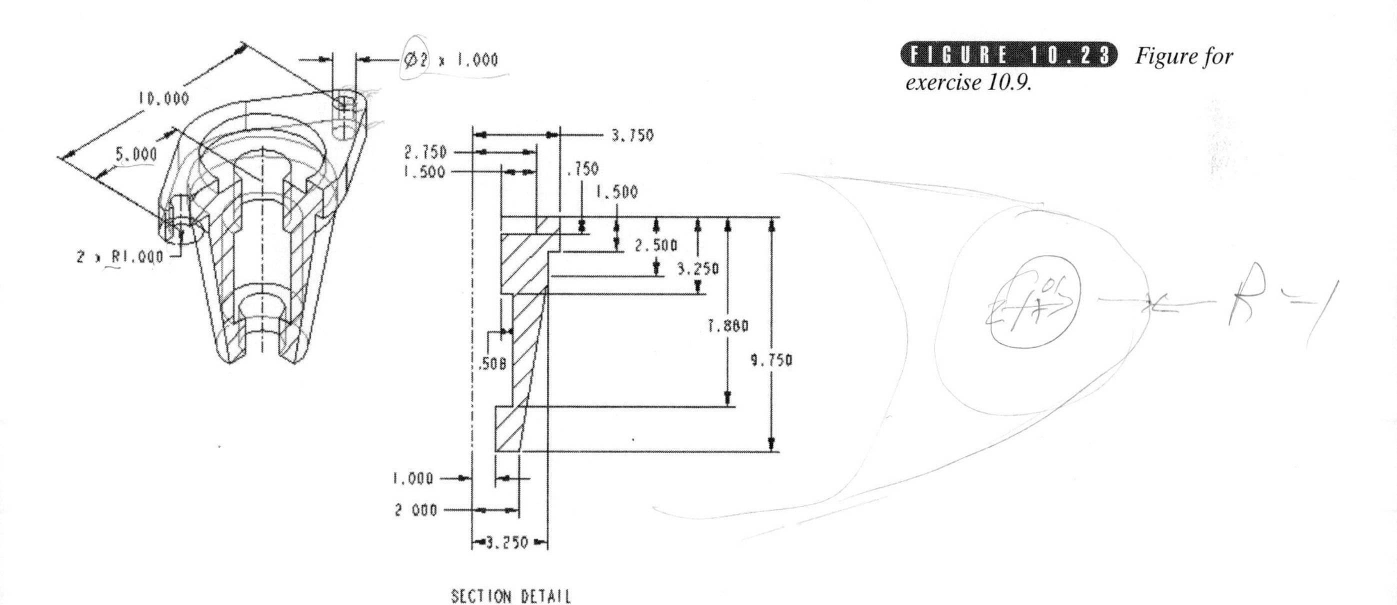


FIGURE 10.23 *Figure for exercise 10.9.*

10.10 Construct the machine wheel, using Figure 10.24 as a reference.

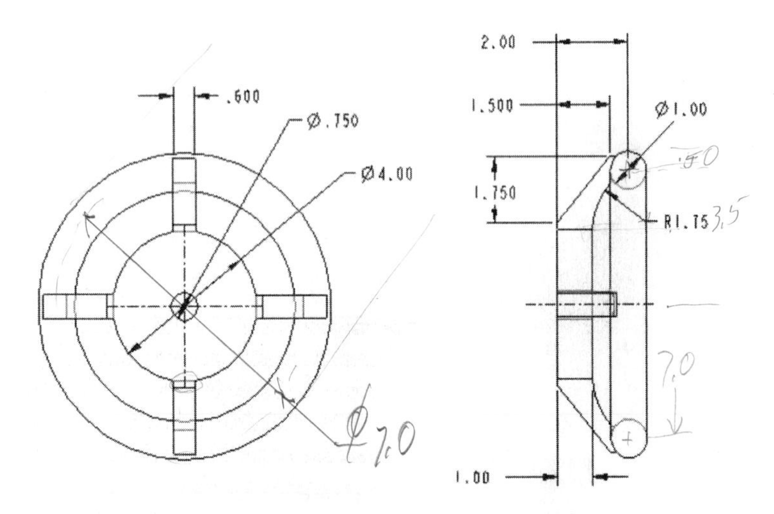


**FIGURE 10.24** *Figure for exercise 10.10.*

10.11 Construct base of the ladle using the *Revolve* option. Use Figure 10.25.

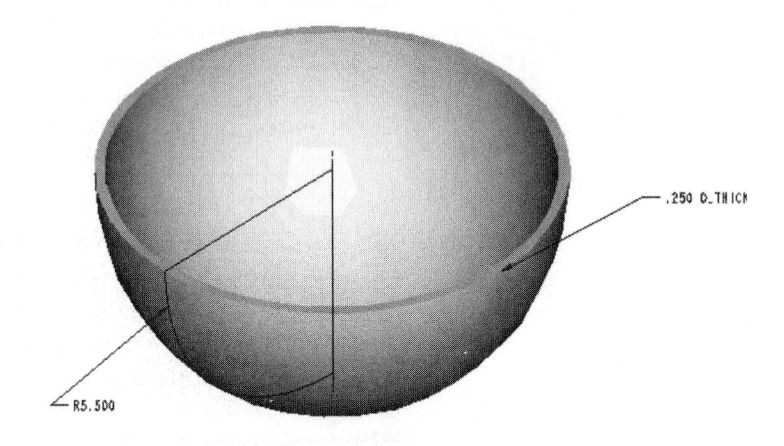


**FIGURE 10.25** *Figure for 10.11.*

10.12 Create the model "FloodLightCover." Use the *Revolve* option and Figure 10.26.

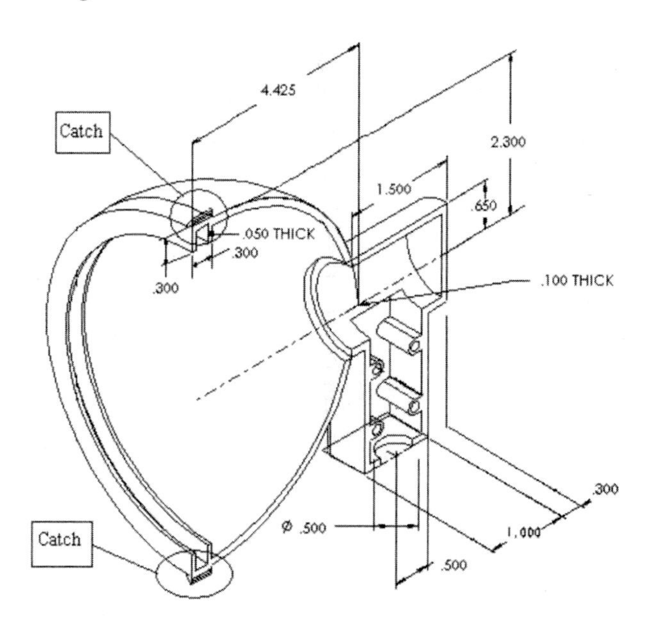


FIGURE 10.26 *Figure for exercise 10.12.*

10.13 Add the "Catch" and the four bosses to the model of the flood light cover as shown in Figure 10.27.

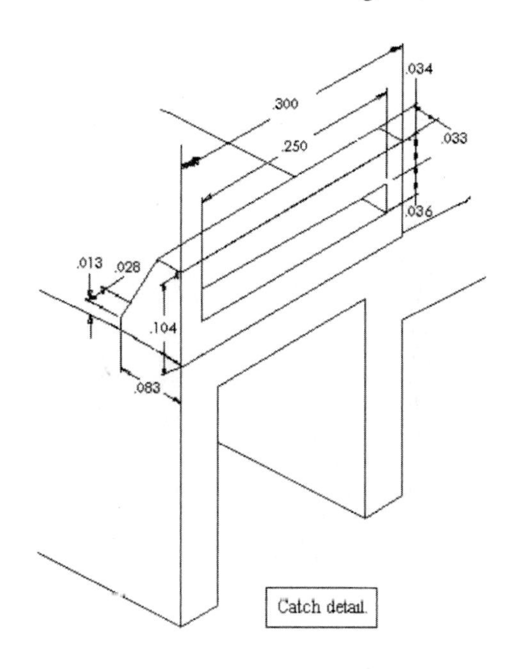


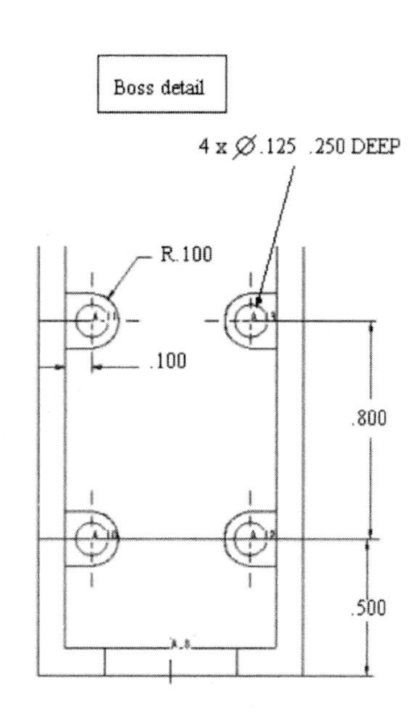


FIGURE 10.27 *Figure for exercise 10.13.*

10.14. Create a mirror copy of the flood light cover as shown in Figure 10.28. Add the counter bore holes and the latch. The geometry of the latch is shown in Figure 10.29.

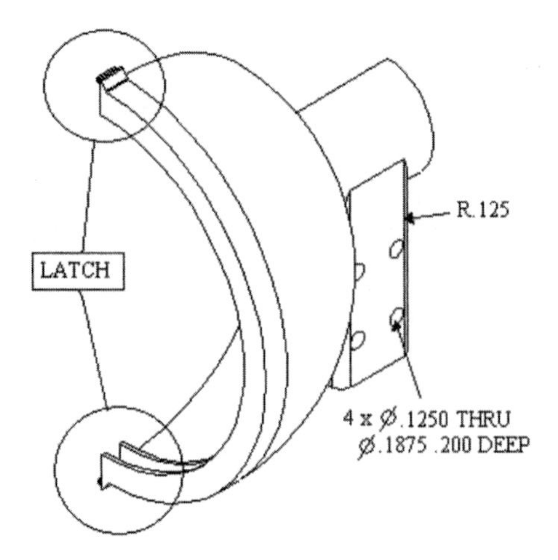


FIGURE 10.28 *Figure for exercise 10.14.*

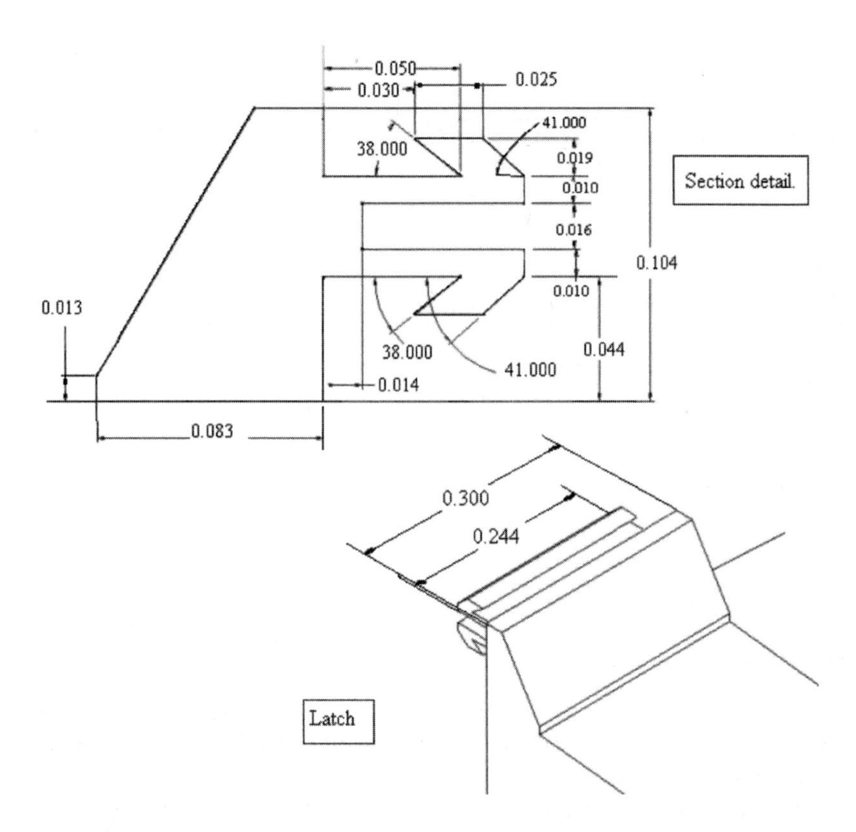


FIGURE 10.29 *An additional figure for exercise 10.14.*

# CHAPTER 11

# FEATURE CREATION WITH SWEEP

## INTRODUCTION AND OBJECTIVES

Many parts have features that are defined along a trajectory. Such features can be created in Pro/E by using the *Sweep* option. In creating a sweep, the user must define the sweep trajectory as well as the section.

In Pro/E, a sweep can be a protrusion (material added to a model), or a cut (material removed from the model). Three-dimensional sweeps may be created by using a spline as the trajectory of the sweep. A *Helical Sweep* can be used to construct springs and parts with similar geometry. A helical sweep is produced by moving a section along a helical trajectory.

The objectives of this chapter are:

1. The reader will develop the ability to construct a sweep trajectory or select a trajectory from suitable features on a model.
2. Using a *Helical Sweep,* the reader will attain the aptitude for constructing springs and threads.
3. The reader will be able to use datum curves as a sweep trajectory.

## 11.1 THE SWEEP OPTION

Sweeps can be used to create features that have a cross section defined along a trajectory, provided the trajectory and the cross section of the protrusion can be easily defined.

The trajectory of the sweep can be defined by either selecting existing edges and/or curves, or by drawing the trajectory (see Figure 11.1). If the

FIGURE 11.1 *The* SWEEP TRAJ *menu used to define the trajectory of the sweep.*

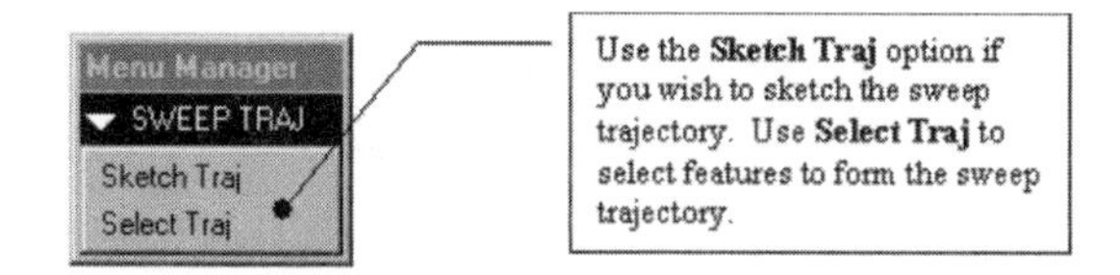


trajectory is sketched, then a sketching plane is required wherein the path or trajectory of the sweep is drawn. The *Sketcher* is used to create the trajectory. After the trajectory is created, the section of the sweep is constructed using the *Sketcher.*

Pro/E chooses the point at which you start your trajectory as the start point for the sweep. In some cases, the start point is required at the end of the trajectory. The start point for the sweep can be changed by using the option *Sec Tools* from the *SKETCHER* menu (Figure 2.2) and *Start Point* from the *SEC TOOLS* menu (Figure 2.11).

On some models with a given geometry of the model and the sweep, the ends of the sweep may not blend into the model because the end trajectory is not perpendicular to the model. Consider the coffee cup shown in Figure 11.2. Notice how the end of the handle does not merge with the cup surface.

If the user wants the sweep to blend into the model, then the *Merge Ends* option must be chosen. If the *Free Ends* option is chosen, the ends are not merged with the surface. The *Merge Ends* and *Free Ends* options are selected from the *ATTRIBUTES* menu shown in Figure 11.3.

### 11.1.1 TUTORIAL # 53: A SIMPLE SWEEP

The handle for the mug from chapter 10 may be created by using the *Sweep* option. Choose *Create, Protrusion, Sweep,* and *Done.* Notice that the sweep can be created as a *Thin* as well as a solid. Choose *Solid* in this case.

We will sketch the trajectory of the sweep so choose *Sketch Traj* from the *SWEEP TRAJ* menu (Figure 11.1). Click on DTM3 to choose that default datum plane as the sketching plane. Choose DTM2 as "*Top*" to define the orientation.

FIGURE 11.2 *Coffee cup shows a sweep constructed with the* Free Ends *option.*

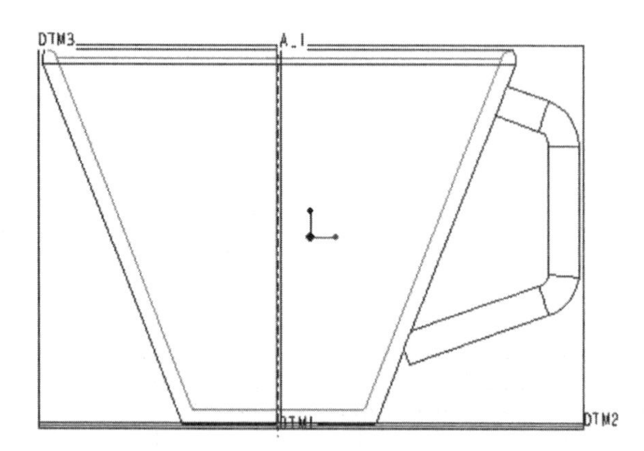

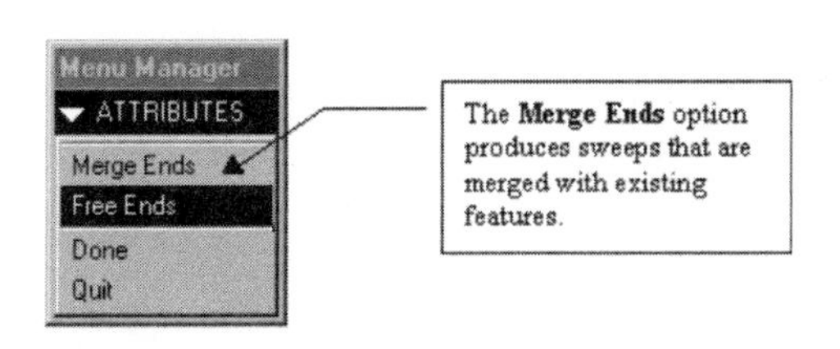


FIGURE 11.3 *The Pro/E* ATTRIBUTES *menu.*

In order to sketch the handle, create a circle as shown in Figure 11.4. Use *Geom Tools* to locate the intersections of the circle with the silhouette of the mug. Delete the unwanted part of the circle. Align the ends of the circle to the mug silhouette; that is, the edge of the mug in the *Sketcher.*

Notice in Figure 11.4, that Pro/E indicates the direction and starting point of the trajectory with a heavy arrow. Locate the center of the circle by aligning the center with the silhouette of the mug. Also, dimension the vertical position of the center with respect to the DTM2 datum plane as shown in Figure 11.5. The position of the center from the DTM2 datum plane, is 2.00 inches. *Dimension* the arc. The radius of the arc is 1.25 inches.

After regenerating, click on *Done.* Choose *Merged Ends* from the *ATTRIBUTES* menu when prompted to do so.

After the trajectory has been defined, Pro/E will reorient the model and ask the user to sketch the section to be swept as shown in Figure 11.6. The intersection of the dotted line and the sketching plane, which in this case is

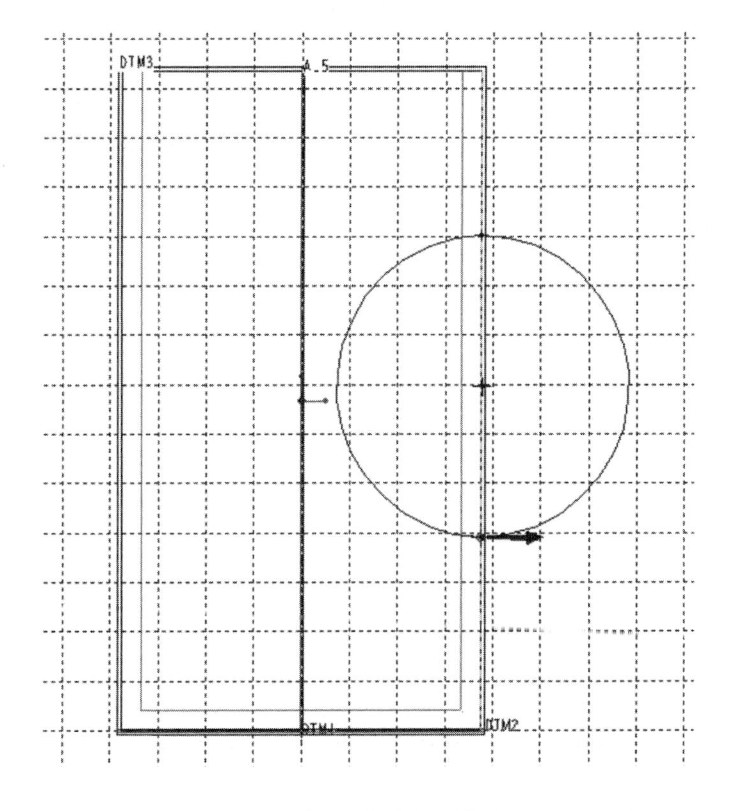


FIGURE 11.4 *The circle used to create the handle of the mug.*

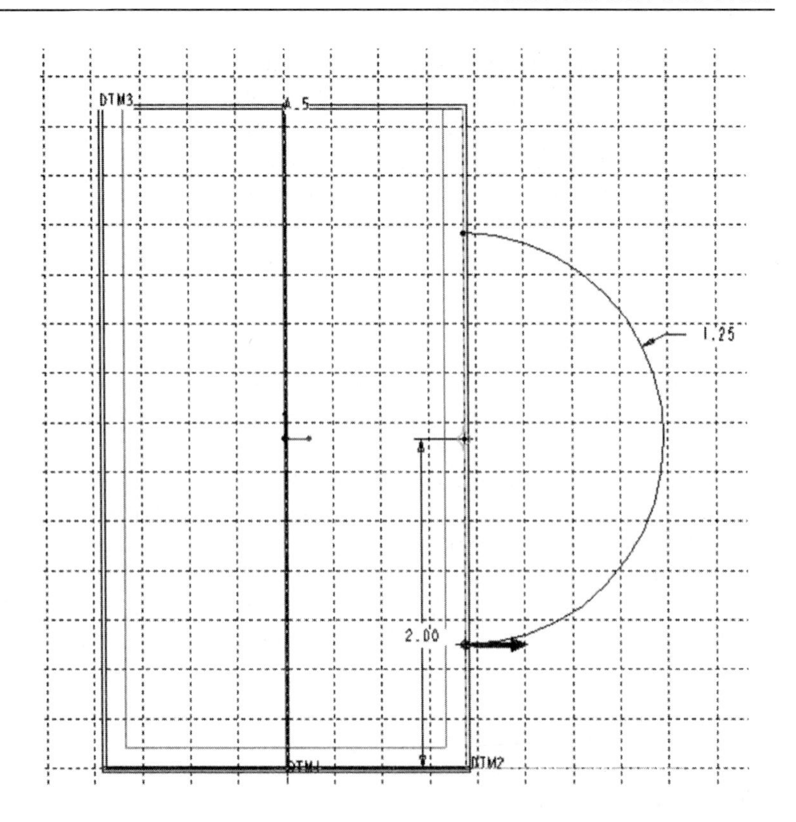


FIGURE 11.5 *The dimensions required for creating the handle trajectory.*

DTM3, forms the starting point for the sweep. Sketch a circle with the center near the starting point. Align the center to DTM3 and the dotted line. Then dimension the circle to have a 0.6-inch diameter. *Regenerate* the sketch and choose *Done.* Select *OK.*

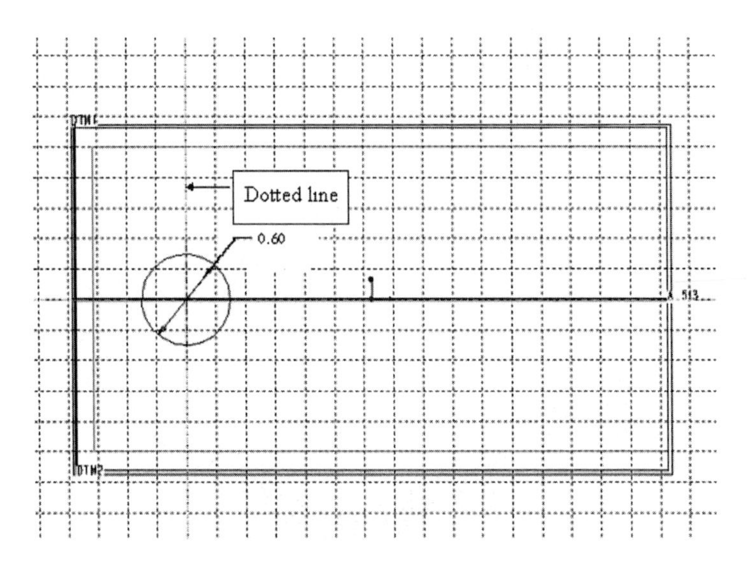


FIGURE 11.6 *Pro/E has reoriented the model to allow the user to sketch the section to be swept. The intersection of the dotted line and the sketching plane (in this case DTM3) forms the starting point.*

The mug is now complete. Add rounds to the inner *and* outer edge of the top of the mug. Use 0.1 for the radius of the rounds. The final version of the mug should appear as in Figure 11.7. Save the part.

Sweeps may fail if the section and trajectory are improperly defined. Often the problem is that during the sweep process, the section intersects itself if the section is too large with respect to the sweep. The problem can be fixed by reducing the size of the section or changing the trajectory.

FIGURE 11.7 *The final version of the mug with the handle created using a sweep.*

## 11.1.2 TUTORIAL # 54: A GASKET

Models that have a complicated cross section running along a trajectory can be easily created with the *Sweep* option. The model, illustrating the topic of this section, is the gasket shown in Figure 11.8.

The part contains symmetry about two axes. Because of this symmetry, only one quarter of the part will be constructed using the *Sweep* option. The rest of the model will be completed by using the *Mirror Geom* option described in section 7.4.

First, create a part, called "Gasket," and add the default datum planes. This part is metric, so the units need to be changed to millimeter. In order to change the units, follow this procedure:

1. Select *Set Up* from the *PART* menu.
2. Then chose *Units* and *Millimeter.*
3. Select *Done* twice.

It is immaterial as to which quarter of the model is constructed using the *Sweep* option. The quarter that we have chosen is reproduced in Figure 11.9. In order to generate the sweep given in the figure, we will draw the cross section on DTM2 and use DTM3 to orient the model in the *Sketcher.*

Select *Create, Protrusion, Sweep, Solid,* and *Done.* Use the *One Side* option. Choose the DTM2 as the sketching plane and DTM3 as the bottom. The trajectory of the sweep is shown in Figure 11.10 and will be sketched.

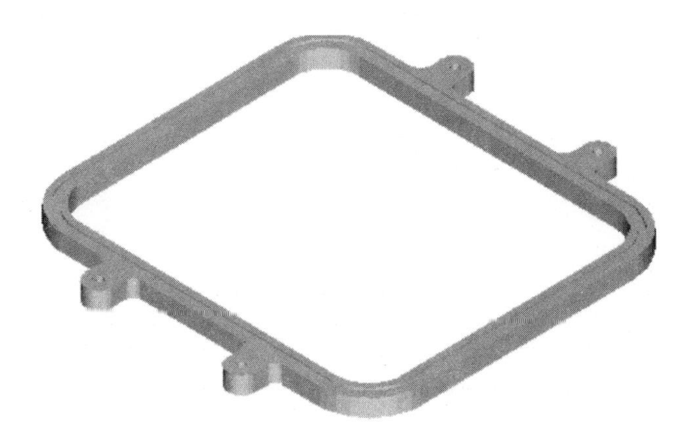

FIGURE 11.8 *A shaded model of the gasket. Note the symmetry of the model. Because of the symmetry, only one-quarter of the model need be constructed. The rest of the model is completed using the* Mirror Geom option.

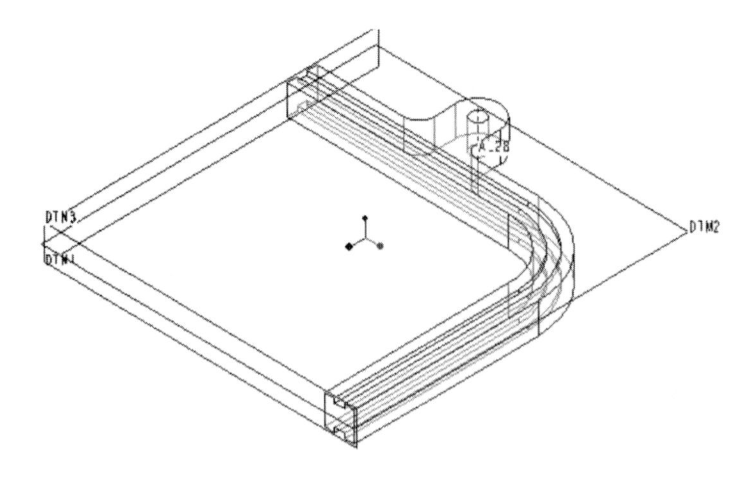


**FIGURE 11.9** *One-quarter of the gasket produced using the* Sweep *option. The alignment tab and hole were added using a protrusion. Note the cross section of the sweep.*

Select the *Sketch Traj* option and sketch the trajectory using the given geometry. *Regenerate* and *Modify* the dimension to the value given in Figure 11.10.

Note the location of the *Start Point* in Figure 11.10, which is indicated by the heavy black arrow. Make sure that the start point is located at the same point on your model. If not, use the option *Sec Tools* followed by *Start Point* to place the start point. After all values have been properly modified and the start point is suitably located, *Regenerate* again.

Now sketch the cross section of the sweep. Use the cross section geometry given in Figure 11.11. *Dimension* and *Regenerate* the sketch. Select *Done* and then *OK*. Pro/E will generate the sweep as shown in Figure 11.12.

We can now proceed to add the alignment tab. The tab is constructed by using the protrusion option and the DTM2 datum plane as the sketching plane. The hole is added after the protrusion is created.

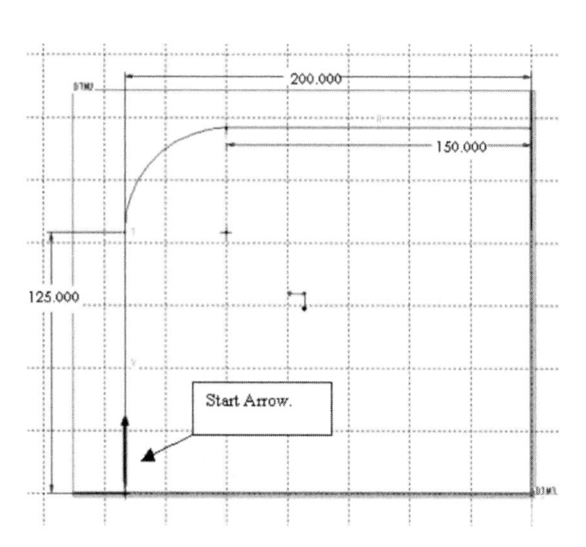


**FIGURE 11.10** *The trajectory for the sweep is sketched using two lines and an arc. Note the location of the start arrow.*

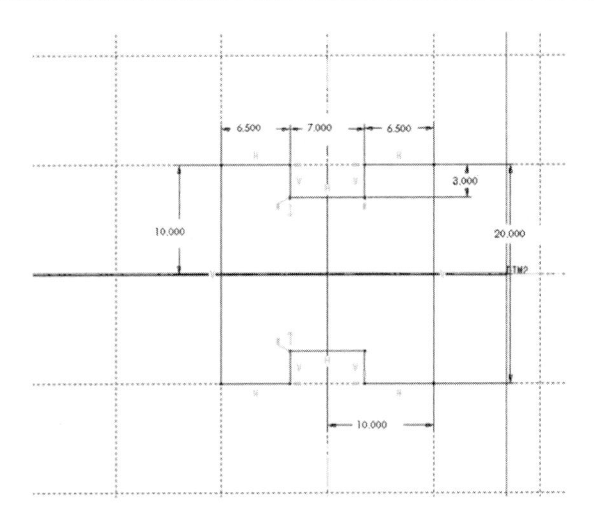


FIGURE 11.11 *Sketch the geometry of the cross section. When dimensioning, be sure to dimension the section to the DTM2 plane and to the trajectory (the dotted line).*

The procedure is:

1. Select *Feature, Create, Protrusion,* and *Extrusion.*
2. Use the *Solid* option and choose *Done.*
3. Use the *Both Sides* option from the *ATTRIBUTES* menu.
4. For the sketching plane, select DTM2. Use DTM3 as the "*Bottom.*"

The geometry for the tab is shown in Figure 11.13. Note that the geometry must be closed by placing a line along the silhouette of the sweep. Regenerate the sketch and modify the parameters to the proper values. Select *Done.* Extrude to a *Blind* depth of 20 mm. The part after the extrusion is shown in Figure 11.14.

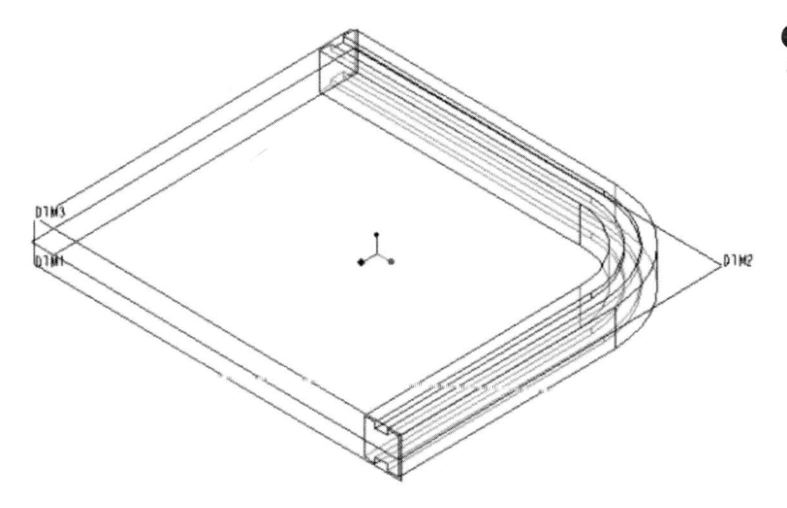


FIGURE 11.12 *The gasket after sweeping the section in Figure 11.11.*

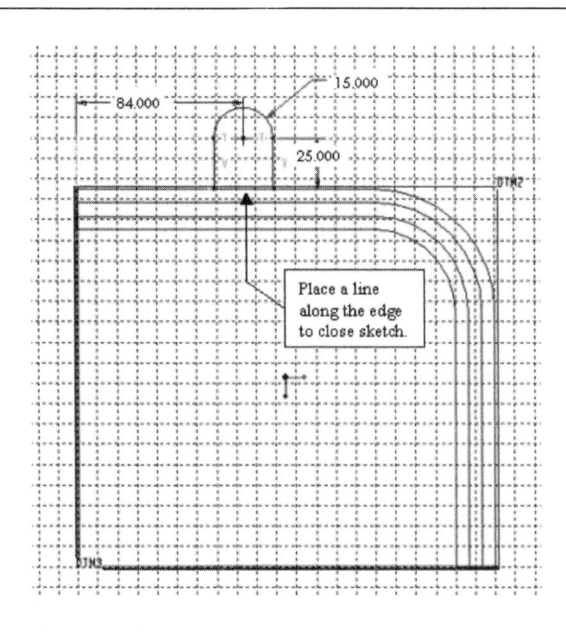


FIGURE 11.13 *The model in the* Sketcher *with the geometry of the alignment tab.*

Place the hole through the tab using the *Hole* option.

1. Select *Feature, Create,* and *Hole* to place the hole.
2. Use the *Coaxial* option and select the axis shown in Figure 11.14.
3. Use the placement plane also shown in Figure 11.4.
4. The depth of the hole is *Thru All* and its diameter is 10 mm.

The model after placing the hole is shown in Figure 11.15.

We will add the two surface-to-surface rounds one at a time. In this case, it does not matter which round is added first. The rounds are each 10 mm in

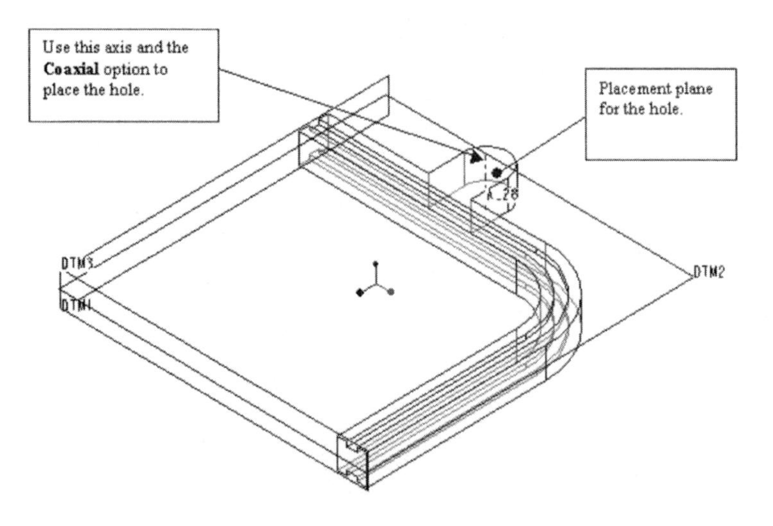


FIGURE 11.14 *The gasket after adding the alignment tab. Surface rounds need to be placed between the sweep surface and the tab.*

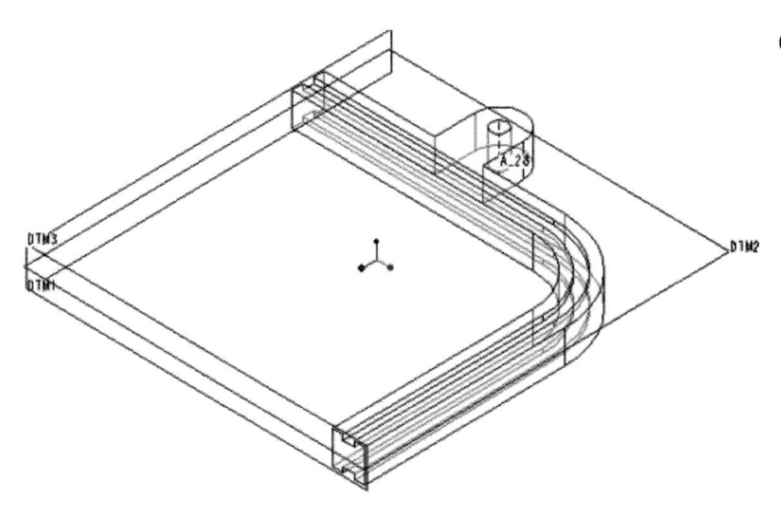


FIGURE 11.15 *The gasket after adding the hole on the tab.*

radius. In order to place the rounds, you need to select the appropriate surfaces. Figure 11.16 shows the model after it has been rotated, to show the appropriate surfaces for placing the rounds.

In order to place the rounds, use this procedure:

1. Use *Feature, Create, Round, Simple,* and *Done.*
2. Select the *Surf-Surf* option and then select the surfaces, as shown in Figure 11.16.
3. Enter the 10-mm radii when asked to do so.

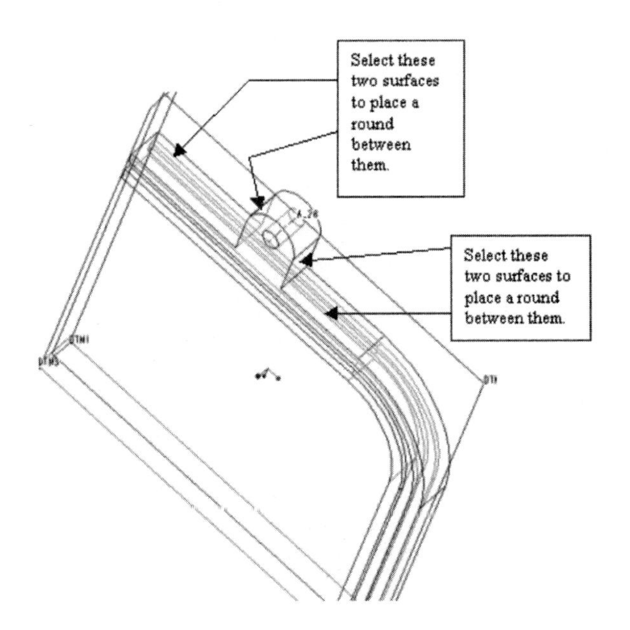


FIGURE 11.16 *The model rotated to show the surface to select when placing the rounds.*

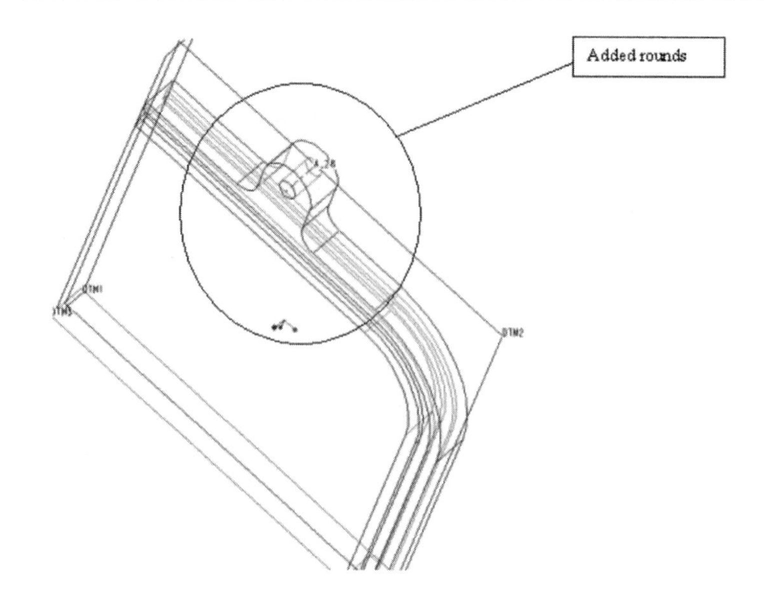


FIGURE 11.17 *The gasket after the surface-to-surface rounds have been placed.*

The same view of the model after the rounds have been added is shown in Figure 11.17.

The model can now be completed by using the *Mirror Geom* option. In constructing the model, we placed the DTM3 and DTM1 in such a way as to make the datum planes suitable as planes of symmetry. By selecting these planes, the feature of the model can be copied by mirroring.

The steps required are:

1. Select *Feature,* and *Mirror Geom.*
2. Click on DTM3.

The software will mirror the features beyond DTM3 and produce the model shown in Figure 11.18.

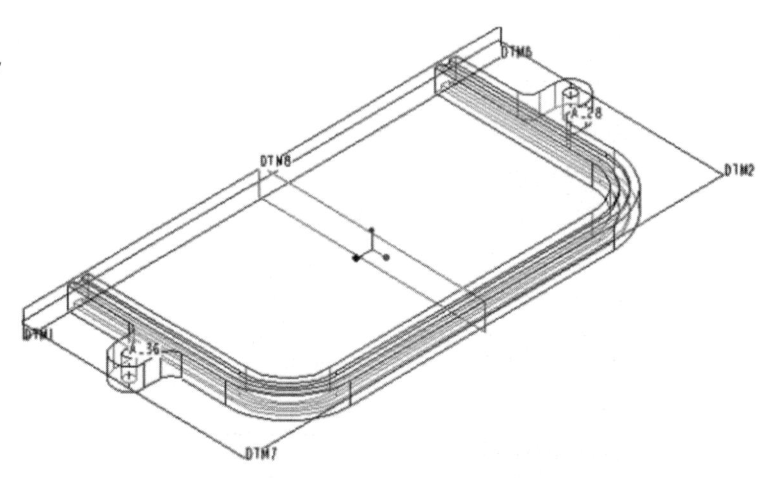


FIGURE 11.18 *The gasket, after using DTM3 to mirror the features. Half of the model is complete. A mirror using DTM1 will complete the model.*

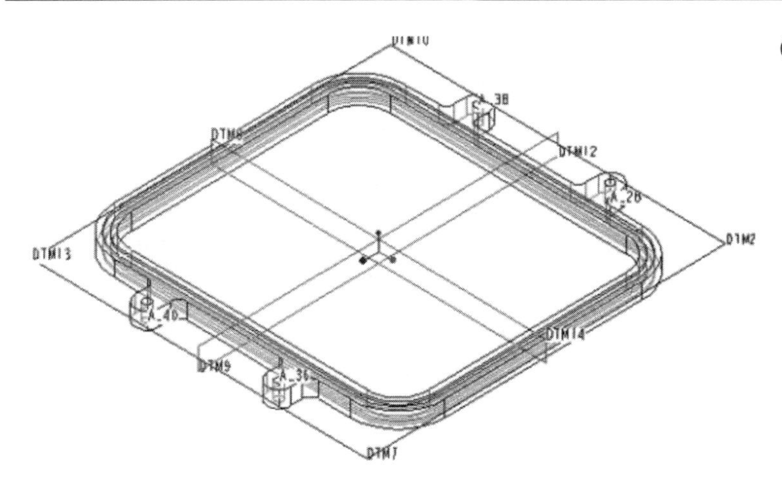


FIGURE 11.19 *After using the* Mirror Geom *option for a second time, this isometric of the completed model of the gasket may be obtained.*

Now,

1. Select *Feature* and *Mirror Geom* again.
2. Then click on DTM1.

Pro/E will complete the model. An isometric of the complete model is shown in Figure 11.19. Save the part.

## 11.2 HELICAL SWEEPS

A *Helical Sweep* is constructed by moving a defined cross section along a helical path. Helical sweeps have practical applications in the construction of springs and thread surfaces.

The *Helical Sweep* option is one of the so-called advanced options in the *SOLID OPTIONS* menu. The *ADVANCED FEATURE OPTIONS* menu, where the *Helical Sweep* option is located, is reproduced in Figure 11.20.

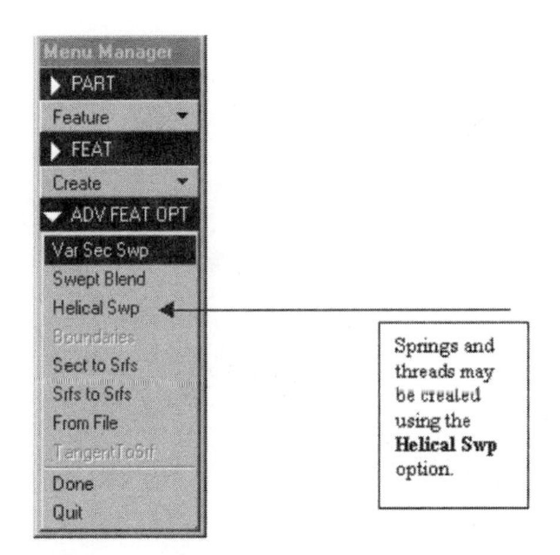


FIGURE 11.20 *The* ADVANCE FEATURE OPTIONS *menu that contains the* Helical Swp *(Sweep) option.*

The distance between the coils of a spring or between the teeth of a thread is called the pitch. In constructing a helical sweep, the value of the pitch must be provided by the user. The other information required in construction of these types of features is entered by selecting from the options in the *HELICAL SWEEP ATTRIBUTES* menu shown in Figure 11.21.

From the menu, it is evident that either right-handed or left-handed sweeps may be constructed. Furthermore, a sweep with a *Constant* or *Variable* pitch may be created. The options *Thru Axis* or *Normal To Traj* refer to the orientation of the sweep cross section.

If a sweep is constructed with the *Thru Axis* option (the default), the cross section is defined so that it lies in a plane that contains the axis of revolution. The *Normal To Traj* option, however, allows the user to create a sweep in which the cross section is normal to the trajectory. Using the *Normal To Traj* option gives the cross section of the sweep a twist relative to the axis of revolution.

### 11.2.1 TUTORIAL # 55: A SPRING CONSTRUCTED WITH A HELICAL SWEEP

Let us construct a spring using the *Helical Sweep* option. Create a new part called "Spring" and load the default datum planes.

Select *Feature, Create, Protrusion, Advanced,* and *Done.* Select *Helical Sweep* from the *ADVANCED FEATURES OPTION* menu (Figure 11.20). We will make a right-handed spring with a constant pitch, so accept the defaults from the *ATTRIBUTES* menu (Figure 11.21) by clicking on *Done.*

Pro/E will inquire about the sketching plane. Select DTM2. Then choose DTM3 as "*Top*" to orient the part in the *Sketcher.*

In the *Sketcher,* place a centerline along the DTM3 edge. Use *Align* and align this line to the DTM3 edge. Sketch a line parallel to the centerline and dimension as shown in Figure 11.22. The dimension will result in a spring with an inner bore of 0.19 inches and a length of 1/2 an inch. Notice the solid black arrow in the figure, which defines the start point of the sweep. *Regenerate* and select *Done.*

Enter a value of 0.1 for the pitch. Now that the trajectory has been defined, we can proceed to sketch the cross section of the spring. We have cho-

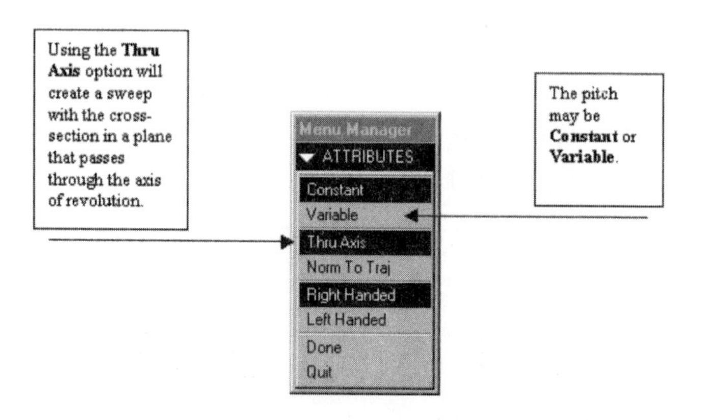


**FIGURE 11.21** *The available attributes for creating a helical sweep.*

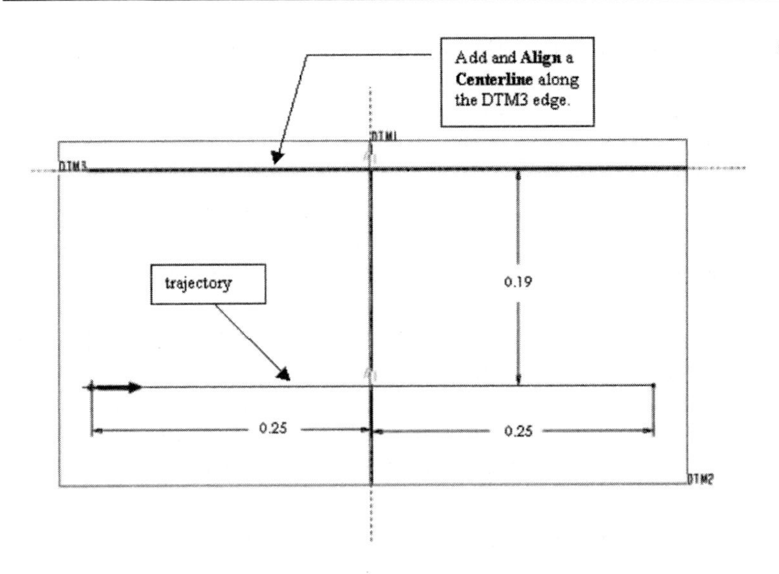


FIGURE 11.22 *The geometry and dimensions required for sketching the trajectory for the spring.*

sen a simple circular cross section of 0.06 inches in diameter. Sketch and dimension the section as shown in Figure 11.23. Regenerate and click on *Done.* Then select *OK.*

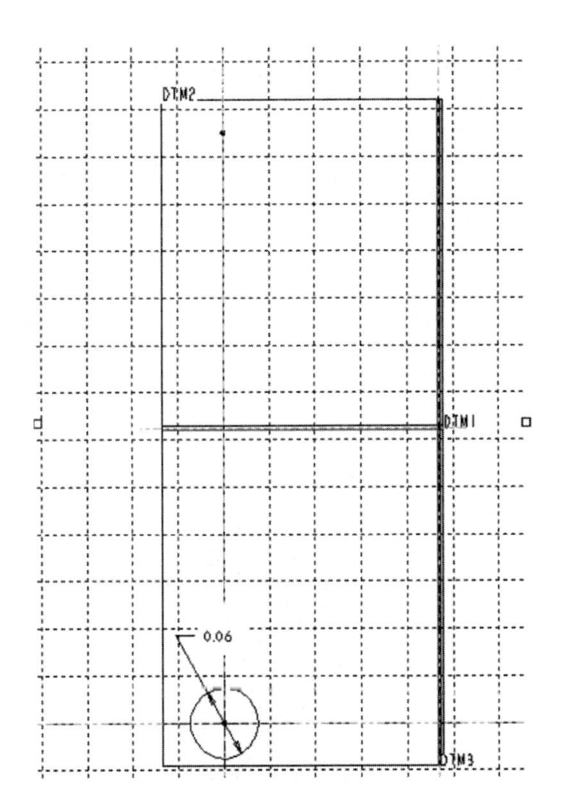


FIGURE 11.23 *Notice in this sketch of the spring cross section, that the center of the circle is aligned to the start point.*

Pro/E will generate the model. Figure 11.24 shows the spring with all its dimensions. A shaded model of the spring is also given in Figure 11.24.

### 11.2.2 TUTORIAL # 56: CONSTRUCTING SCREW THREADS USING A HELICAL SWEEP

As we will see in chapter 14, Pro/E has a built-in option for creating screw threads. However, this option is a cosmetic option; that is, Pro/E creates a surface that represents the threads, but does not actually draw them. One advantage is that that model regeneration time is minimized. However, if the actual threads are desired, they can be created using a helical sweep.

Let us place a thread on the adjustment screw from chapter 5. The threads to be placed on the screw are ANSI standard 0.25 pitch square threads. The surface to be threaded is shown in Figure 11.25.

The threads can be constructed by placing the trajectory of the sweep along the edge of the surface to be threaded. This process will require aligning the trajectory with the silhouette of the surface. We will see how it works shortly.

After retrieving this part, select *Feature, Create, Protrusion, Advanced,* and *Done.* Select *Helical Sweep.* The thread is right-handed with a constant pitch. Use the *Thru Axis* option. Orient the part in the *Sketcher.* Suggestion: use DTM2 as the sketching plane and DTM3 as the top.

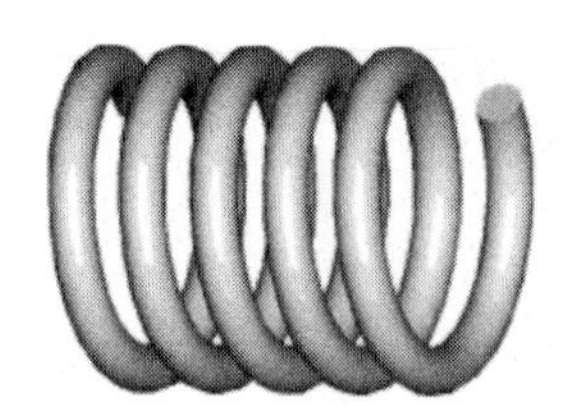

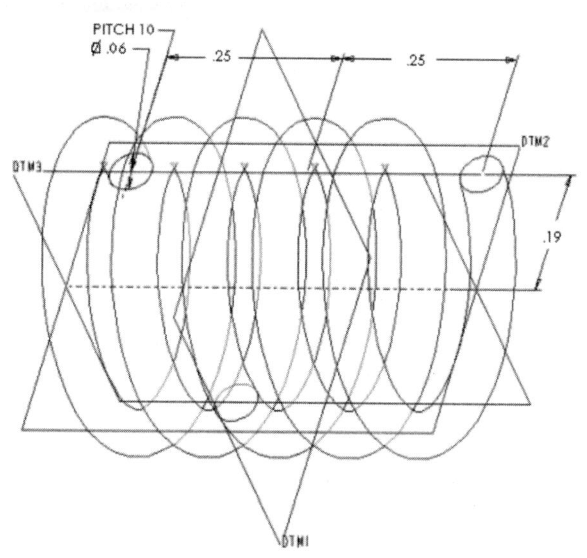


FIGURE 11.24 *Wireframe and shaded models of the spring.*

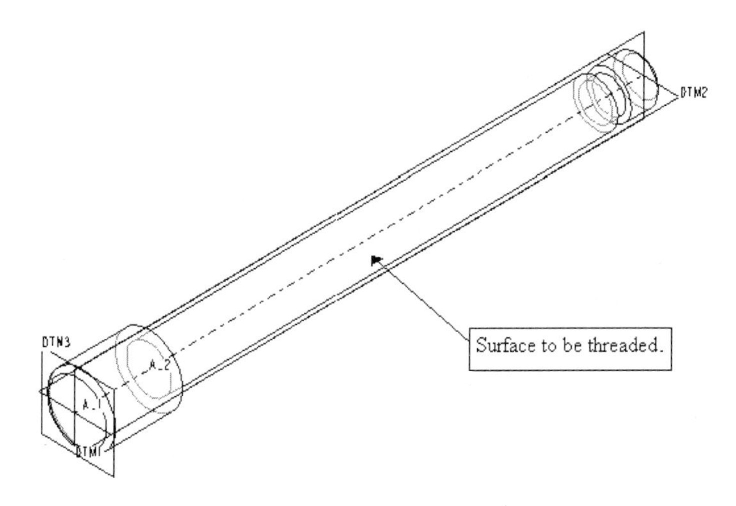


**FIGURE 11.25** *The adjustment screw with the surface to be threaded.*

Figure 11.26 shows the screw in the *Sketcher.* Place a centerline along the DTM1 plane edge and align it to the edge. Then sketch a line as shown in the figure. The starting point of the line is indicated by the big black arrow. The direction of the sweep is in the direction of the arrowhead. Note that the trajectory is 0.5 inches from one end and 0.25 inches from the other to prevent the screw from tightening against the head of a vise. *Regenerate* and select *Done.* Enter a value of 0.25 for the pitch.

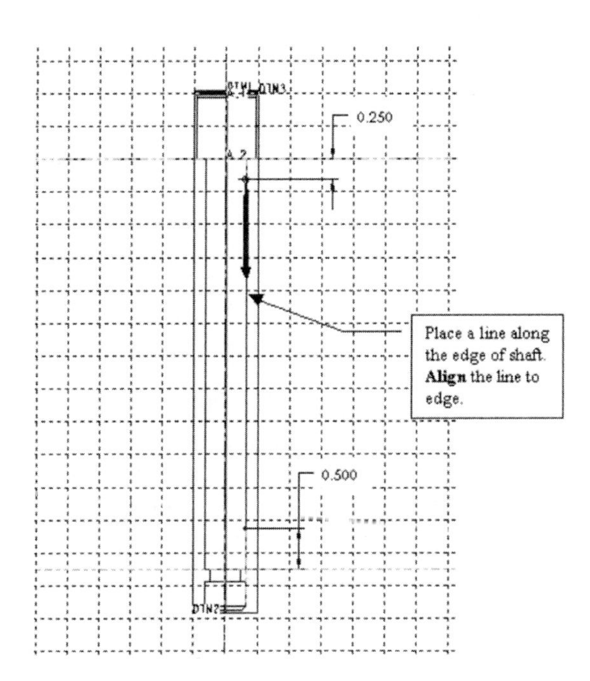


**FIGURE 11.26** *The adjustment screw in the* Sketcher *showing the geometry for the trajectory of the sweep.*

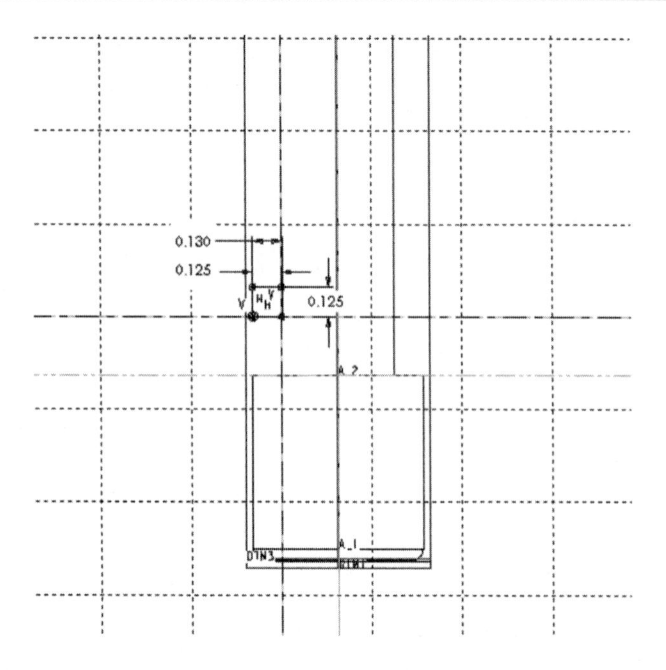


FIGURE 11.27 *An enlarged sketch of the thread section. Note that the rectangle is drawn slightly into the adjustment screw to ensure that the helical sweep intersects the part.*

Pro/E will now inquire about the cross section of the sweep. The cross section is shown in Figure 11.27. The size of the thread is 0.125 by 0.125 inches. In order to ensure that the tooth geometry intersects the adjustment screw during the sweep, the width of the thread is drawn slightly larger, say to a value of 0.130.

Use the *Rectangle* option and sketch the profile. Then *Dimension* the sketch as shown. *Regenerate* the model and choose *Done.* Select *OK.*

Pro/E will now generate the sweep. Be patient. The sweep is somewhat involved and depending on the computer, model generation may take some time. Eventually, Pro/E will display the model. It should appear similar to the one shown in Figure 11.28.

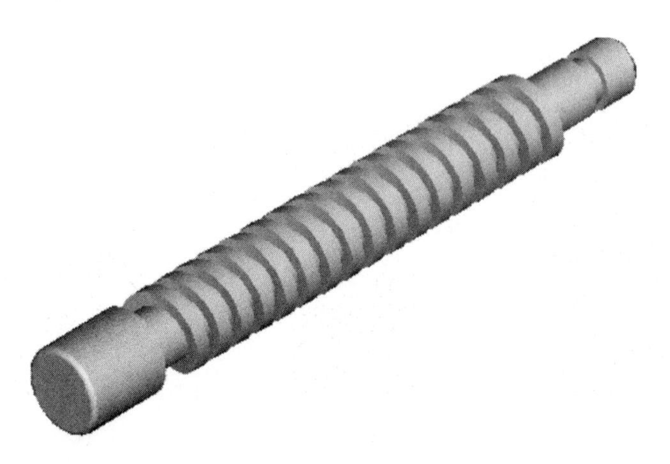

FIGURE 11.28 *A shaded model of the adjustment screw with added square threads.*

In Figure 11.29, we have reproduced a top view of the screw that shows the spacing of the threads more clearly. Notice how the threads end at the appropriate points.

FIGURE 11.29 *The model has been rotated to readily illustrate the threads.*

## 11.3 SWEEPS ALONG A DATUM CURVE

A sweep may be constructed along a datum curve by using the *Select Traj* option (Figure 11.1) and picking the datum curve as the sweep trajectory with the appropriate options from the *CHAIN* menu, shown in Figure 11.30a. The options from this menu are:

1. *One By One.* Create a chain of curves and edges by selecting each of the features one at a time.
2. *Tangnt Chain.* The user selects an edge. The software selects all the edges tangent to the chosen edge and completes the trajectory.
3. *Curve Chain.* Define the chain by selecting a curve. The *CHAIN OPT* menu shown in Figure 11.30b allows the user to select the endpoints of the chain. The *Select All* option selects *all* the curves that are connected to the curve selected by the user. The *From-To* option provides the user with the ability to create a composite chain by selecting vertices or ends of the features.

*Bndry Chain* and *Surf Chain* are used to construct a chain from the edges of a quilt (see chapter 15, for a discussion of a *Quilt*). *Unselect* may be used to disable a choice.

In situations where the chain forms a closed loop, Pro/E can add top or bottom faces to the section as it sweeps along the trajectory. This function is the *Add Inn Fcs* option shown in Figure 11.30c. However, when using this option, the section must be an open section. For closed sections use the default option, *No Inn Fcs.*

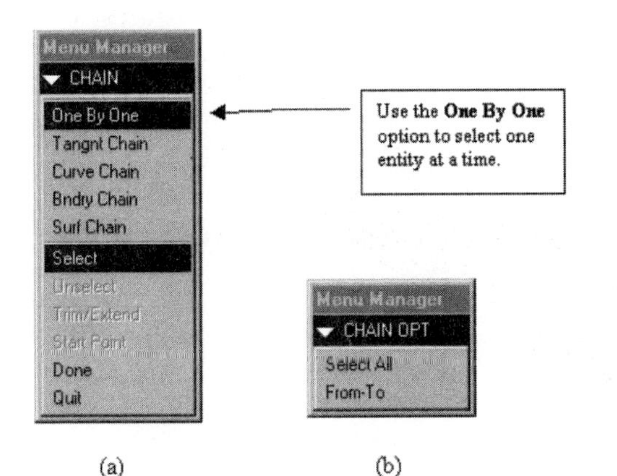

FIGURE 11.30 *The* CHAIN *and* CHAIN OPT *(Options) menus.*

### 11.3.1 TUTORIAL # 57: COMPLETING THE TWO-DIMENSIONAL FRAME

In chapter 9, we constructed a datum curve passing through five datum points. Recall that the name of the part is "2dFrame." Retrieve this part. We will now complete the part by sweeping a circular section along the curve.

After the part has been loaded, select *Feature, Create, Protrusion, Sweep,* and *Done.* From the *SWEEP TRAJ* menu (Figure 11.1), choose *Select Traj.* The *Chain* menu (Figure 11.30a) will appear, which allows the trajectory of the sweep to be chosen. Because we are going to select a datum curve, choose the *Curve Chain* option from the menu.

With your mouse, select the datum curve. The *CHAIN OPT* menu will appear. Choose the *Select All* option and then *Done* to ensure that the entire curve is chosen for the trajectory. A red arrow will appear. This arrow defines the upward direction for the sweep direction. Make sure it points towards you. Pro/E will load the part into the *Sketcher.*

Sketch the circular cross section of the frame as shown in Figure 11.31. The diameter of the circle is 0.04. Regenerate. Upon successful regeneration, select *Done.* Pro/E will sweep the cross section along the datum curve. The final part is shown in Figure 11.32. Save the part.

### 11.3.2 TUTORIAL # 58: COMPLETING THE KING POST TRUSS

The geometry for the king post truss considered in chapter 9 was created using datum points. The *Spline* option was used to place several datum curves through the datum points. Unlike the "2Dframe" from section 11.3.1, this option allows the user to construct a part using sweeps, without having to define a radius at the bends.

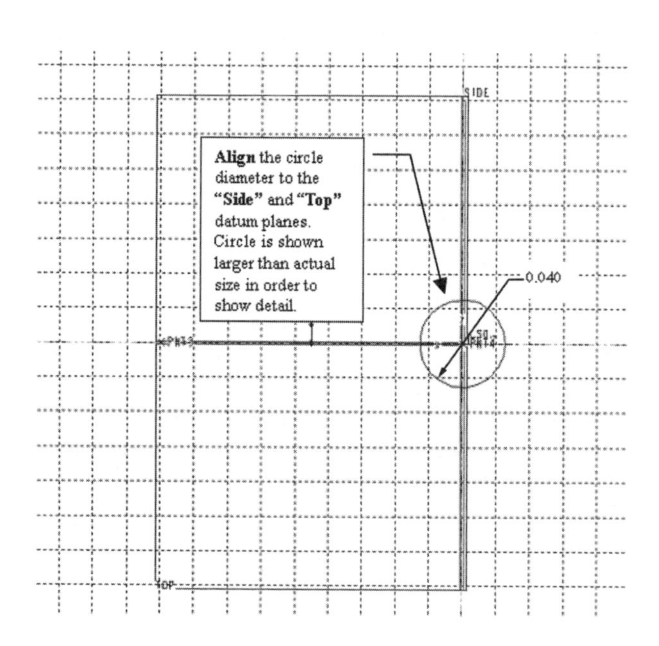


FIGURE 11.31 *The cross section of the two-dimensional frame.*

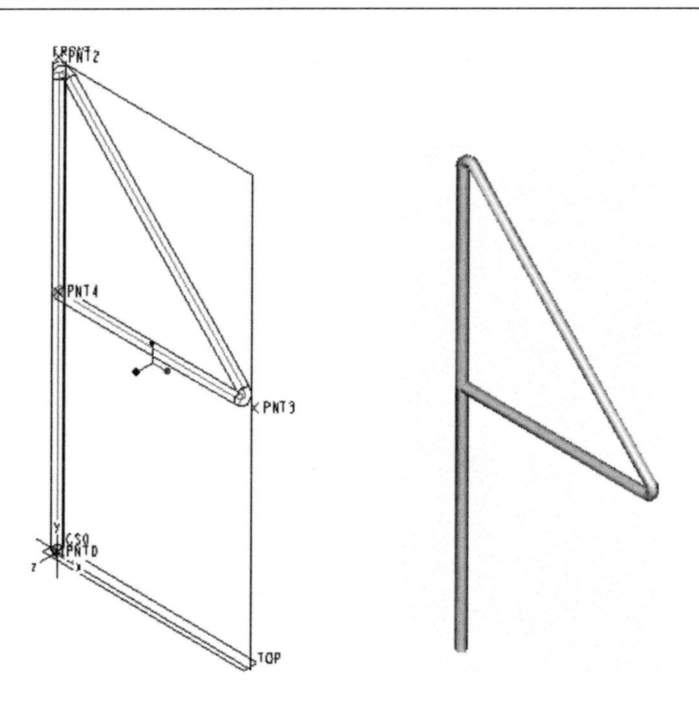


FIGURE 11.32 *The two-dimensional frame after sweeping the circular cross section along the datum curve.*

The truss is made of 2 × 4 dimensional lumber, (actually 1.5 × 3.5 due to shrinkage during drying). The elements were modeled using datum curves. As long as the selected curves form a loop, the structural members may be modeled by sweeping a rectangular section with the proper dimensions along the datum curves. The loop does not have to be closed. In completing the truss, we will use two loops. These loops are shown in Figure 11.33.

Retrieve the part called "KingPostTruss." We now proceed to construct the outer loop made up of the three datum curves shown in Figure 11.33.

Select *Feature, Create, Protrusion, Sweep, Solid,* and *Done.* Choose *Select Traj* and *One By One.* Pick the datum curves shown in Figure 11.33 to form the chain. Pro/E will inquire whether to add inner faces (*Add Inn Fcs* option, Figure 11.30c). Because the section is closed, you must select *No Inn Fcs.*

Notice the arrow in Figure 11.33. Make sure that the arrow on your screen is pointing in the same direction. If not, use the *Flip* option to change the direction of the arrow.

After the software has loaded the model in the *Sketcher,* sketch the rectangular geometry shown in Figure 11.34. *Dimension* as shown. Select *Relation* and *Add.* Enter the following relations (note the parameter names may be different):

$$sd2 = sd3/2$$

$$sd4 = sd5/2$$

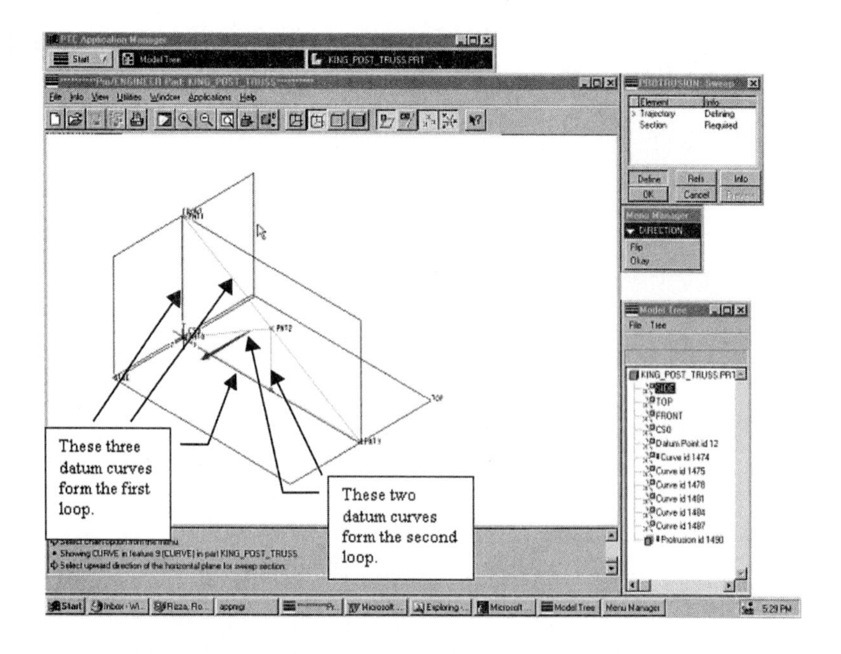


FIGURE 11.33 *The king post truss is shown with the two loops.*

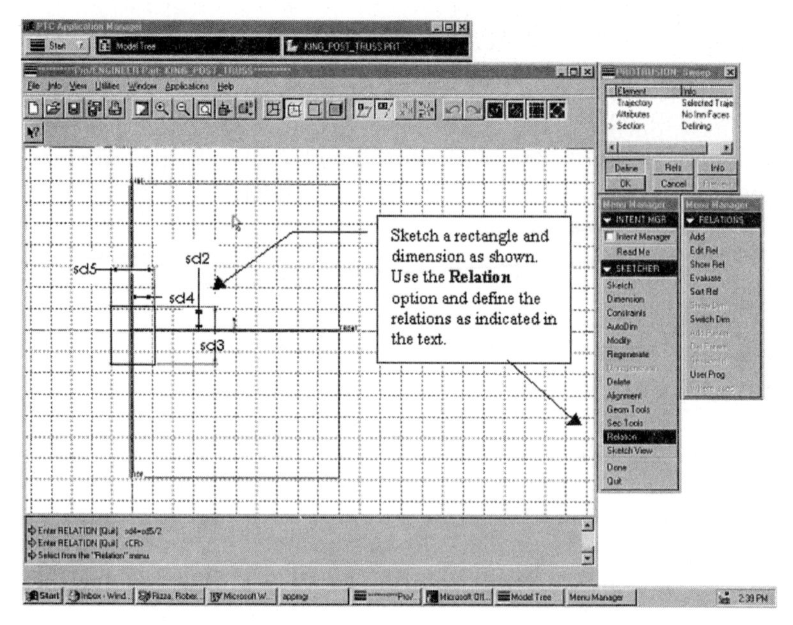


FIGURE 11.34 *The cross section of the sweep is a rectangular section.*

These relations will properly orient the section. *Regenerate* and *Modify* to the values:

$$sd3 = 3.5/12$$

$$sd5 = 1.5/12.$$

We divide by 12 because the units of the model are in feet.

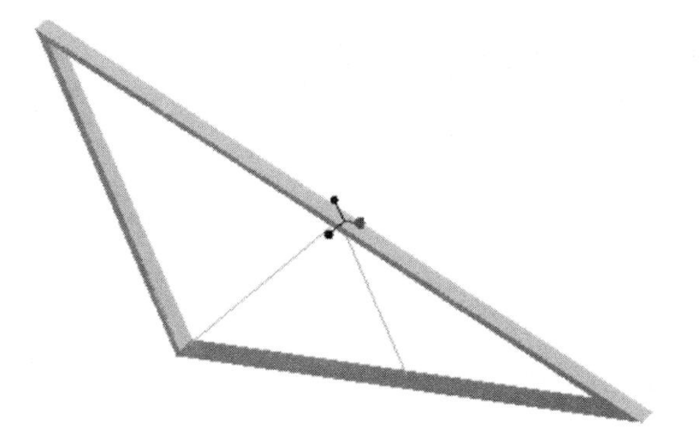

FIGURE 11.35 *The model of the king post truss is shown after the first sweep.*

Choose *Regenerate* and *Done.* Then select *OK.* The software will sweep the section. After this section has been swept, your model should appear similar to the one shown in Figure 11.35.

In order to add the second sweep:

1. Select *Feature, Create, Protrusion, Sweep, Solid,* and *Done.*
2. Choose *Select Traj* and *One By One.*
3. Use Figure 11.33 and pick the shown datum curves to form the chain.
4. Again, make sure the direction arrow points in the direction, as shown in Figure 11.33.
5. Use the *Merge Ends* option.
6. In the *Sketcher,* sketch and *Dimension* the section shown in Figure 11.36.
7. Select *Relation* and *Add.* Using the parameters from Figure 11.35, set sd2 = sd3/2 and sd4 = sd5/2.
8. *Regenerate* and *Modify* to sd3 = 3.5/12 and sd5 = 1.5/12.
9. Choose *Done* and *OK.*

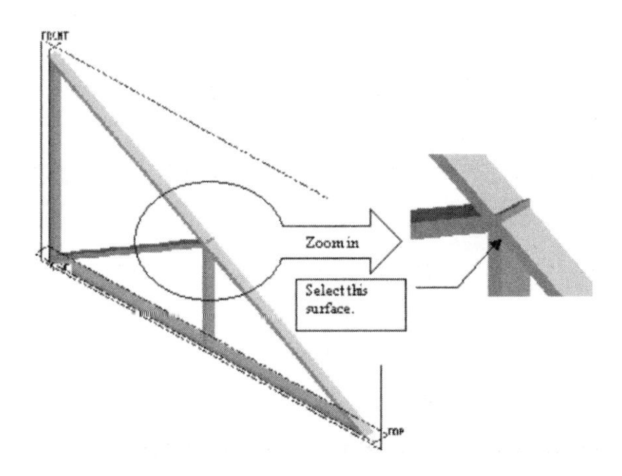


FIGURE 11.36 *The truss is shown after completing the second sweep. Because of the geometry of this sweep, a small triangular feature protrudes beyond the first sweep.*

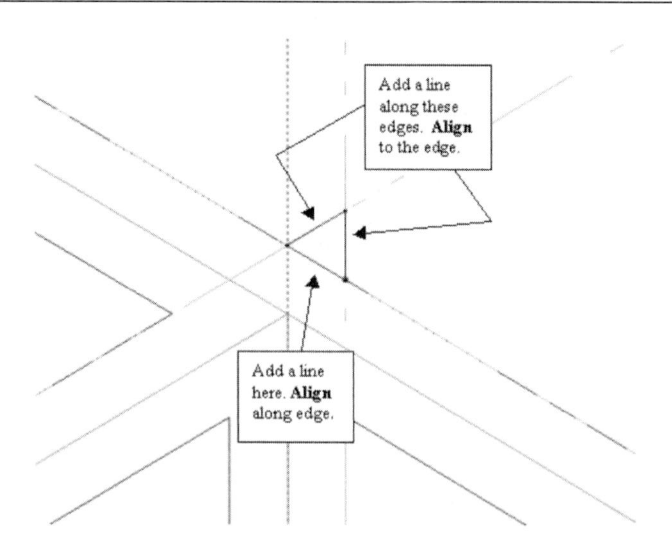


FIGURE 11.37 *The section needed for the cut with a "zoomed in" view of the intersection.*

The model with the second sweep is shown in Figure 11.36. Notice that even after using the *Merge Ends* option, which merges two features at their line of intersection, a small triangular feature protrudes beyond the first protrusion. This triangular shape is due to the geometry of the second sweep and may be removed using a simple cut.

In order to remove the unwanted triangular shape, select *Feature, Create, Cut, Extrude, Solid,* and *Done.* Use the *One Side* option and select the surface shown in Figure 11.36. To orient the model in the *Sketcher,* choose the "*Top*" option and select the default datum plane called "*Top.*" Make sure the orientation arrow points into the screen.

In the *Sketcher,* use the sketch shown in Figure 11.37 for the geometry of the cut. Notice that lines must be aligned to the geometry of the triangular section. *Regenerate* the sketch. Choose *Done* and *Thru All* for the depth of the cut. The effect of the cut on the model is shown in Figure 11.38.

The rest of the model may be completed by using the *Mirror Geom* option. Choose *Feature* and *Mirror Geom.* Then select the hidden surface

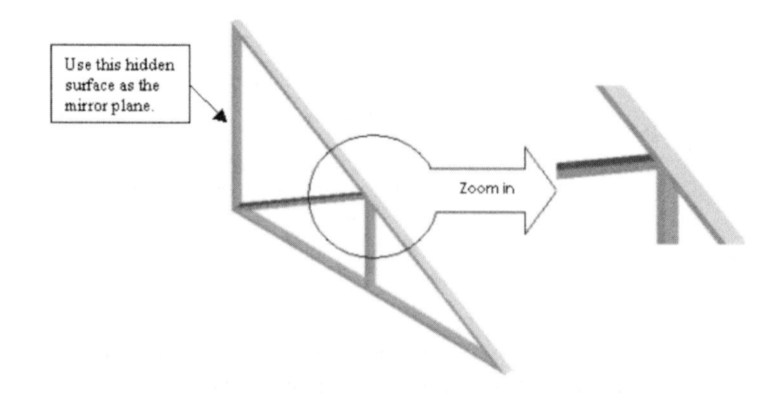


FIGURE 11.38 *After the use of the cut, the intersection under consideration takes on the appropriate appearance.*

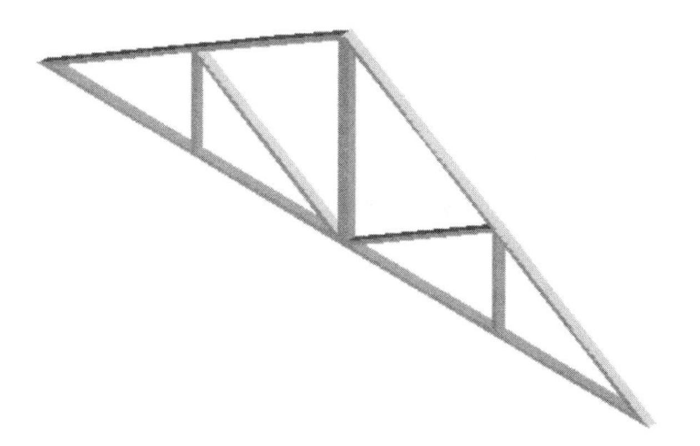

FIGURE 11.39 *The king post truss after the use of the* Mirror Geom *option. Other than the placement of nail plates at the joints, the model is complete.*

shown in Figure 11.38. The model, after using the *Mirror Geom,* option is reproduced in Figure 11.39. Save the model.

## 11.4 SUMMARY AND STEPS FOR USING THE SWEEP OPTION

The *Sweep* option may be used to construct features that have a constant cross section defined along a trajectory. In creating a sweep, the user must define the trajectory as well as the cross section. The trajectory may be sketched using the *Sketcher* (*Sketch Traj* option) or constructed as a chain or entities that already exist on the model (*Select Traj*).

Datum curves may be used as the trajectory for a sweep by selecting the datum curve. Sweeps along datum curves are useful in creating models of structures such as trusses. A chain containing several curves may be created by using the *Curve Chain* option from the *CHAIN* menu (Figure 11.30a). In order to select individual datum curves, use the *One By One* option.

In general, a sweep is constructed by following these steps:

1. Select *Feature, Create, Protrusion* (or *Cut), Sweep, Solid* (or *Thin*), and *Done.*
2. From the *SWEEP TRAJ* menu (Figure 11.1) choose *Select Traj* or *Sketch Traj.*
3. If you choose the *SKETCH TRAJ* option, choose a sketching plane and an orientation plane. Then, sketch the trajectory and dimension. *Regenerate* and select *Done.* If you are selecting the trajectory, pick the appropriate entities to form the trajectory. Then choose *Done.*
4. Sketch the cross section of the sweep. *Dimension, Modify,* and *Regenerate* to the appropriate values.
5. Select *Done.*
6. Choose *OK.*

A *Helical Sweep* is constructed by using a helical trajectory. Such sweeps are used to create springs and threads on fasteners.

The basic approach that may be used to create this type of sweep is:

1. Select *Feature, Create, Protrusion* (or *Cut*), *Sweep* (or *Thin*), *Advanced,* and *Done.*
2. Choose *Helical Sweep* from the *ADV FEAT OPT* menu (Figure 11.20).
3. From the *ATTRIBUTES* menu (Figure 11.21), select the desired options, for example, a right-handed helix versus a left-handed helix.
4. Choose *Done.*
5. Pick a sketching and an orientation plane.
6. In the *Sketcher,* sketch and *Dimension* the profile of the surface of revolution. *Dimension* the profile to the axis of revolution.
7. Select *Done.*
8. Enter the value of the pitch.
9. In the *Sketcher,* draw and dimension the cross section. *Regenerate* and *Modify* to the appropriate values. Choose *Done.*

## 11.5 ADDITIONAL EXERCISES

11.1 Construct the handle shown in Figure 11.40 using the *Sweep* option. Note that this is a metric part.

FIGURE 11.40 *Figure for exercise 11.1.*

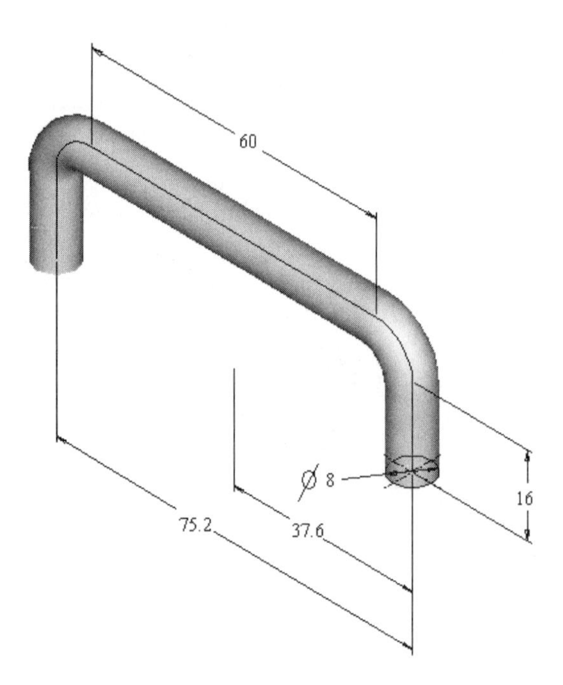

11.2 Use the *Sweep* option to construct the spring clip shown in Figure 11.41.

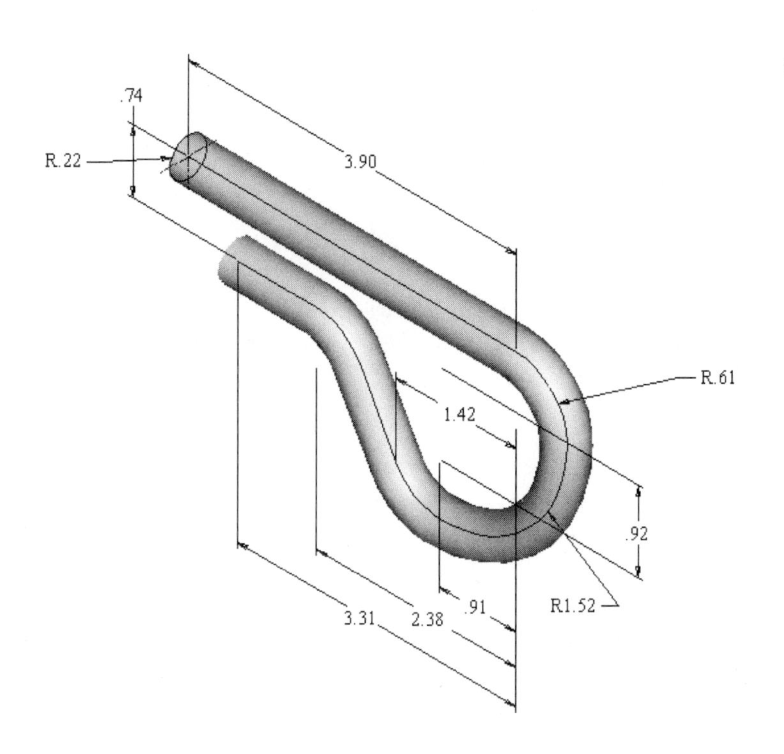


FIGURE 11.41 *Figure for exercise 11.2.*

11.3 Using the *Sweep* option, create the metric model of the paper clip. Use Figure 11.42 as a reference.

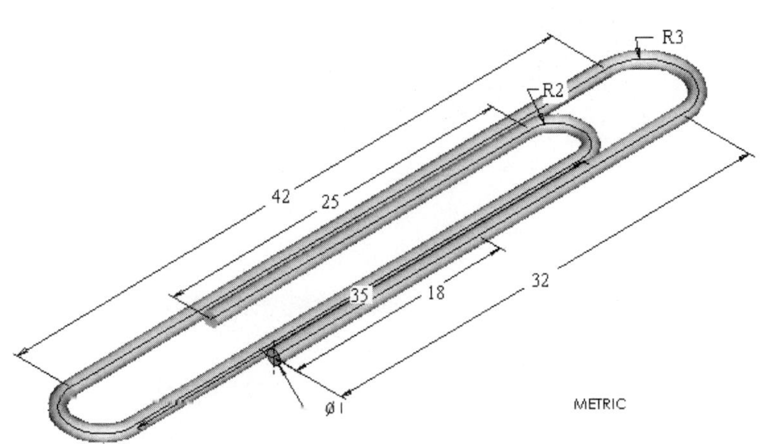


FIGURE 11.42 *Figure for exercise 11.3.*

11.4 Complete the model of the fink truss using the *Sweep* option. Use 2 × 4 (1.5 × 3.5 actual) dimensional lumber for the elements of the truss. The base model was created in exercise 9.8. The completed model is shown in Figure 11.43.

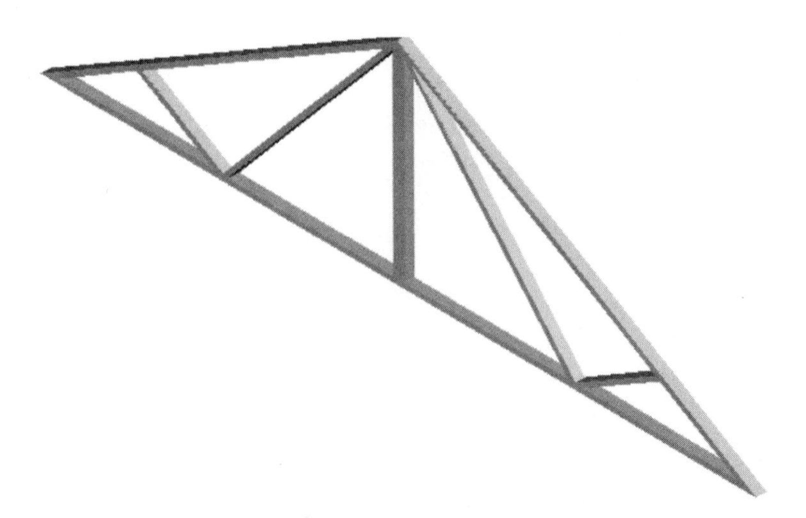

FIGURE 11.43 *Figure for exercise 11.4.*

11.5 Complete the bridge truss, whose base feature was originally constructed in exercise 9.9. Use 8 × 8 (7.5 × 7.5 actual) dimensional lumber for the elements of the truss. The completed model is shown in Figure 11.44.

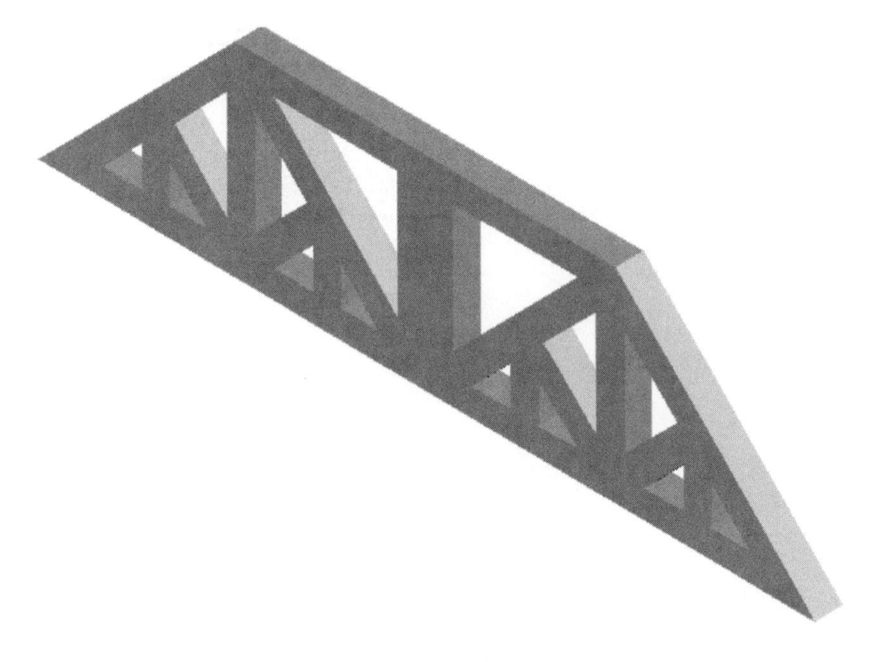

FIGURE 11.44 *Figure for exercise 11.5.*

11.6 Complete the highway sign frame using the datum points and curves generated in exercise 9.2. The cross section of the sweep is a 4–inch diameter circle. Hint: use the *Sweep* option to create the protrusion through the frame. Then use *Mirror Geom* with DTM3 as the mirror plane. Add the cross members by sketching on the DTM3. Extrude *Blind* with the *Both Sides* option. For the depth of the extrusion, use the *UpTo Pnt/Vtx* option. The completed model is shown in Figure 11.45.

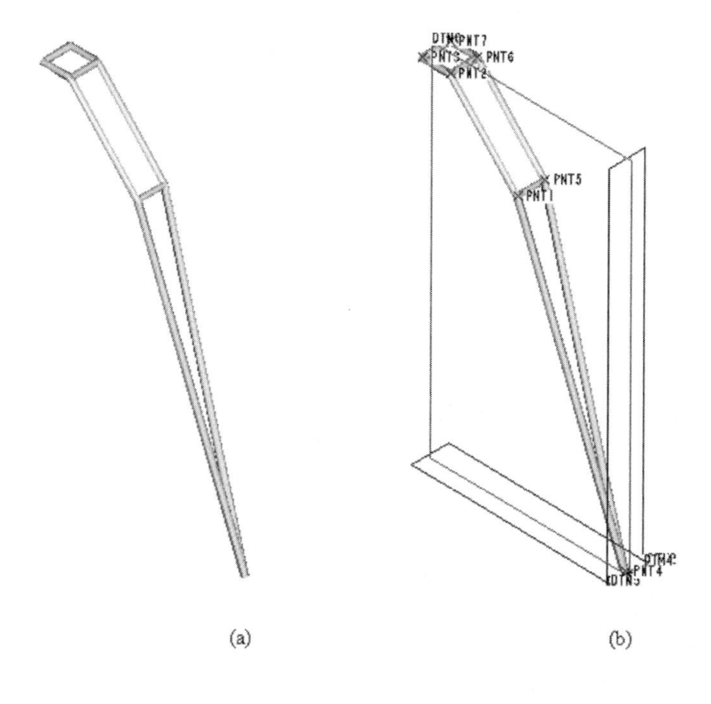


FIGURE 11.45 *Figure for exercise 11.6.*

11.7 Create a model of the hook spanner using the geometry shown in Figure 11.46.

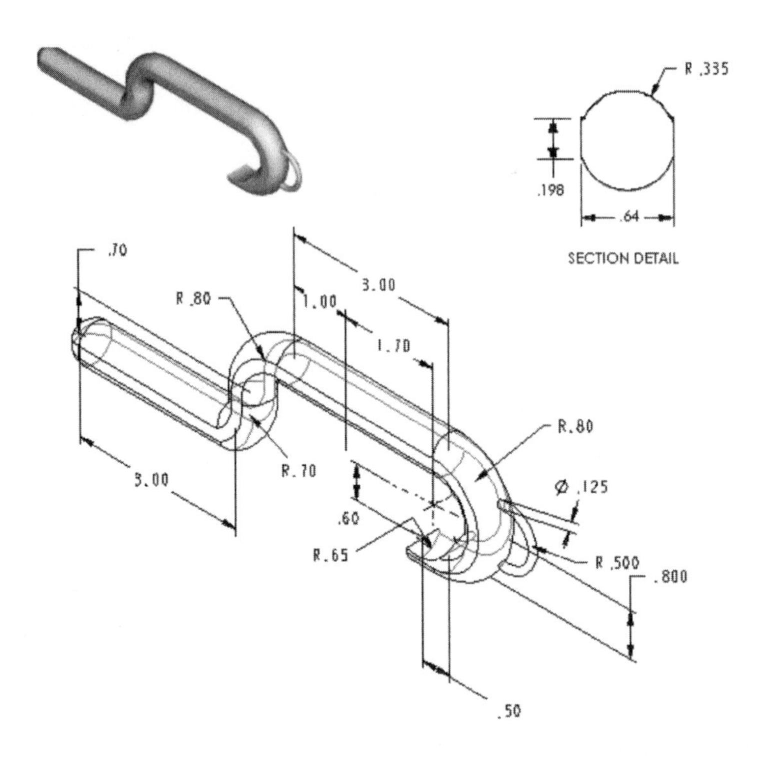


FIGURE 11.46 *Figure for exercise 11.7.*

11.8 Construct a model of the shaft hanger using the appropriate Pro/E options and the geometry in Figure 11.47.

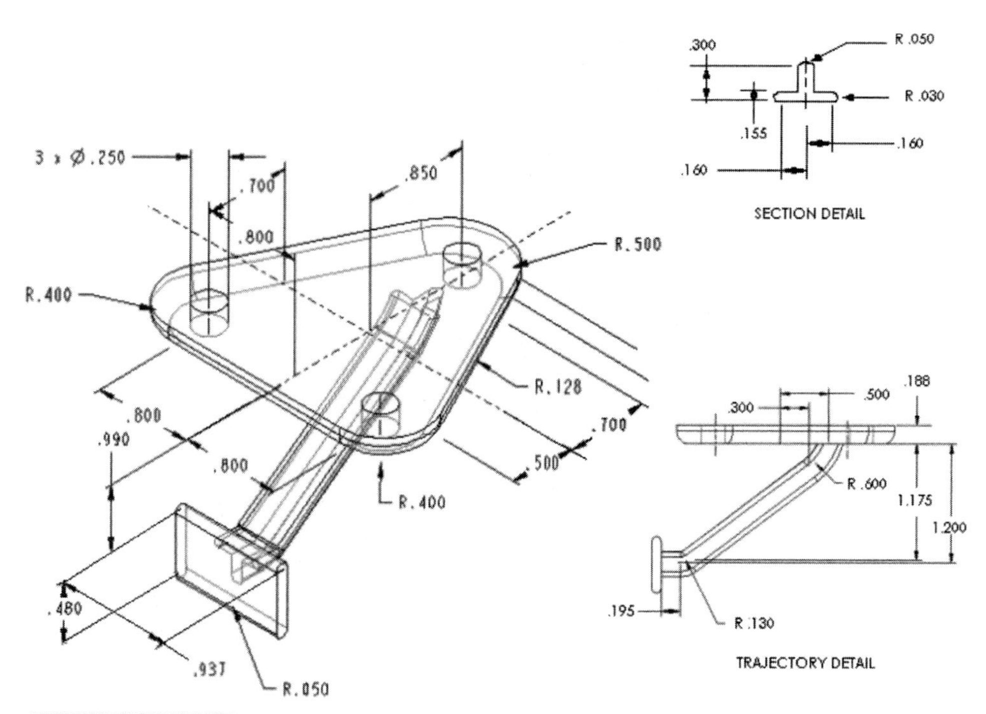


FIGURE 11.47 *Figure for exercise 11.8.*

11.9 Create a model of the hand rail support. For the cross section of the sweep, use the given section in Figure 11.48.

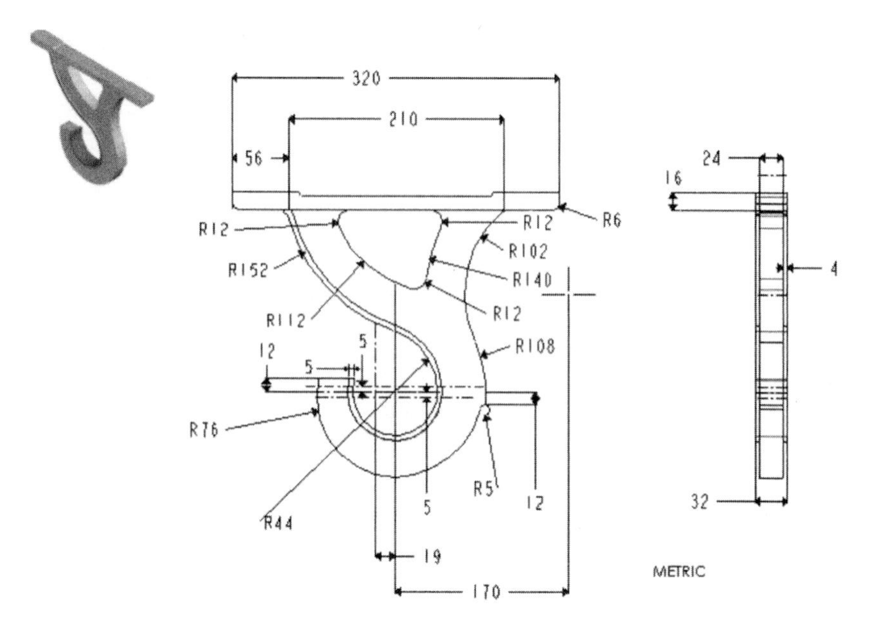


FIGURE 11.48 *Figure for exercise 11.9.*

11.10 Construct the model of the dash pot lifter. Use the appropriate Pro/E options. The geometry of the part is given in Figure 11.49.

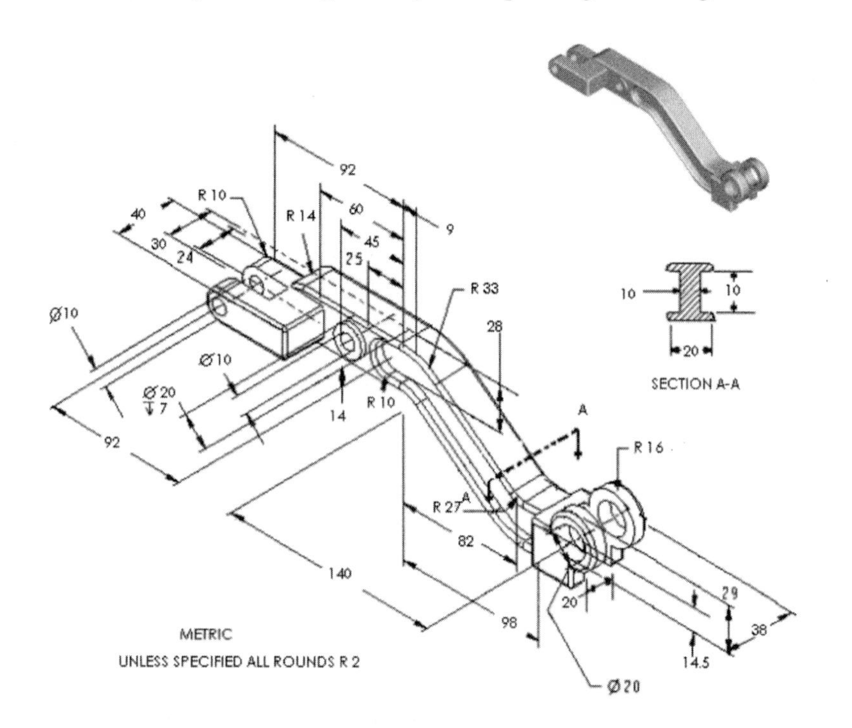


FIGURE 11.49 *Figure for exercise 11.10.*

11.11 Add a handle to the ladle constructed in exercise 10.11. Use the *Sweep* option and Figure 11.50.

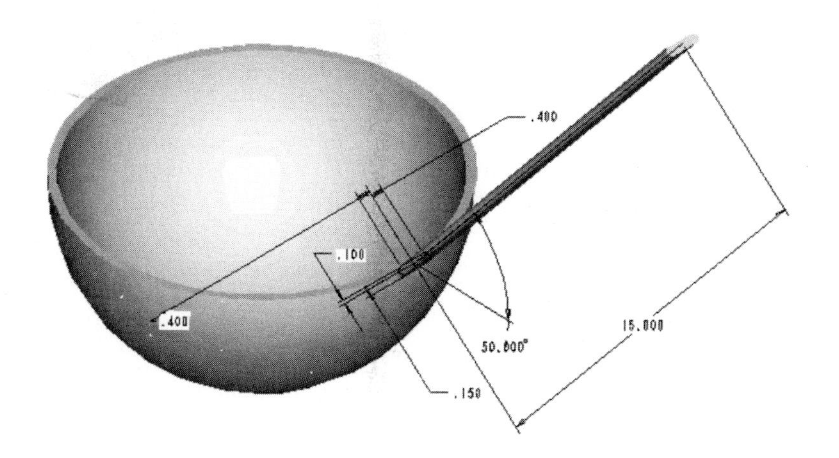


FIGURE 11.50 *Figure for exercise 11.11.*

11.12 Add a grip to the ladle from exercise 11.11. Use the geometry in Figure 11.51.

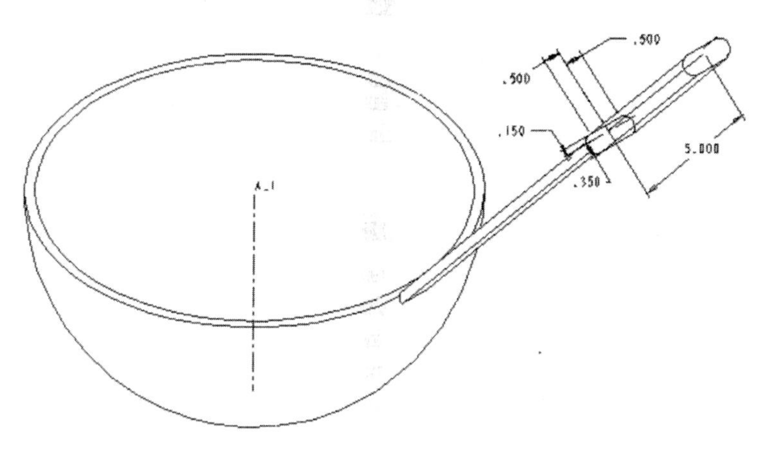


FIGURE 11.51 *Figure for exercise 11.12.*

# BLENDS

## INTRODUCTION AND OBJECTIVES

A *Blend* is a feature constructed of more than one section joined together by transitional surfaces. These surfaces may be smooth or straight. The three types of blends are *Parallel, Rotational,* and *General.*

In a *Parallel* blend, all the sections lie on parallel planes. *Rotational* blends are created if the sections are rotated. Pro/E allows a maximum of 120° rotation. The last type of blend a *General* blend, allows both rotation and translation of the sections.

A blend along a trajectory may be created by using the *Swept Blend* option. This option is a combination of the *Sweep* and *Blend* options in which the user sketches the trajectory and sections. This option is useful for constructing features that would normally be created using the *Sweep* option, but whose cross sections are not constant.

The objectives of this chapter are:

1. The software user will attain a proficiency in constructing blends.
2. The reader will be able to create a model using the *Swept Blend* option.

Pro/E assumes that all sections in a blend have the same number of entities. This assumption presents a problem when, say, the user wishes to blend two sections that do not contain the same number of entities. An example is the blend between a circular and a rectangular section. Because the software considers a circle as one entity and the rectangle as four, the blend cannot be done unless the circle is divided into four elements. Of course, the circle must be properly divided. This requirements leads to an additional objective:

3. The user will attain experience in properly dividing a section so that it may be blended with another section containing a different number of entities.

## 12.1 The Blend Option

The *Blend* option is used to create a blended feature. A blended feature is constructed of at least two planar sections that are joined together by a continuous feature. This feature can be either a smooth blend or a straight blend. In a smooth blend, the blend is accomplished by joining the sections with smooth curves. In a straight blend, the blend is performed by joining the sections with straight lines.

The blend type depends on the orientation of the planar sections. If all the planar sections are parallel, then a *Parallel* blend can be performed. Sections that are rotated about the y-axis lead to *Rotational* blends. Finally, sections that are rotated and translated about the x- and y-axes are blended using the *General* blend option.

In creating a blend, the user is asked to sketch, dimension, and regenerate each subsection before proceeding to the next one. In order to sketch the next subsection, the previous subsection must be grayed after successful regeneration by selecting *SEC TOOLS* (Figure 2.9) and then *Toggle* from the *SEC TOOLS* menu (Figure 2.18).

After clicking on the *Toggle* option, the previous subsection is grayed and the *Sketcher* can be used to sketch the next subsection by clicking on *Sketch* in the *SKETCHER* menu. The new subsection is drawn, dimensioned, and regenerated. The process is repeated as many times as the number of subsections. Only after the final subsection has been successfully regenerated is the *Done* option selected in the *SKETCHER* menu.

Pro/E blends the subsections by connecting the vertices of each subsection to the next. Thus, all the subsections must have the same number of entities. For example, a blend cannot be performed between a circle and a square because the circle contains only one entity while the square contains four. In such a case, the circle must be broken into four arcs.

During the creation of the subsection, the sketch must be dimensioned and regenerated. If datum default planes are used, then the subsections can be dimensioned with respect to the datum planes. If, on the other hand, default datum planes are not used, then the subsections must be dimensioned with respect to one another.

It is also important to keep track of the start point for each section. If the start point is not the same for all the subsections, then a part with a twisted blend is obtained.

### 12.1.1 Tutorial # 59: HVAC Takeoff

We have chosen an HVAC takeoff as an example of the blend. The takeoff under consideration has two subsections, the first being a 6-inch × 6-inch square, and the second, a 6-inch diameter circle. Because the number of entities for each subsection is not the same, we will need to divide the circle into four equal length arcs by first sketching and regenerating the square. Then, diagonal lines will be drawn connecting the corners of the square and dividing the circle into the equal length arcs.

Create a part called "HeatTakeoff." Create the default datum planes. Select *Feature, Solid, Protrusion,* and *Blend.* Notice that the blend can be performed on a solid as well as a thin feature. Choose *Thin.* Then select *Done.*

From the *BLEND OPTS* menu (Figure 12.1), select *Parallel* and *Regular Sec,* because the subsections will be parallel to one another. After selecting *Done,* the *ATTRIBUTES* menu will appear as reproduced in Figure 12.2. Click on *Straight.* Choose DTM3 as the sketching plane and DTM2 as the "*Top*" plane to orient the model in the *Sketcher.*

Draw the square as dimensioned and as shown in Figure 12.3. Now, select the *Relation* option from the *SKETCHER* menu and

1. Choose *Add.*
2. Enter sd1 = sd0/2. Hit *Return.*
3. Enter sd3 = sd2/2. Hit *Return.*
4. Hit *Return* one last time.
5. *Regenerate. Modify* the dimensions so that sd1 = sd3 = 6.
6. *Regenerate* the model again.

Click on *Sec Tools* in the *SKETCHER* menu (Figure 2.9) and then on *Toggle* from the *SEC TOOLS* menu (Figure 2.18). The *Toggle* option will gray the square.

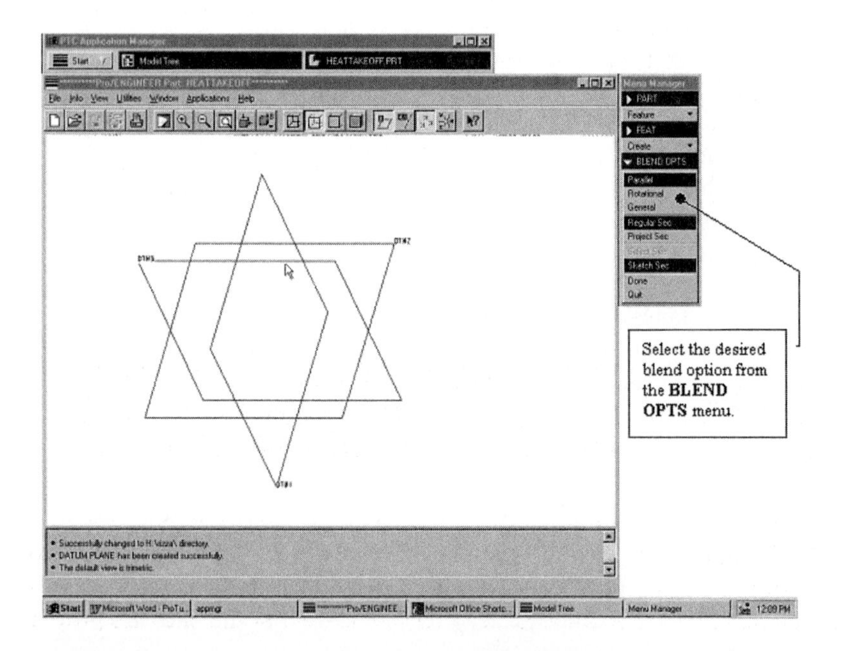


FIGURE 12.1 *The* BLEND OPTS *menu.*

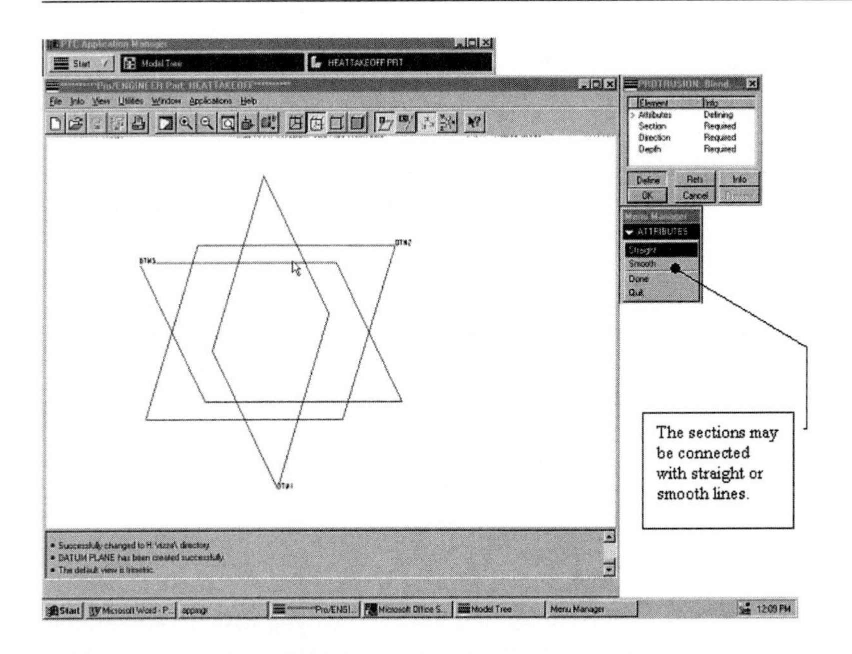

The sections may be connected with straight or smooth lines.


FIGURE 12.2 *The* ATTRIBUTES *menu showing the* Blend *options* Straight *and* Smooth.

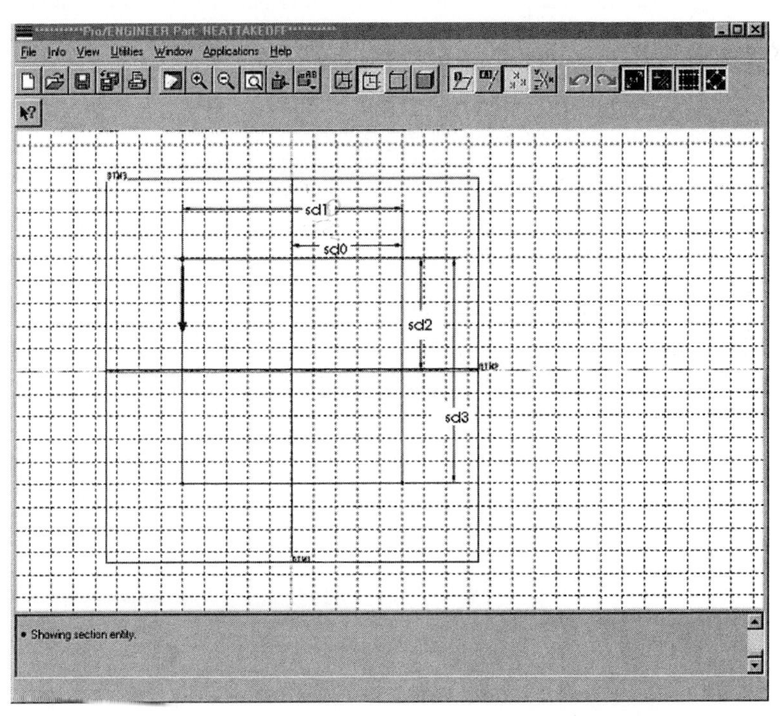


FIGURE 12.3 *The first subsection with added dimensions.*

Click on *Sketch* and draw the circle. Align the center of the circle to the DTM1 and DTM2 datum plane edges. Draw diagonal lines through the corners of the square. The lines will be 45° with the horizontal and will divide the circle into four equal arcs. In order to ensure that the lines are properly positioned, dimension the lines with respect to the DTM2 edge. As shown in Figure 12.4, the angular dimensions should have values of 45°. If they do not, *Modify* the values.

Using *Geom Tools* and *Intersection,* click near the intersections of the circle and the lines. By doing so, you will obtain four intersecting points, as shown in Figure 12.4. After obtaining these points, use *Delete* to remove the lines.

Notice in Figure 12.3 that the start point for the square is in the upper left-hand quadrant formed by the datum planes. We want the start point for the circle to be in the same quadrant as that for the square. Most likely, the start point for the circle moved after you deleted the lines. If it moved, place the start point for the circle in the same quadrant as for the square by selecting *Sec Tools* and then *Start Point.* Then click on the intersection point in the proper quadrant. The placement of the start point for the circle is shown in Figure 12.5.

Pro/E will need to know the location of the intersections on the circle. *Dimension* the points as shown in Figure 12.5. Because the lines were properly located, the points will be, also. *Regenerate* the sketch.

As Pro/E begins to recognize the position of the circular subsection with that of the square subsection, it will highlight dimensions that are no longer needed in red. You may delete these dimensions if you wish. However, it is not necessary to do so. Choose *Done.* Fill the thickness of the part *inward* with a thickness of 0.05 inches. Enter a depth of 12.0 inches. Then select *OK.*

The takeoff should appear as in Figure 12.6. Save the part.

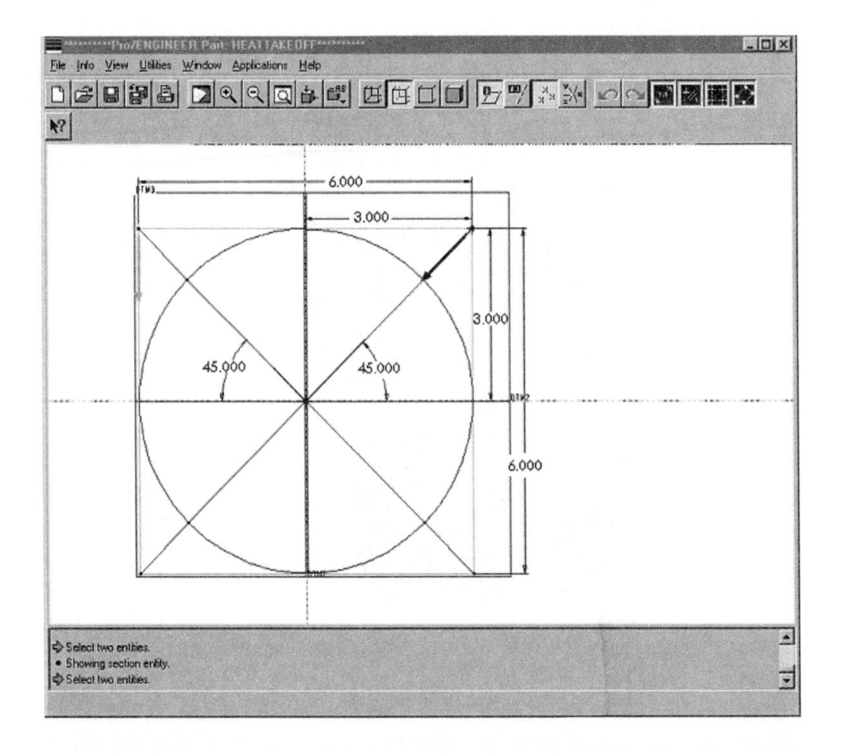


**FIGURE 12.4** *The circular subsection divided into four arcs by using two diagonal lines and the* Geom Tools *option.*

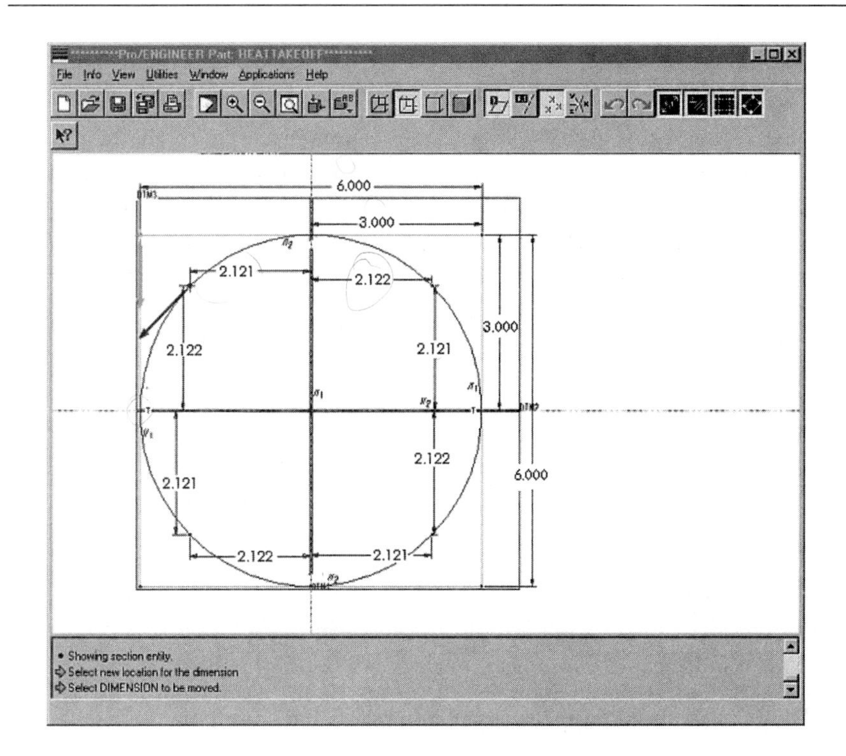


FIGURE 12.5 *Notice the proper placement of the* Start Point *in the circular subsection.*

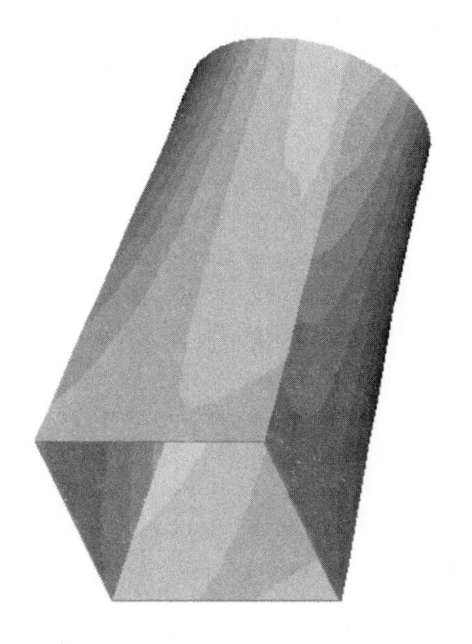

FIGURE 12.6 *After constructing with a* Blend, *this shaded solid model of the HVAC takeoff may be obtained.*

## 12.2 SWEPT BLENDS

A *Swept Blend* is a blend along a defined trajectory. Multiple sections may be placed along the trajectory and their geometry defined. The trajectory may be closed or open. In the case of an open trajectory, the section *must* be sketched at the endpoints of the trajectory. For closed trajectories, the sections must be drawn at the start point and one other point along the trajectory. As with a regular sweep, the trajectory may be sketched or selected from predefined datums, curves, or edges.

After defining the trajectory, Pro/E will place points on the sketch where the section is to be sketched. The section can be rotated about the z-axis of the given coordinate system. The value of the rotation must be between +120° and −120°. The section may be placed normal to the trajectory (*Nrm to Spine* option) or normal to trajectory as viewed from a pivot point (*Pivot Dir* option). These options may be chosen from the *BLEND OPTS* menu shown in Figure 12.7.

### 12.2.1 TUTORIAL # 60: A CENTRIFUGAL BLOWER

The model of the centrifugal blower shown in Figure 12.7 is constructed from a revolution (for the body) and a swept blend (for the discharge). The first phase in constructing the model is to make the body of the blower.

One side of the body has a circular hole that allows the fluid to enter the blower. Certainly, this hole can be added to the model later. However, the hole may be created by properly sketching the cross section for the revolution. The cross section is shown in Figure 12.8.

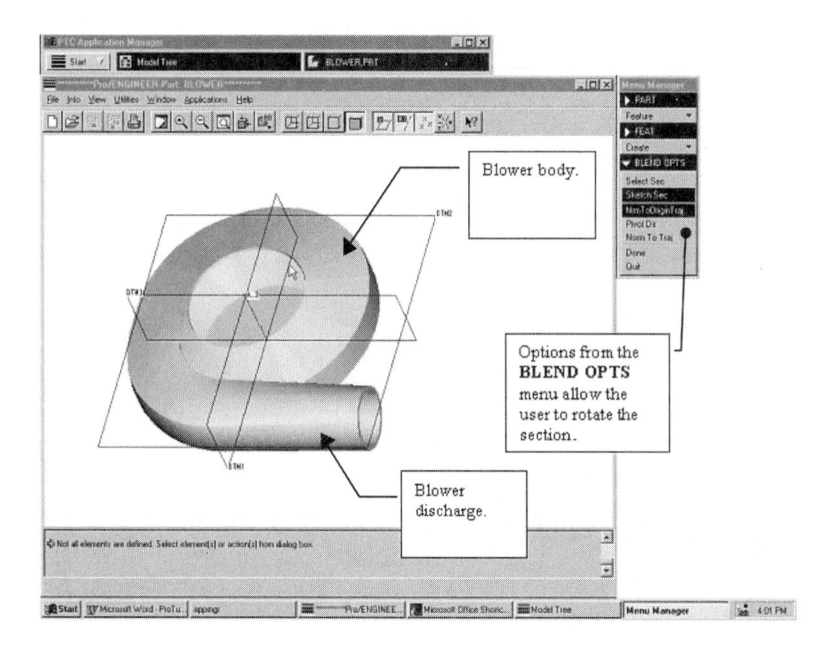


FIGURE 12.7 *The options in the* BLEND OPTS *menu are used to define the orientation of the section and trajectory (spine).*

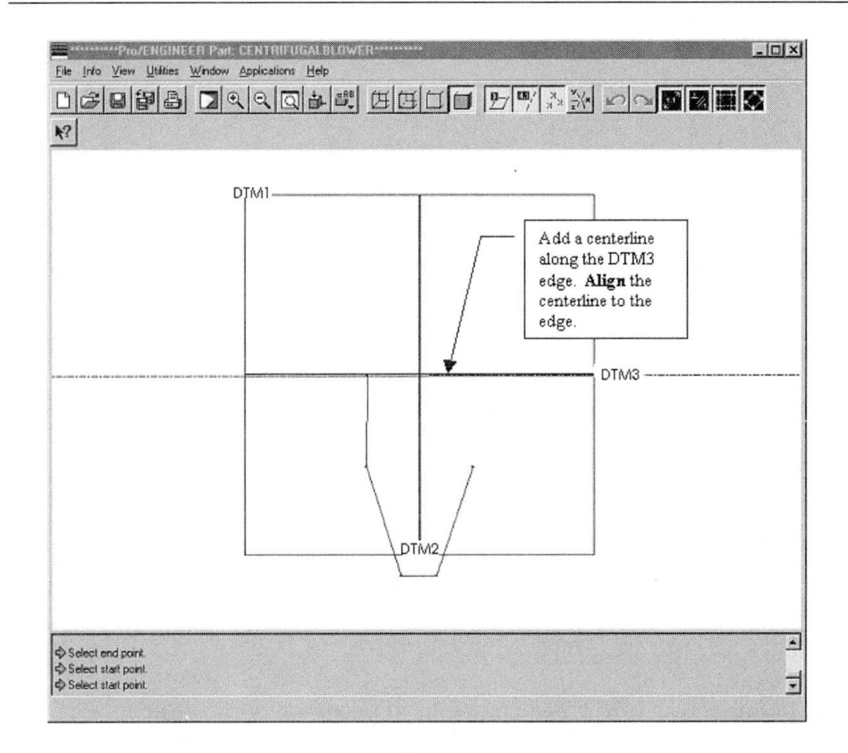


FIGURE 12.8 *The geometry necessary for constructing the body of the blower.*

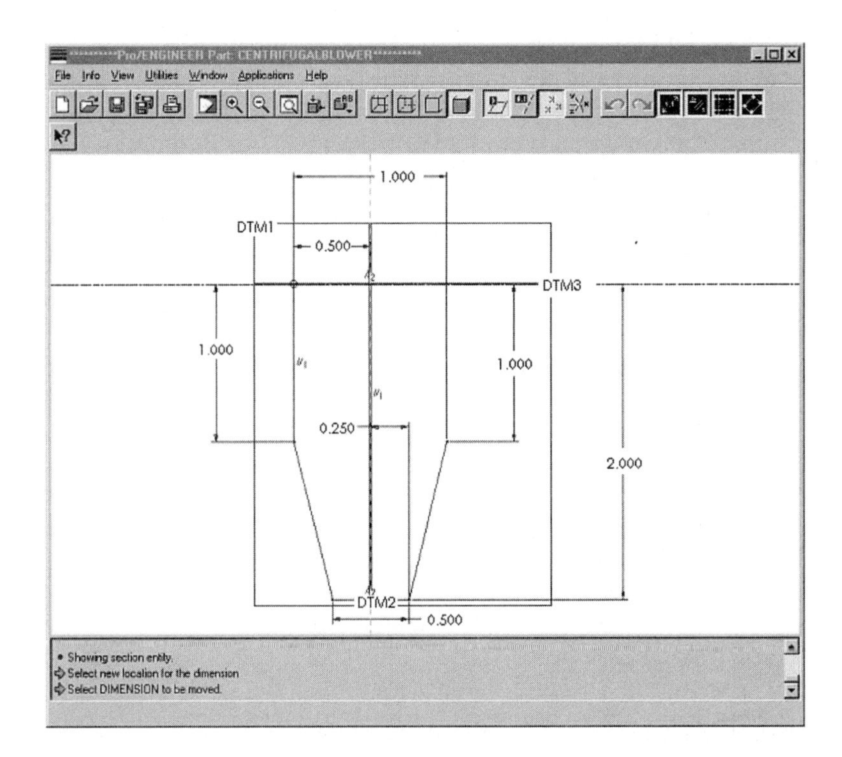


FIGURE 12.9 *Dimension the geometry as shown.*

Let us construct the body of the blower. The procedure is:

1. Create a part file called "Blower" and add default datum planes.
2. Select *Feature, Create, Protrusion, Revolution,* and *Thin.*
3. Use the *Both Sides* option.
4. Select DTM1 as the sketching plane and DTM3 as the "*Top.*"
5. These choices will properly orient the model in the *Sketcher.* Sketch the section shown in Figure 12.9.
6. Select *Line* and *Centerline.* Add an axis of revolution along the DTM3 edge.
7. Align the centerline to the DTM3 edge.
8. For the thickness, grow the part inward using a value of 0.05 for the thickness.
9. Revolve the cross section 360°.

Construct the swept blend by selecting *Feature, Create,* and *Protrusion.* From the *SOLID OPTS* menu (Figure 2.6) select the *Thin* and *Advanced Feature* options. Choose the *Nrm to Spine* option from the *BLEND OPTS* menu (Figure 12.7)

We will be sketching the trajectory. Use DTM2 as the sketching plane and select DTM3 as the "*Top*" to reorient the model. Now sketch vertical and horizontal lines as shown in Figure 12.10. Add an arc between the two lines. *Dimension* as shown in the figure.

Notice the position of the start point in Figure 12.10. Make sure that the start point is located in the same place on your part. If not, use *Sec Tools* and *Start Point* to move the start point to its proper location. If the start point is properly located, choose *Done.*

Now Pro/E will inquire as to where you want to sketch the section. It will query you by highlighting a point. Notice that the trajectory has four points. Because it is an open section, the section must be defined at the endpoints. Thus, Pro/E will begin by highlighting one of the interior points. Use the *Next* option (if necessary, use the *Previous* option) and scroll through the points on the trajectory until Pro/E highlights the endpoint where the start arrow is located. Choose *Accept* when it has done so.

Notice that Pro/E has placed an x-y-z coordinate system at the point. The software will inquire as to the rotation angle. Enter a value of zero.

Pro/E will reorient the sketch as shown in Figure 12.11. Notice the coordinate system indicating the point. Sketch a circle with the center of this circle aligned to the origin of the coordinate system, as illustrated in Figure 12.12. Sketch two lines roughly at ±45° to the horizontal. The purpose of the lines is to divide the circle into four equal arcs to match the next section, which will contain four entities.

Use *Geom Tools* and find the intersection of the circle with the lines, which yield four points as shown in Figure 12.13. Delete the lines, because

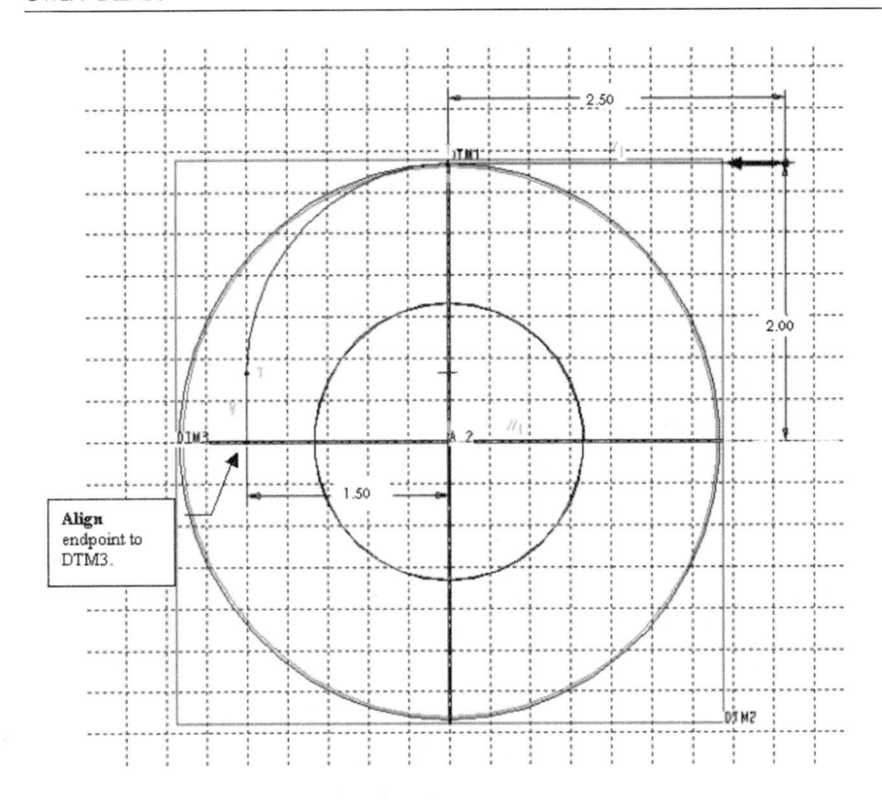


FIGURE 12.10 *The blower in the* Sketcher *with the geometry needed to define the trajectory of the swept blend nozzle.*

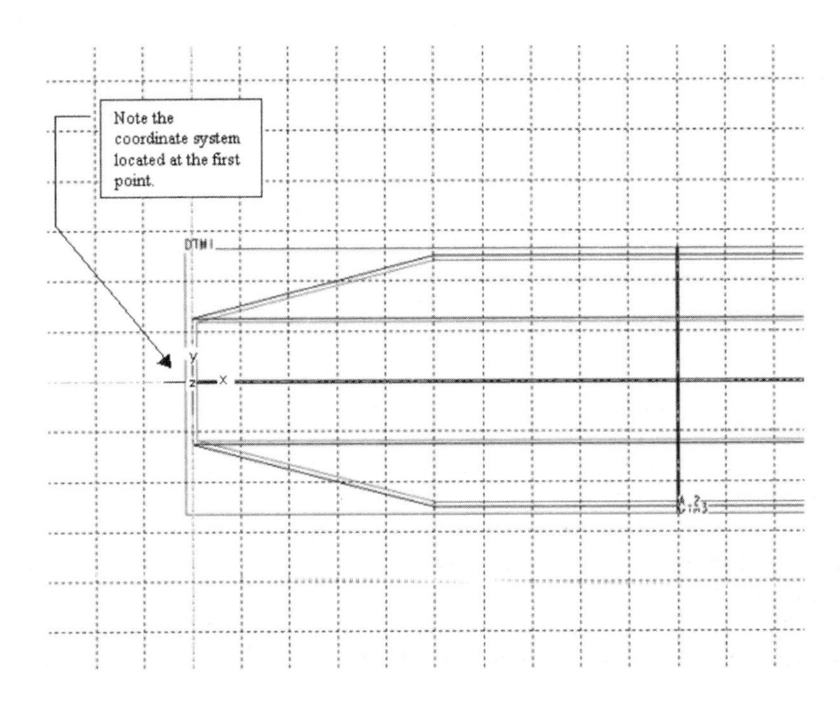


FIGURE 12.11 *The coordinate system on the blower is locating the first point of the trajectory.*

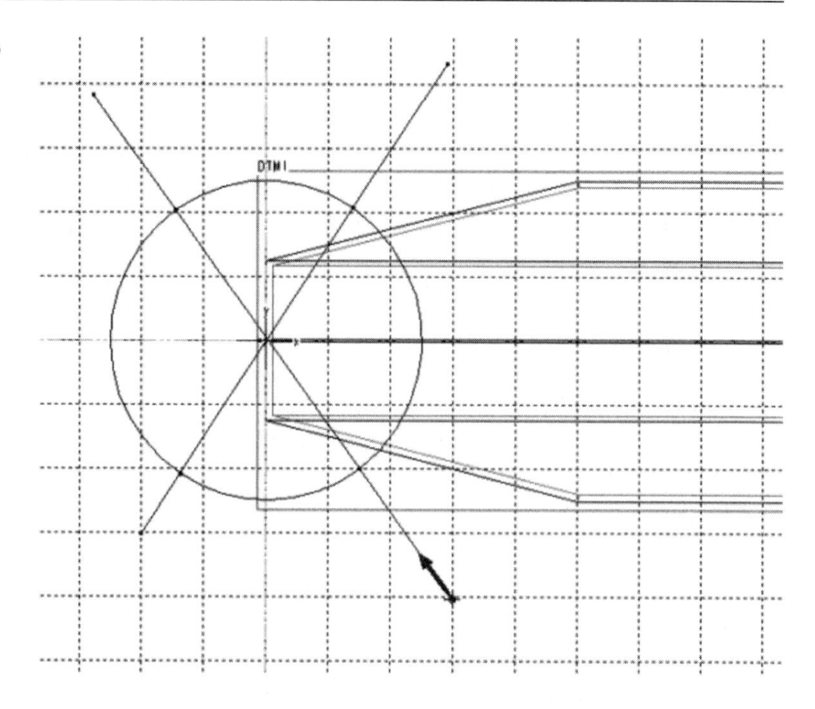


**FIGURE 12.12** *The circle bisected by two lines. These lines will be used to divide the circle into four equal arcs.*

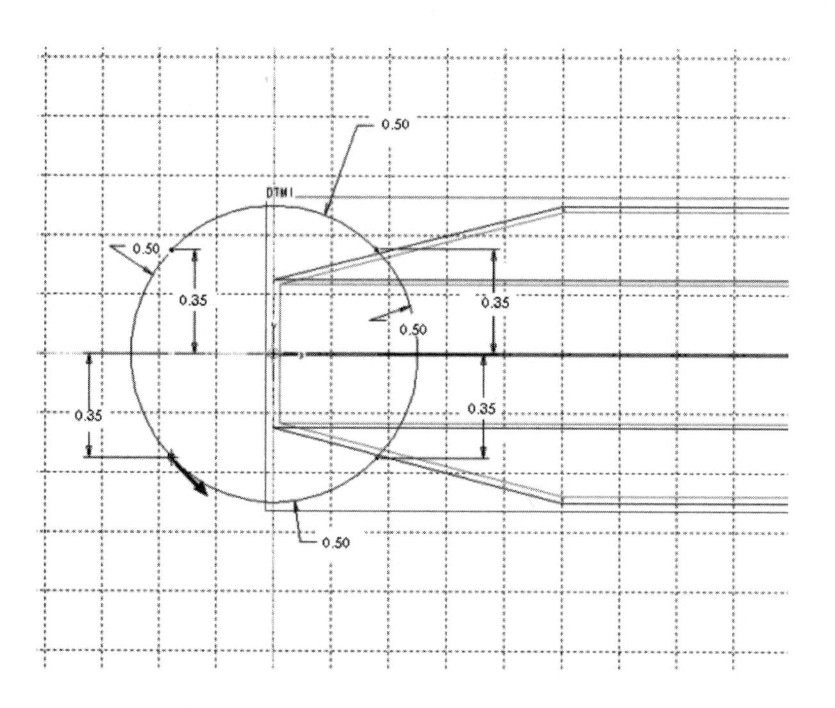


**FIGURE 12.13** *After dividing the circle into four equal arcs, the section can be sized using the shown dimensions.*

they are no longer needed. *Dimension* the four arcs and the location of the points. Note that the vertical dimension of the points is $0.5 \times \sin 45 = 0.35$.

*Regenerate.* Before moving on to the next section, make sure that the start point for this section is located as shown in Figure 12.13. It is immaterial on which of the four points the start point is located, but it must be located in the same quadrant as the one for the next section. After the start point is located, select *Done.*

Grow the thin section inward. Pro/E will inquire about the rotation for the next section; enter a value of 0.00. Again, the software will reorient the model and place a coordinate system as shown in Figure 12.14 indicating the second point. The second section is an outline of the silhouette of the part body and a vertical line. This section is given in Figure 12.15.

In Figure 12.15, notice the location of the start point for the second section. Make sure that the start point is located at the same place in your model. Also notice that the total number of entities for this section is four (four lines), which explains the need to break the first section (the circle) into four equal arcs. In Pro/E, the number of entities must be the same for each section, unless the blend is created using the *General* option.

Dimension the second section as shown in Figure 12.15. Then *Regenerate.* Select *Done* and grow the part inward. Enter a value of 0.05 for the thickness. Select *OK* from the part creation box. Pro/E will generate the feature. A shaded model of the blower is given in Figure 12.7.

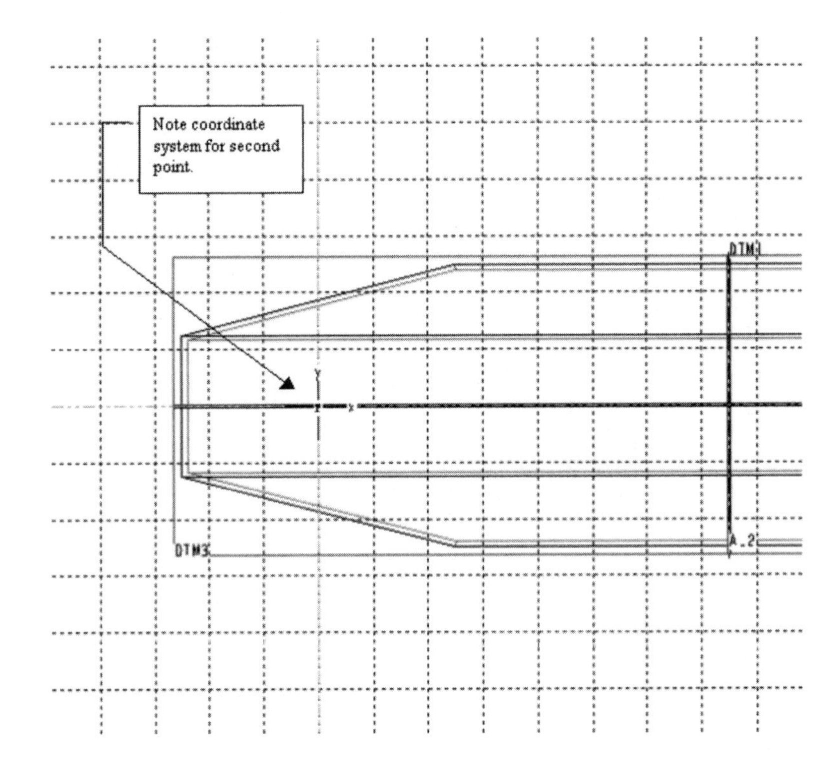


FIGURE 12.14 *The blower with the second point locating the second section.*

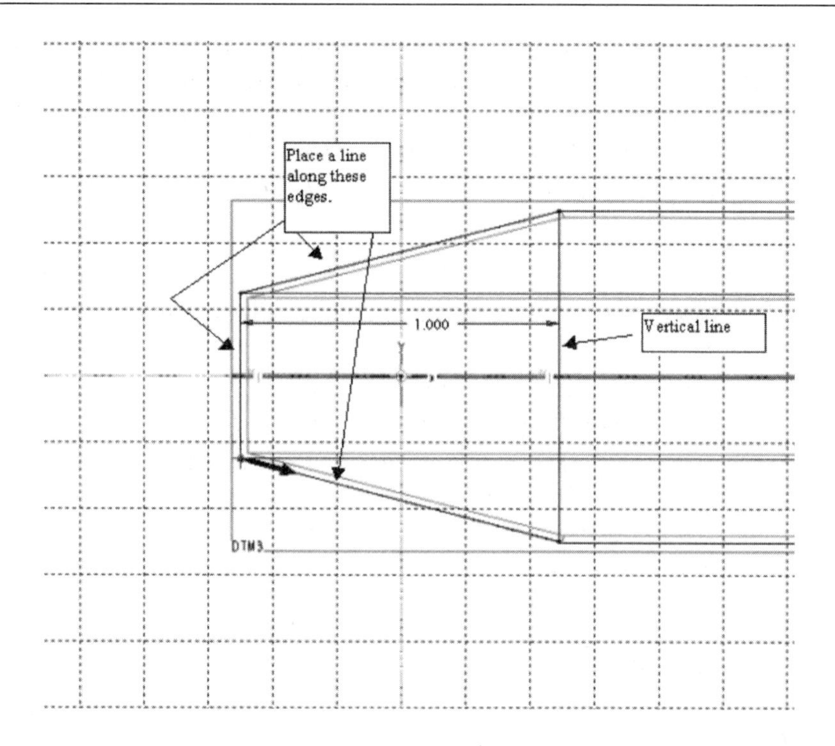


FIGURE 12.15 *The geometry of the second section includes a vertical line.*

## 12.3 SUMMARY AND STEPS FOR CREATING BLENDS

A blend is a feature constructed by joining more than one section using smooth or linear transitional surfaces. The blend is created by sketching and dimensioning or by selecting each section.

The software blends the individual sections by connecting the vertices of each subsection. Therefore, in order to avoid a "twisted blend," the start arrow for each subsection must be located at the same vertex.

The general sequence of steps for creating a *Blend* is:

1. Select *Feature, Create, Protrusion, Solid* (or *Thin*), and *Done.*
2. Choose one of these options: *Parallel, Rotational,* or *General.* In addition, select from *Regular Sec* or *Project Sec.*
3. Select *Done.*
4. Select either *Straight* or *Smooth.* Then select *Done.*
5. Choose a sketching and an orientation plane.
6. *Sketch, Dimension,* and *Regenerate* the first subsection.
7. Locate the start point by using *Sec Tools* and *Start Point.*
8. Select *Sec Tools* and *Toggle* to gray the first subsection.
9. Repeat steps 6 through 8 for the additional subsections. For the last subsection, choose *Done.*
10. Provide the depth of the blend.

A blend may be performed along a trajectory, which is called a *Swept Blend.* As with a regular blend, multiple sections may be defined. However, in this case, the sections are defined for discrete points along the trajectory. In general, the following steps may used to create a *Swept Blend:*

1. Select *Feature, Create, Protrusion, Solid* (or *Thin*), *Advanced Feature,* and *Done.*
2. Choose *Swept Blend* and *Done.*
3. Select *Sketch Sec* in order to sketch the sections or *Select Sec* to pick the sections.
4. The trajectory needs to be defined by sketching (*Sketch Traj*) or by selecting the appropriate entities (*Select Traj*).
5. Depending on your response in step 4, either sketch or pick the trajectory. If you have chosen to sketch the trajectory, then the software will need a sketching plane and an orientation plane.
6. Sketch the subsection for each desired location. For an open trajectory, the section *must* be sketched at the endpoints of the trajectory.

## 12.4 ADDITIONAL EXERCISES

12.1 Construct the model of the stop using the *Blend* option and Figure 12.16.

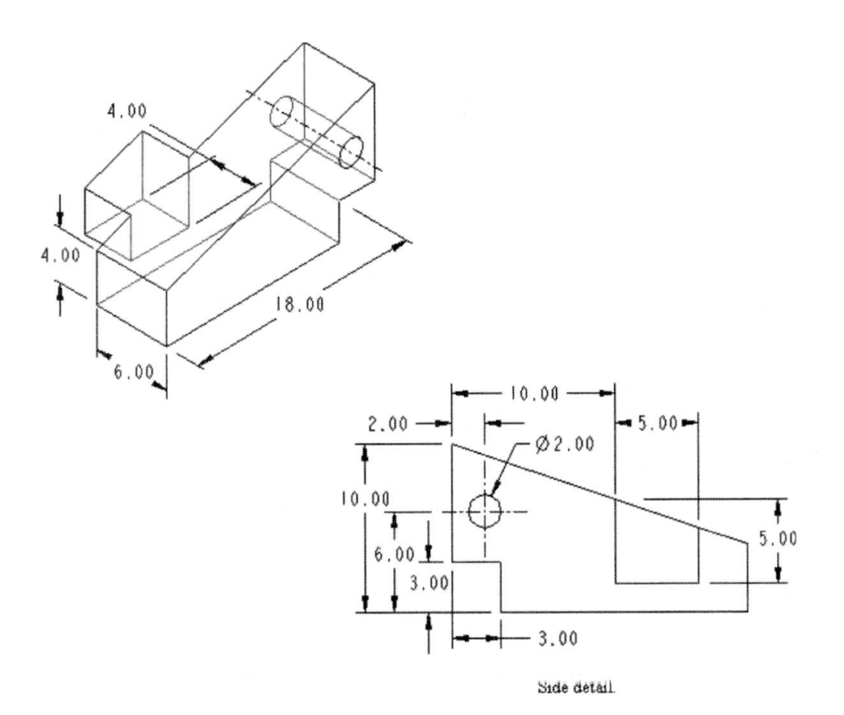


FIGURE 12.16 *Figure for exercise 12.1.*

12.2 Create the model of the holder using the *Blend* option. Use Figure 12.17 as a reference.

FIGURE 12.17 *Figure for exercise 12.2.*

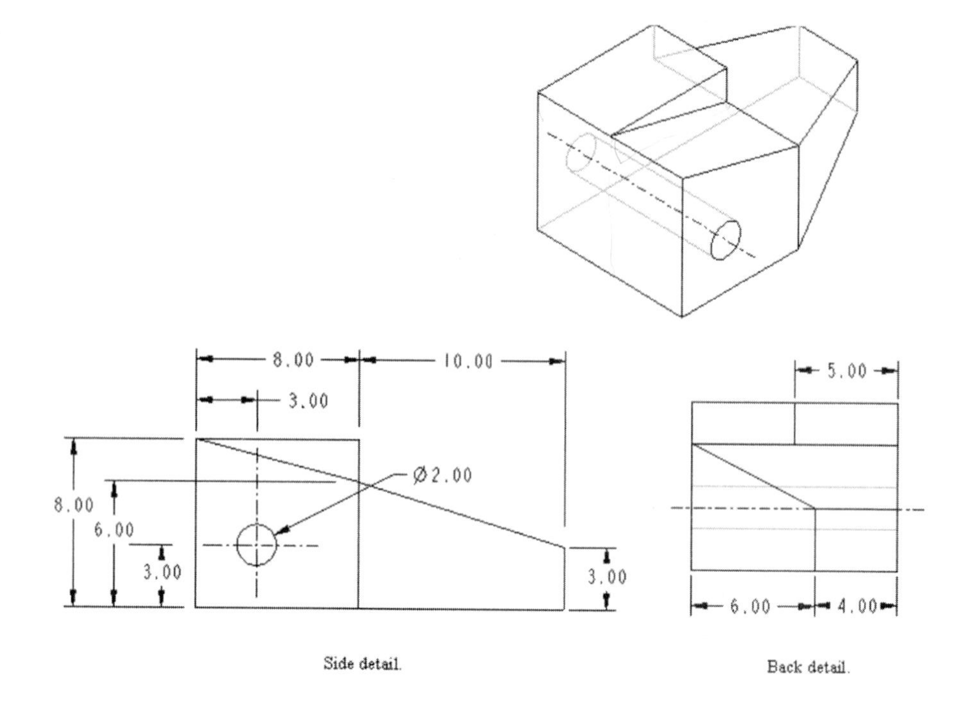


12.3 Using the *Blend* option and Figure 12.18, construct the model of the support.

FIGURE 12.18 *Figure for exercise 12.3.*

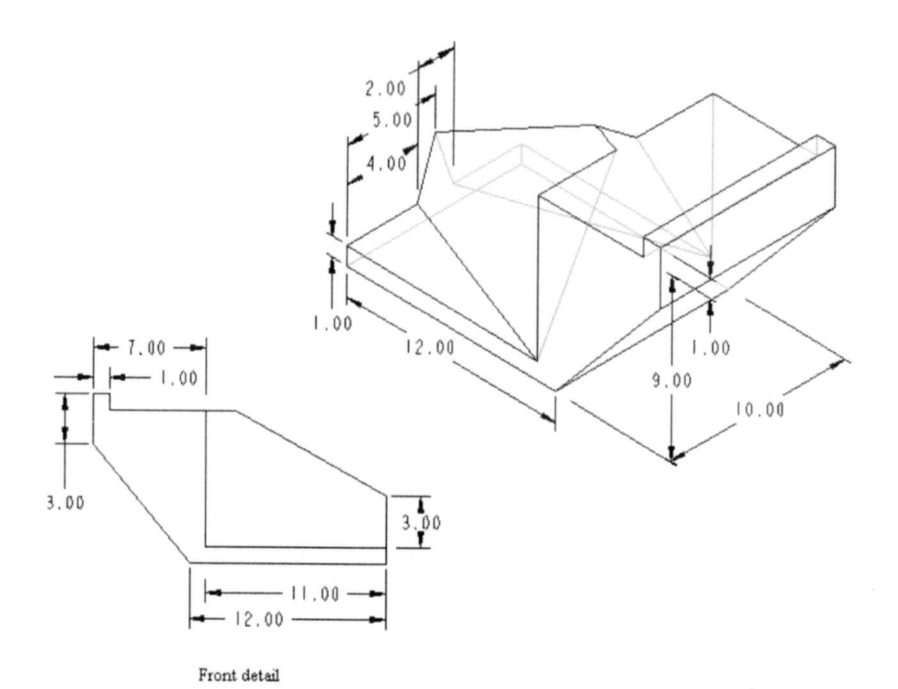

12.4 Add a spout to the ladle from exercise 11.12. Use the *Swept Blend* option and Figure 12.19.

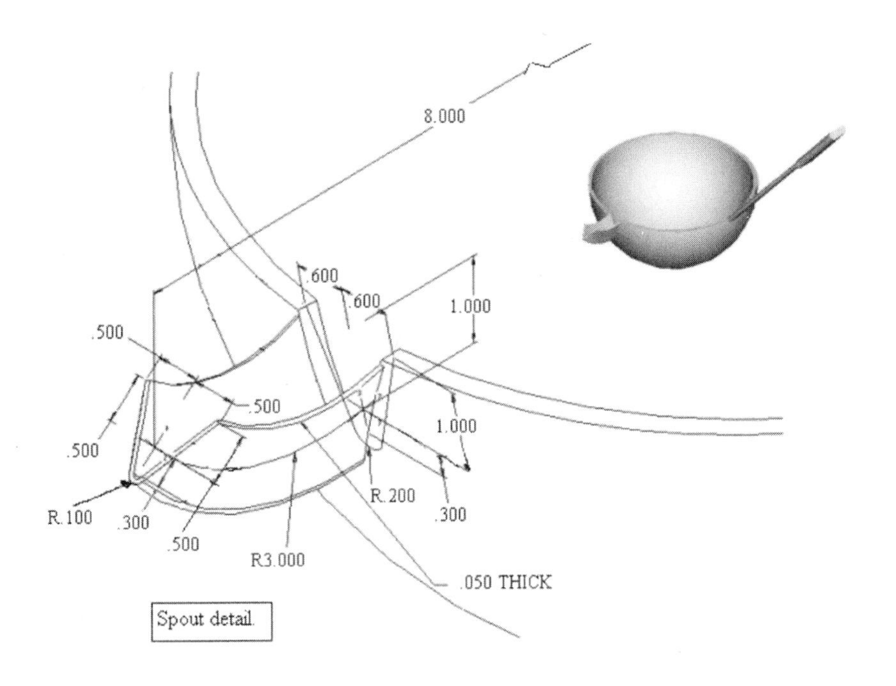


FIGURE 12.19 *Figure for exercise 12.4.*

12.5 Create the faucet. Use the *Swept Blend* option. In Figure 12.20, note the four sections along the sweep trajectory.

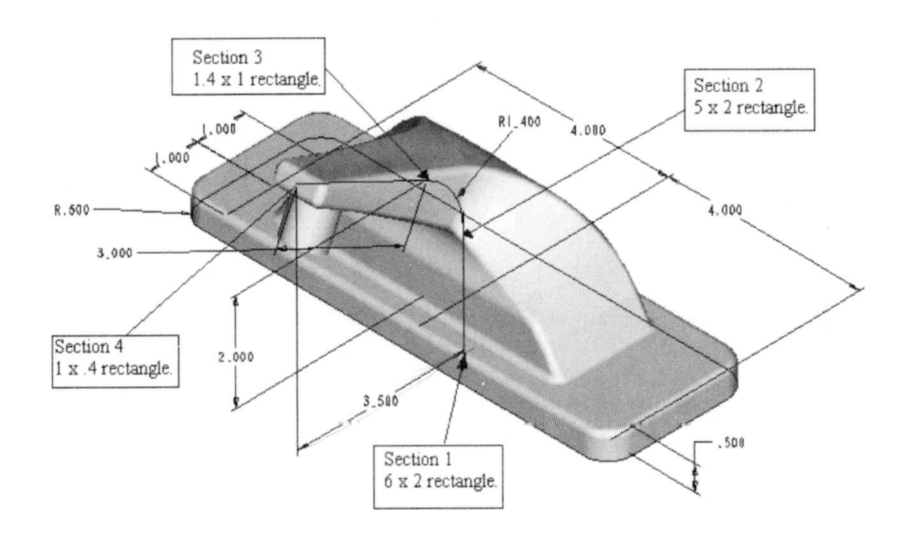


FIGURE 12.20 *Figure for exercise 12.5.*

12.6 Add rounds and an aerator to the model of the faucet as shown in Figure 12.21. All rounds are 0.125.

FIGURE 12.21 *Figure for exercise 12.6.*

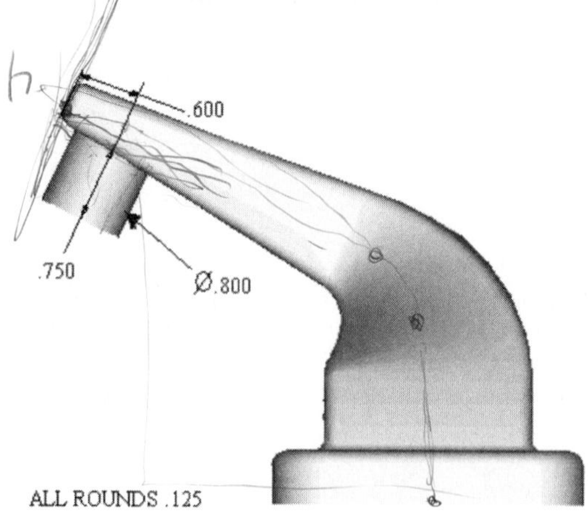

# CHAPTER 13

# SOME OPTIONS FOR MANAGING FEATURES

## INTRODUCTION AND OBJECTIVES

The options in this chapter are from various menus that may be used to group, modify, reorder, or hide existing features.

From the *FEAT* menu, the options under consideration are *Redefine, Reorder, Reroute,* and *Suppress.* Perhaps the most helpful of these options is *Redefine,* which allows the users to change a feature by stepping through the elements defined to construct it. The *Reorder, Reroute,* and *Suppress* options allow the user to control the generating and displaying of features, respectively. The option, *Resume,* restores the visibility of suppressed features.

A part, drawing, or assembly may contain multiple levels. These levels are called layers by Pro/E. Often different types of features are placed on different layers. Features in a layer are treated as a group. Operations may be performed on the group. For example, say a user wishes to turn off the display of all the rounds in a part. The rounds may be placed on a layer called "All_Rounds" and then the visibility of this layer turned off. Obviously, this capability saves time, because without the use of layers, the visibility of each round would have to be deactivated. With names given by the user, layers allow the user to group features in a logical manner. The *Layer* option is found in the *PART* menu.

The last option that will be considered in this chapter is the *Resolve* option. This option does not appear on any of the standard menus. It becomes active only after a failure occurs in feature construction. The option allows the user to diagnose and fix the failure.

The objectives of this chapter are as follows:

1. The user will attain experience in creating *Layers* and the ability to add features to a layer.
2. The software user will obtain proficiency in using the following feature management options: *Redefine, Reroute, Reorder, Suppress,* and *Resume.*
3. With the use of *Resolve,* the user will diagnose and fix a failure in creating a feature.

## 13.1 THE REDEFINE OPTION

The *Redefine* option in the *FEAT* menu can be used to reconstruct a model because of a desired change, or if the model was created incorrectly. The *Redefine* option is used in conjunction with the *Define* button in the Feature Creation box (Figure 13.1). Any of the elements defining the feature may be modified by going through the steps defining the feature and by selecting the element and then choosing the *Define* option.

### 13.1.1 TUTORIAL # 61: REDEFINING A FEATURE

We have chosen the spring from chapter 11 as the part to illustrate the use of the *Redefine* option. Retrieve the part "Spring.prt."

You may recall that the spring was created using a *Helical Sweep.* The cross section of the spring was a circular section. The trajectory was parallel to the axis of revolution. Of course, any of the elements of the spring, including the cross section and trajectory, may be redefined. In this tutorial, we have decided to redefine the trajectory so that a tapered spring is constructed.

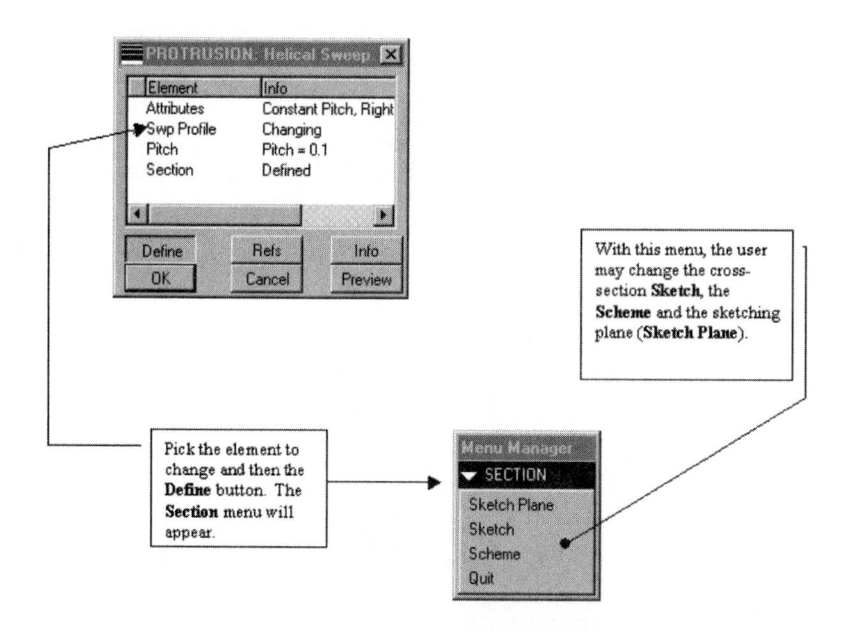


FIGURE 13.1 *The Feature Creation box. Any of these elements may be redefined by selecting the element and then* Define.

From the *FEAT* menu, select *Redefine.* Pick the spring with your mouse pointer. The Feature Creation box will appear as shown in Figure 13.1. We wish to change the sweep profile (*Swp Profile*). Select this element in the box, then choose the *Define* button. Choose *Modify* and *Done.*

The *SECTION* menu will appear as shown in Figure 13.1. With this menu, the user may change the *Sketch, Scheme,* and the sketching plane of the feature.

Select *Sketch.* The software will load the trajectory in the *Sketcher.* Delete the line on the screen and sketch a new line as shown in Figure 13.2. *Dimension, Regenerate,* and *Modify* the sketch to the values given in the figure.

Choose *Done.* Then select *OK.* The software will construct the new model, which is shown in Figure 13.3.

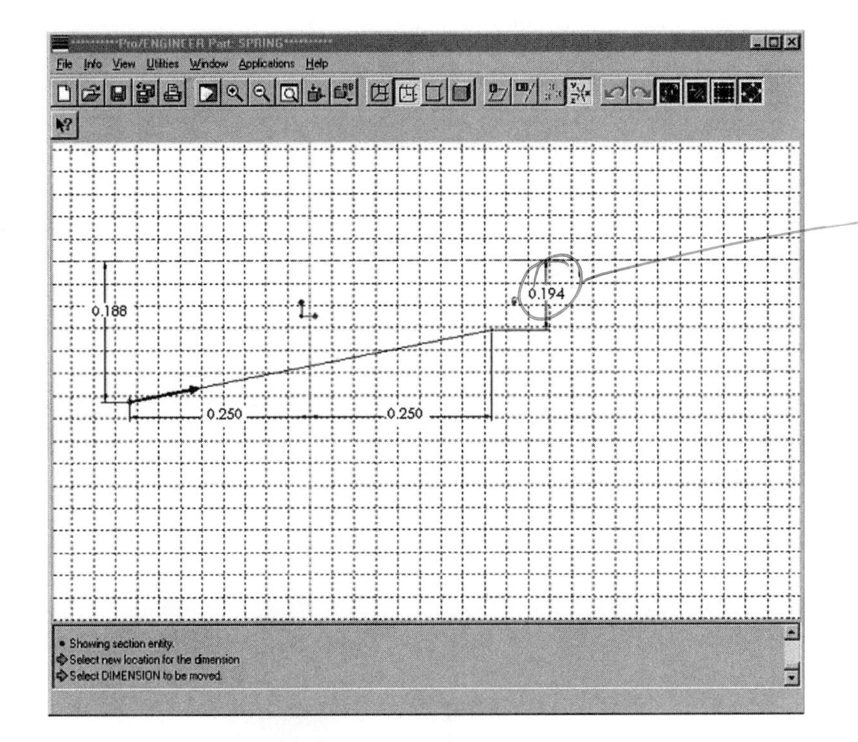


FIGURE 13.2 *Change the trajectory of the helical sweep to the one shown.*

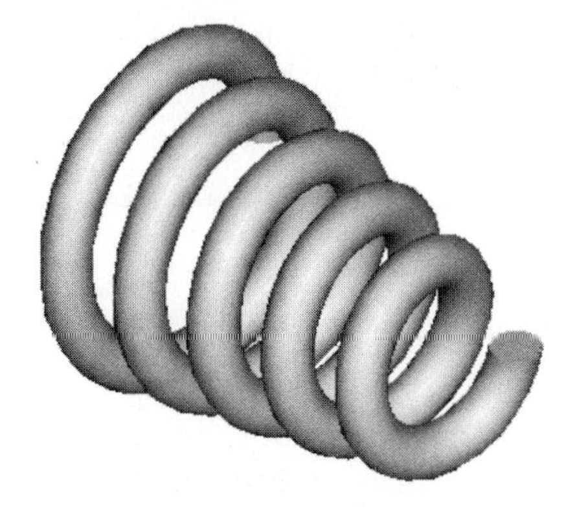

FIGURE 13.3 *After regenerating, a part with a redefined feature is obtained.*

Save the part under a new name by using *File* and *Save As.* Enter the name "TaperedSpring." Then, using *Erase,* remove the part "Spring" from memory.

## 13.2 THE REORDER AND REROUTE OPTION

The order in which a feature or set of features is created is important. It is possible that after working on a part, you discover that a feature should have been added to the model before several other features. So what do you do? Do you delete all the features, add the new feature, and then reconstruct those that you delete? Of course not. The correct solution is to use the *Reorder* option to place the new feature before the other features.

When using the *Reorder* option, the user must keep in mind the relationship between the features. If a feature is a child of another feature, then it cannot be reordered so that it appears in the regeneration list *before* its parent. A child may be made independent of its parent by using the *Reroute* option. This option allows the user to define new, sketching, and dimensioning features.

### 13.2.1 TUTORIAL # 62: AN EXAMPLE USING REORDER

Take, for example, the emergency light cover from chapter 5. Recall that in constructing the cover, we used the *Shell* option to remove material from the base feature after holes had been located in the base feature. Because the shelling process was performed after the holes were added, the holes were shelled as well as the rest of the base, leaving bosses to be used in the assembling process.

Suppose that it was our intention not to shell the holes; that is, the bosses are not desired. Can the model be modified with this change in mind? Yes, if the shelling process was performed before the holes were placed. The way to make the changes is to use the *Reorder* option.

Retrieve the part "EmergencyLightBase." Notice the hierarchy of the features in the *Model Tree* that is reproduced in Figure 13.4. The feature to

FIGURE 13.4 *The* Model Tree *for the emergency light cover.*

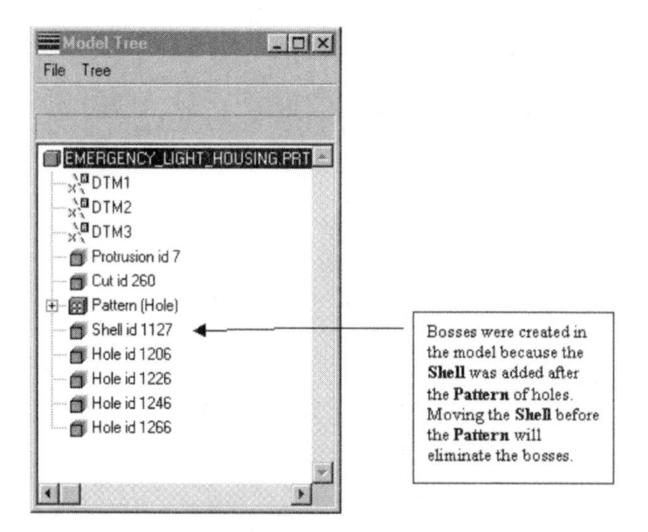

be moved is the pattern of the holes or the shell. If the pattern of the holes is moved, it must be moved to occur *after* the shell. If the shell is to be moved, it must be done so that it occurs *before* the pattern of the hole.

Let us move the shell feature before the pattern of the hole. From the *FEAT* menu, select *Reorder.* Click on the Shell id 1127 in the *Model Tree* or on the model itself.

Pro/E should recognize that the only place that the shell feature can be moved is before the pattern of the hole (features 5–6) or after the other holes (features 12–14). Notice that it cannot be moved before the protrusion (id 7), because the shell is a child of the protrusion.

Pick the *Pattern* feature in the *Model Tree.* Pro/E will place the shell before the pattern of the hole and the model will be updated. The model should appear as in Figure 13.5. The holes are no longer shelled. Notice the updated *Model Tree* reproduced in Figure 13.6. You do not need to save this version of the model.

## 13.3 LAYERS AND FEATURES

In Pro/E, a *Layer* is used to blank, that is, hide certain features during the construction of the part. Often datums, coordinate systems, and dimensions are placed on a different layer because they tend to clutter the drawing. If these features are placed on a layer separate from the rest of the model, then their display can be turned off by using *Blank* from the *Layer Display* box (Figure 13.7). The layer can also be *Blanked* by using the *Set Display* option from the *LAYERS* menu shown in Figure 13.8.

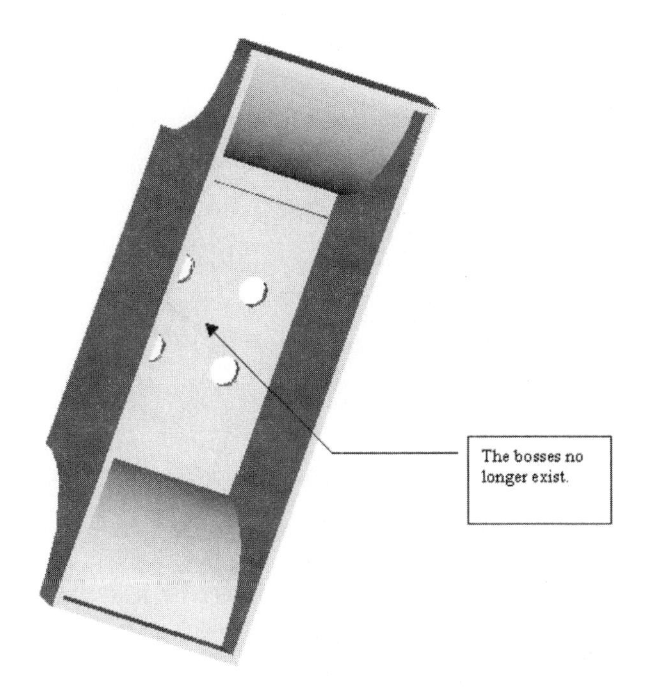


FIGURE 13.5 *The base after changing the order of the shell feature.*

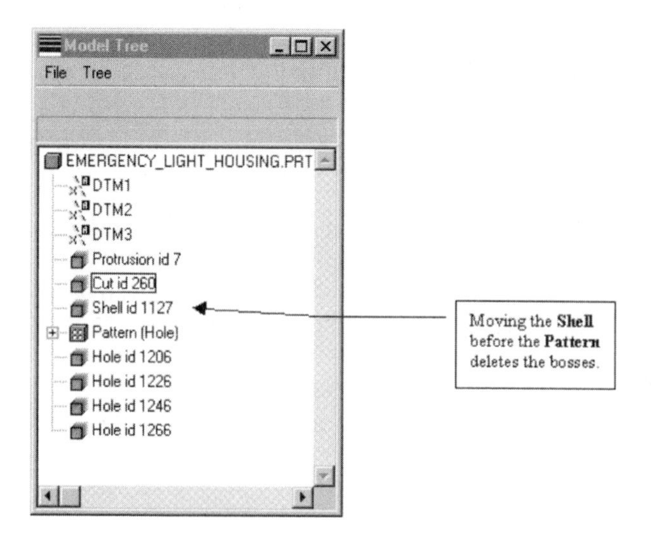


FIGURE 13.6 *The* Model Tree *of the cover after changing the order of the shell feature.*

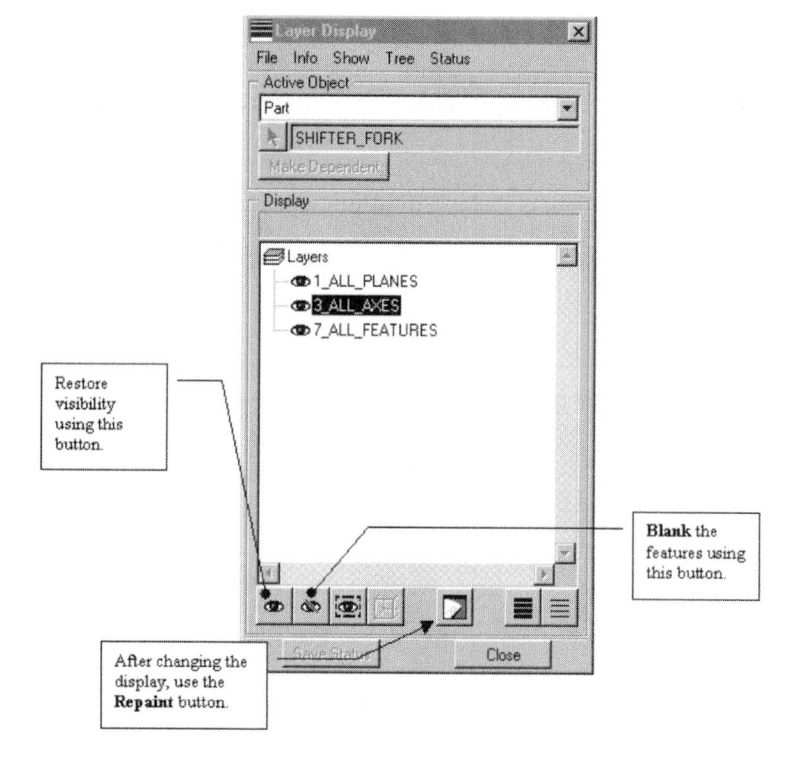


FIGURE 13.7 *The* Layer Display *box may be used to turn the visibility of different layers on and off.*

Layers are named upon creation and can be deleted or *Renamed* as shown in Figure 13.8.

Solid features such as cuts, holes, etc., may be placed in a layer. However, such features cannot be *Blanked;* rather the *Suppress* option is used to turn their visibility off. Rounds, fillets, and chamfers are often placed on different

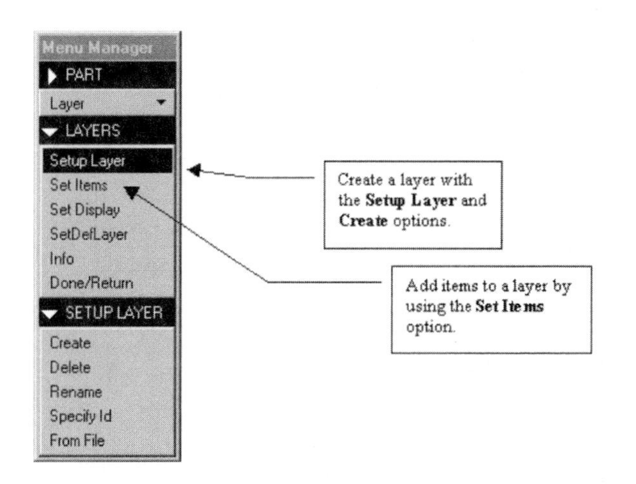


FIGURE 13.8 *The options in the* LAYERS *menu may be used to create and manipulate the features in a layer.*

layers so that their visibility can be suppressed. The features may be redisplayed using the *Resume* option.

Items may be added to a layer with the *Set Items.* The appropriate layer must be selected. To add the items choose *Add Items* after selecting the layer. In the part mode, only items from the active model can be added to a layer.

### 13.3.1 TUTORIAL # 63: CREATING A LAYER AND ADDING FEATURES TO THE LAYER

Retrieve the part "Shifter_Fork" from chapter 3. This part is incomplete. We wish to add several features, including two coaxial holes. After the holes have been added to the model, a layer will be created and the holes added to the layer.

The holes may be constructed by using the *Coaxial* option according to the following procedure:

1. Select *Feature, Create, Straight,* and *Hole.*
2. Choose *Coaxial* and *Done.*
3. Pick axis A_1 and the associated placement plane shown in Figure 13.9.
4. Enter a diameter of 1.000.
5. Repeat steps 1 and 2.
6. Pick axis A_3 and the corresponding placement plane in Figure 13.9.
7. Enter a diameter of 1.000.
8. Select *OK.*

Now that the holes have been created, they may be added to a different layer. First, we must create the layer. Select *Layer* from the *FEAT* menu. Choose *Setup Layer* and *Create.* Enter "All_Holes" for the name of the layer. Hit the *Return* key twice.

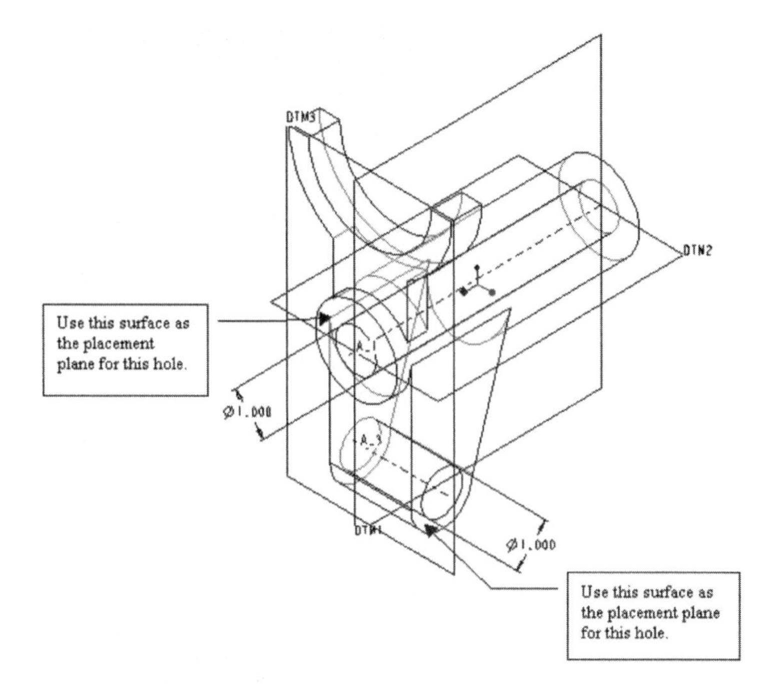


FIGURE 13.9 *The model with the added holes.*

In order to add the holes to the layer, select *Set Items* and then *Add Items.* Choose the layer called "All_Holes" in the list. Select *Done Sel* and *Feature.* Using *Query Sel,* choose the two holes. Then select *Done Sel* and *Done/Return* twice.

Check to see whether the holes have been added to the layer by selecting *Layer, Set Display,* and *Show.* Choose the *Layer Items* option and then double click on "All_Holes." The two holes should appear in the list as shown in Figure 13.10.

Save the part. In the next tutorial, we will use this part as the example for the *Suppress* and *Resume* option.

## 13.4 SUPPRESS AND RESUME

The *Suppress* option may be used to hide one or more features. The *Resume* option is used to restore any suppressed features. Both these options may be found in the *FEAT* menu.

Suppressing a feature is helpful when the design intent is not clear, especially when a particular feature interferes with another one. The feature causing the interference can be suppressed and the model analyzed for the design intent. Later, the suppressed feature can be restored.

The *Suppress* option can also be used when a feature fails to regenerate. In such cases, by suppressing the troublesome feature, the scheme used in generation of the feature and its interaction with the model can be more easily addressed.

Features may be suppressed and resumed by layer as shown in Figure 13.11. The *Range* option may be used to select the desired features by number.

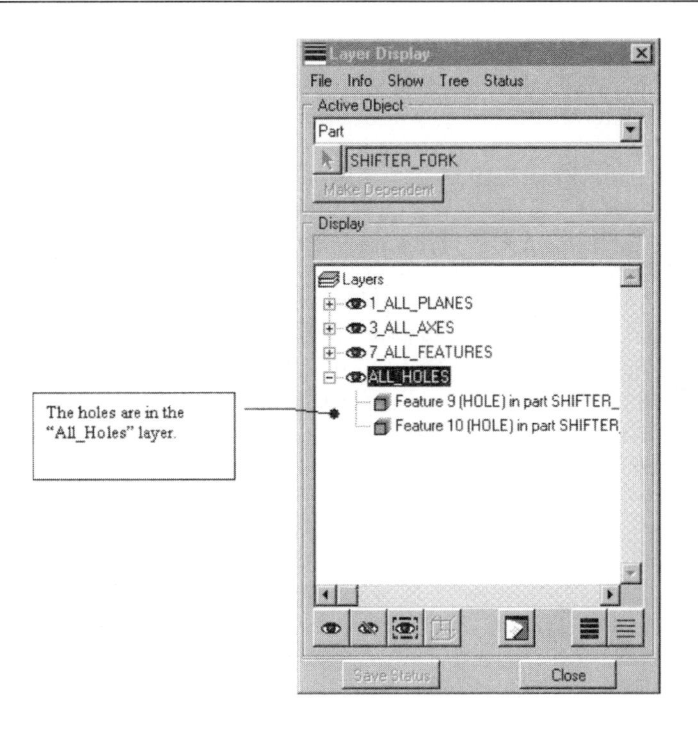


FIGURE 13.10 *The list includes the names of the two holes indicating that they have been placed in the layer "All_Holes."*

## 13.4.1 Tutorial # 64: Suppressing and Resuming Features

In some cases, a feature might be so complicated that it takes some time to regenerate. However, the feature may not be especially important to the overall model as in the case for the adjustment screw. Recall that in chapter 11, we added threads to the model using a helical sweep.

Let us retrieve this model and suppress the threads. After the part has been retrieved, select *Feature* and *Suppress.* Now, choose the threads created with the helical sweep. You may use your mouse and the *Query Sel*

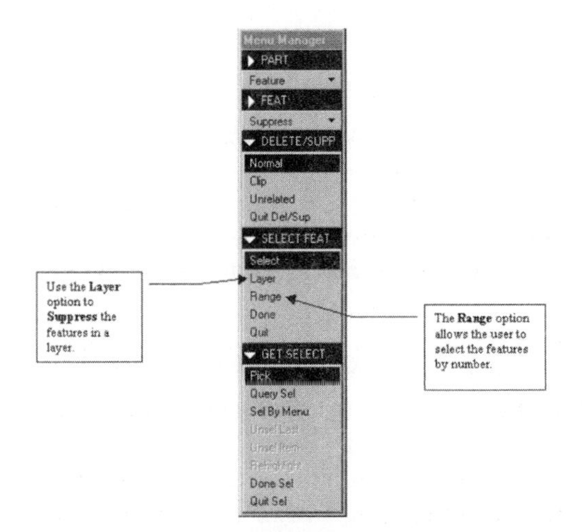


FIGURE 13.11 *Features may be suppressed by layer by using the* SELECT FEAT *menu.*

option or select the threads by clicking in the *Model Tree.* Choose *Done Sel* and *Done.*

Pro/E will regenerate the model without the threads. Notice that the identification for threads no longer appears in the model tree.

The threads can be restored by selecting *Resume* from the *FEAT* menu. In our case, we only have one layer, so choose *All* and *Done* from the *RESUME* menu. The threads will be restored.

### 13.4.2 TUTORIAL # 65: SUPPRESSING FEATURES IN A LAYER

Solid features may be suppressed by layer by selecting the desired layer. Retrieve the part "Shifter_Fork." Then, select *Suppress* from the *FEAT* menu. Choose *Layer.* Pick the layer "All_Holes" in the list. Select *Done Sel.* Choose *Done.* The software will hide the holes.

In order to redisplay the holes, select *Resume* and *Layer.* Again, choose "All_Holes" from the list. Choose *Done Sel* and *Done.* The software will redisplay the holes.

## 13.5 THE RESOLVE OPTION

The *Resolve* option is used to diagnose and fix a failed regeneration. The option becomes available after the user selects *Preview* or *OK* from the Feature Creation box. Please note that the failure *must* be resolved before continuing any further.

The options contained in the *RESOLVE FEAT* menu (Figure 13.12) may be used to resolve the problem. The options contained in the menu are used as follows:

1. *Undo Changes.* Remove the changes made to the model and return to the previous successful regeneration.
2. *Investigate.* Using the *Investigate* menu, attempt to find the cause of the failure. The *Investigate* menu contains the option *Show Ref.* This option shows the references of the failed feature.
3. *Fix Model.* Fix the model by using various options.
4. *Quick Fix.* Fix the model by redefining the elements of the failed feature.

The *Failure Diagnostics* box shown in Figure 13.12 lists the failed feature. In some cases, it may provide a hint as to how to fix the failure.

### 13.5.1 TUTORIAL # 66: RESOLVING A FAILED REGENERATION

The part "EmergencyLightHolder" from chapter 6 is incomplete. Several features need to be added to the model including rounds. Let us add rounds to the part using Figure 13.13 as a reference.

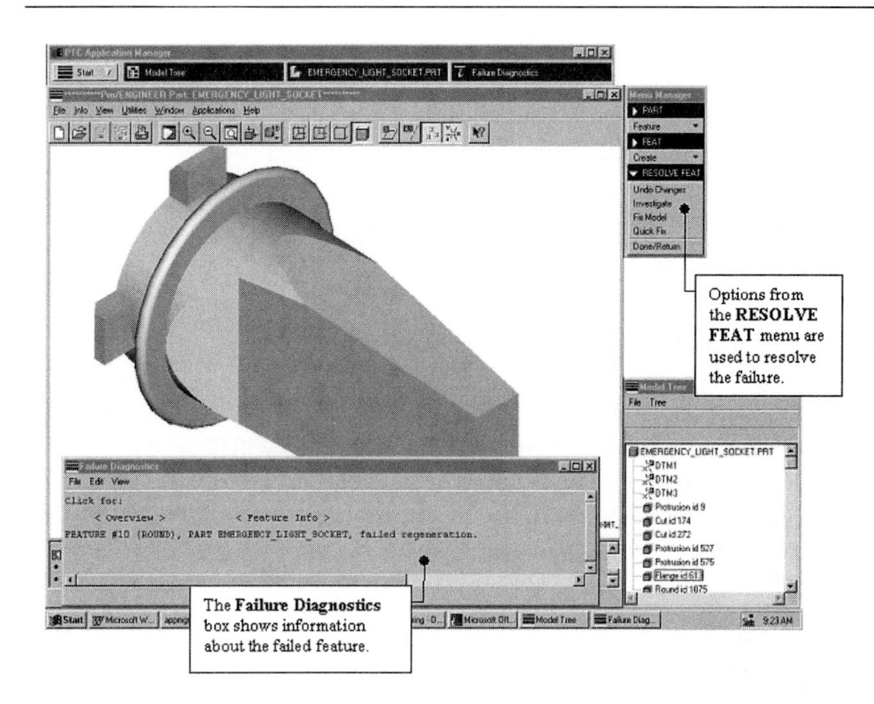


**FIGURE 13.12** *The* RESOLVE FEAT *menu may be used to resolve a failed regeneration.*

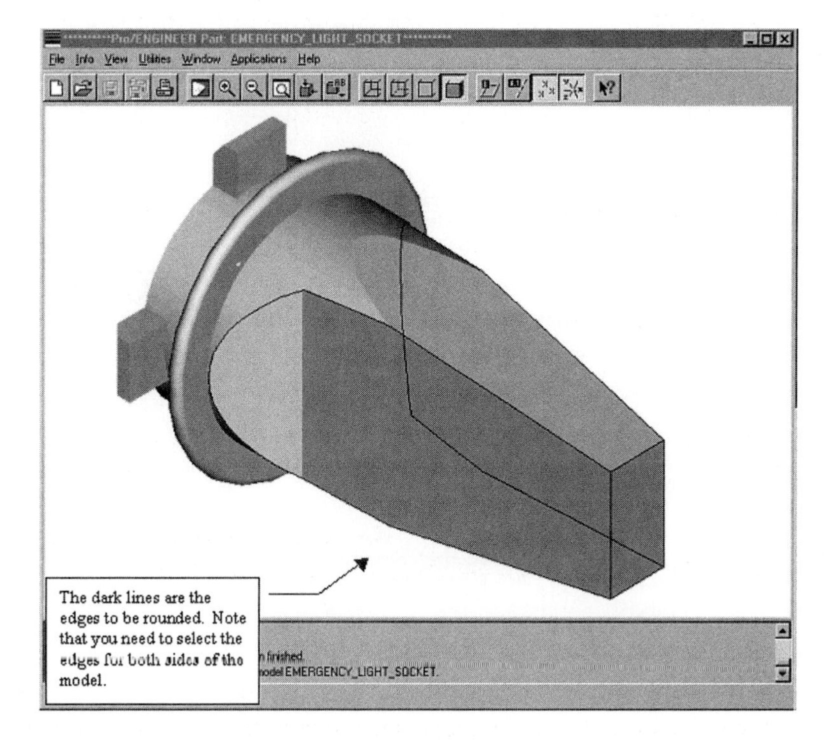


**FIGURE 13.13** *Use the shown edges as the edges for the round.*

The procedure for adding the rounds is as follows:

1. Select *Feature, Create,* and *Round.*
2. Choose *Simple* and *Done.*
3. Select *Constant, Edge Chain,* and *Done.*
4. Use the *One By One* option and select the edges shown in Figure 13.13. Note that the edges are the same for both sides of the model.
5. Choose *Done* and enter a value of .125. Notice that this value is much larger than the suggested value.
6. Choose *Preview* or *OK.*

Creation of the round will fail. The software will load the *SHOW ERRORS* menu and draw a curve around the failure. In this case, failure occurs at more than one edge. Select *Done/Return* and choose the *Resolve* button.

Because we entered a value for the round much larger than the suggested value, our first thought might be that the radius of the round is causing the failure. Unfortunately, the *Failure Diagnostics* box does not provide any further information.

Choose *Quick Fix, Redefine,* and *Confirm.* Then select the *Radius* element from the feature. Choose *Define.* Enter a value of .005. Then select *Preview.*

Again, the round will fail. Select *Resolve.* From the *RESOLVE FEAT* menu, choose *Investigate* and *Failed Geom.* In this case, the software will only show a failure along both curved edges as shown in Figure 13.14.

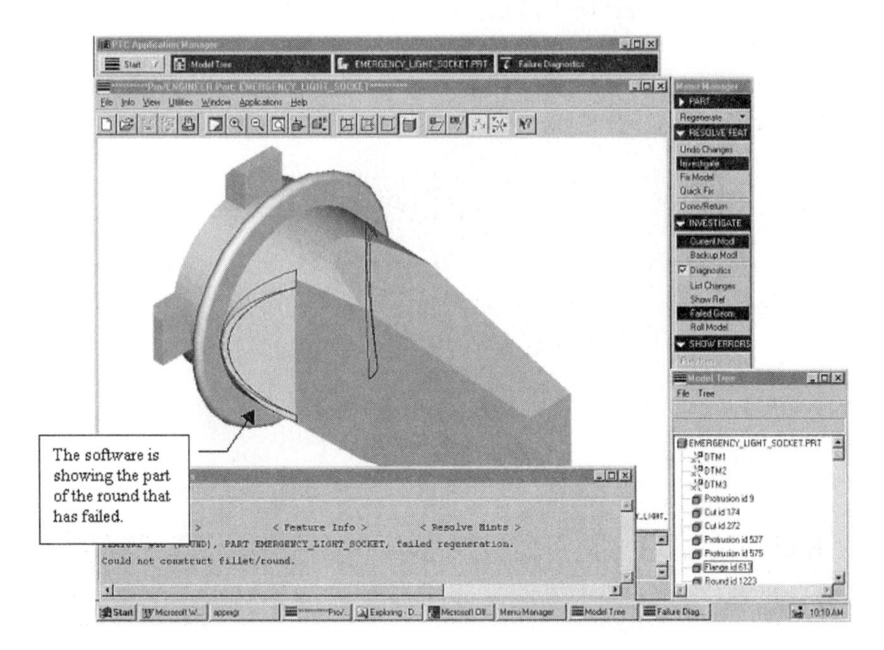


FIGURE 13.14 *The software highlights the failed feature(s).*

The round fails along the curved edge because of the geometry of the part. Creation of the round interferes with the round along the other edges and possibly with the flange. Obviously, creation of the round is failing along different edges for diverse reasons. Because the failure was fixed in a different location by reducing the radius of the round, the round should be placed on the curved edges separately from the other edges. Then, different values for the radius can be used for each chain of edges.

Select *Quick Fix, Redefine,* and *Confirm.* Choose the *References* element and then the *Define* button. Choose the *Unselect* option. Use *Query Sel* and pick the curved edges to remove the curves from the edge chain. Choose *OK.* The round will be created successfully. The part, after the addition of the round, is shown in Figure 13.15.

The round may be added on the curved edges by using the following procedure:

1. Select *Feature, Create,* and *Round.*
2. Choose *Simple* and *Done.*
3. Select *Constant, Edge Chain,* and *Done.*
4. Use the *Tangnt Chain* option and select the two curves.
5. Choose *Done* and enter a value of 0.005.
6. Select *OK.*

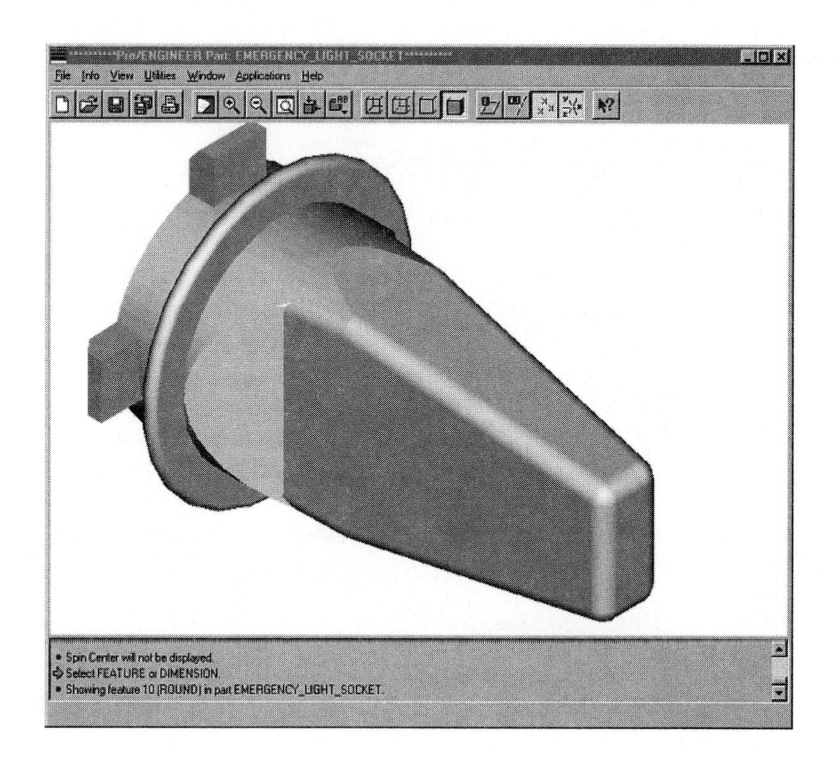


FIGURE 13.15 *The part after the addition of the round.*

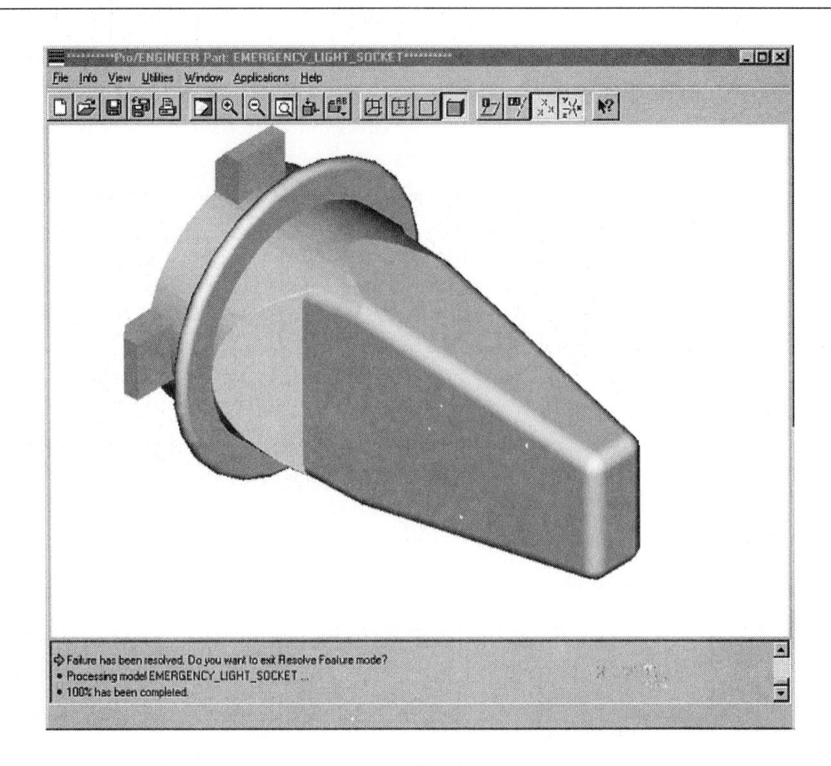


FIGURE 13.16 *The model with both rounds added.*

The second round will be added to the model. The round is shown in Figure 13.16. Save the part.

## 13.6 SUMMARY AND STEPS FOR USING THE FEATURE MANAGEMENT OPTIONS

Feature management is the subject of this chapter with regard to the options *Redefine, Reorder, Reroute, Layer, Suppress, Resume,* and *Resolve.* Using the *Redefine* option, features may be changed by refining the elements that are used to define the feature. Features may be placed on a different level within the model with the *Layer* option. A feature or set of features may be removed from the generation list with the *Suppress* option and placed back on the list using the *Resume* option. The *Reorder* option may be used to change the order in which a feature is generated, with respect to the other features in the model. If a feature is a child of another feature, it may be moved ahead of its parent by using the *Reroute* option. Finally, the *Resolve* option may be used to correct a failed regeneration by diagnosing the failure.

In general, the following steps may be used to *Redefine* a feature:

1. Select *Feature* and *Redefine.*
2. Choose the feature to redefine.
3. Select the element to change from the Feature Creation box.

4. Choose *Define.*
5. Modify the element as necessary.
6. Select *OK.*

The general sequence of steps to follow for reordering a feature is:

1. Select *Feature* and *Reorder.*
2. Choose the feature to reorder from the screen or from the *Model Tree.*
3. Choose the insertion point in the generation list. Pick this object from the screen or *Model Tree.*

The following sequence of steps may be used to create a layer:

1. Select *Layer* from the *FEAT* menu.
2. Choose *Setup Layer* and *Create.*
3. Enter a name of the layer.
4. Hit the *Return* key twice.

After a layer has been created, features may be added to the layer using the following procedure:

1. Select *Set Items* and then *Add Items.*
2. Choose the desired layer in the list.
3. Select *Done Sel* and *Feature.*
4. Pick the desired features.
5. Select *Done Sel* and *Done/Return* twice.

Solid features may be removed from the generation list by using the *Suppress* option. The features may be in a specific layer. In order to suppress one or more features use the following procedure:

1. Select *Feature* and *Suppress.*
2. Now, choose the desired feature or features to hide.
3. Choose *Done Sel* and *Done.*

If the desired features are contained in a layer, then use the following procedure:

1. Select *Suppress* from the *FEAT* menu.
2. Choose *Layer.* Pick the name of the layer in the list. Select *Done Sel.* Choose *Done.*

Pro/E will regenerate the model without the suppressed features. The features may be placed back in the generation list by using the *Resume* option. In general, this option may be used as follows:

1. Select *Resume* from the *FEAT* menu.
2. Choose *All* and *Done.*

In order to resume features in a layer, use the following sequence of steps:

1. Select *Resume* and *Layer.*
2. Choose the layer by name from the list.
3. Choose *Done Sel* and *Done.*

The model will be regenerated and the suppressed feature restored.

## 13.7 ADDITIONAL EXERCISES

13.1 Create a layer called "Hole_Layer" for the "EmergencyLightHolder." Add a 0.250 diameter, 0.500 deep coaxial hole to the model. Then place the hole in the layer. Use Figure 13.17 as a reference.

FIGURE 13.17 *Figure for exercise 13.1.*

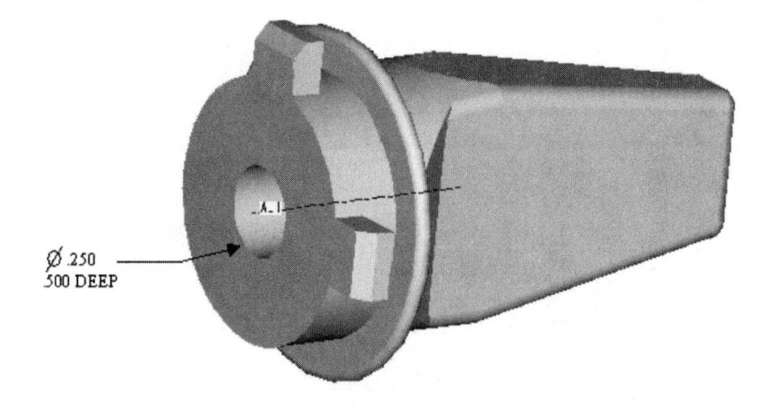


13.2 For the "ShifterFork" model, create two layers, call the first layer "All_Rounds" and the second layer "Cut."

13.3 Create 0.50 radii rounds as shown in Figure 13.18 for the part "ShifterFork," and add these rounds to the layer "All_Rounds."

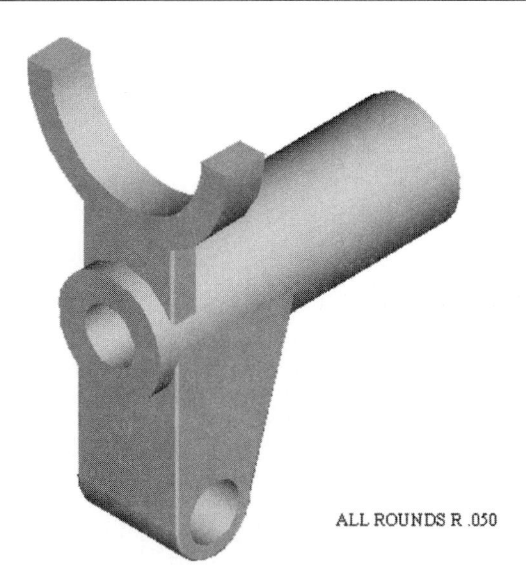


FIGURE 13.18 *Figure for exercise 13.3.*

13.4 Complete the support in the model "ShifterFork" using a 1-inch wide by 2.25-inch high rectangular cut. Add this feature to the layer "Cut." The geometry of the cut is shown in Figure 13.19.

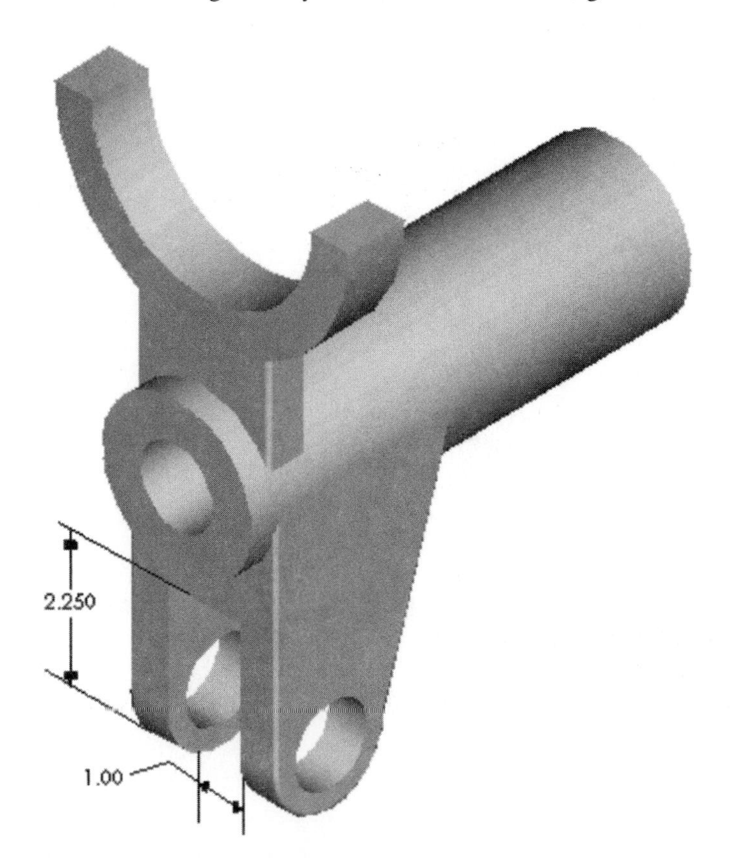


FIGURE 13.19 *Figure for exercise 13.4.*

13.5 Use the *Redefine* option and change the rectangular protrusion in the part "BaseOrient" to a circular protrusion of diameter 3.000 as shown in Figure 13.20.

FIGURE 13.20 *Figure for exercise 13.5.*

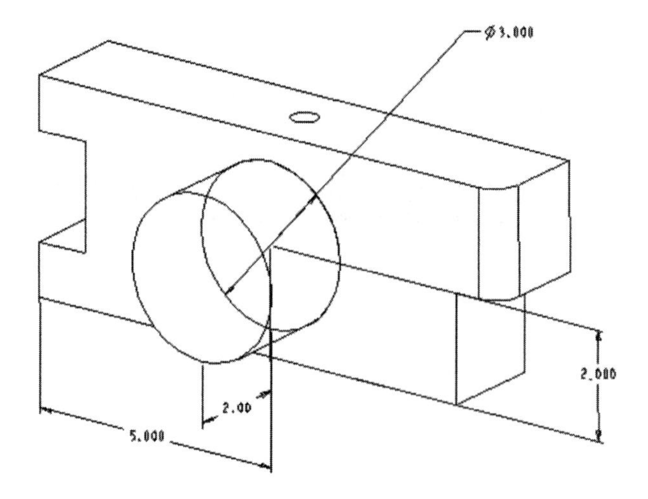

# CHAPTER 14

# COSMETIC FEATURES

## INTRODUCTION AND OBJECTIVES

The *Cosmetic* option is used to place special features, including text for company logos and serial numbers, projected sections, and threads. The various cosmetic options are accessed via the *COSMETIC* menu shown in Figure 14.1.

The cosmetic *Groove* is constructed by projecting a sketch onto a surface. The cosmetic *Groove* is useful in the manufacturing process *Groove,* where the groove is used to define the tool path.

The option *Sketch* is used to place logos and serial numbers. The *Thread* option is used to create a surface on the part that represents the threads. Pro/E does not actually draw the individual threads. If you want to draw the individual threads, see the section on helical sweeps in chapter 11.

The objectives of this chapter are as follows:

1. The user will be able to place cosmetic threads on a surface.
2. The reader will attain an ability to create cosmetic text on a part.

## 14.1 PRO/E COSMETIC OPTIONS

Using the *Cosmetic* feature, threads may be placed by defining the thread. In this case, the surface to be threaded (*Thread Surf*), the position to start threading (*Start Thread*), the depth to which the feature is threaded (*Depth*), and the major diameter of the thread (*Major Diam*) must be defined.

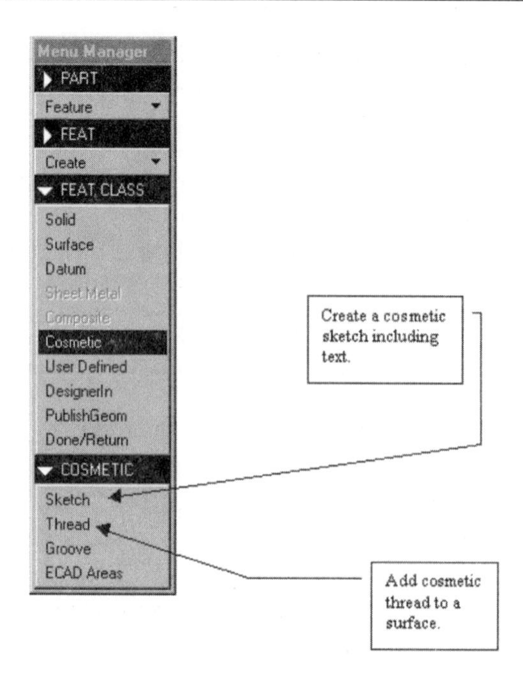


**FIGURE 14.1** Menu Manager *with the* Cosmetic *options.*

Pro/E will create both internal and external threads depending on the surface on which the thread is located. If the thread is placed on a hole, then an internal thread is created. If the thread is placed on a cylinder, then an external thread is created.

Pro/E sizes all threads 10% larger or smaller than the governing surface size. For example, if the thread is placed on a hole, then the thread is sized 10% larger than the size of the hole. If the thread is placed on a shaft, then the thread is created 10% smaller than the diameter of the shaft.

The parameters of the thread may be accessed and modified by the *FEAT PARAM* menu shown in Figure 14.2. When the *FEAT PARAM* menu appears, select *Mod Params.* Pro/E will load the Pro/Table editor. The editor will list the parameters of the thread. Change the parameters of the thread as desired.

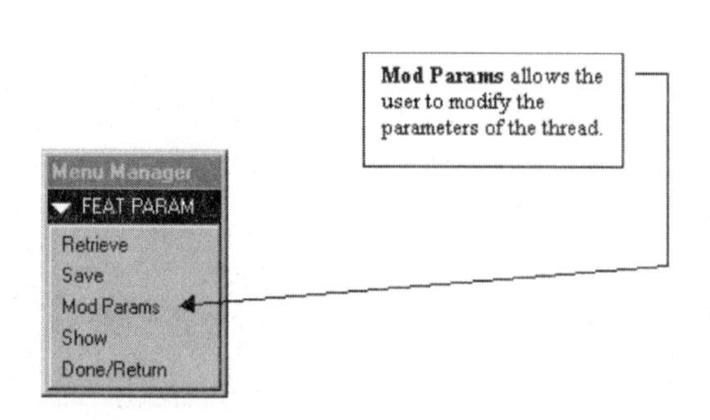


**FIGURE 14.2** *The* FEAT PARAMS *menu may be used to access the parameters defining the thread.*

Another interesting use of the *Cosmetic* option is to place a logo or text on a part. The text is often a serial or part number. Because the *Sketcher* is used to construct the text or logo, the cosmetic placed on the part can be quite complex and elaborate.

The style and size of the text can be modified by selecting *Modify* and then the text. Pro/E will load the *MODIFY SEC TEXT2* menu, reproduced in Figure 14.3. Selecting the *Text Style* option from this menu will bring up the *Sketcher Text Style* Box shown in Figure 14.3. The available fonts are shown in Figure 14.4 and are as follows:

1. *font3d,* a text that appears as three-dimensional.
2. *Font,* an ASCII-type font.
3. *leroy,* a Leroy-type font.
4. *cal_grak,* a Calcomp Greek character font.
5. *cal_alf,* a Calcomp alphanumeric font.

The text, like any section created in the *Sketcher,* must be properly located with respect to the part by dimensioning the base of the sketch.

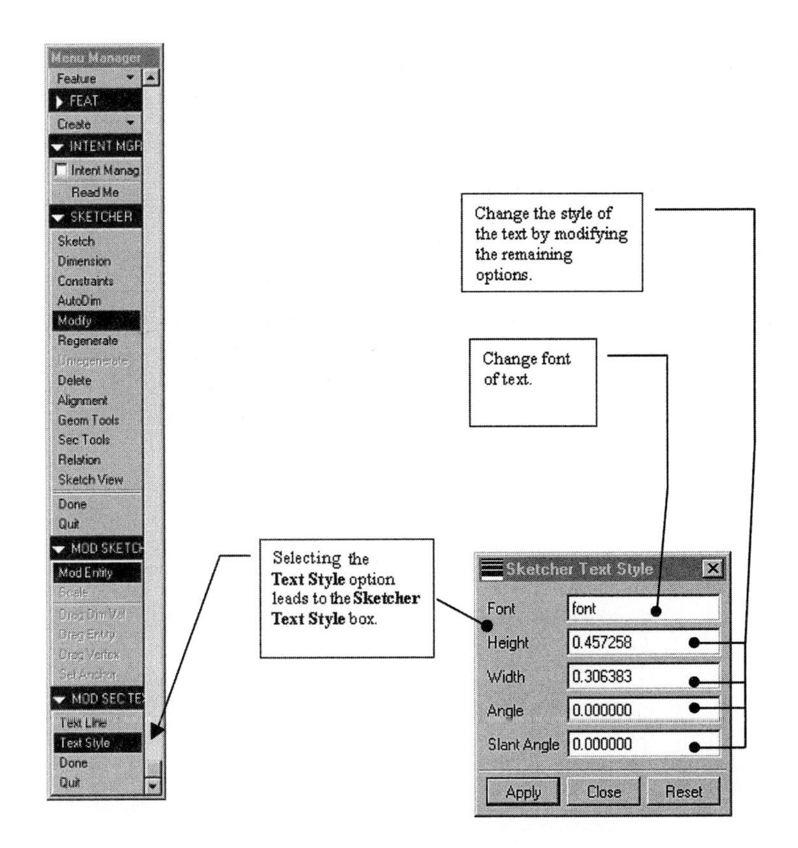


**FIGURE 14.3** *The* MOD SEC TEXT2 *menu and the* Sketcher Text Style *box are used to modify cosmetic text.*

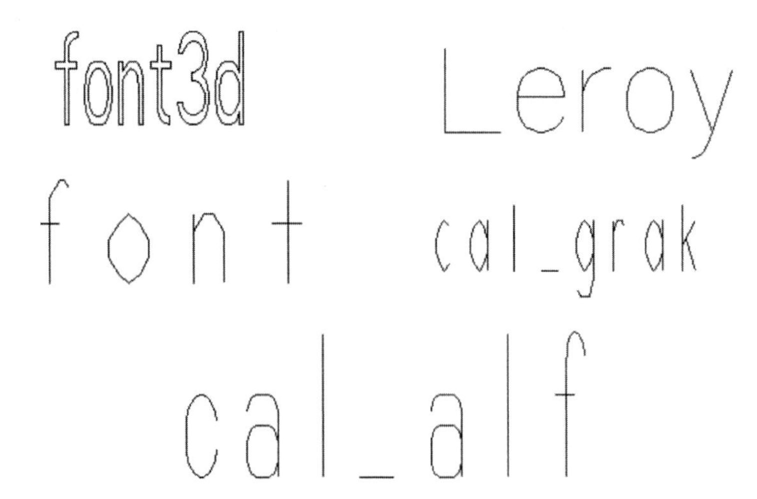


FIGURE 14.4 *Several different font types are available for cosmetic text.*

### 14.1.1 TUTORIAL # 67: COSMETIC THREADS FOR THE NUT

Retrieve the part called "Nut" constructed in chapters 7 and 8. Then select *Feature, Create,* and *Cosmetic.* The *COSMETIC* menu will appear as shown in Figure 14.1. Click on *Thread.*

Using your mouse, pick on the surface to thread. Pick the cylindrical surface, which forms the surface of the hole through the nut as shown in Figure 14.5. Next, you need to pick the starting surface. Use *Query Select* and pick the starting surface again using the figure for reference. Pro/E will display an arrow indicating the direction of the thread creation. The direction of the arrow should be the same as in Figure 14.5.

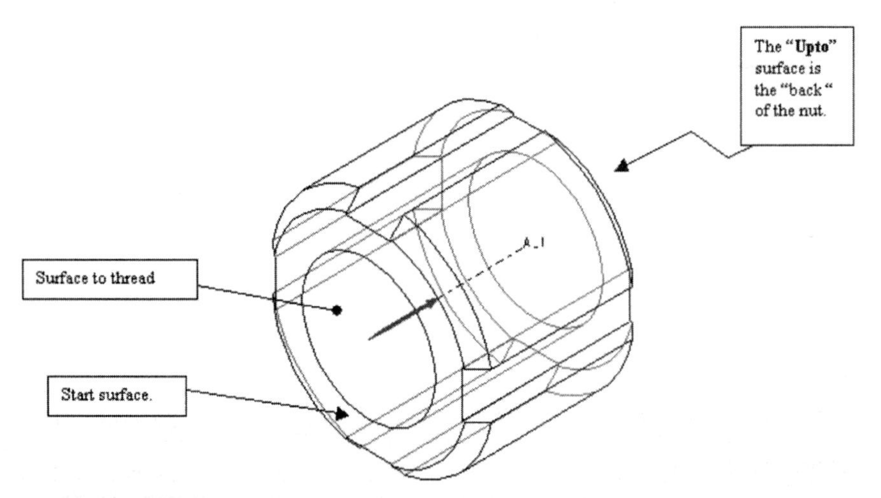


FIGURE 14.5 *The arrow indicates the direction for the creation of the threads on the nut. The base of the arrow is attached to the* Starting Surface.

Flip the arrow if it is not in the same direction as the one shown in the figure. Choose *OK.*

We want the threads to extend along the entire surface of the hole, so from the *SPEC TO* menu, pick the "back" of the nut as shown in Figure 14.5.

For the thread diameter, Pro/E will calculate 10% of the size of the hole and present the result as the default value, in this case 0.33 inches. In order to make the thread a standard size, enter the value .375 (3/8"). The ANSI standard pitch for a 3/8–inch major diameter UNC thread is 16 threads per inch. We can change the pitch of the thread by selecting the *Mod Params* option from the *FEAT PARAM* menu shown in Figure 14.2.

Pro/E will load the Pro/Table editor. The editor will list the parameters of the thread. The parameters for this thread are reproduced in Figure 14.6. Click on the value for the pitch and enter 16. Also, make sure that the metric option is set to *False.* Then select *File* and *Exit.* From the *FEAT PARAM* menu, choose *Done/Return.* Then select *OK.* Pro/E will load the threads but will not draw the individual teeth. The threads will be indicated by an additional surface. This additional surface is illustrated in Figure 14.7. In this case, the surface is coincident with the surface forming the hole. Save the part.

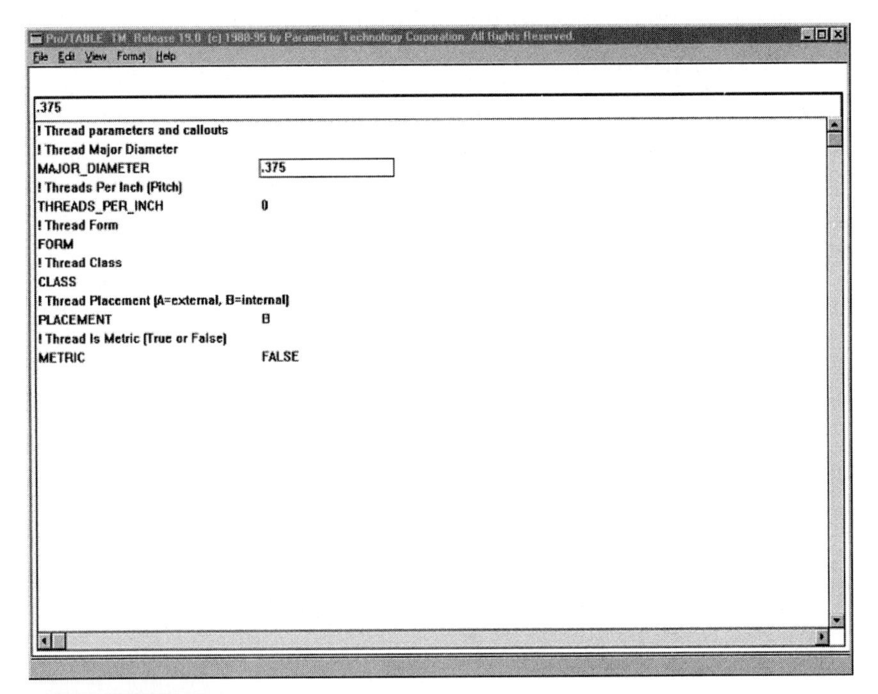


**FIGURE 14.6** *The Pro/Table editor showing the parameters for the thread on the nut. Change the* Threads per Inch *to read 16. Also, make sure that the* Metric *option is set to False.*

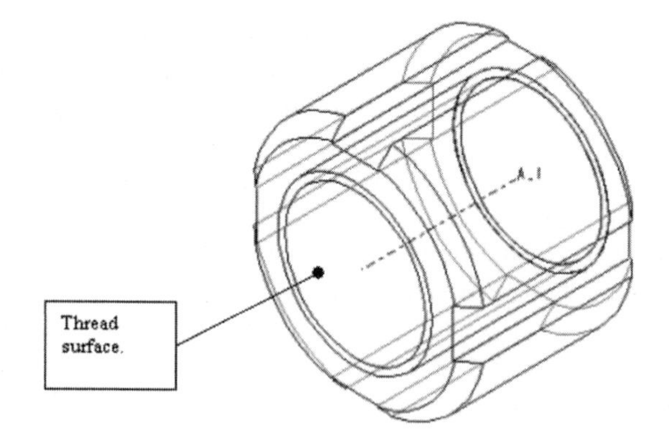


**FIGURE 14.7** *The nut after adding the cosmetic thread. Pro/E does not draw the individual thread teeth; rather, it inserts a surface that represents the threads. Compare with Figure 14.5.*

## 14.1.2 TUTORIAL # 68: COSMETIC TEXT FOR THE COVER PLATE

Let us see how to place cosmetic text on the part. The part is a simple plate that is used in the emergency light as a cover. Before adding the text, we need to construct the part as follows:

1. Select *File* and *New.*
2. Create a file called "CoverPlate."
3. Add the default datum planes.
4. Select *Feature, Create, Protrusion, Extrude, Solid,* and *Done.*
5. *One Side* and *Done.*
6. Choose DTM2 as the sketching plane. DTM1 as "*Bottom.*"
7. In the *Sketcher,* draw a rectangle that is evenly spaced with respect to the DTM1 and DTM3 edges (consider Figure 14.8).
8. *Dimension* as shown in Figure 14.8. *Regenerate.*

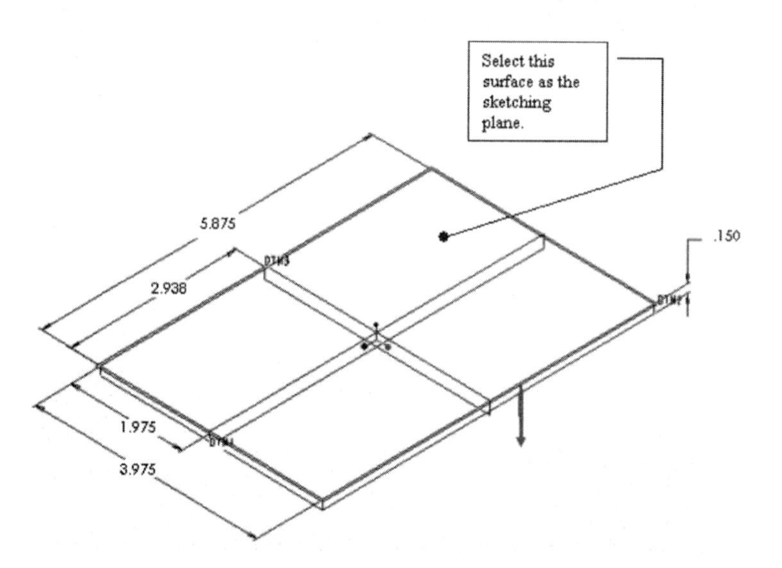


**FIGURE 14.8** *Dimensions for the model "CoverPlate."*

Now, we can add the text. Select *Feature, Create Cosmetic,* and *Sketch.* The section can be placed onto a surface (*Regular Sec*) or projected from a sketching plane onto another surface (*Project Sec*). The menu, reproduced in Figure 14.9, lists these options. For this model, we will place the cosmetic text on the surface itself, so select *Regular Sec.* Click on *Done.*

Select the top of the plate using *Query Sel.* Make sure that the arrow points in the direction shown in Figure 14.8. Use DTM1 as "*Top*" to reorient the model in the *Sketcher.* Pro/E will open the *Sketcher.*

Choose *Advanced Geometry* and then *Text.* Enter the text: Emergency Light. In order to place the text, you must define a box to contain the text. With your mouse, drag a box across the base.

*Dimension* the section as shown in Figure 14.10. *Regenerate* the sketch and select *Done.* Pro/E will add the cosmetic text to the model. Save the part.

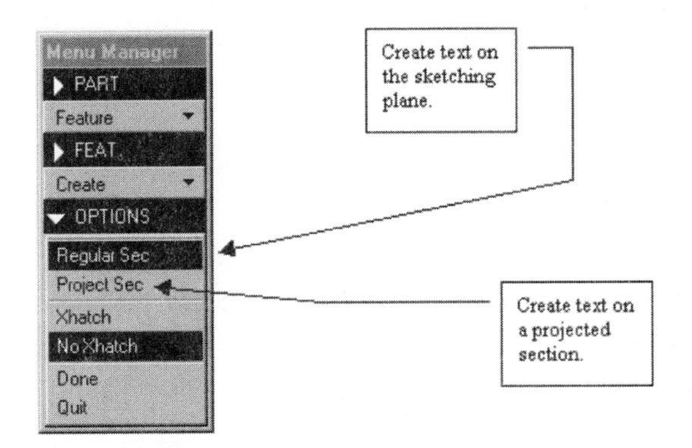


FIGURE 14.9 *Regular or projected text may be created with the* OPTIONS *menu.*

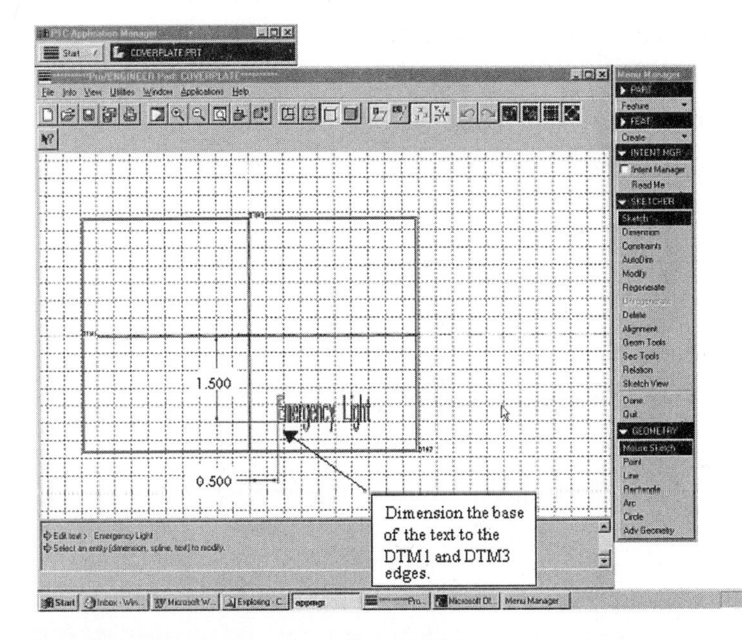


FIGURE 14.10 *Dimension the base of the text to the datum planes as shown.*

## 14.2 SUMMARY AND STEPS FOR ADDING COSMETIC FEATURES

The *Cosmetic* option may be used to place logos and text, such as serial numbers and threads, on a part. When Pro/E creates a *Cosmetic Thread,* it constructs a surface that represents the threads. A cosmetic *Sketch* is added to a part by using the *Sketcher.*

The general steps for adding a cosmetic sketch to the part are as follows:

1. Select *Feature, Create, Cosmetic,* and *Sketch.*
2. Choose the sketching surface and an orientation plane.
3. Sketch the geometry.
4. *Dimension* the geometry to the part and *Regenerate.*
5. Select *Done.*

In general, cosmetic threads may be added to a surface by using the following procedure:

1. Choose *Feature, Create, Cosmetic,* and *Thread.*
2. Select the surface to be threaded.
3. Choose the feature from which the threading is to begin.
4. Select the direction of the thread.
5. Indicate the depth of the thread.
6. Enter the parameters of the thread including its diameter.
7. Choose *Done/Return.*

## 14.3 ADDITIONAL EXERCISES

14.1 Add the *Cosmetic* thread to the handle (exercise 11.1). Use the geometry in Figure 14.11.

FIGURE 14.11 *Figure for exercise 14.1.*

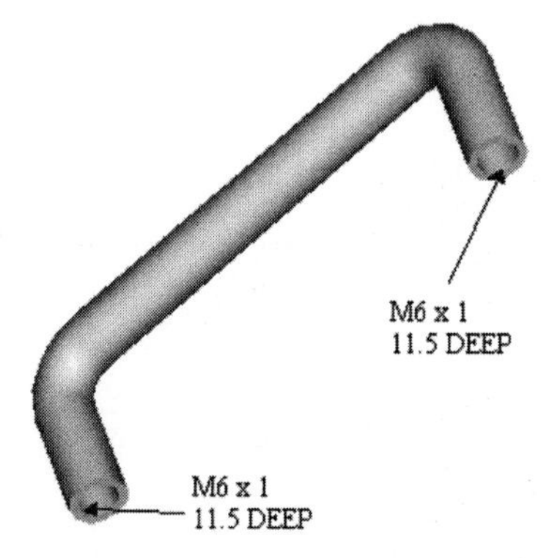

14.2 For the model of the coupler from exercise 6.9, add the *Cosmetic* thread as shown in Figure 14.12.

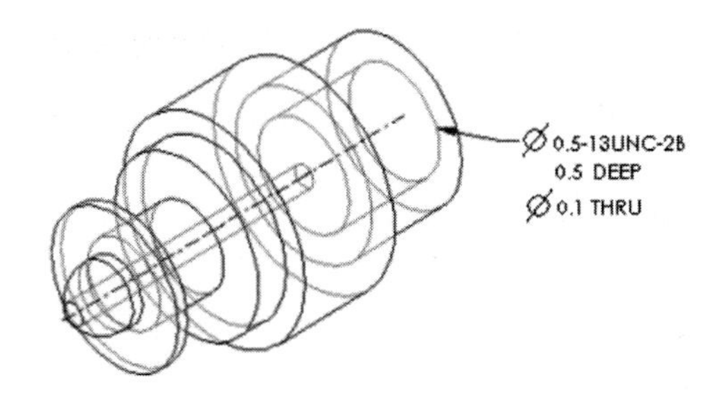


FIGURE 14.12 *Figure for exercise 14.2.*

14.3 Add *Cosmetic* threads to the holes in the model of the pipe clamp bottom (exercise 8.6). Use Figure 14.13 for the geometry of the threads.

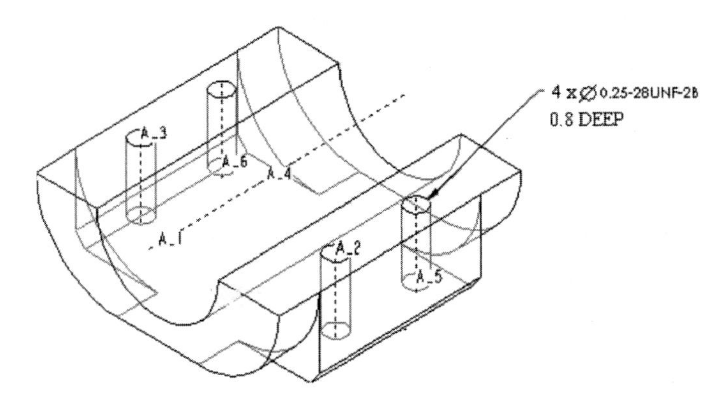


FIGURE 14.13 *Figure for exercise 14.3.*

14.4 For the model of the offset support from exercise 4.6, consider Figure 14.14 and add the *Cosmetic* thread.

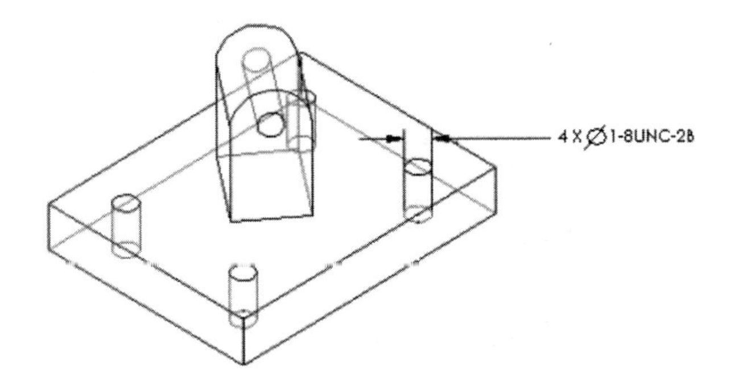


FIGURE 14.14 *Figure for exercise 14.4.*

14.5 Add the threads to the holes on the side support model (exercise 8.7). Use Figure 14.15 for the geometry of the threads.

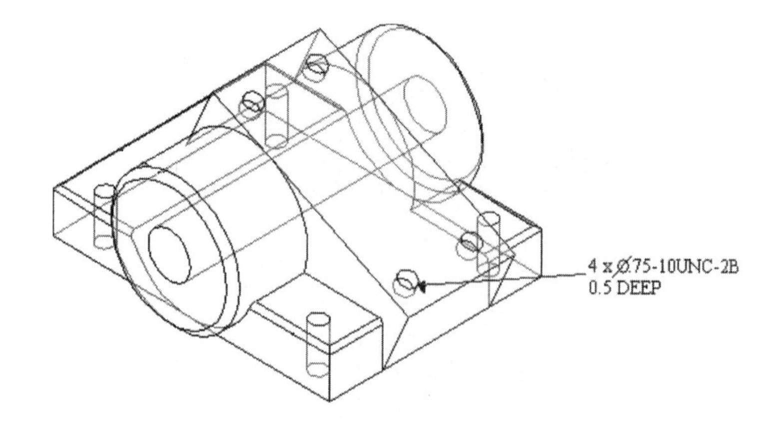


FIGURE 14.15 *Figure for exercise 14.5.*

14.6 Add the *Cosmetic* thread to the hole in the angled block (exercise 4.9). Information on the thread is given in Figure 14.16.

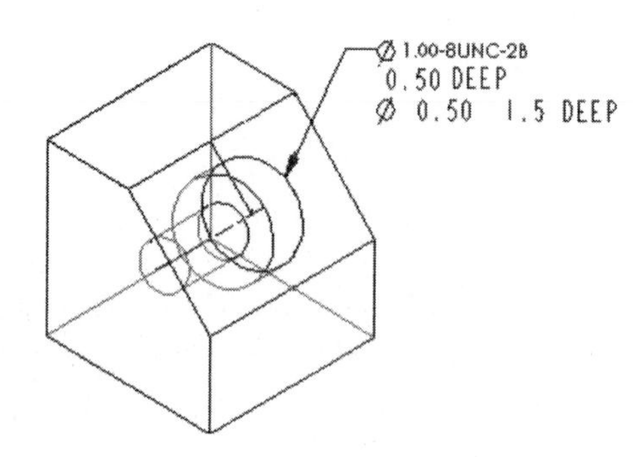


FIGURE 14.16 *Figure for exercise 14.6.*

14.7 Add the text "RoboArm" to the model of the robot arm housing (exercise 5.9). The location of the text on the model is shown in Figure 14.17.

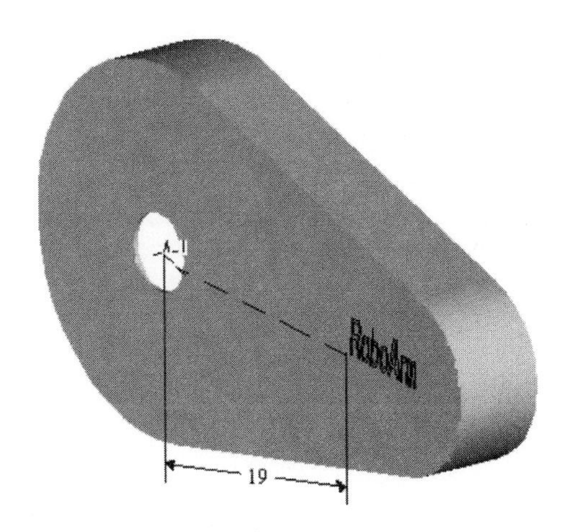


FIGURE 14.17 *Figure for exercise 14.7.*

14.8 For the model of the electric motor cover, add the text "EMotors." Use Figure 14.18 for the location of the text.

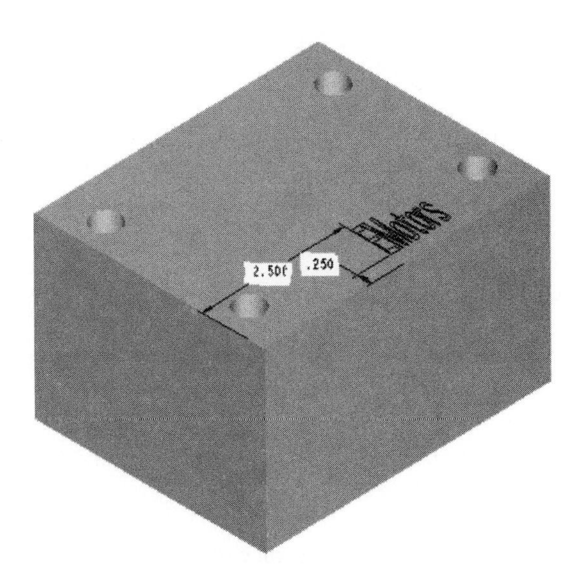


FIGURE 14.18 *Figure for exercise 14.18.*

# CHAPTER 15

# QUILTS AND SOME OPTIONS FROM THE TWEAK MENU

## INTRODUCTION AND OBJECTIVES

The *Tweak* option is obtained by selecting *Create* from the *FEAT* menu. The *TWEAK* menu is obtained by choosing the *Tweak* option. This menu is shown in Figure 15.1.

The options contained in the *TWEAK* menu are specialized options. Among the options in the menu, you will find *Draft, Ear,* and *Lip* to be especially useful.

The *Draft* option is used to add a draft angle on an existing feature. The angle must be between −15° and +15°. A *Lip* is a small protrusion often used to interlock one part with another. An *Ear* is a protrusion extending from a surface that is bent at its base.

With regard to the options in the *TWEAK* menu, the objective of this chapter is the following:

1. The user will be able to use the *Draft, Lip,* and *Ear* options to place specialty protrusions on a part.

A *Quilt* is a nonsolid surface, that is, a surface that has no thickness. A quilt is a feature created using the same fundamental options such as *Extrude, Revolve, Sweep,* and so on. Quilts may be manipulated. The *QUILT SURF* menu shown in Figure 15.2 is used to manipulate a quilt.

A surface is useful when the design intent is to analyze the surface not the interior of an object. For example, in aerodynamic studies, the surface geometry of a wing is analyzed. Therefore, only a model of the surface is required to calculate the lift and drag forces on the wing.

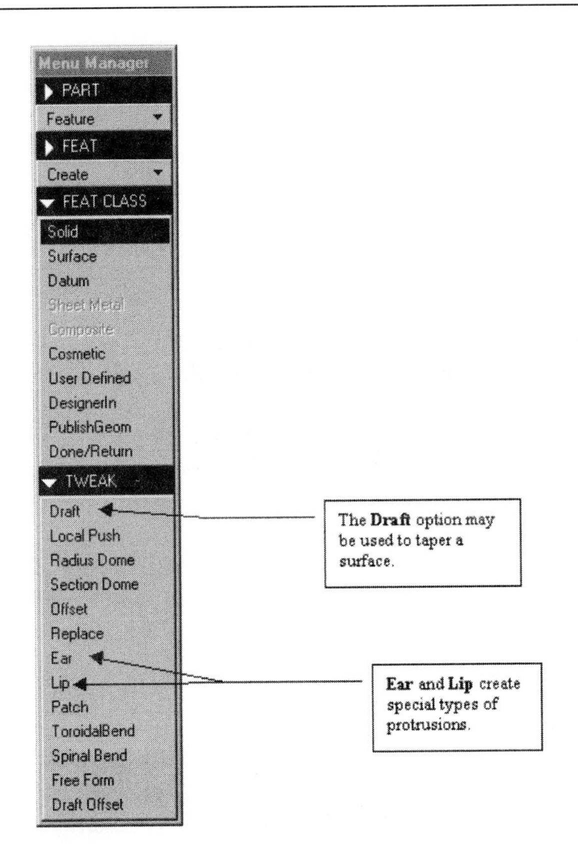


FIGURE 15.1 *The* TWEAK *menu contains advanced options for modifying a feature.*

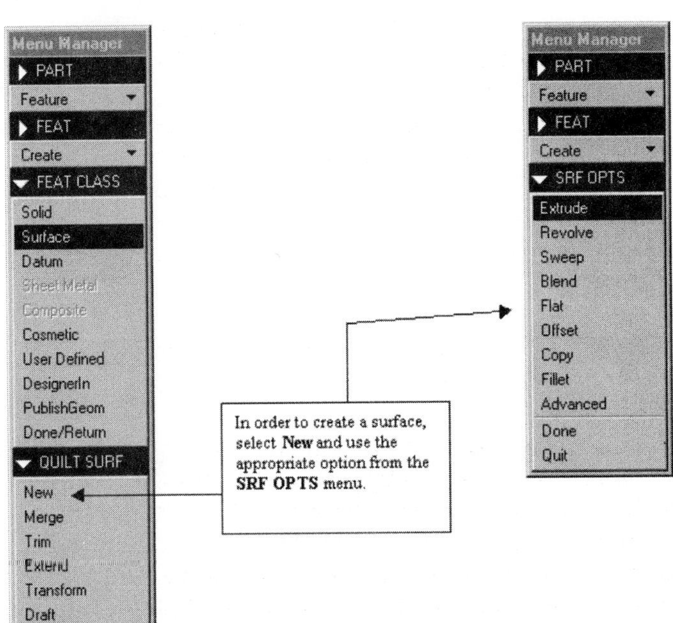


FIGURE 15.2 *A quilt may be created by using the* New *option from the* QUILT SURF *menu.*

In this chapter, discussion of surfaces in this chapter is intended to be an introduction about the subject. Thus, we may add to our list of objectives for this chapter:

2. The user will be able to create and manipulate a quilt.

## 15.1 ADDING A DRAFT TO A BASE FEATURE

A draft is constructed by moving a surface or surfaces through the given angle about a neutral plane (*Neutral Pln* option) or curve (*Neutral Crv* option). The neutral plane or curve must be indicated by the user. The surfaces that are to be moved are also chosen by the user.

The only surfaces that can be drafted are surfaces of a cylinder or plane. Pro/E divides the cylindrical surface into two separate surfaces. If a draft is to be performed on the entire cylindrical surface, then both surfaces must be chosen by the user.

Fillets pose a special problem in constructing a draft, because a draft cannot be performed on the surface edge. Thus, if a feature is to have both a draft and a fillet, the draft must be performed first, and then the fillet can be added.

The surface to be drafted is chosen using the *SURF OPTIONS* menu shown in Figure 15.3. Surfaces chosen for drafting can be split at a defined curve (*Split at Crv* option), surface (*Split at Srf* option), plane, or at a sketched geometry. This split allows the user to define a different draft angle for each desired region. If the surface is not split, then either a constant or a variable angle may be provided. The default is a constant value.

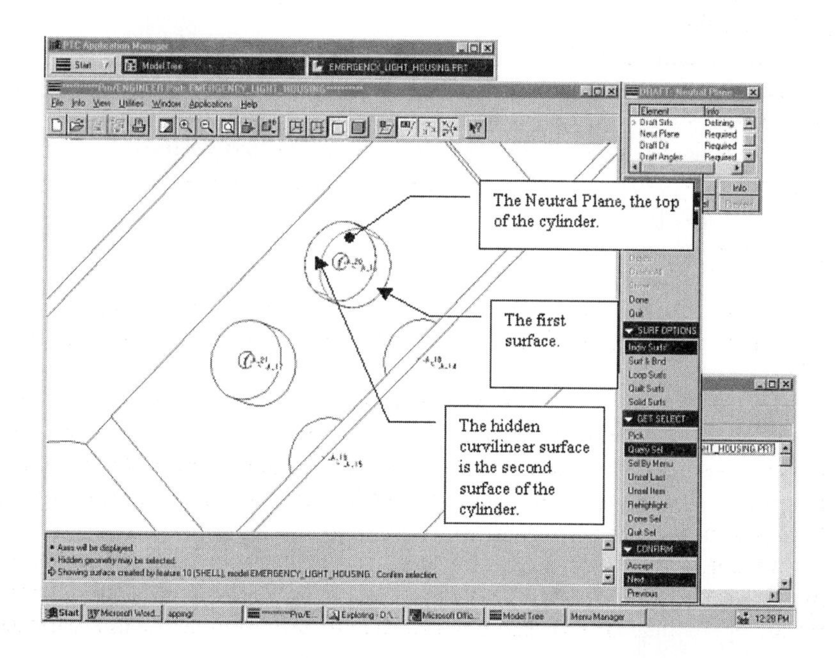


FIGURE 15.3 *The desired surfaces on one of the bosses. Note that the surface of the cylinder is actually two surfaces. Both of these surfaces must be selected.*

The draft direction is normal to the reference plane chosen by the user. The draft angle is measured from this normal.

Drafts are placed on features that are cast so that they can be removed from molds. For example, a boss, such as on the emergency light cover in chapter 6, is often given a small draft.

### 15.1.1 TUTORIAL # 69: A DRAFT FOR THE EMERGENCY LIGHT COVER

Let us place a 5° draft on the bosses of the emergency light cover. Retrieve the part called "EmergencyLightCover." Notice that the surfaces to be drafted are the cylindrical surfaces created around the holes by the shell. Also, notice that only the original holes are included in the pattern, not the surfaces around these holes. Thus, a draft will have to be performed individually on each of these bosses.

Select *Feature, Create,* and *Tweak.* The *TWEAK* menu will become available. Select *Draft.* From the *DRAFT OPTS* menu, keep the *Neutral Pln* default by choosing *Done.* Also, select *Done* from the *ATTRIBUTES* menu to keep the *No Split* option.

For the surface, select one of the bosses and choose the outermost cylindrical surface. Note that the cylindrical surface has two parts, as shown in Figure 15.3. Use *Query Sel* to help in picking the correct surfaces. After you have chosen *both* surfaces, click on *Done Sel.* Select *Done.*

The neutral plane is the top of the cylinder, as shown in Figure 15.3. A direction must be given for the draft to be constructed. Select the *Crv/Edg/Axis* option and click on either one of the axes running through the boss. You will probably need to use *Query Sel* to pick the axis from the rest of the entities.

Enter a value of −5 for the draft angle. Pro/E will add the draft. Your boss should look like the one shown in Figure 15.4. Compare Figure 15.4 to Figure 15.3.

Repeat the process, adding a draft to the remaining three bosses. The final version of the cover should appear as in Figure 15.5. Save the part.

## 15.2 EARS

An *Ear* is a small protrusion that is bent at its base. In Pro/E, the amount by which the protrusion is bent is determined by the user. Ears are used to align parts.

The section of an ear is sketched using the *Sketcher* after a sketching plane has been identified. The length of the ear may be one of two types. For a *Variable* ear, the length of the sketch includes the bent part of the ear. For a *Tab* ear, the length of the sketched section includes the *projection* of the bent part of the ear.

In creating an ear, the software requires that the sketching plane be normal to the surface to which the ear is attached. Furthermore, the sketch must be an open section with the ends of the section aligned to the attached

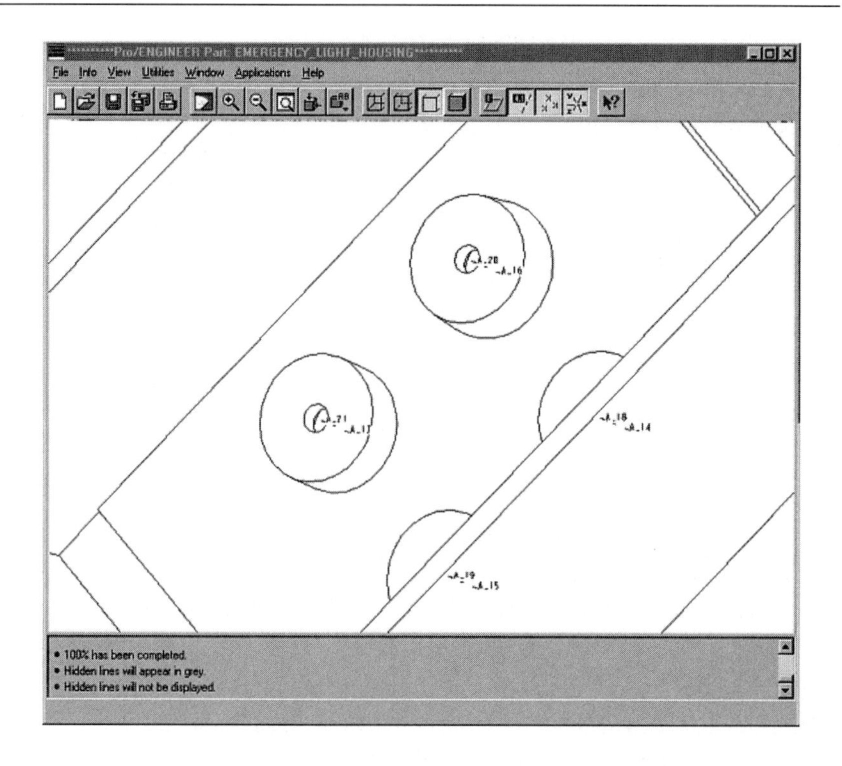


**FIGURE 15.4** *The cover with the boss shown in Figure 15.3 tapered with a* Draft.

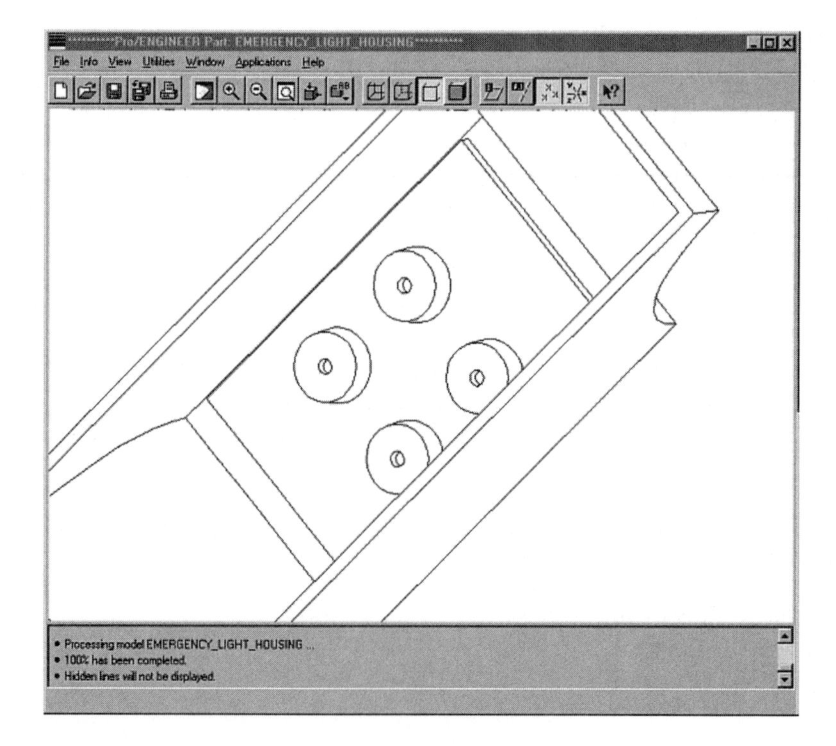


**FIGURE 15.5** *The cover with tapered bosses.*

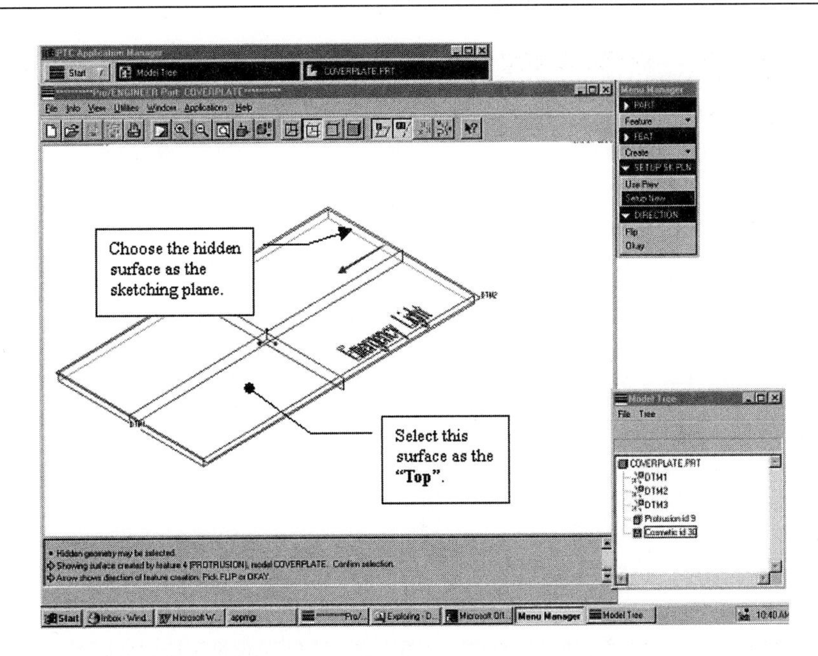


FIGURE 15.6 *Use the given surfaces as the sketching and orientation planes.*

surface. While the section of the ear may have any geometry, the part of the geometry that is attached to the surface must have edges that are parallel to each other and long enough to form the bend.

### 15.2.1 TUTORIAL #70: EARS FOR THE COVER PLATE

The part "CoverPlate" from chapter 14 is used to hide the fasteners used to attach of the emergency light cover. The cover plate fits into a rectangular recess. It is aligned with ears.

Retrieve the part "CoverPlate." Then select *Feature, Create, Tweak,* and *Ear.* The ear will be bent 90° to the attachment surface, so choose *90 deg tab* and *Done.*

Select the sketching plane shown in Figure 15.6. Notice the direction of the arrow in the figure. Orient the model in the *Sketcher* by choosing the surface with the cosmetic text as the "*Top.*"

In the *Sketcher,* draw the open section shown in Figure 15.7. *Dimension* as shown and *Modify* to the proper values. Then select *Done.*

Enter 0.05 for the depth of the ear. Key in the same value for the bend radius. The software will create the ear.

A second ear may be added to the part using the same procedure or the *Pattern* option. By using the *Pattern* option, the second ear may be created faster. The procedure is as follows:

1. Select *Feature* and *Pattern.*
2. Select the ear.
3. Use *General* and *Done.*

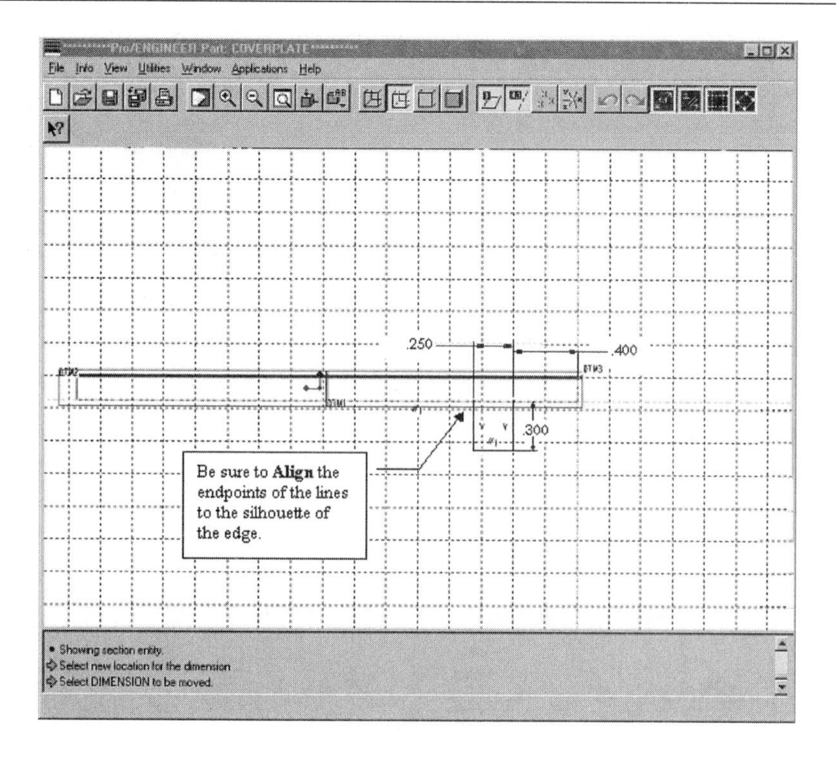


FIGURE 15.7 *Draw an open section in the* Sketcher.

4. Choose the dimension .400 as the first direction.
5. Enter the formula $3.95 - 0.80 - 0.25 = 2.90$ as the increment. Note the value 3.95 is the depth of the plate. The value 0.80 will give an ear equidistant from either side of the plate. The value 0.25 is included in the formula in order to take into account the depth of the ear.
6. *Done.*
7. Enter 2 for the total number of instances.
8. *Done.* The plate with both ears is shown in Figure 15.8.

Because of the ears, two small slots must be added to the model of the emergency light cover. These slots will not be added here. In order for the cover to lock in position, a latch must be added to the plate. This latch will be constructed using a two-sided protrusion. The procedure for adding the latch to the plate is as follows:

1. Select *Feature, Create,* and *Protrusion.*
2. Choose *Extrude, Thin,* and *Done.*
3. *Both Sides* and *Done.*
4. Choose DTM1 as the sketching plane and DTM2 as "*Top.*"
5. Sketch the section shown in Figure 15.9.

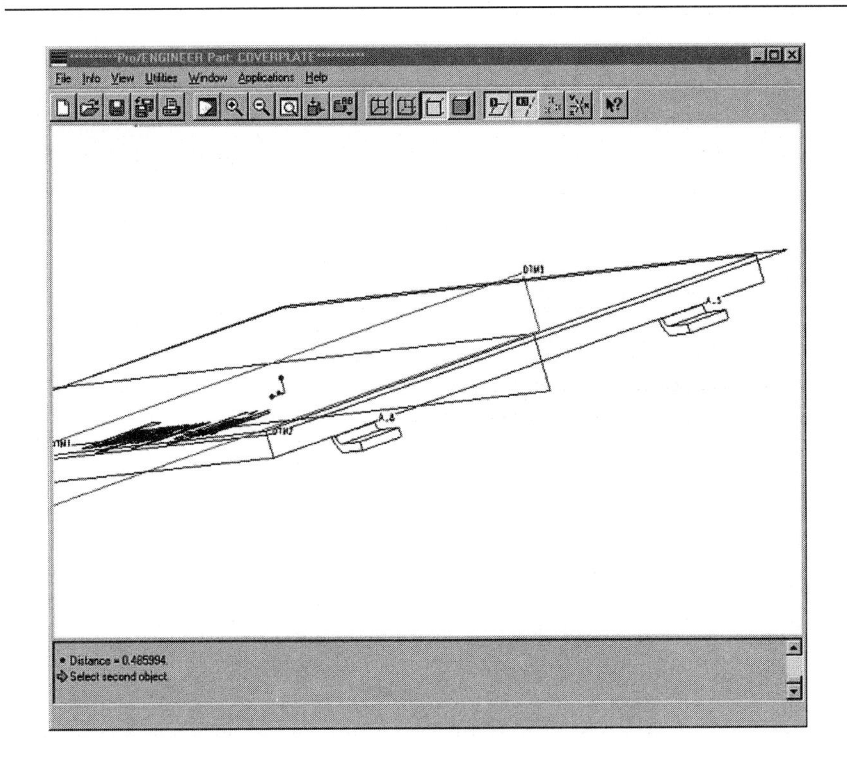


FIGURE 15.8 *The model with both ears.*

6. *Dimension, Regenerate,* and *Modify,* to the proper values.
7. Select *Done.*
8. Choose *Blind* and *Done.*
9. Enter a thickness of 0.050 for the *Thin.*
10. Enter a depth of 2.00.

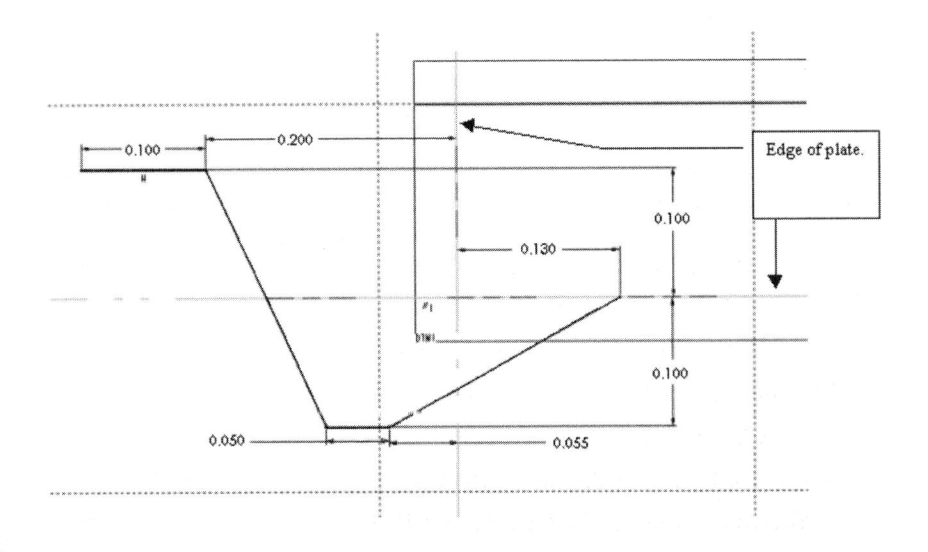


FIGURE 15.9 *Use this geometry to Sketch and dimension the latch.*

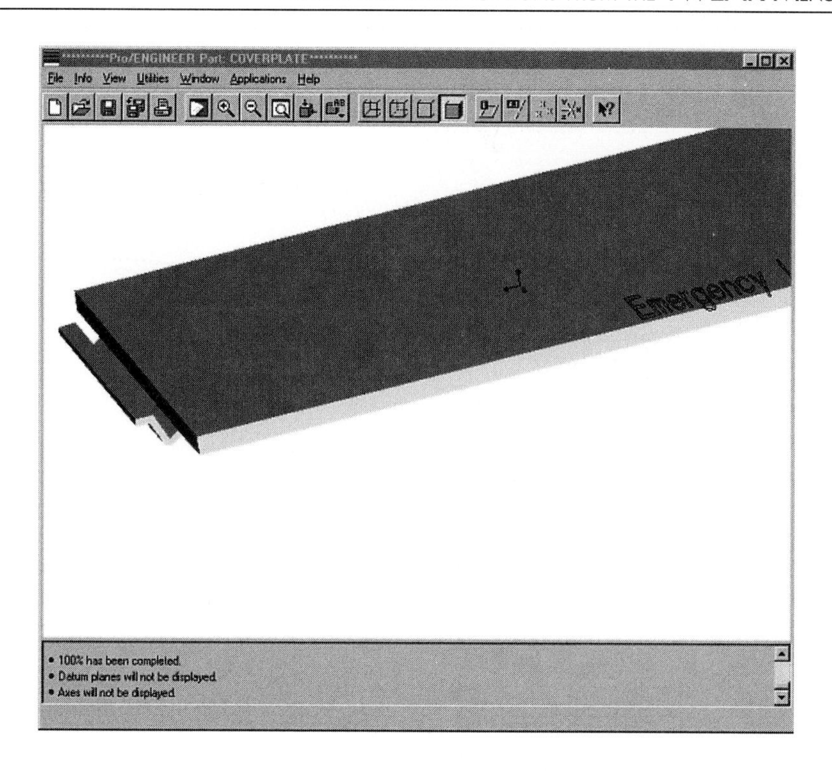


FIGURE 15.10 *The model after the addition of the latch.*

The plate with the latch is shown in Figure 15.10. Save the part.

## 15.3 THE LIP OPTION

A *Lip* is a feature used to interlock two parts, as shown in Figure 15.11. One part contains the protrusion, and the other part, a cut corresponding to the protrusion.

A lip is created by offsetting a given surface called the Mating Surface. The offset distance is a parameter that must be provided by the user. The lip may have a sloped side at a prescribed angle. The side offset is also a parameter of the lip.

### 15.3.1 TUTORIAL # 71: ADDING A LIP TO A PART

A lip is used to interlock parts. A part that has such a feature is the cover of a remote control. In this section, we will create this cover and add a lip.

Use the following procedure to create the part:

1. Create a part called "RemoteControlCoverLower."
2. Add the default datum planes.
3. Select *Feature, Create,* and *Protrusion.*
4. Choose *Extrude, Solid,* and *Done.*

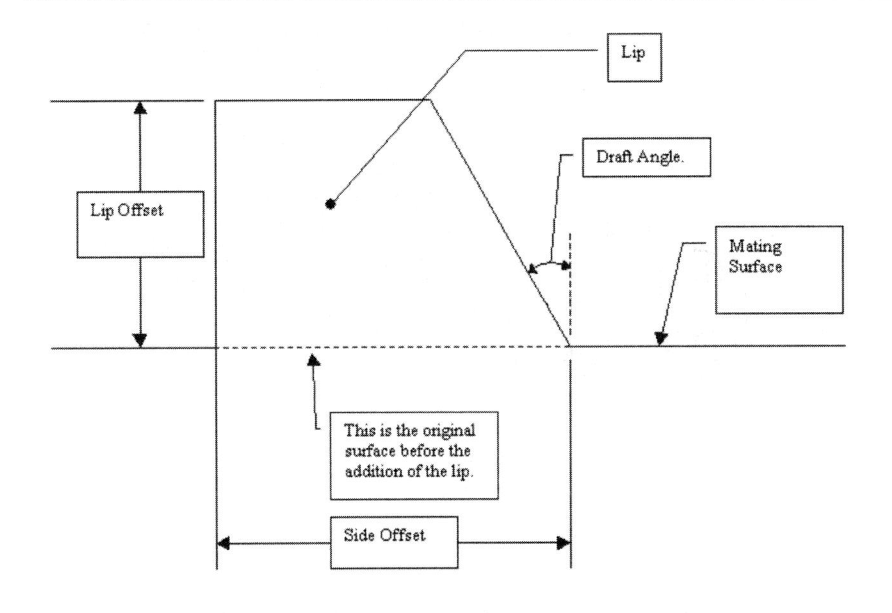


**FIGURE 15.11** *A* Lip *is created by offsetting a surface by a certain amount.*

5. *Both Sides* and *Done.*
6. Select DTM3 as the sketching plane and DTM2 as the "*Top.*"
7. In the *Sketcher,* draw the section shown in Figure 15.12.

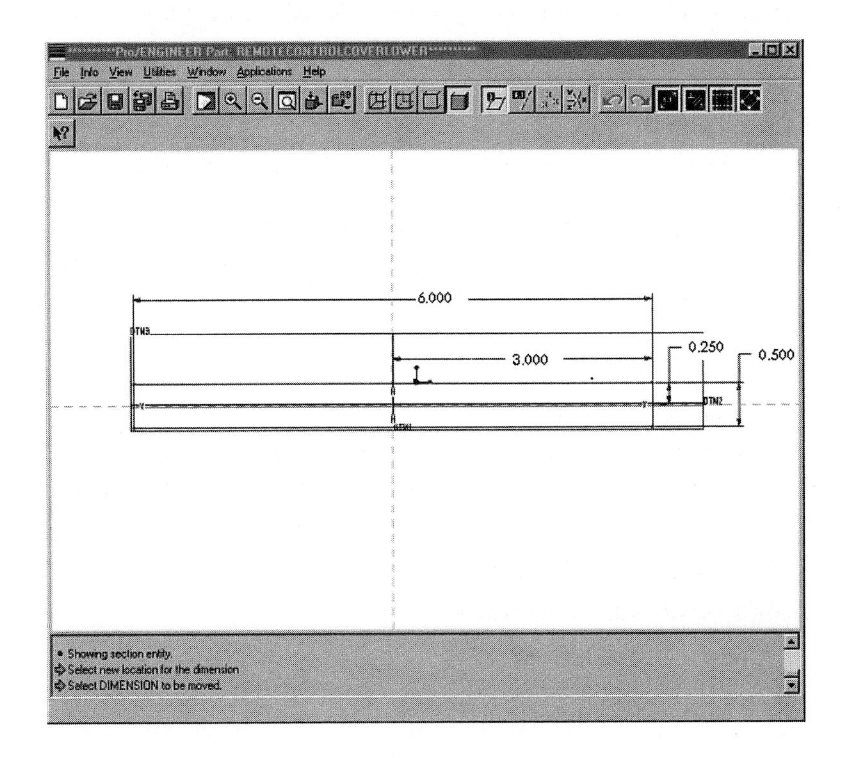


**FIGURE 15.12** *Use this geometry to construct the remote control.*

8. *Dimension* and *Modify* to the values shown in Figure 15.12.
9. *Done.*
10. Extrude *Blind* to a depth of 2.50 inches. The model at this point is shown in Figure 15.13.

We need to shell the model. This task may be accomplished by using the following sequence of steps:

1. Select *Feature, Create,* and *Shell.*
2. Choose the plane shown in Figure 15.13 as the surface to remove.
3. Choose *Done Sel* and *Done Refs.*
4. Enter 1/8 for the thickness of the shell.

The lip is to be added on the inner portion of the shell thickness using the edges shown in Figure 15.14. In order to add the *Lip,* select *Feature, Create, Tweak,* and *Lip.* Then, using the *Single* option, select the edges shown in Figure 15.14. Choose the surface shown in Figure 15.14 as the mating surface. Enter an offset of 0.125 and a side offset of 0.125/2. The draft angle is 0°. The model with the lip is shown in Figure 15.15. Save this part.

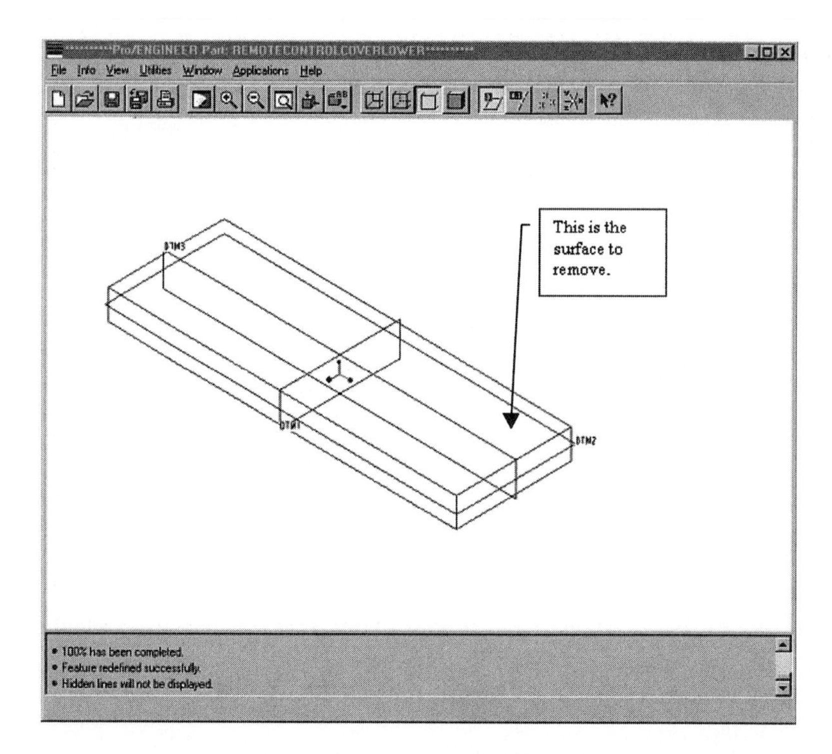


FIGURE 15.13 *This model after the creation of the protrusion.*

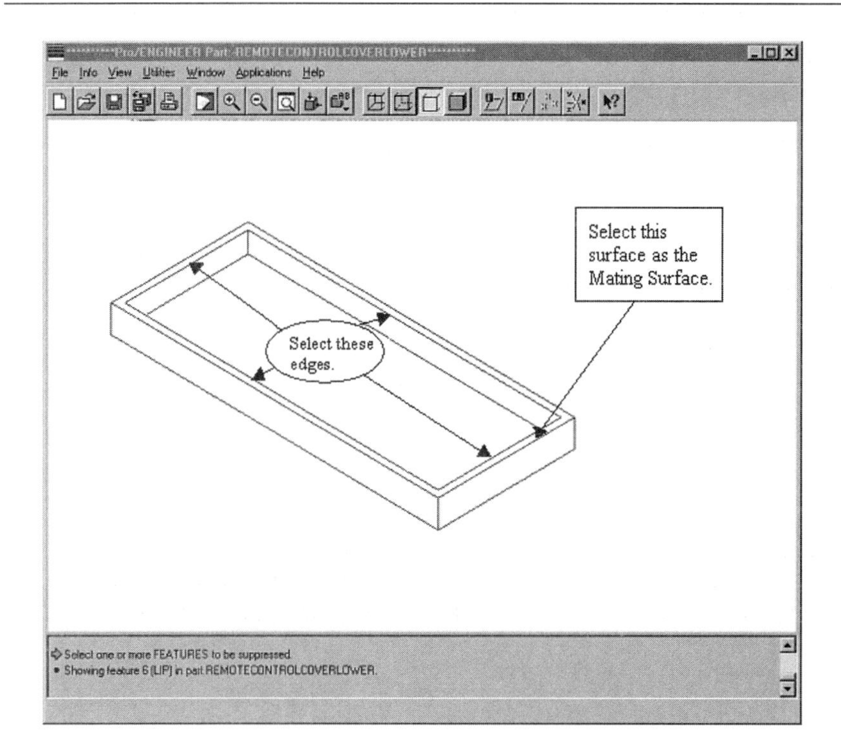


**FIGURE 15.14** *Shows the model after the use of the* Shell *option. The* Lip *may now be added to the model.*

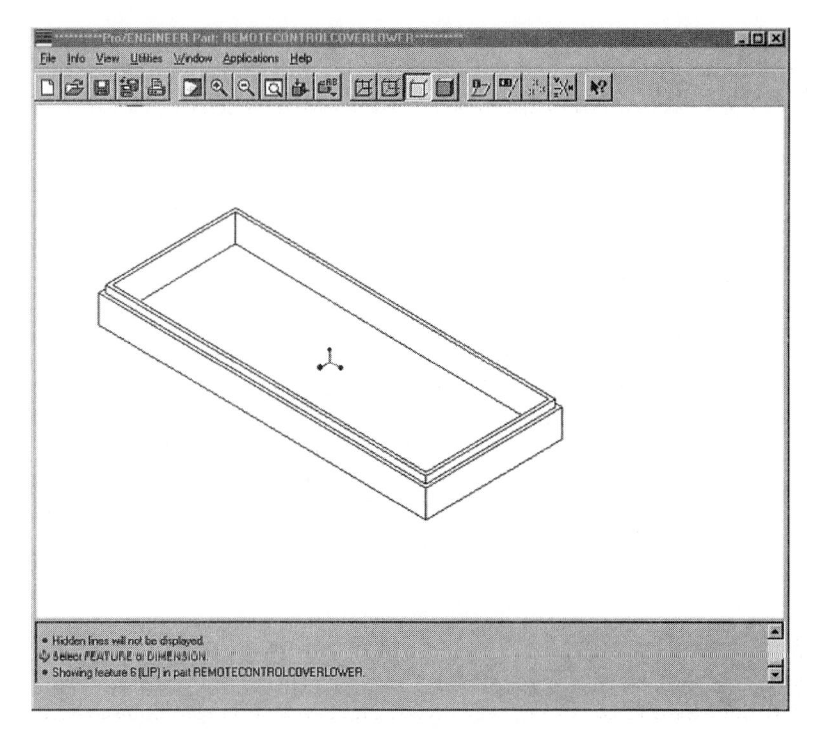


**FIGURE 15.15** *A lip may be added to the model. Compare this figure with Figure 15.14.*

## 15.4 QUILTS

A *Quilt* is a surface feature that is created by extending one or more curves. In Figure 15.16, we see a practical application of the *Quilt* option. The shape of the wing is determined by the NACA airfoil section. By extending the datum curve, a three-dimensional surface is created, which models the surface of the wing.

A *Quilt* may be created and manipulated by using the options in the *QUILT SURF* menu (Figure 15.2). The *New* option is used to create a quilt. By using the *Transform* option, the user may *Move, Copy,* and *Mirror* a quilt. A *Fillet* quilt may be used to construct a surface-to-surface round, where the elementary method discussed in chapter 8 fails. A quilt may be placed in a layer and blanked.

### 15.4.1 TUTORIAL # 72: AN AIRCRAFT WING

In chapter 9, we created a part containing datum points and curves. These points and curves model the geometry of the wing root (see Figure 15.16). Using the options in the *QUILT SURF* menu, it is possible to generate a surface model of an aircraft wing.

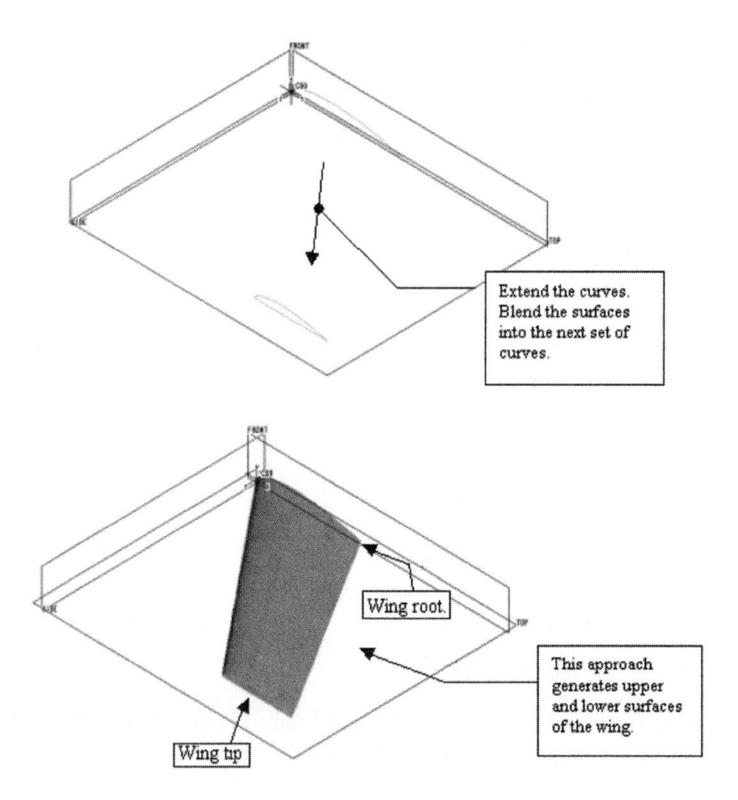


FIGURE 15.16 *By extending a curve, a surface may be constructed. In this case, multiple curves are used, and the surface is blended into both sets of curves.*

Retrieve the part "Airfoil." The part contains two datum curves created in chapter 9. Let us call the upper curve "Curve 1" and the lower curve "Curve 2."

Before proceeding to add the surfaces, we need to add some additional datum curves. These curves will model the geometry of the wing tip as shown in Figure 15.16.

The datum points are contained in the files called "cs1.pts" and "cs2.pts" and may be found on the disk supplied with this book. You will need to have access to these files before proceeding any further.

Add the datum points as follows:

1. Select *Feature, Create, Datum,* and *Point.*
2. Choose *Offset Csys.*
3. Select the coordinate system "Cs0" on your screen and then choose the *Cartesian* option.
4. Use the *With Dims* option. Select *Read Points.*
5. Find the file "cs1.pts" and double click.
6. Choose *Done.*

Zoom into these points. Consider Figure 15.17. Notice that the points may be used to form two curves. Let us place a curve through the upper points and a separate curve through the lower points. The procedure for adding datum curves through the points is as follows:

1. Select *Feature, Create, Datum,* and *Curve.*
2. Choose *Thru Points* and *Done.*
3. Select *Spline* and *Single Point.*
4. Now pick the upper points as shown in Figure 15.18. Start with point #69 then #36, #37, and so on. The last point is point #52.
5. Choose *Done* and *OK.*

Let us call this curve "Curve 3." Shortly, we will add a curve to the lower points by using the same procedure.

Add the datum curve to the lower points by doing the following:

1. Select *Feature, Create, Datum,* and *Curve.*
2. Choose *Thru Points* and *Done.*
3. Select *Spline* and *Single Points.*
4. Now pick the lower points; start with point #69 and proceed to point #68, point #67, and so on. Finish with point #52.
5. Choose *Done* and *OK.*

Call this curve "Curve 4." Turn off the display of the datum points. The curves should appear as in Figure 15.16.

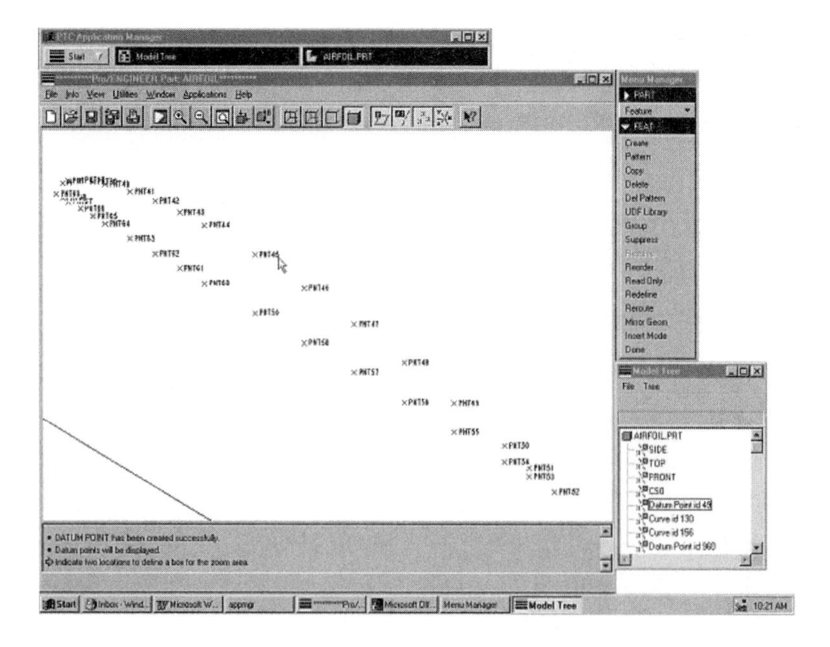

FIGURE 15.17 *The datum points from file "cs1.pts" may be used to create the geometry of the wing tip.*

We need to add another set of points in order to construct a curve for the end of the wing. The points are found in the file "cs2.pts." Add these points to the model as shown in Figure 15.19. Then create a datum curve through the points. Again, use the options *Spline* and *Single Point.* Start with point #70, proceed to point #71, point #72, and so on. The last point is

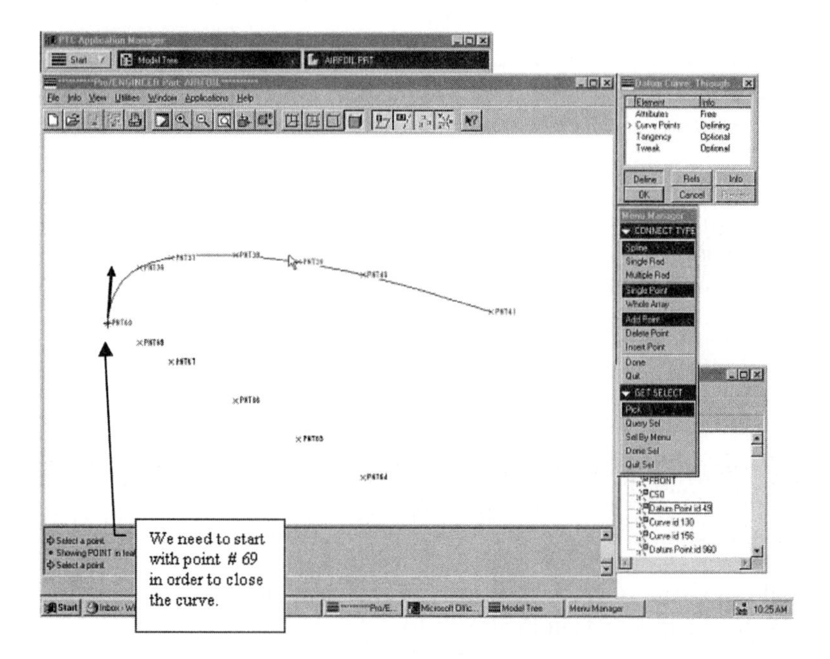


FIGURE 15.18 *Add a datum curve through the upper points by using the options* Spline *and* Single Points.

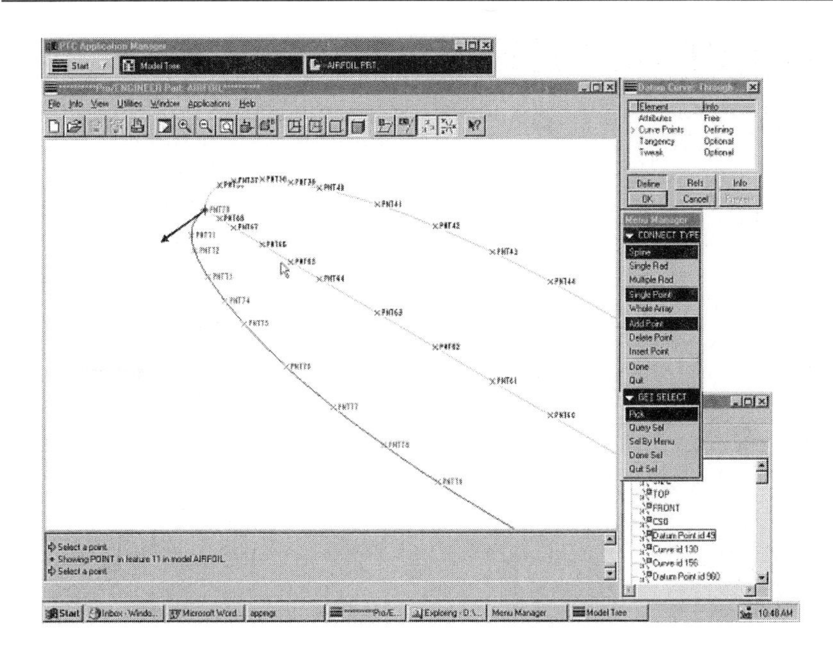


**FIGURE 15.19** *Add a datum curve through the points from file "cs2.pts."*

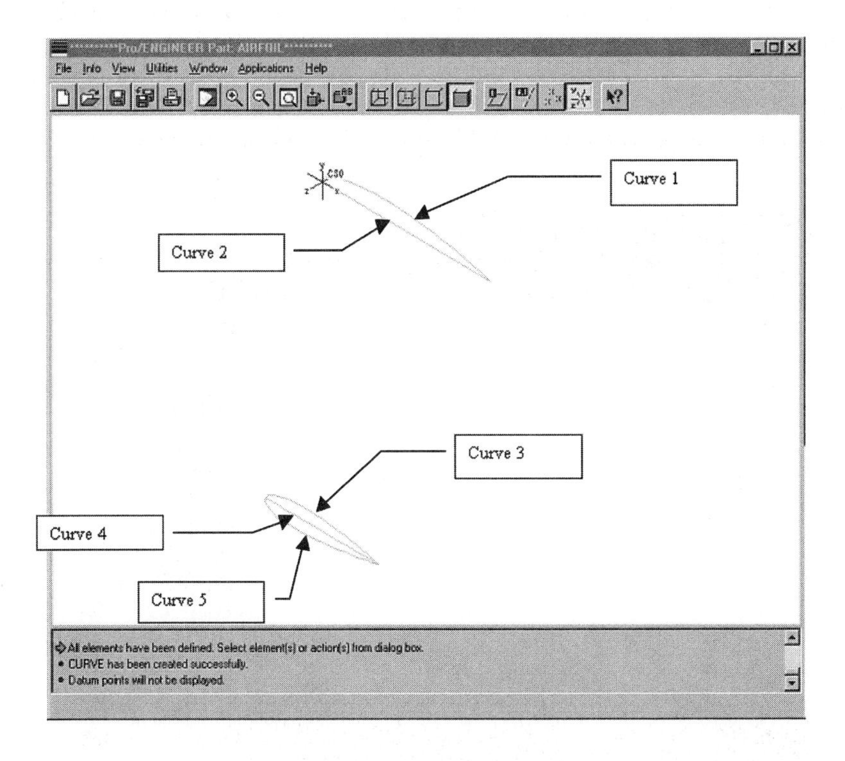


**FIGURE 15.20** *The model after the addition of "Curve 5."*

#87. Let us call this curve "Curve 5." The model should appear as it does in Figure 15.20.

One set of curves must be added to the model in order to generate the support structure of the wing. This part is attached to the fuselage. We can create these two datum curves, say, Curves 6 and 7, by copying Curves 1 and 2. Because Curves 1 and 2 form a surface, we may simply transform this surface to obtain the desired goal. The procedure is as follows:

1. Select *Feature, Create, Surface,* and *Transform.*
2. Choose *Copy* and *Done.*
3. Select Curves 1 and 2. Choose *Done Sel.*
4. Choose coordinate system Cs0.
5. Enter zero for the transformation in the $x$ direction.
6. Enter zero for the transformation in the $y$ direction.
7. For the $z$ direction, enter −0.5.
8. Choose *Done/Return.*

The software will create a new surface containing Curves 6 and 7, as shown in Figure 15.21. We are now ready to create the surface of the wing. The curves will be used as boundaries to generate the various surfaces.

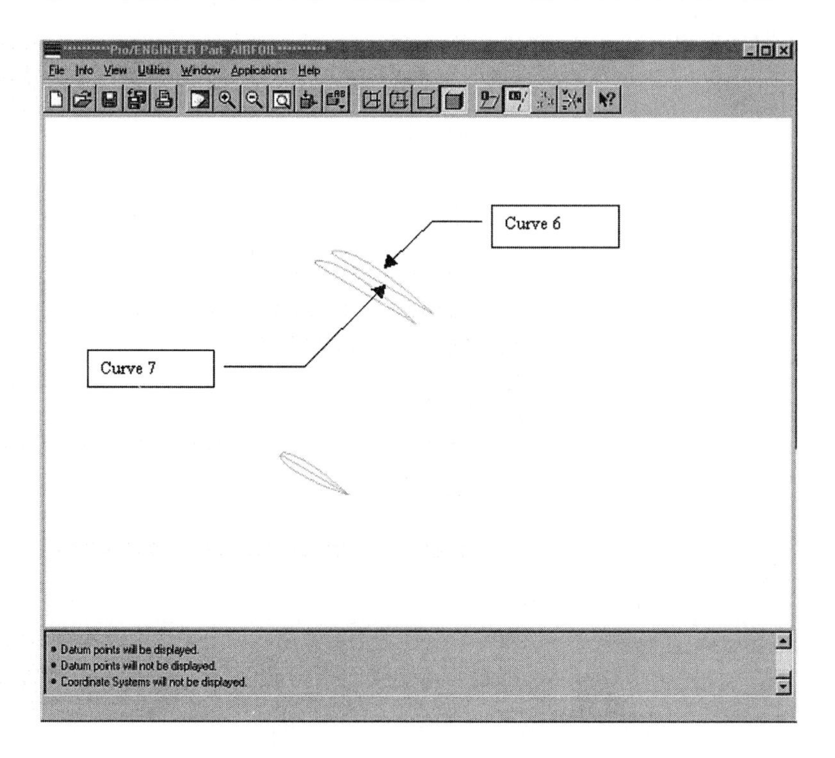


FIGURE 15.21 *The model with the two new curves.*

The surfaces are added to the model by using the *Boundaries* option and the *Blended Surf* option. We begin by adding a surface between Curve 1 and Curve 3. The model, after adding this surface, is shown in Figure 15.22. The procedure is as follows:

1. Select *Feature, Create, Surface,* and *New.*
2. Choose *Advanced* and *Done.*
3. Select the *Blended Surf* option and *Done.*
4. Pick Curve 1 and Curve 3. Choose *Done Sel* and *Done Curves.*
5. Select *OK.*

The software will add the surface. Consider Figure 15.22. We need to repeat the process for the remaining surfaces. The steps for creating the surfaces are the same, except that a different set of curves must be chosen for each surface. Follow this procedure in order to add the remaining quilts:

1. Repeat the steps. Use Curves 2 and 4.
2. Repeat the steps with Curves 3 and 5.
3. Repeat the steps with Curves 4 and 5.
4. Repeat the steps with Curves 1 and 6.
5. Repeat the steps one last time with Curves 2 and 7.

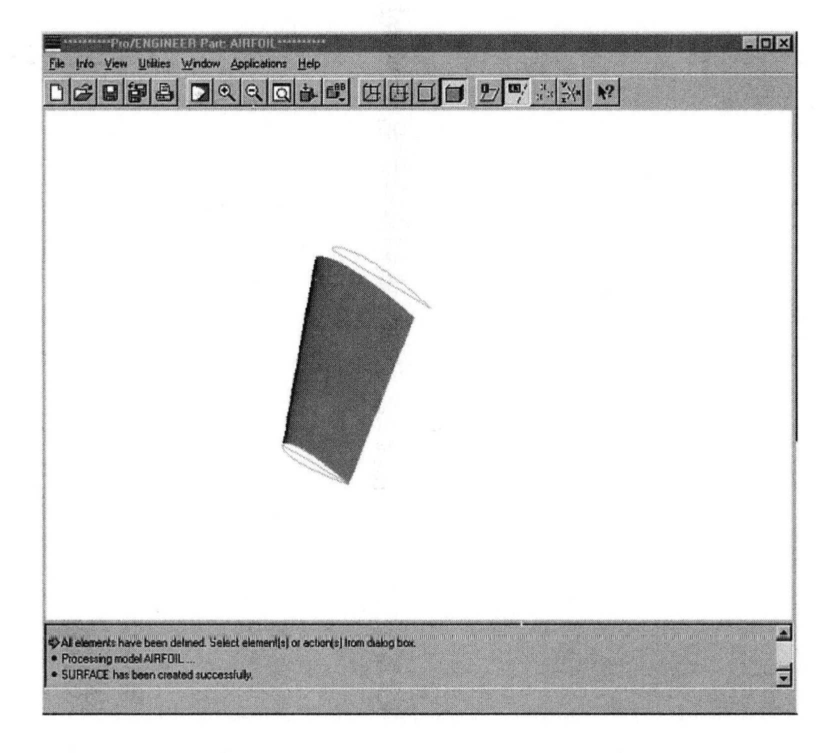


**FIGURE 15.22** *The model with a surface whose boundaries are Curve 2 and Curve 3.*

**FIGURE 15.23** *The rest of the surfaces may be added to the model by selecting the correct boundaries.*

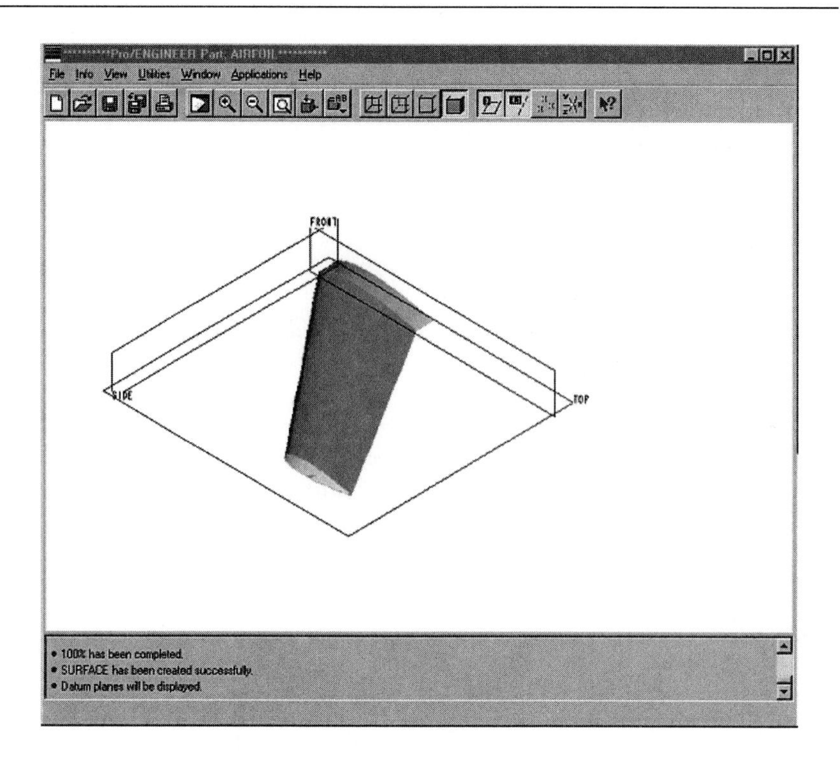


The model at this point is shown in Figure 15.23.

Because the wing is symmetric, we may create the other half of the wing by simply using the *Mirror* option. However, before rushing to do so, we must create a mirror plane. This plane goes through Curves 6 and 7 by selecting *Feature, Create,* and *Datum.* Then use *Offset* and select the datum "Front." Enter a value of −0.5.

Mirror the geometry by selecting *Feature* and *Mirror Geom.* Choose the new plane (DTM1). The model is shown in Figure 15.24. Save the model.

## 15.5 SUMMARY AND STEPS FOR THE TWEAK OPTIONS

The scope of this chapter is to provide introductory material on the options contained in the *TWEAK* menu and the creation of surfaces using quilts. From the *TWEAK* menu, the options *Draft, Ear,* and *Lip* were considered.

The *Draft* option is used to taper surfaces. The surfaces may be tapered as much as ±15°. In general, a *Draft* may be added to a surface by using the following steps:

1. Select *Feature, Create, Tweak,* and *Draft.*
2. Choose either *Neutral Pln* or *Neutral Crv.* Then, select *Done.*
3. Choose any split surfaces, edges, or curves and *Done.*
4. Select the surface(s) to draft. Choose *Done Sel* and *Done.*

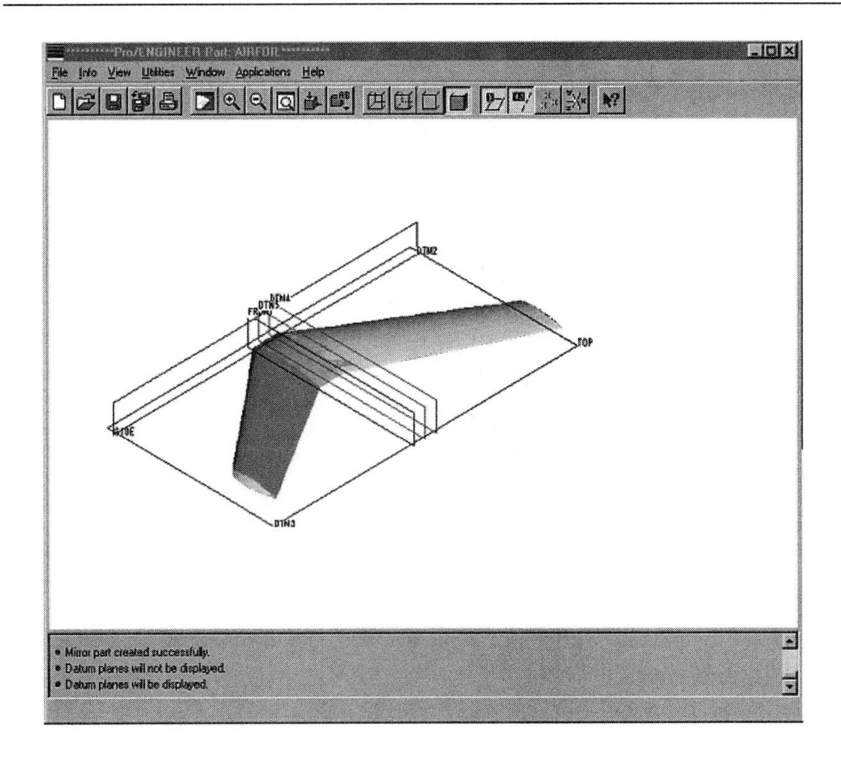


FIGURE 15.24 *The completed model of the wing.*

5. Choose the neutral plane.
6. Pick the normal direction.
7. Enter the draft angle.

An *Ear* is a protrusion that is attached to a base feature and bent at its point of attachment. An ear may be of *Variable* type or *90 deg tab.* In the case of a *Variable* type ear, the user may enter the bend radius. The general sequence of steps for creating an *Ear* is as follows:

1. Select *Feature, Create, Tweak,* and *Ear.*
2. Choose the type of ear, that is *90 deg tab* or *Variable.*
3. Pick the sketching plane.
4. Select the orientation plane.
5. In the *Sketcher,* draw the geometry of the ear.
6. *Dimension* and *Modify* to the proper values. Select *Done.*
7. Enter the depth of the ear.
8. Key in the bend radius.
9. If the ear is of type *Variable,* enter the angle of the bend.

A *Lip* is a protrusion that is used to interlock multiple parts. The protrusion is constructed on one part and on the mating part, a cut with the same geometry as the lip is created. A lip may be added to a surface using the following general procedure:

1. Select *Feature, Create, Tweak,* and *Lip.*
2. Choose the edge defining the boundary of the mating surface.
3. Select the mating surface.
4. Enter the offset distance from the mating surface.
5. Key in the side offset.
6. Enter the draft angle.

A *Quilt* is a surface. Quilts are created by extending a curve or edge. A surface may be created between one or more boundaries. The surface may be blended to the boundary using the *Blended Surf* option.

In general, a surface quilt may be created by using the following sequence of steps: *Feature, Create, Surface,* and *New.* Quilts may be constructed by using the same options as for a protrusion, including *Extrude, Revolve, Sweep,* and so on. If the boundaries of the quilt are known, then the surface may be created by using these steps:

1. Select *Feature, Create, Surface,* and *New.*
2. Choose *Advanced* and *Done.*
3. Select the *Blended Surf* option and *Done.*
4. Pick the curves that form the boundary.
5. Choose *Done Sel* and *Done Curves.*
6. Select *OK.*

## 15.6 ADDITIONAL EXERCISES

15.1 Using the geometry of the Howe roof truss in exercise 9.9 and the *Transform* option, create a set of trusses ten meters apart. The result is given in Figure 15.25.

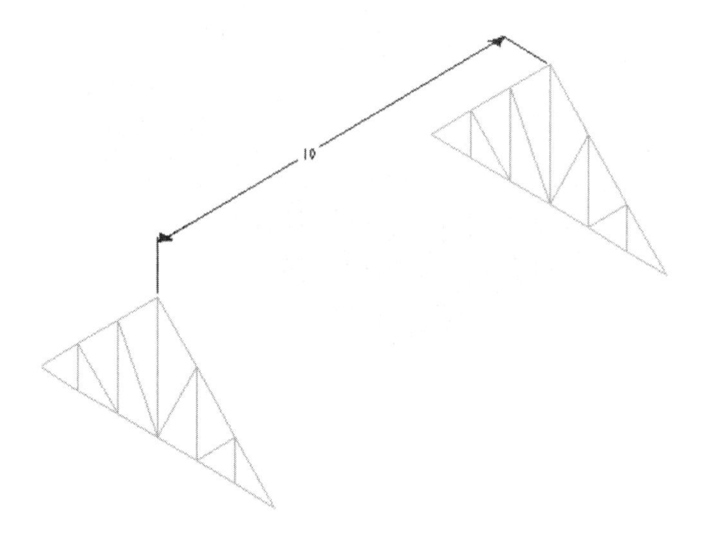


FIGURE 15.25 *Figure for exercise 15.1.*

15.2 Use the results from exercise 15.1 to create surfaces representing the roof. Use Figure 15.26 for reference.

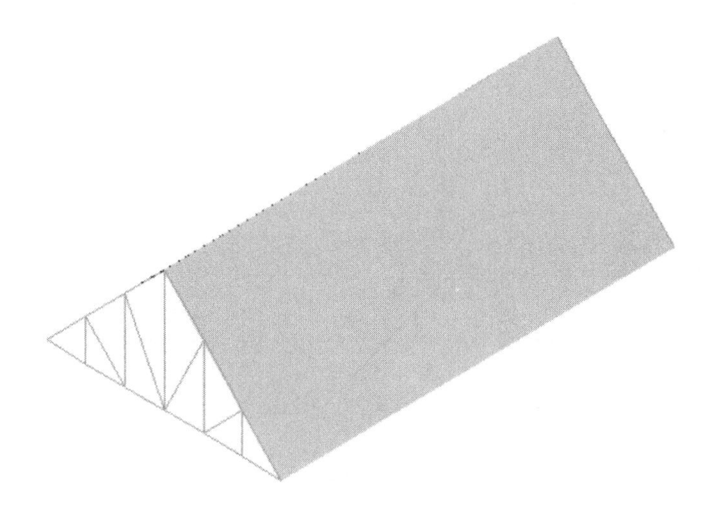

FIGURE 15.26 *Figure for exercise 15.2.*

15.3 Add walls to the results from exercise 15.2 by using *Feature, Create, Surface, New,* and *Extrude.* The "walls" are shown in Figure 15.27.

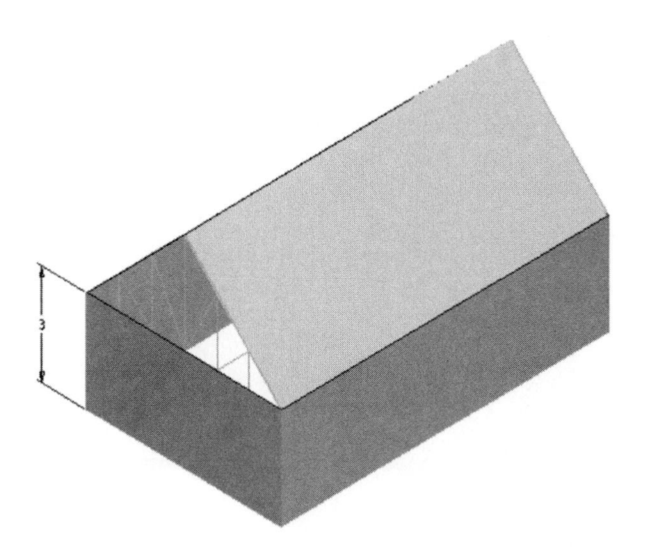


FIGURE 15.27 *Figure for exercise 15.3.*

15.4 Finish off the ends of the house from exercises 15.1 through 15.3, as shown in Figure 15.28. Hint: use the options *Feature, Create, Surface, New, Extend,* and *Along Dir* and extend the roofline.

FIGURE 15.28 *Figure for exercise 15.4.*

15.5 Create the 14 × 14-inch oil pan illustrated in Figure 15.29. Use a draft angle of 5°.

FIGURE 15.29 *Figure for exercise 15.5.*

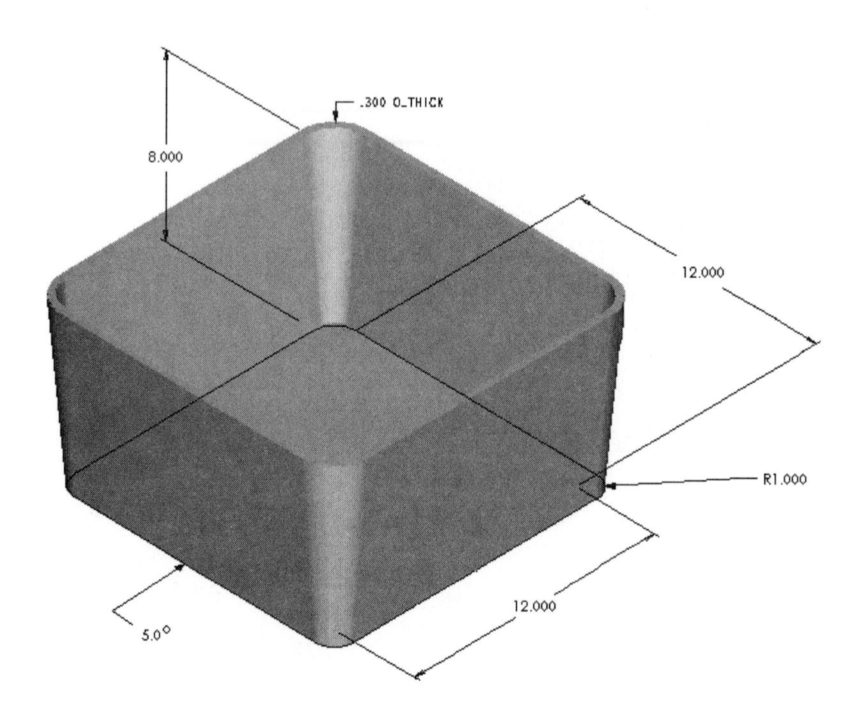

15.6 Add the second protrusion and lip shown in Figure 15.30 to the oil pan from exercise 15.5.

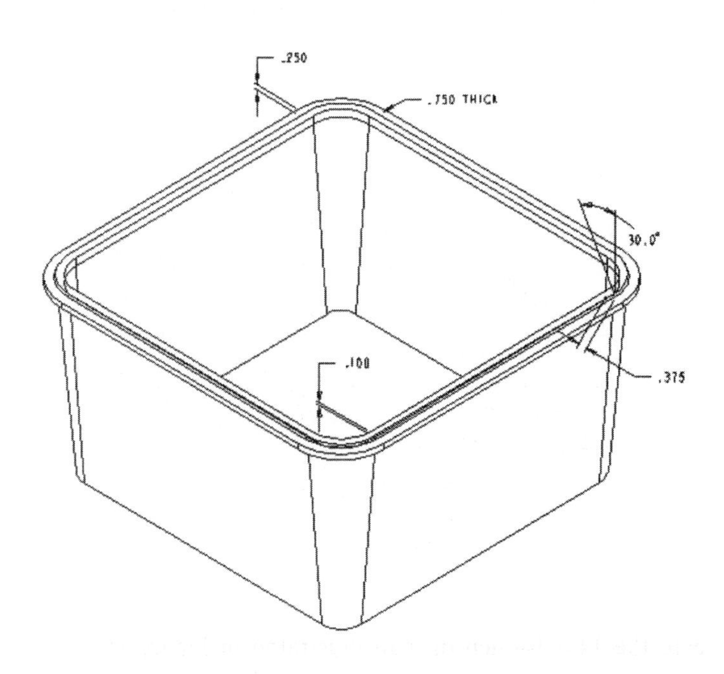


**FIGURE 15.30** *Figure for exercise 15.6.*

15.7 Create the cover shown in Figure 15.31. Add the ears.

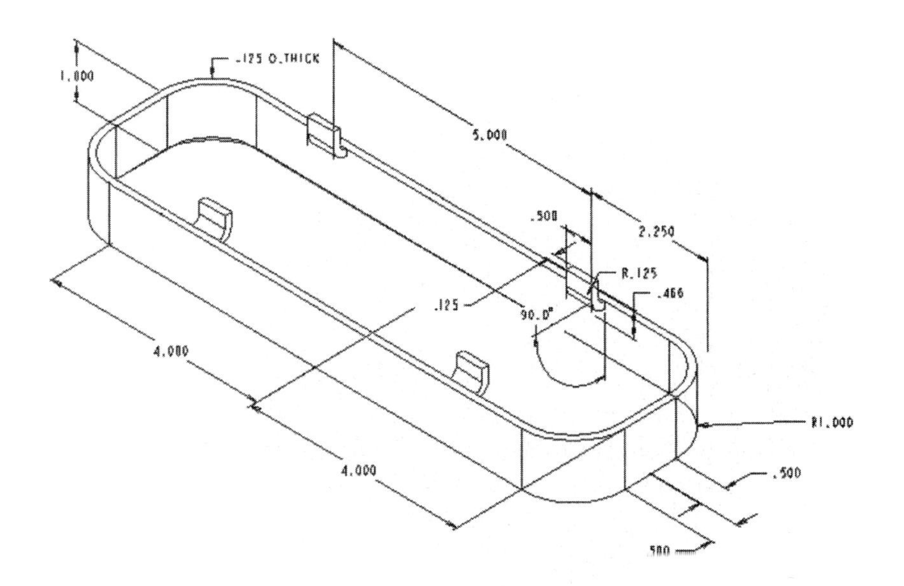


**FIGURE 15.31** *Figure for exercise 15.7.*

# CHAPTER 16

# THE DRAWING MODE

## INTRODUCTION AND OBJECTIVES

So far, we have only studied options available in order to create models using Pro/E. However, special views of the model are desired in order to convey its features. Obtaining special views can be done in several ways. A standard method is the use of multiviews.

Typically, multiview drawings are obtained of the top, front, and side of the model. The views are fully dimensioned and tolerances are specified. If desired, a pictorial, such as an isometric, is often added to further clarify the features inherent in the model.

The module for creating multiviews in Pro/E is the *Drawing* mode. Similar to the approach used for part files, the software allows the user to create drawing files. A standard format may be loaded during the creating of the drawing file.

The objectives of this chapter are the following:

1. The software user will become familiar with the drawing mode.
2. The user will be able to retrieve or create a format.
3. The user will attain an ability to create a multiview drawing and add dimensions and notes to the views.

## 16.1 DRAWINGS AND FORMATS

A drawing is a layout containing various views of a model. Most drawings contain front, top, and a right-hand side view of the model. Often, the drawing may contain a section of the model created by slicing the model with a plane and displaying one side of the cut model.

All drawing files are designated with the extension ".drw" by Pro/E. In order to create a drawing, select either the *File* and then *New* (Figure 1.6) or the *Create New Object* (Figure 1.11) icon. As shown in Figure 16.1, choose the *Drawing* option and enter the name of the drawing.

The size and orientation of the drawing may be defined by the *New Drawing* dialog box illustrated in Figure 16.1. If desired, enter the name of the model in the cell provided. You may also give the name of the model when the view is added to the drawing. Select the desired paper size or retrieve a format using the *Retrieve Format* option. In general, we suggest that the *Landscape* orientation be used.

A format is nothing more than a boundary for the drawing, with spaces for your name, date, name of the model, etc. The Pro/E software is bundled with standard formats. Any of these formats may be chosen by simply selecting the paper size. However, a simple format contained in the file "Aformat.frm" may be found on the disk supplied with this book. This simple format is provided so that the user may complete the tutorials. The format is designed for A size paper and *Landscape* orientation. You will need to make sure that you have access to this file.

Custom formats may be created with the *Format* mode. The geometry of the format may be created by sketching the entities. In order to create a

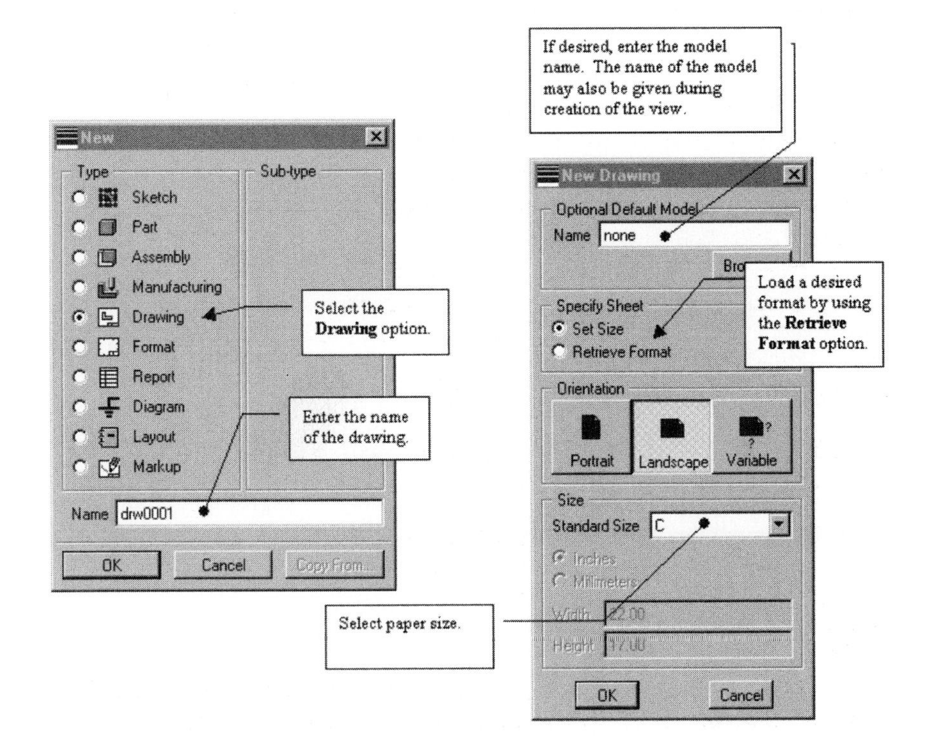


**FIGURE 16.1** *Create a drawing by using the* New *and* New Drawing *dialog boxes.*

format, choose *File* and *New,* or select the *Create New Object* icon. Then, choose the *Format* option, as shown in Figure 16.2. Give the format a name and select *OK.* Select the paper size and orientation.

In adding views to a drawing, the model must be defined, and Pro/E must have access to the model. If the part file is not located in the default directory, then the directory must be changed or the file moved.

When adding the first view, the drawing may be scaled, if desired. If no scale is given, then Pro/E scales the initial view so that additional views may be added to the drawing. The scale size is written at the bottom of the drawing window and may be modified by clicking on *Modify* and then on the scale with your mouse pointer.

The first view added to the drawing is a *General* view. Subsequent views may be *Projections, Auxiliary, Revolved,* and *Detailed* views. Most often, we will be using the *Projection* option to create additional views. A *Detailed* view shows a portion of the model. Detail views are helpful in highlighting special features inherent in the model. *Auxiliary* views are special views. Such views are usually used to illustrate incline or oblique surfaces. A *Revolved* view is a view created by revolving a section 90 degrees around a cutting plane. Revolved views are helpful in illustrating cross sections.

The angle bracket and the connecting rod will be used as the models for this chapter. These models must be complete in order to proceed.

### 16.1.1 TUTORIAL # 73: CREATING A FORMAT

In this tutorial, we will create a format for B size paper (11 inch × 17 inch) that may be used in cases where "Aformat" is unsuitable. Begin the creation

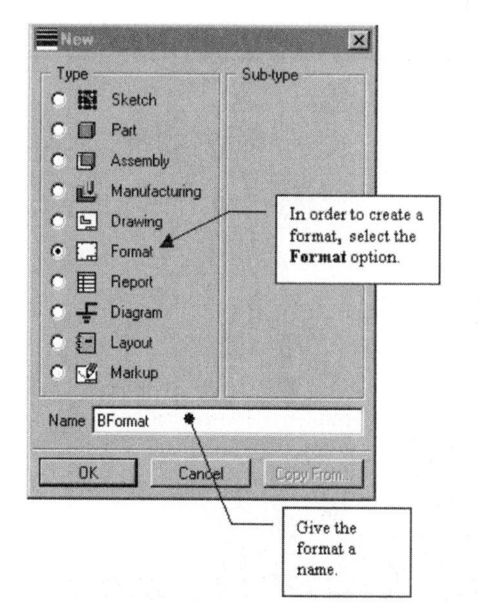


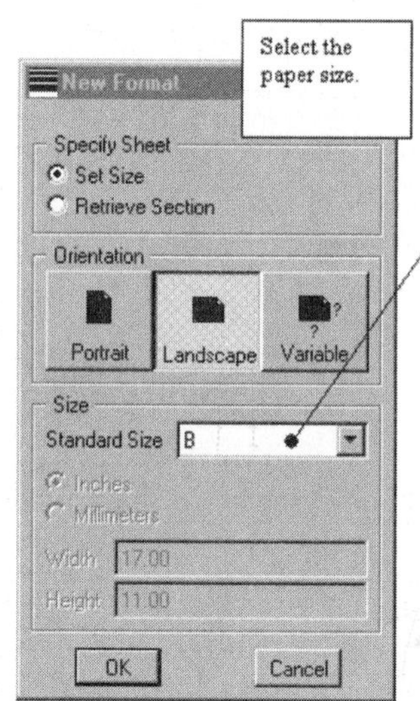


**FIGURE 16.2** *Create a format by using the* New *and* New Format *dialog boxes.*

of this format by selecting *File* and *New.* In the *New* dialog box, (Figure 16.2), choose the *Format* option and enter the name "Bformat." Select the *OK* button.

In the *New Format* dialog box choose *Landscape* orientation and B size paper. Then, press the *OK* button. The software will display the outline of the sheet of paper as shown in Figure 16.3. Note that the paper size is indicated below this outline.

We will create a simple format with several spaces. The *Sketch* option in the *DETAIL* menu (Figure 16.3) may be used to create the required geometry.

First, let us change some of the parameters describing the format. Select *Set Up* and *Modify Val.* The software will display the drawing configuration file for this format. Change drawing_text_height to 0.3 and *draw_arrow_style* from *Closed* to *Filled.* These changes will make the text larger than the default value, as well as change the style of the arrowhead. Then select *File* and *Exit.* Hit the *Repaint* button to activate the changes.

A grid may be displayed for making the geometry construction easier. For better visibility, the spacing of the vertical and horizontal lines may be changed. Select *Modify* and *Grid.* The spacing in the *x* and *y* direction may be set to separate values. For our purposes, we can set the spacing in both directions to the same values. Choose *Grid Params* and *X&Y Spacing.* Enter a value of 0.5. Hit the *Return* key and then select *Done/Return.*

Turn on the grid shown in Figure 16.3 by selecting *Grid On* and *Done/Return.* In order to exit the *Modify* mode select *Done/Return.*

Now, let us add a half-inch border on all four sides of the format by adding vertical and horizontal lines. These lines are created by locating the endpoints

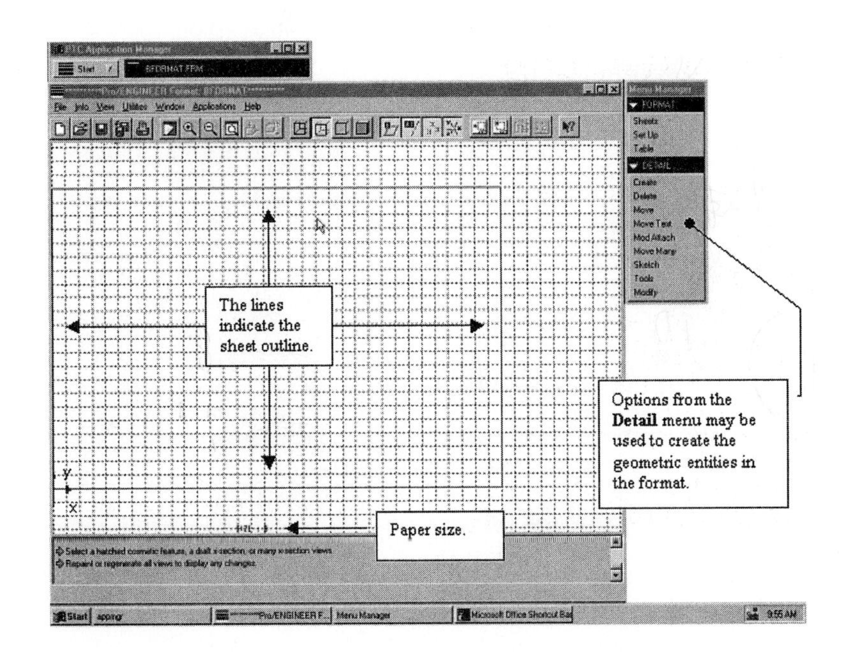


FIGURE 16.3 *Create a custom format by using the* DETAIL *menu and Sketch option.*

of the lines, which can be done in several ways. One approach is to use the grid and the mouse to locate the lines. A far easier and much more accurate approach is to enter the coordinates of the endpoints of the lines.

The coordinates will be entered with reference to the absolute coordinate system shown in Figure 16.3. Using the values in Table 16.1, the procedure is as follows:

1. Select *Sketch, Line,* and *2 Points.*
2. Choose *Abs Coord.*
3. Enter the $x$ coordinate of the first point. Hit *Return.*
4. Enter the $y$ coordinate of the first point. Hit *Return.*
5. Choose *Abs Coord.*
6. Enter the $x$ coordinate of the second point. Hit *Return.*
7. Enter the $y$ coordinate of the second point. Hit *Return.*

Add the four lines to the format by using the outlined procedure. The four lines are shown in Figure 16.4. Then, add the remaining horizontal and ver-

TABLE 16.1 COORDINATES OF THE LINES NEEDED TO COMPLETE THE FORMAT.

| LINE # | AXIS | FIRST POINT | SECOND POINT |
|---|---|---|---|
| 1 | x | .5 | .5 |
| | y | 10.5 | .5 |
| 2 | x | .5 | 16.5 |
| | y | 10.5 | 10.5 |
| 3 | x | 16.5 | 16.5 |
| | y | 10.5 | .5 |
| 4 | x | .5 | 16.5 |
| | y | .5 | .5 |
| 5 | x | .5 | 16.5 |
| | y | 1.5 | 1.5 |
| 6 | x | 4.5 | 4.5 |
| | y | 1.5 | .5 |
| 7 | x | 10.5 | 10.5 |
| | y | 1.5 | .5 |
| 8 | x | 11 | 16 |
| | y | 1 | 1 |
| 9 | x | .5 | 10 |
| | y | 1 | 1 |

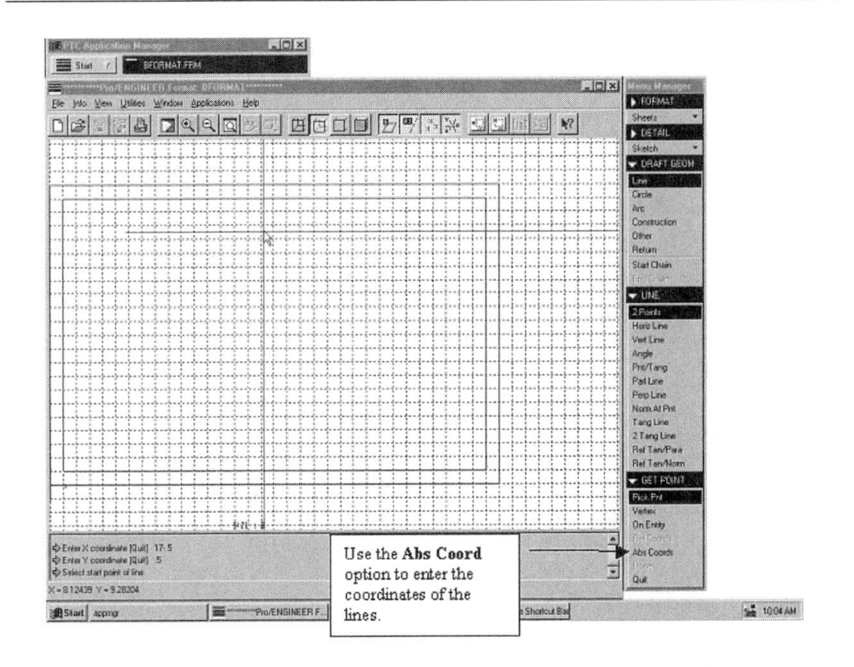


FIGURE 16.4 *The format with the four lines forming the boundary of the drawing.*

tical lines using the same procedure and the remaining data from Table 16.1. The format with all the lines added is shown in Figure 16.5.

The format is completed by adding text to the appropriate spaces. In a drawing, text is called a *Note* by Pro/E. A note may be added anywhere on the drawing by simply clicking at the desired location with your mouse pointer or by entering the coordinates of the text.

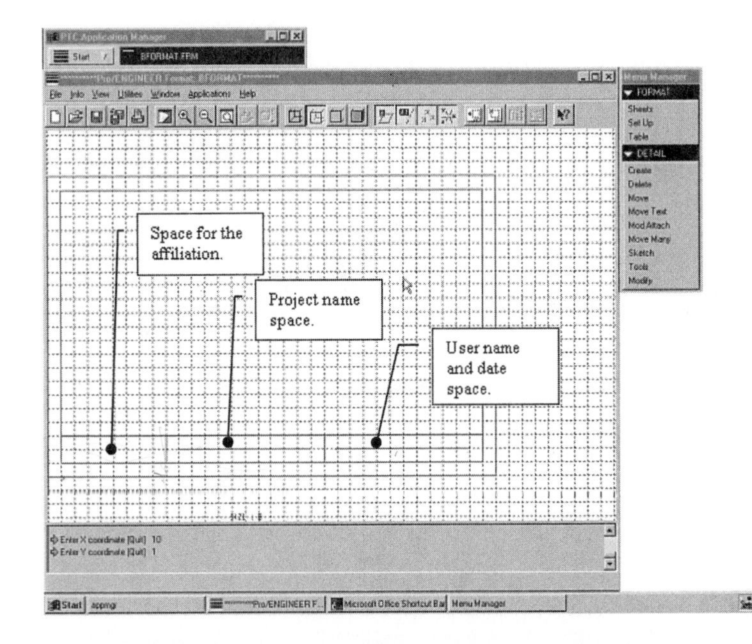


FIGURE 16.5 *The remaining lines are added to the format, which creates the spaces for the necessary information.*

The procedure for adding a note is as follows:

1. Select *Create* and *Note.*
2. Choose the text options from the *NOTE TYPES* menu. In particular, choose the orientation and justification of the text. In this case, select the defaults *No Leader, Horizontal, Standard,* and *Default* justification.
3. Select *Make Note.*
4. Pick the location of the note. In this case, choose *Abs Coord.*
5. Enter the coordinates of the text.
6. Select *Done/Return.*

Add the text given in Table 16.2 by the using the outlined procedure. The format with the added text is shown in Figure 16.6.

Add the name of your university or company in the affiliation space (see Figure 16.5). Because the size of this text varies from organization to organization, we cannot outline a standard procedure for this task. However, it may be possible to add the text using the *Abs Coord* option. The space available is 4 inches wide by 1 inch high.

Another approach is to locate the text by clicking near the desired location with your mouse. Then, the option *Move Text* may be used to move the text as desired.

Save the format. Select *File* and *Erase* to remove the format from memory. Then load the format "Aformat" and add the name of your university or company in the space provided. When you are done, save "Aformat."

## 16.1.2 TUTORIAL #74: CREATING A DRAWING

Let us create a multiview drawing of the angle bracket. In a subsequent tutorial, we will add dimensions and tolerances to the drawing. Select *File* and

TABLE 16.2 TEXT AND THE LOCATION OF THE TEXT NEEDED TO COMPLETE THE FORMAT

| TEXT # | AXIS | COORDINATES | TEXT |
|---|---|---|---|
| 1 | x | 11 | NAME: |
| | y | 1.2 | |
| 2 | x | 11 | DATE: |
| | y | .8 | |
| 3 | x | 5 | TITLE: |
| | y | 1.2 | |

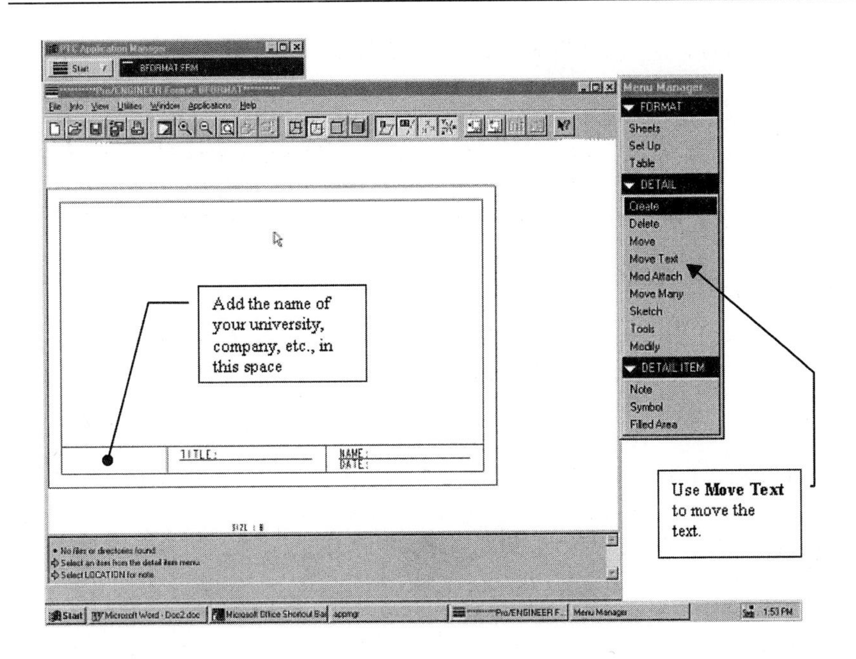


FIGURE 16.6 *The format is reproduced with the added text.*

*New* (or click on the *Create New Object* icon). Choose the drawing option and enter the name: "AngleBracket." Click *OK*.

Enter the model name and then select the *Retrieve Format* option. Enter "Aformat" in the cell provided. Choose *OK*.

The software will load the format and display it on the screen. We can now add the first view to the drawing. The placement of the first view is important because projections will be created from the first view. Because the subsequent projections depend on the first view, the orientation of the first view must be properly defined.

First, make sure that the orientation is set to *Isometric.* Choose *Utilities, Environment,* and change the default orientation to *Isometric.* Select *OK* to save the change.

Add the first view by selecting *Views* and *Add View.* Accept the defaults shown in Figure 16.7 (the *VIEW TYPE* menu) by selecting *Done.* Click somewhere in the lower left-hand portion of the screen. Pro/E will place an isometric view of the model. Reorient the model so that it appears as in Figure 16.7.

Obviously, the view needs to be rescaled. Select *Done/Return* and then *Modify* from the *DETAIL* menu. Select the scale on the screen, (the value 1.000), and enter a value of 0.5. Then choose *Done/Return.* The resized view is shown in Figure 16.8.

By the way that the first view has been reoriented, a projection made of this view to the right will generate a right-hand side view. Any projection constructed above the first view will generate a top view.

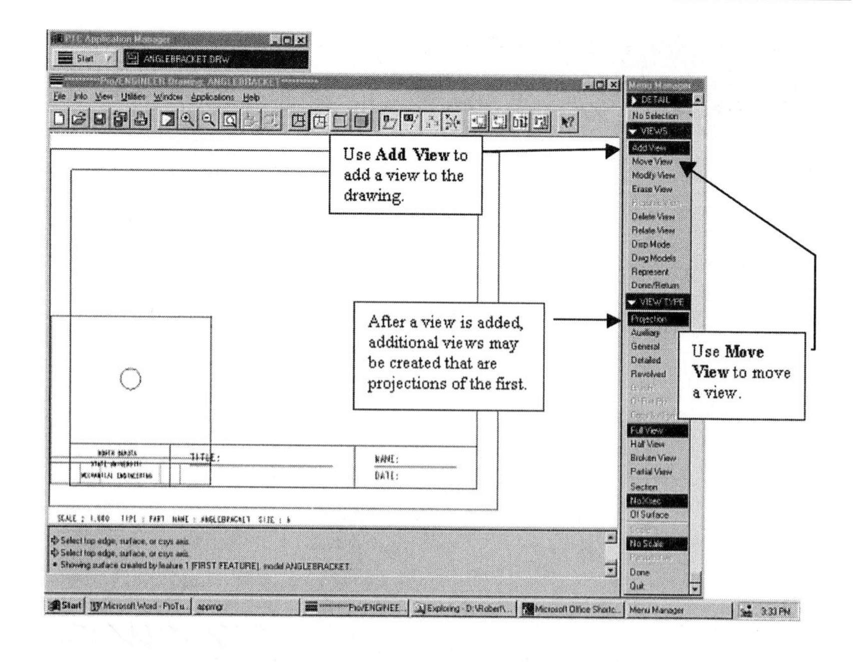


**FIGURE 16.7** *The* VIEW TYPE *menu may be used to place a view in the drawing.*

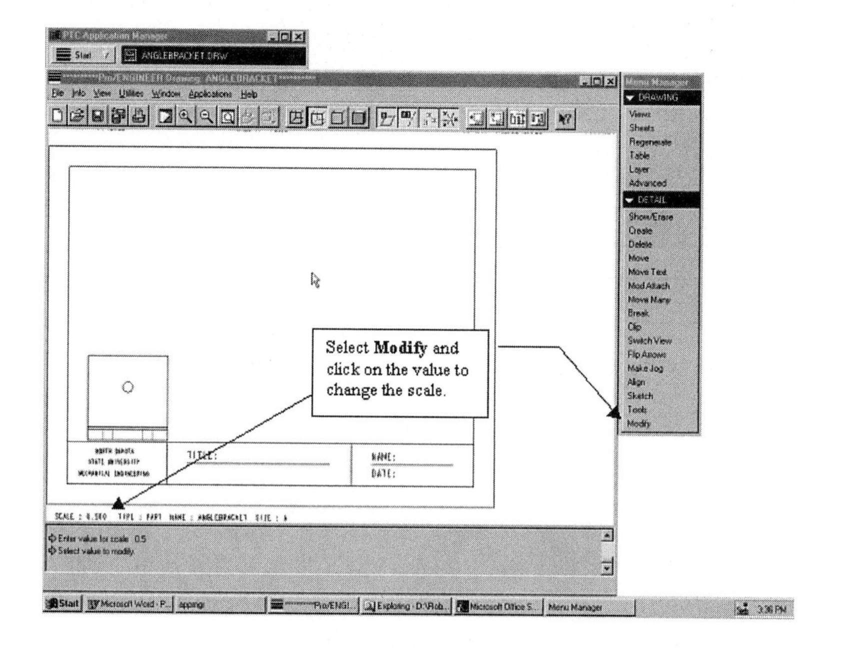


**FIGURE 16.8** *The* Modify *option from the* DETAIL *menu may be used to resize a view.*

Choose *Views, Add View, Projection,* and *Done.* Then click in the space above the first view. A top view will be added. Choose *Add View* and click in the space to the right of the first view. A right-hand side view will be added. Your drawing should now appear as in Figure 16.9.

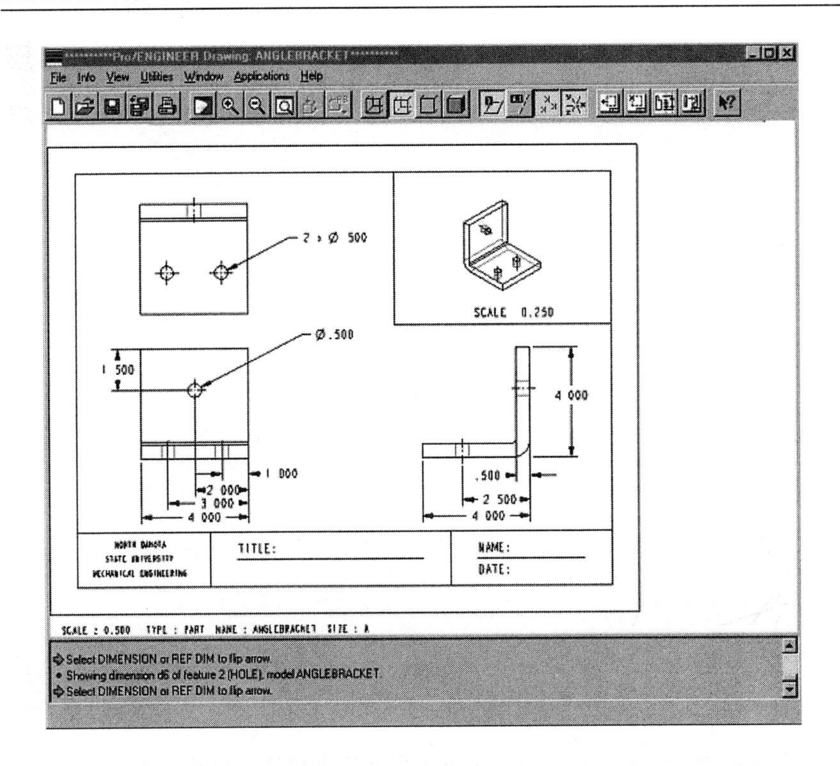


FIGURE 16.9 *Additional views may be created as projections of the first.*

Your views might need to be centered. Select *Views* and *Move View.* Click on the first view and move it to create equal borders around the views. Move the other view if necessary using *Move View.*

Many drawings contain a pictorial (*General* view) of the model. We may add this additional view by doing the following:

1. Select *Views* and *Add View.*
2. Choose *General* from the *VIEW TYPE* menu in Figure 16.7.
3. In comparison to the other views, this view will be too big to fit on the drawing. Therefore, we need to scale the view. Choose *Scale.*
4. Select *Done.*
5. Click in the upper right-hand side of the screen.
6. Enter a value of, say, 0.25 for the scale (this number can always be changed).
7. Select *OK* from the *Orientation* dialog box.
8. Choose *Done/Return.*

As shown in Figure 16.10 use *Views* and *Move View* to reposition the new view. Use *Move Text* from the *DETAIL* menu and reposition the scale of the view below the view.

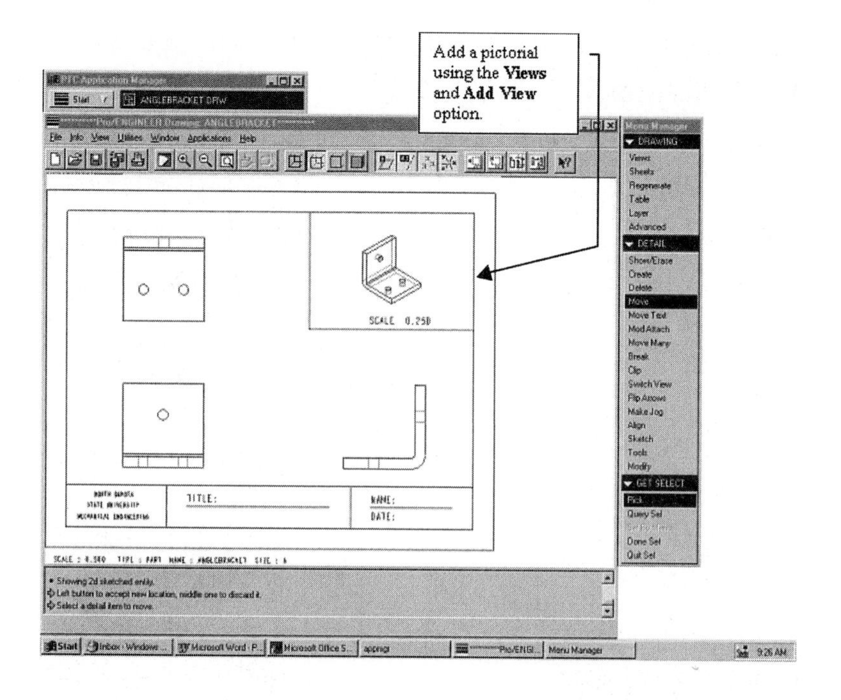


**FIGURE 16.10** *An isometric pictorial may be added to the drawing. Lines may be sketched to separate this view, which is of different scale, from the rest of the drawing.*

Because the new view is of different scale from the rest, it is convenient to sketch a border around this view by using the *Sketch* option in the *DETAIL* menu. Following this procedure:

1. Select *Sketch* from the *DETAIL* menu.
2. Choose *Line, 2 Points,* and *Pick Pnt.*
3. Then sketch a horizontal and vertical line, as shown in Figure 16.10.
4. Choose *Return.*

As with all other geometric entities in a drawing (except for the views), the lines may be deleted by using the *Delete* option from the *DETAIL* menu. Use this option if you wish to erase the lines and start again. Likewise, the lines may be moved by using the *Move* option from the *DETAIL* menu.

Save your drawing. In the next tutorial, we will add dimensions.

## 16.2 ADDING DIMENSIONS AND TEXT TO A DRAWING

Dimensions that were used to construct a model (driven dimensions) may be displayed on the drawing by using the *Show/Erase* command from the *DETAIL* menu. After selecting the *Show/Erase* command, the dialog box given in Figure 16.11a appears. The buttons across the top of the box are from left to right: dimension, reference dimension, geometric tolerance, note, balloon, axis, symbol, surface finish, datum plane, and cosmetic feature. If you place your mouse pointer on any of these buttons and leave it there, the name of the button will be displayed.

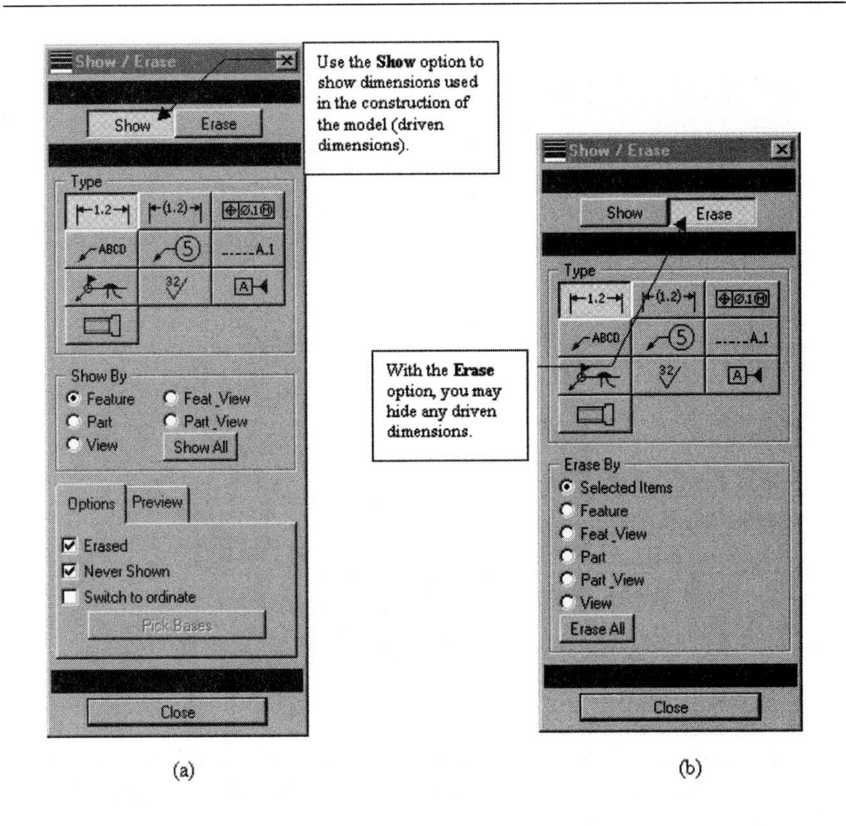


FIGURE 16.11 *Driven dimensions may be displayed or hidden by using the* Show *and* Erase *options, respectively.*

Notice that any of these options can be activated per view, for each feature or for the entire part. When activating the option per feature, you will need to click on the desired feature. Selection of multiple features is allowed; simply click on each feature in succession.

Often when using the *Show* option in placing dimensions, Pro/E places the dimension in an inconvenient location and dimensions may be on top of or too close to another dimension. Luckily, the option *Move Text* can be used to reposition the text or the option *Move* can be used to move the entire dimension. Furthermore, the option *Switch View* from the *DETAIL* menu may be used to switch a dimension from one view to another.

The *Erase* command may be used to hide any driven dimension. The corresponding dialog box is shown in Figure 16.11b.

You can create your own dimensions by clicking on *Create* in the *DETAIL* menu and then on *Dimension.* The dimension is constructed by selecting the appropriate entities.

In placing the dimension values between the leaders, you may find the *Text Style* option (Figure 16.12) useful. By changing the height or width of the text, you can make the text fit between the leaders. The *Text Style* box may be used in conjunction with the options in the *MODIFY TEXT* menu to change the appearance and orientation of any text on the drawing. The sequence for activating these options is *Modify, Text,* and *Text Style.*

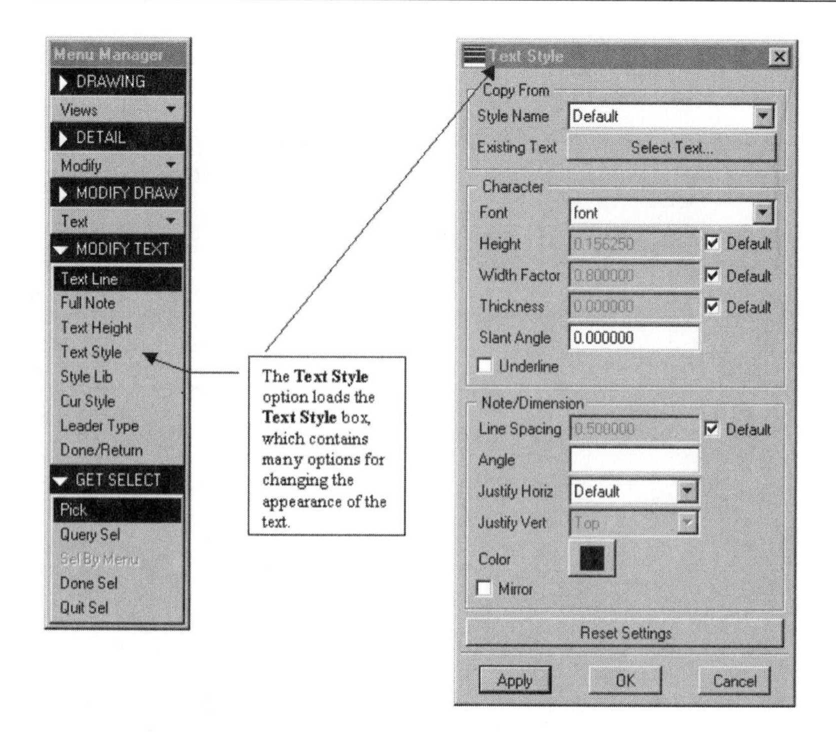


**FIGURE 16.12** *The* Text Style *box and the* MODIFY TEXT *menu are shown. This dialog box is retrieved after clicking on the* Text Style *option in the* MODIFY TEXT *menu.*

The style of the dimensions, including the arrowhead type, can be changed by selecting *Modify* from the *DETAIL* menu and *Dim Params* (Figure 16.13*)*. Pro/E supports eight different arrowhead styles. These styles are:

1. *Default.* The default style depends on the type of dimension.
2. *Arrowhead.* A single triangle is drawn on the end of the line. Whether the arrow is filled or closed is controlled by the *draw_arrow_style* option in the drawing setup file.
3. *Dot.* A dot is placed at the end of the line.
4. *Double Arrow.* A double triangle is drawn at the end of the line. This symbol is used to indicate a clipped dimension.
5. *Slash.* A slash is drawn at the line's end. The slash is constructed by drawing a line 45° to the dimension line.
6. *Integral.* An integral sign is used for the end of the dimension line.
7. *Box.* The line is completed with a rectangular box.
8. *None.* No ending is placed on the line.

Among the other options in the *DIM PARAMS* menu is the option *Scheme.* This option may be used to change a dimension by redefining the feature. Thus, you do not have to exit the *Drawing* mode and load the *Part* mode in order to redefine a feature.

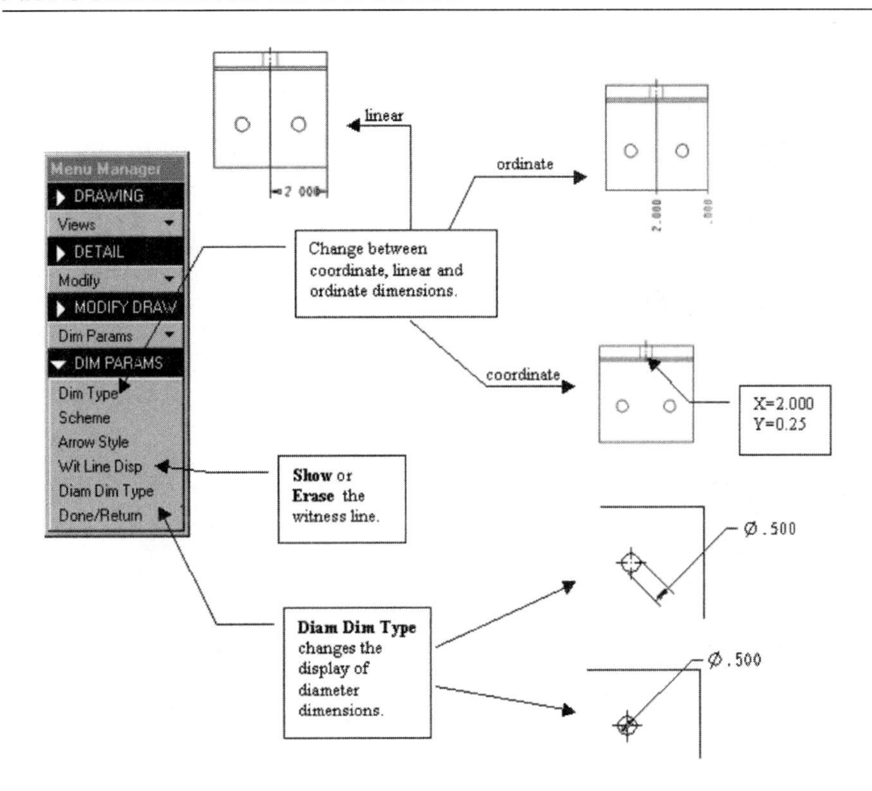


FIGURE 16.13 *The* DIM PARAMS *menu contains options for modifying the various aspects of the dimension.*

### 16.2.1 TUTORIAL # 75: DIMENSIONS AND TEXT FOR THE DRAWING OF THE ANGLE BRACKET

Before proceeding to add dimensions and text to the drawing, check to see that the arrowhead style is set to "*Filled.*" In order to check style, select *Advanced* from the drawing menu, *Set Up* and then *Modify Val.* The software will load the drawing setup file into the PROTAB editor. Scroll down until you find the option *draw_arrow_style.* If the option is set to "*Closed,*" change it to "*Filled.*"

Notice from Figure 16.11a, that all the dimensions may be shown at the same time by using the *Show All* option. It is the opinion of the author that it is more practical to display the dimensions for one feature at a time. In this tutorial, we will follow this approach.

It really does not matter which feature we dimension first. We will begin with the one shown in the front view.

Select *Show/Erase* from the *DETAIL* menu. Then pick the *Show* option. In addition, select the *Dimension* and *Axis* buttons. Set the toggle to *Feature* and click on the hole. The software will load the driven dimensions and axes for the hole, as shown in Figure 16.14. We find that it is better to hide the axis name label. Turn off the display of these labels by selecting the Datum axis button (see Figure 16.14).

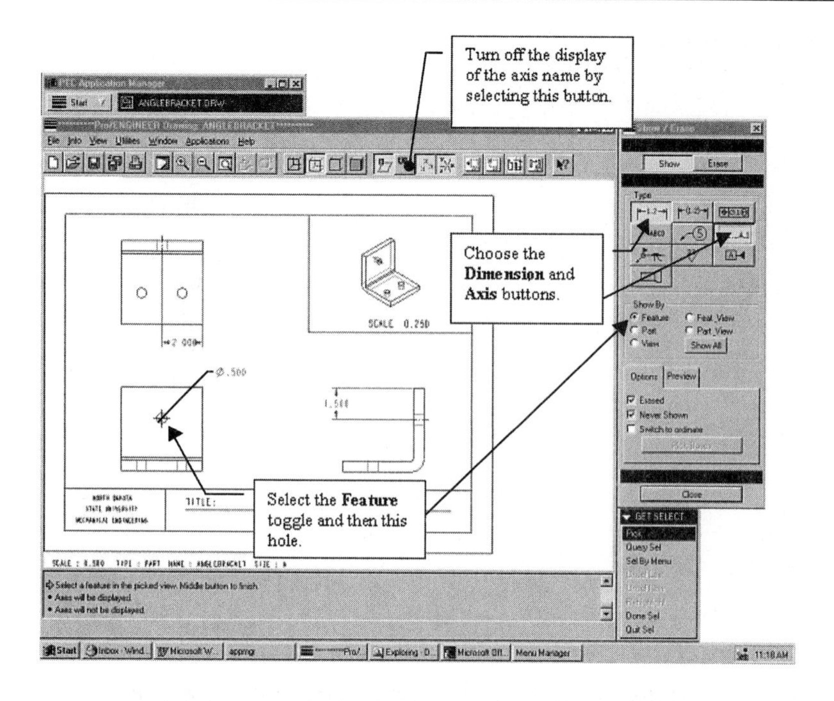


**FIGURE 16.14** *The drawing after displaying the dimensions of the hole in the front view.*

At this point, you may add dimensions to the remaining features; however, we suggest doing one feature at a time. With this in mind, select *Close.* Then choose *Switch View.* Select the 2.000 and 1.500 dimensions and *Done/Sel.* Then click on the front view to move the dimensions so that your drawing appears similar to the one shown in Figure 16.15. Also, select *Flip* arrows and then click on the .500 dimension to change the direction of the arrows to a more suitable appearance.

Let us add some more dimensions. Select *Show/Erase.* Set the toggle to *Feature* and click on the model in the top view. Be sure to avoid either hole. Pro/E will add the dimensions defining the width, height, and thickness of the bracket. Your drawing should look like the one shown in Figure 16.16.

The thickness dimension should be moved to the right-hand side view. In fact, it is best to redefine this dimension as a *Linear* dimension.

In this case, let us erase the thickness dimension and create a new one in the right-hand view. Click on *Erase* and then on the *Dimension* button and set the option to *Selected Items.* With your mouse, click on the thickness dimension and then click on *Done Sel.*

To add the dimension in the proper place, close the *Show/Erase* dialog box. Choose *Create* from the *DETAIL* menu. Then select *Dimension.* Make sure that the option *On Entity* is highlighted. With your mouse, click on the two vertical lines defining the thickness of the bracket in the right-hand view. Then click your middle mouse button.

Pro/E will place the dimension. Use *Flip Arrows* and change the direction of the arrows for this dimension. Your drawing should have the added dimension shown in Figure 16.17.

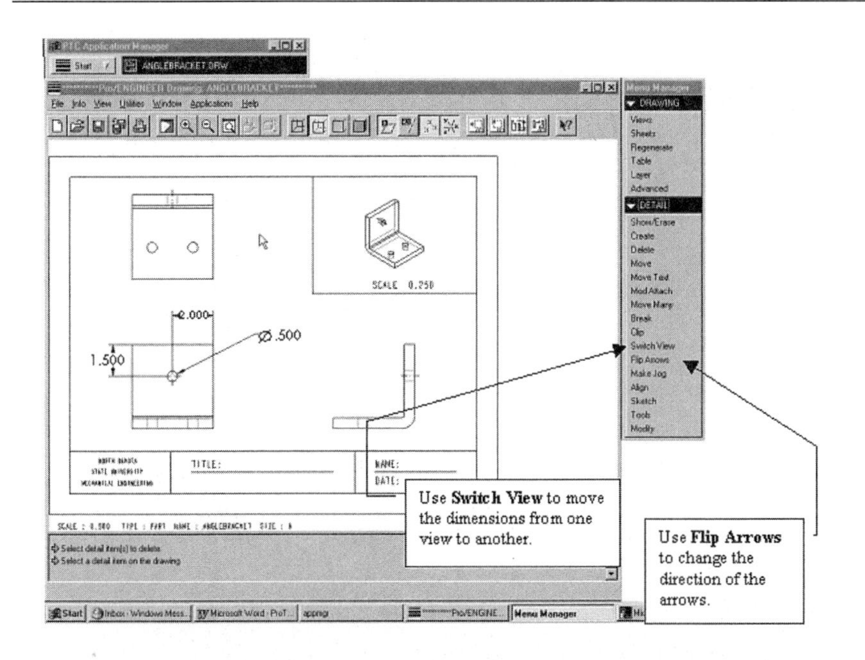


FIGURE 16.15 *Move the dimensions to roughly the same positions shown.*

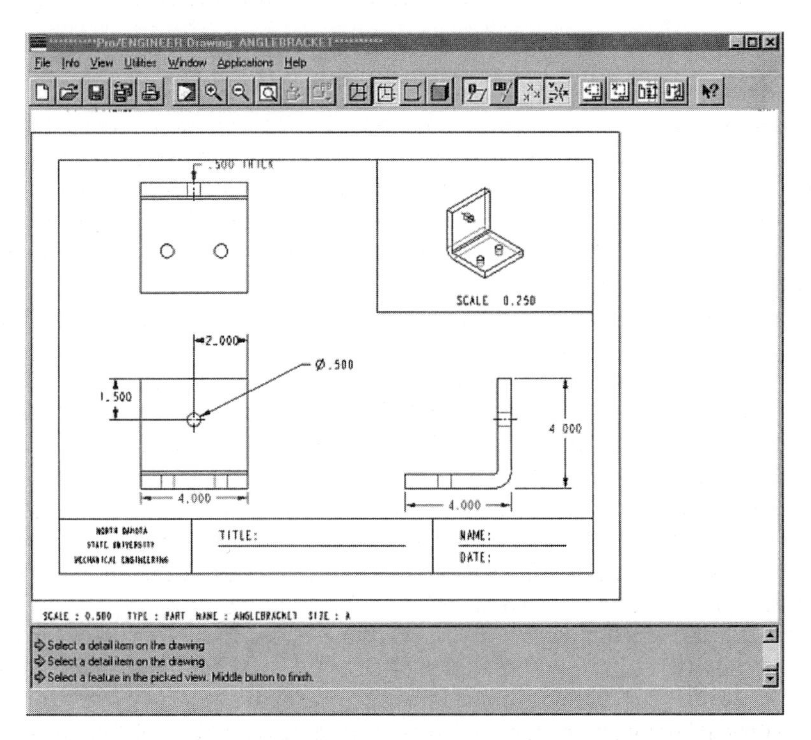


FIGURE 16.16 *Add dimension for the overall size of the model.*

Move the remaining dimensions so that they are grouped as shown in Figure 16.17. You will probably need to move the views in order to create some space for the dimensions by selecting *Views* and *Move View*.

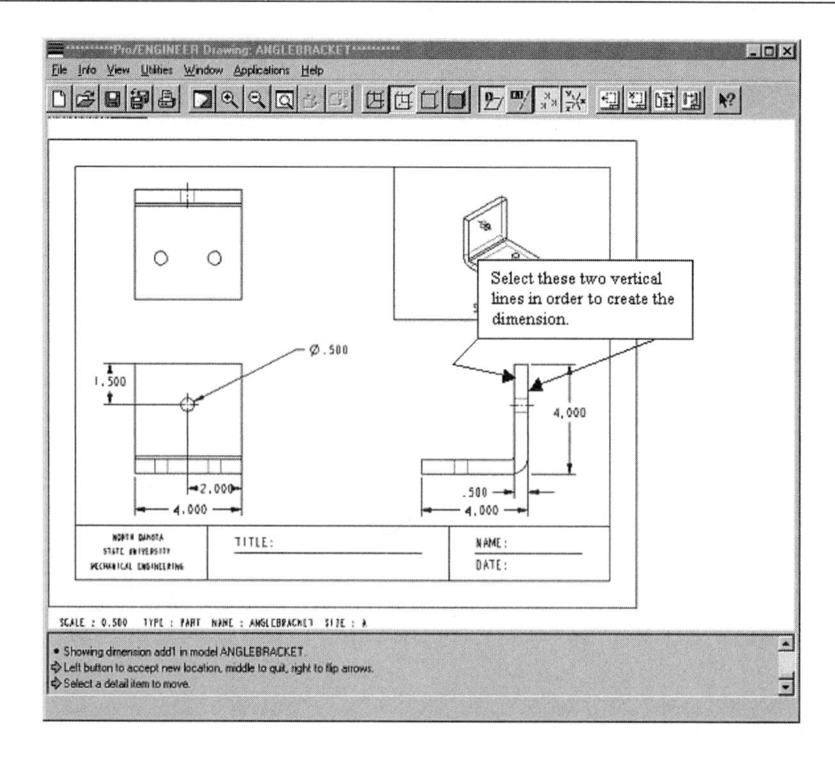


**FIGURE 16.17** *The dimension indicating the thickness of the part may be recreated in a different view.*

Now, add the dimension to the two holes in the top view. The procedure to follow is:

1. Select *Show/Erase.*
2. Choose *Dimension* and *Axis.*
3. Select both holes in the top view.

Consider Figure 16.18. Because both holes are the same size, we can erase one of the diameter dimensions and modify the text of the other. This format will achieve the same result as displaying both dimensions.

Erase the second diameter dimension. Then select *Close.* From the *DETAIL* menu select *Modify* and then *Text.* Click on the remaining diameter dimension. In the Message Window, as shown in Figure 16.18, the software will load code describing the dimension. Scroll to the left and insert the text "2 x" before the diameter symbol. Then hit *Return.*

The remaining dimensions may be either moved into position or erased and recreated using the *Create* and *Dimension* options. Consulting Figure 16.19, redefine the dimensions as necessary.

Add the dimension for the rounds by setting the option to *Feat & View* and then click on the arcs in the right-hand side view. You will need to click on each arc separately. Move the dimensions to the appropriate position as shown in Figure 16.20.

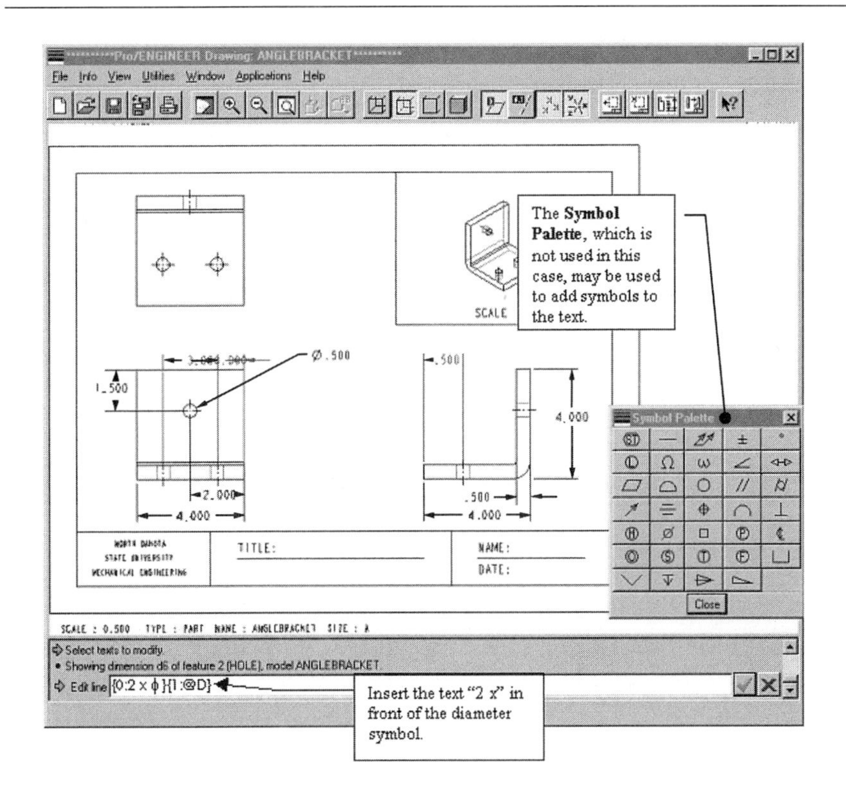


**FIGURE 16.18** *Modify the text of the dimension by adding "2 x" in front of the diameter symbol.*

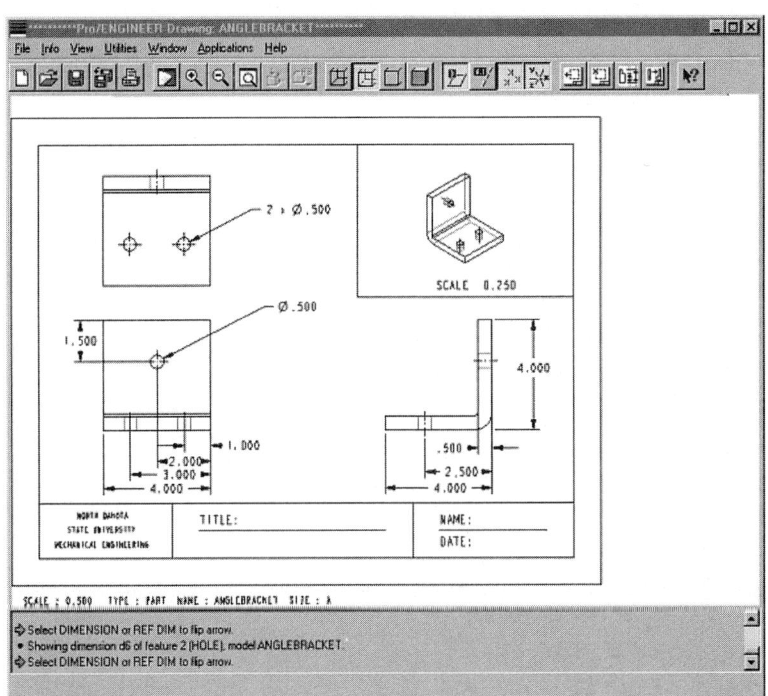


**FIGURE 16.19** *With all the dimensions properly located, your drawing should appear as shown.*

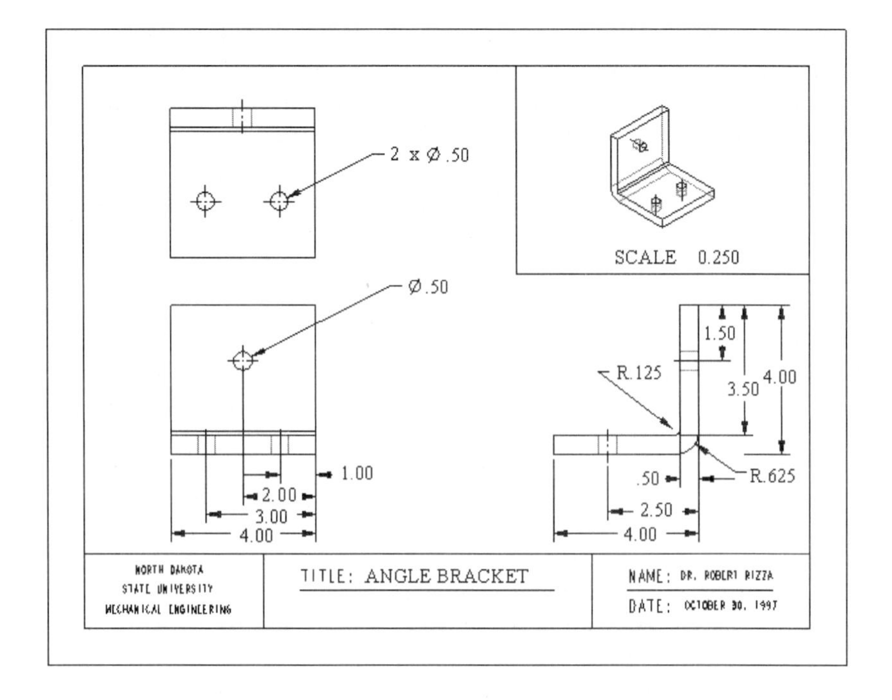


**FIGURE 16.20** *The drawing with dimension and appropriate notes added.*

Finally, this drawing need not display three decimal points on most of the dimensions. Select *Modify* and *Dimension.* Click all the dimensions except for those related to the rounds. Then choose *Done Sel.* The *Modify Dimension* box will appear. Change the number of digits from three to two. Select *OK.*

Create the remaining dimension shown in Figure 16.20. Then, add your name, title (shown in Figure 16.20), date, and scale to the drawing. The procedure is as follows:

1. Select *Create* and *Note.*
2. Accept the defaults *No Leader, Horizontal, Standard,* and *Default* justification.
3. Choose *Make Note.*
4. Pick the location of the note with your mouse pointer.
5. Enter the note.
6. Hit *Return.*

Save the drawing. You may print a copy of the drawing by selecting *File* and *Print.*

## 16.3 AUXILIARY VIEWS AND PRO/E

An auxiliary view is constructed from an image plane that is not a frontal, horizontal, or profile plane. Auxiliary views are useful in visualizing an oblique plane; that is, a plane that does not appear true size and true shape in any orthogonal view. An example of a plane that is oblique is shown in Figure 16.21. In order to visualize the hole on plane A in true size and shape, a projection must be constructed on a plane parallel to surface A and then revolved.

An auxiliary view can be created with Pro/E by using the *Drawing* mode after a model has been created. The auxiliary view is then constructed from a frontal, horizontal, or profile view.

### 16.3.1 TUTORIAL #76: CREATING AUXILIARY VIEWS IN A DRAWING

Let us make a drawing of the rod support that contains the primary orthogonal view as well as an auxiliary view of surface A. The auxiliary view can either be a partial or full view.

Begin the creation of the drawing by entering the *Drawing* mode. Create a drawing called "RodSupport" and load the format "Aformat." Add a general view to the top left-hand side of your screen and reorient the view to create top, front, and right-hand side projections. These views are shown in the final version of the drawing given in Figure 16.22.

The procedure for adding an auxiliary view is simple. The auxiliary view is generated from a previously placed and oriented view. This view is called the *Parent* view by Pro/E. For this example, the parent view is the front view.

Create a full auxiliary view by choosing *Add* view from the *VIEWS* menu. Select *Auxiliary* and *Done.* Click slightly above the front view to locate the auxiliary view. This location is the center point of the view.

The auxiliary view is, in essence, a revolved view, and Pro/E will ask for an edge or axis around which the revolution is desired. In this case, the edge of choice is edge A, shown in Figure 16.21.

Pro/E will place the auxiliary view. If the view is too close to the front view, use *Move View* to relocate the view. The next step would be to add dimensions to the views, using the auxiliary view to dimension surface A and the hole it contains.

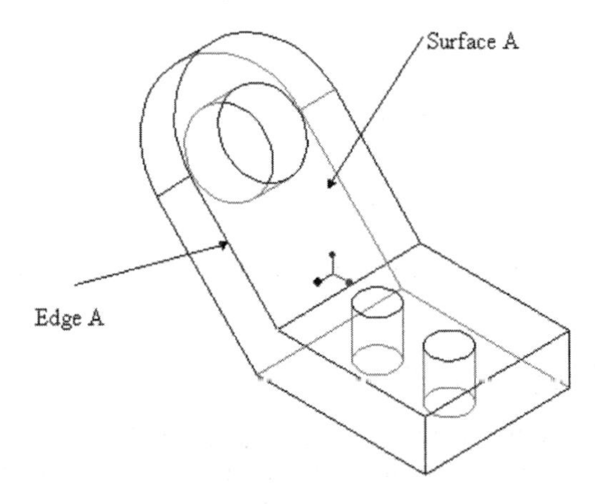


FIGURE 16.21 *An example of a part with an incline oblique plane. The plane A will not appear true size or true shape on any orthogonal view.*

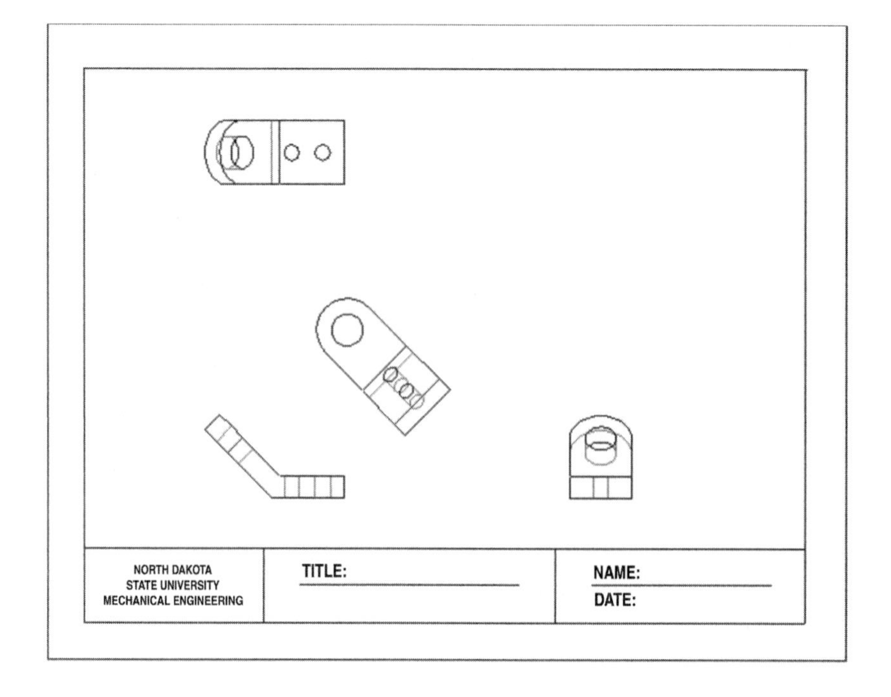


FIGURE 16.22 *A drawing containing the three primary orthogonal views and a total (full) auxiliary view of surface A is shown. The drawing is incomplete because dimensions would normally be added.*

Unless instructed otherwise, do not keep this drawing. Use *Views* and *Erase View* and remove the auxiliary view from the drawing. We will now proceed to add a partial auxiliary view.

The procedure for adding the partial auxiliary view is the same for adding the total view except that the option *Half View* must be chosen. However, a cutting plane is required. Unfortunately, the model was created without such a suitable plane. Thus, before proceeding, we must construct this plane.

Choose *Create, Datum,* and *3D Datum.* The software will load the Datum dialog box shown in Figure 16.23. Enter the name DTM5 and select the *Define* button as shown. Then using the options from the Datum plane menu, construct the datum so that it passes through the hidden line (use *Query Sel* to pick the line) and at an angle to DTM4.

Choose *Done.* Select *Enter Value* and enter a value of 90. Choose *OK* and *Done/Return.* The software will add the datum.

Select *Add View* from the *VIEWS* menu and select *Auxiliary* and *Half View.* Then click on *Done.* Locate the half view by clicking on the screen near the front view. Again, choose edge A. Pro/E will place the view with all the datum planes showing. The partial auxiliary view will contain only the part of the model above the DTM5 plane. Click on the DTM5 when asked to do so. Arrows will appear indicating the direction of the view to be saved. These arrows should point toward the hole. If they do, choose *OK;* otherwise flip the arrows.

Your drawing should appear as in Figure 16.24. Add the appropriate text and dimensions. You may need to switch some dimensions from one view to

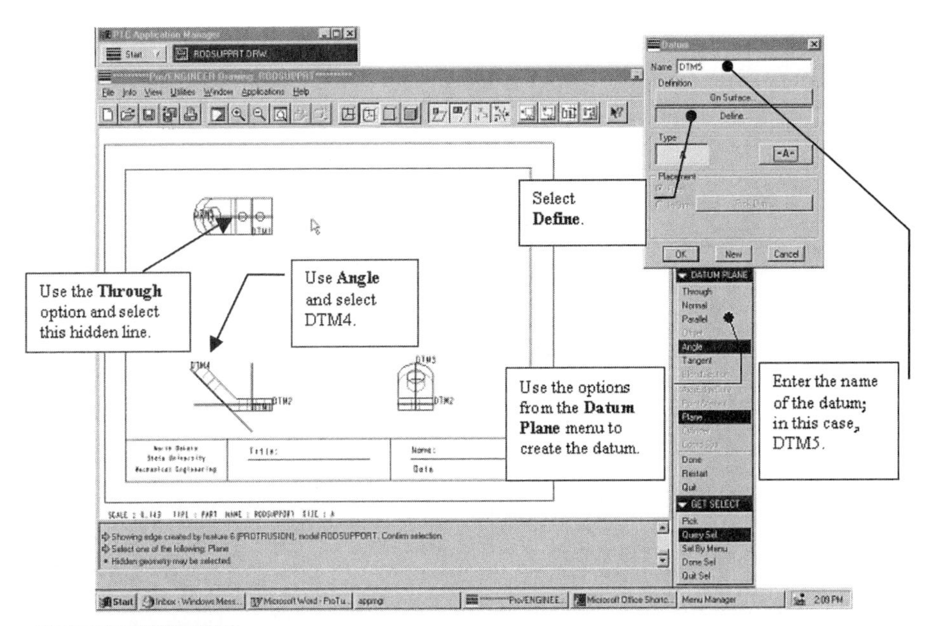


FIGURE 16.23 *A datum may be created in the drawing mode by using the Datum dialog box.*

another. This requirement will mostly likely occur in the dimensions defining the features in the auxiliary view. In order to switch a dimension from one view to another, use the *Switch View* option from the *DETAIL* menu. Pick the dimension with your mouse pointer. You can pick as many dimensions as desired. Choose *Done Sel* when finished picking the dimensions. Then click on the desired view. The dimensions will be switched to the desired view automatically. Save the drawing when you are finished.

## 16.4 SUMMARY

The drawing containing various multiviews of a model may be created by using the *Drawing* mode. The software creates drawing files in a manner analogous to the creation of part files.

A standard format may be added to the drawing, but it must be added during the creation of the drawing file by using the *Retrieve Format* option. Format files may be created by using the *Format* mode. The geometry in the format is created by sketching entities.

After a drawing is created, views may be added to the drawing by using the *Views* and *Add Views* options. The first view added to the drawing is a *General* view. Subsequent views may be *Projections, Auxiliary, Revolved,* and *Detailed* views.

Dimensions that were used to construct the model may be displayed in the drawing by using the *Show/Erase* option. Such dimensions often need to be moved in the drawing. Moving dimensions is accomplished with the *Move*

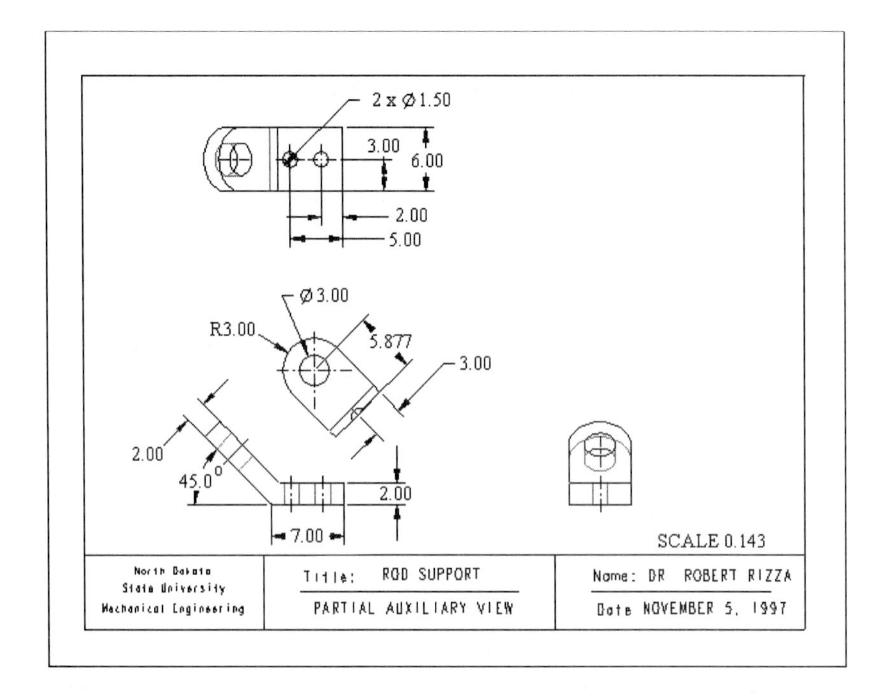


FIGURE 16.24 *A drawing of the rod support containing a partial auxiliary view.*

and *Move Text* options. Additional dimensions may be added to the drawing by using the *Create* and *Dimension* options. The user needs to select the entities defining the dimension.

## 16.4.1 STEPS FOR CREATING A FORMAT

In general, a format may be created by using the following sequence of steps:

1. Choose *File* and *New,* or select the *New Object* icon.
2. Select the *Format* option.
3. Enter a name for the format and select *OK.*
4. Choose the paper size and orientation. Select *OK.*
5. Draw the entities in the format using the *Sketch* option.
6. Save the format.

## 16.4.2 STEPS FOR CREATING AND ADDING VIEWS TO A DRAWING

The general procedure for creating a drawing is as follows:

1. Choose *File* and *New,* or select the *New Object* icon.
2. Select the *Format* option.
3. Enter a name for the format and select *OK.*

4. Choose the paper size and orientation.
5. Select *Retrieve Format* and enter the name of the format.
6. Choose *OK.*

Views may be added to the drawing with the *Add View* option by following this procedure:

1. Select *Views* and *Add Views.*
2. Select the view type from the *VIEW TYPE* menu. Note that the first view placed in the drawing must be of type *General.*
3. Choose *Done.*
4. Reorient the view if desired.
5. If desired, *Modify* the scale of the view.
6. Move the view using *Move View.*

### 16.4.3 STEPS FOR ADDING DRIVEN DIMENSIONS TO THE VIEWS

The driven dimensions, dimensions used in constructing the model, may be displayed on the drawing by using the *Show/Erase* option. The procedure is as follows:

1. Select *Show/Erase* from the *DETAIL* menu.
2. Choose *Show* and *Dimension.*
3. Choose how to show the dimensions. The possible options are by *Feature, Part, View, Feat_View* (feature and view), and *Part_View* (part and view). We suggest displaying the dimensions by feature.
4. Select *Close.*

Likewise, the display of the driven dimensions may be hidden. The procedure is:

1. Select *Show/Erase* from the *DETAIL* menu.
2. Choose the *Erase* button.
3. Select the *Dimension* button.
4. Choose how to erase the dimensions. The available options are *Selected Items, Feature, Feat_View, Part_View,* and *View.* We suggest by *Select Items* that will allow you to erase the dimensions one at a time.
5. Select *Close.*

## 16.5 ADDITIONAL EXERCISES

For the exercise selected, create a fully dimensioned multiview drawing of the model.

16.1 Cement trowel (see also exercise 8.1).
16.2 Side guide (see also exercise 8.2).
16.3 Control handle (see also exercise 8.3).
16.4 External lug (see also exercise 8.4).
16.5 Pipe clamp top (see also exercise 8.5).
16.6 Pipe clamp bottom (see also exercise 8.6).
16.7 Side support (see also exercise 8.7).
16.8 Guide plate (see also exercise 8.8).
16.9 Single bearing bracket (see also exercise 8.9).
16.10 Double bearing bracket (see also exercise 8.10).
16.11 Roller (see also exercise 10.1).
16.12 Baseball bat (see also exercise 10.2).
16.13 Create a drawing containing an auxiliary view of the slotted support (see also exercise 4.12). The drawing must contain at least the views shown in Figure 16.25.

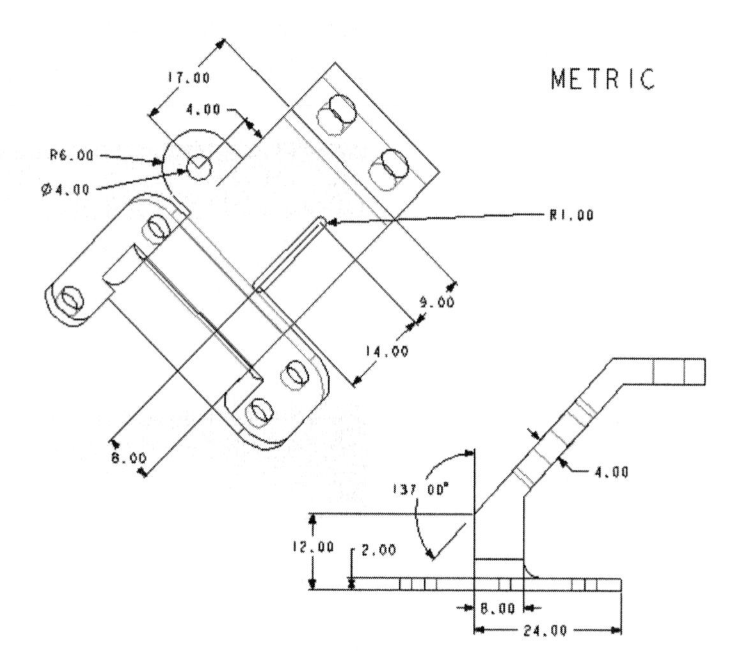


FIGURE 16.25 *Figure for exercise 16.13.*

16.14 Create a drawing of the clamp base (see also exercise 4.13) containing the views shown in Figure 16.26.

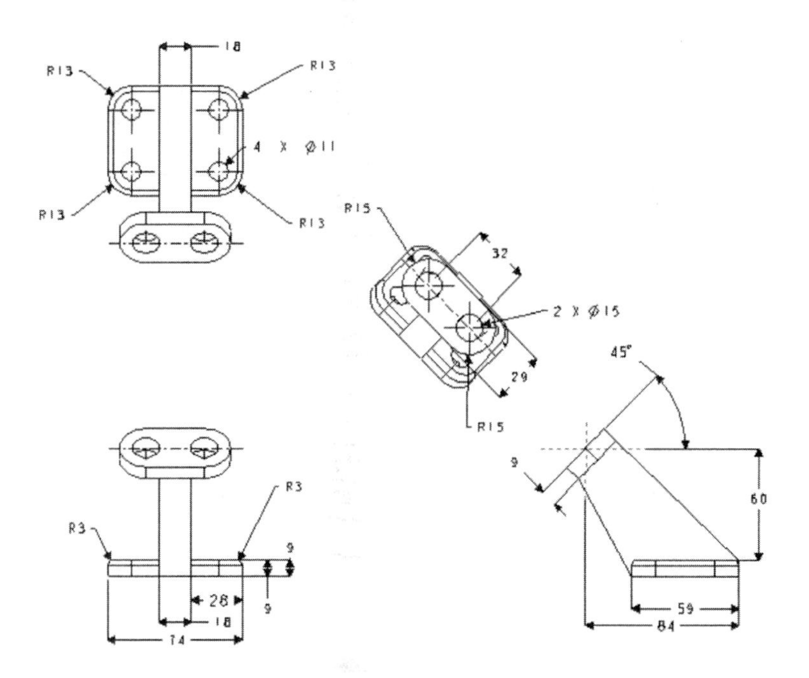


FIGURE 16.26 *Figure for exercise 16.14.*

# CREATING A SECTION

## INTRODUCTION AND OBJECTIVES

The drawing module may be used to create views containing either full or partial sections. Pro/E allows the user to create a section in the part, assembly, or drawing modes and then place the section in the drawing. In this chapter, creation of all sections will be taking place in the drawing mode. The sections will be constructed from previously created models.

In general, sections can be full, half, or partial sections, depending on how the cutting plane is passed through the part. In this chapter, we will consider how Pro/E may be used to create a full, offset, half, or revolved section. Thus, we have the first objective:

1. The user will become familiar with the procedure required to create a full, offset, half, or revolved section.

The Pro/E software allows multiple pages (sheets) in a drawing. This option is helpful when several different drawings of a part are required. The following is the second objective:

2. The software user will be able to use the *Sheets* option to add multiple pages to a drawing.

## 17.1 CREATING AND PLACING A SECTION

As was seen in chapter 16, when adding a view to a drawing, the user is given the option to create a section (see Figure 16.7, the *VIEWS* menu). In doing so, the user must choose the *Section* option. After choosing this option, the user is presented the cross-section type menu (*XSEC TYPE*). This particular menu is reproduced in Figure 17.1.

The default type of section is *Full*. A *Full* section is created by passing a cutting plane entirely through the model. Half sections are used to show a section, wherein half of the model is cut with the cutting plane. The part of the model that is not cut is represented in a normal fashion. Examples of *Full* and *Half* sections are shown in Figure 17.2.

If the *Local section* option is chosen, a section of the model within a defined boundary is created. The *Full & Local* option is used to create a full section with local cross sections.

In creating sections, the preferred standard is to show not only the surface in contact with the cutting plane, but also the edges that become visible on the model after the cut has been made. In Pro/E, this edge visibility can be achieved by choosing the *Total Xsec* option. If the edges are not desired, then the *Area Xsec* option may be selected. In this book, the *Total Xsec* option will always be used.

The *Total Align* option creates a total (*Total Xsec*) type section that is revolved and aligned. The *Total Align* option is useful when making offset sections. For an *Offset* section, the cutting plane is varied so that it goes through

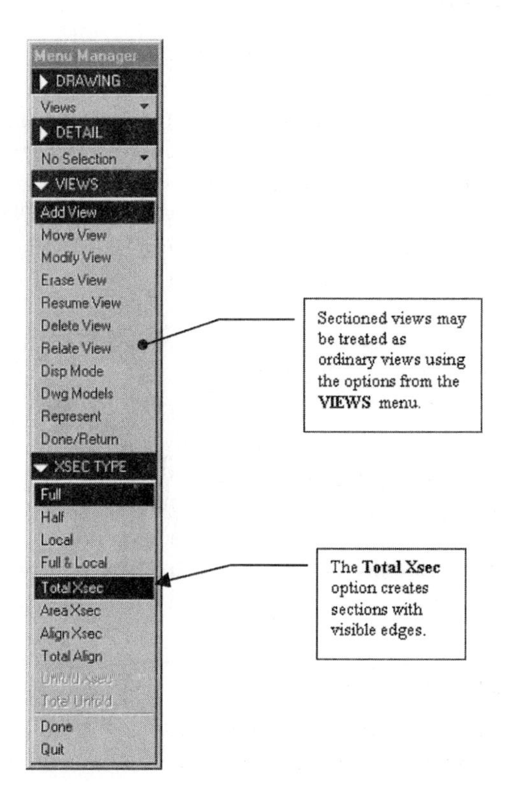


FIGURE 17.1 *The* XSEC TYPE *menu is used to define the type of section desired.*

**FIGURE 17.2** *Examples of* Full, Half, *and* Offset *sections. The model for these sections is given next to the offset section. (a) A* Full *section is shown of the surface containing the holes. The drawing above the section is the view without the added section. (b) A* Half *section of the upper part of the model is given. The view without the section is shown directly above. (c) An offset section.*

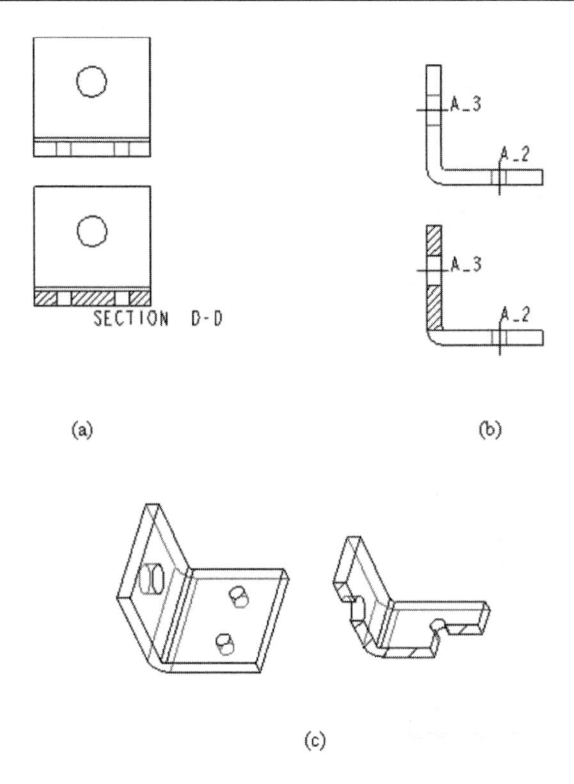


important features. The section produced is often difficult to visualize unless it is revolved.

The *Unfold Xsec* and *Total Unfold* options are also useful in offset cross sections, because they can be used to construct a projection of the offset. The projection is drawn flat. Both options are only available if the *Full* cross-section type is chosen.

An offset section is shown in Figure 17.2, along with the original model. The cutting plane was created by using the *Sketcher* to define the edge of the cutting plane. A sketching plane parallel to the two holes was used.

In defining the offset cutting plane, an open section is drawn through the desired features. After the contour of the plane has been sketched, it must be located with respect to the part and regenerated. Often the *Alignment* option can be used to locate the sketch with respect to the part.

If the cross section was created in another mode, say the *Part* or *Assembly* mode, then the cross section can be retrieved. Otherwise, the cross section must be created using the *Create* option from the *XSEC ENTER* menu shown in Figure 17.3.

Upon creating a section, the section is given a name. After the section is displayed, the name of the section is displayed as well.

A section may be created using the options shown in Figure 17.4. Included in these options are the *Planar* and *Offset* options familiar to students of en-

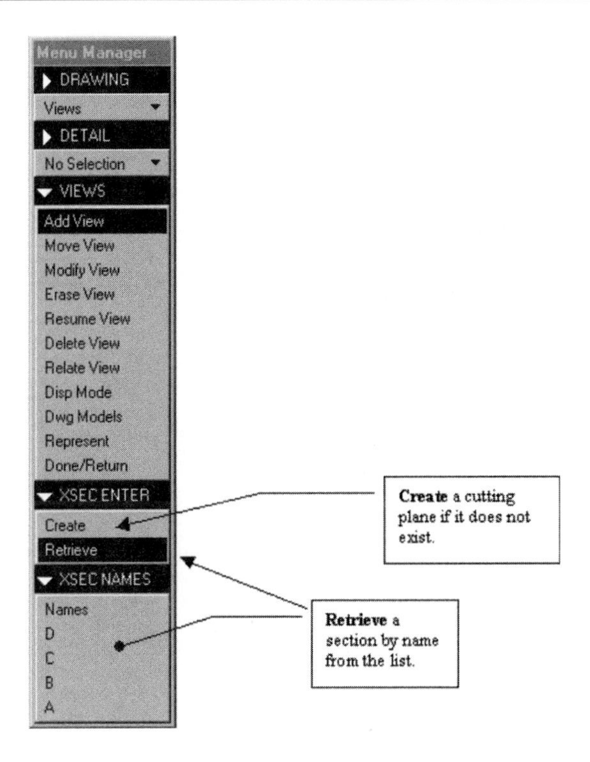


FIGURE 17.3 *If the cross section already exists, it can be retrieved by name, otherwise it must be created and named.*

gineering graphics. In Pro/E, when creating a planar section, the user must define the cutting plane. This plane must be parallel to the screen. Therefore, the view from which the plane is defined must be properly chosen and oriented. Often default datum planes can be used for this purpose. If no suitable plane exists, then a plane may be constructed by using the *Make Datum* options.

The crosshatch in a section may be modified by using the option *Modify*. As shown in Figure 17.5, the *Spacing, Angle* and *Line Style* are among the features that can be changed. The standard for crosshatch spacing is 1/16" to 1/8" (1.5 mm to 3 mm) apart.

### 17.1.1 Tutorial # 77: A Full Section of the Angle Bracket

Let us create a full section of the angle bracket from chapters 2, 4, and 8. The part should be complete with holes and rounds. Begin by creating a drawing called "AngleBracketSection." Load the format "Aformat." Add a general view of the bracket. Click on the top of the drawing, centered from left to right. *Reorient* the view so that the double holes are parallel to the screen. Use the final version of the drawing shown in Figure 17.6 for reference.

Add a second view below the first. This view will be sectioned, so choose the *Section* option from the *VIEWS* menu (Figure 16.5). Select *Full* and *Total Xsec* from the XSEC TYPE menu shown in Figure 17.1.

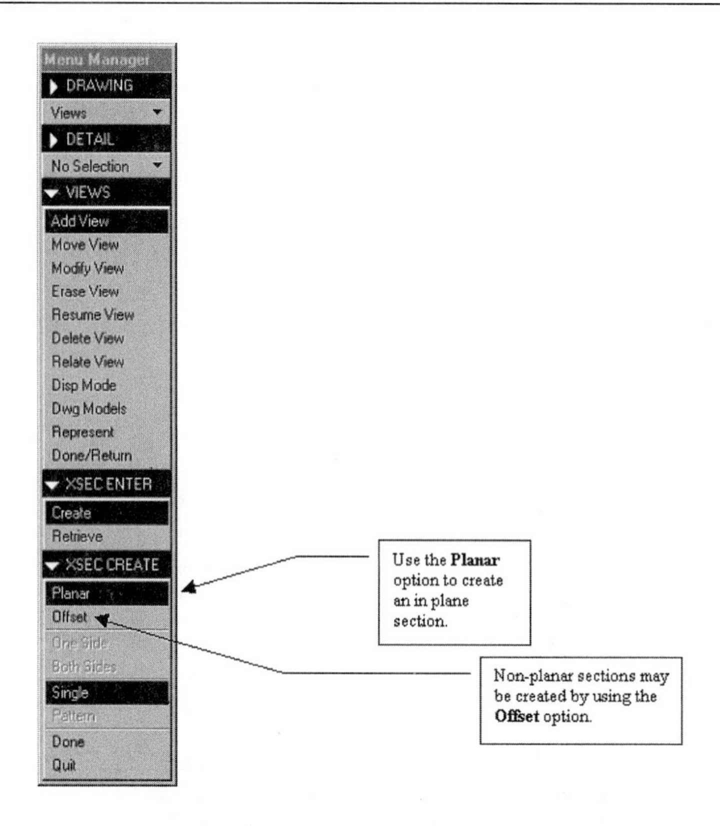


FIGURE 17.4 *The options for creating a cross section.*

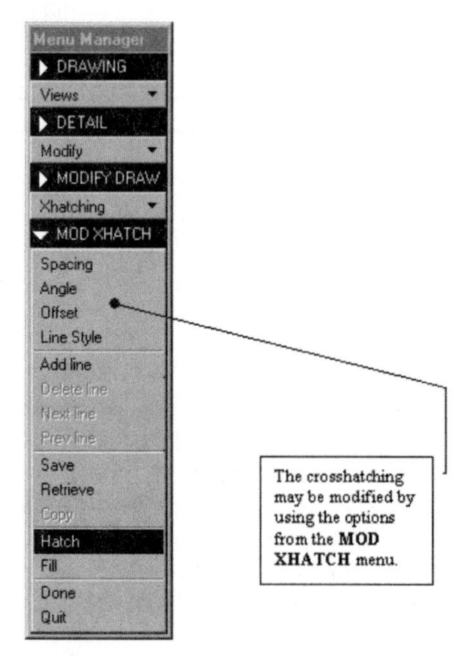


FIGURE 17.5 *The options available to modify the crosshatching.*

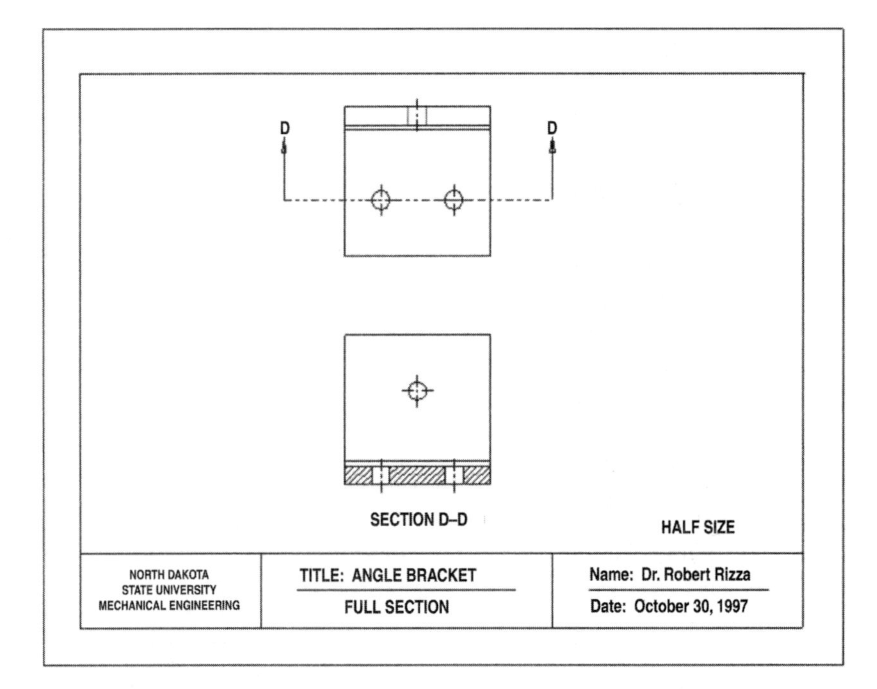


FIGURE 17.6 *This drawing of the angle bracket shows a section of the part in the front view.*

We will need to create the section. In order to do so, choose the *Create* option from the *XSEC ENTER* menu in Figure 17.2. Accept the *Planar* and *Single* defaults. Enter the name D.

We need to use *Make Datum* to create the cutting plane. The plane will be defined such that it goes through the center of one of the double holes in the first view placed. Actually, because both holes are aligned, the plane will pass through both the holes. In addition, we will construct the plane so that it is parallel to the front of the model.

Thus, in order to create this plane, select *Through* and pick on the center of one of the circles. Then select *Parallel* and click on the front of the model.

We want the arrows that define the location of the cutting plane to reside in the top view. Click on the top view to locate the arrows.

Pro/E will create the section. You may need to move the views, so that the drawing looks uniform.

Change the spacing of the hatch lines to 1/16" (0.0625). Use the options *Modify, Xhatching, Spacing,* and *Value.* Enter 1/16 and hit the *Return* key.

Dimensions may be added to the drawing. However, for our purposes in this tutorial, it is not required. Therefore, do not add any dimensions unless asked to do so by your instructor.

Add the necessary text as needed to complete the drawing. Change the scale of the drawing to half scale. Select *Modify* then click on the scale with your mouse pointer. Enter 0.5.

Use the *Show/Erase* option and place the axis for each circle and centerline for each hole. Turn off the axis labels by selecting the *Axis* icon in the toolbar (see Figure 1.11). When you are done, save the drawing.

### 17.1.2 TUTORIAL # 78: AN OFFSET SECTION OF THE ANGLE BRACKET

We will now create the offset section of the angle bracket. The part should be complete with all required holes, fillets, and rounds. If it is not complete, take a few moments to add the holes or the fillets and rounds.

Our first exercise will be to create the offset section shown in Figure 17.2c. Then we will proceed to an offset section located on a projection of the top view.

Instead of creating a new drawing, we can add a new page (sheet) to the drawing created in section 17.1.1. If the drawing Angle_bracket_section is not in memory, retrieve the file by selecting *File* and *Open.* After the file is in memory, select *Sheets* from the *DRAWING* menu (Figure 16.6). Then choose *Add.* Notice that the software has loaded the format on the new page of the drawing.

Select *Views* and *Add View.* Add a *General* view near the top of the screen, centered from left to right. When adding this view, choose the *Section* option and then *Done* from the *VIEWS* menu (Figure 5.5). Select *Full* and *Total Xsec.* Then click on the screen to place the view. Do not reorient the view. Choose *OK.*

Select *Create* from the *XSEC ENTER* menu (Figure 17.3). Then choose the *Offset* option from the *XSEC CREATE* menu shown in Figure 17.4. For an offset section, the *Both Sides* option must always be used so accept the default. For the section name, enter A.

At this point, the program prompts the user to choose a sketching plane. In choosing this plane, you want a plane or surface where the cutting plane would appear as an edge. In this case, the cutting plane will pass through the holes and will be parallel to the L-shaped profile of the part. Thus, any surface perpendicular to the L-profile can be chosen as the sketching plane. With your mouse pointer, click on the top part of the model as shown in Figure 17.7.

In order to orient the model in the *Sketcher,* choose the "*Right*" option and then click on the face shown in Figure 17.7. The model should appear as in Figure 17.8.

The offset cutting plane will pass through the center axis of the large hole and the center of the rear smaller hole in Figure 17.8. Select the *Sketch* and draw the edge of the plane as shown in Figure 17.9.

The sketch must be located with respect to the part before being regenerated. In this case, no dimensions are required if the endpoints of the line are aligned to the silhouettes of the edge of the bracket. Also, align the endpoints of the lines where they intersect at the center of the small circle with the center of that circle. *Regenerate.* After the sketch has been successfully regenerated, choose *Done.*

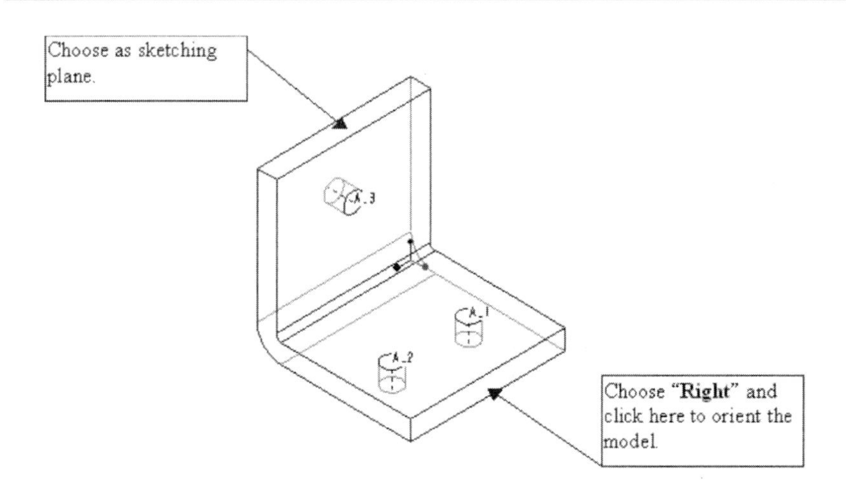


FIGURE 17.7 *The angle bracket model with the appropriate surfaces for the sketching plane and the orientation plane.*

Because we are going to create a section on the general view, arrows are not desired. Click your middle mouse button to unselect the placement of the arrows.

After the section is generated, use the *Modify* option and change the spacing of the hatch lines to 1/16". The section should appear as shown in Figure 17.10.

Our purpose in generating the offset section using the general orientation was to help the user to better visualize the sectioning. However, in multiview

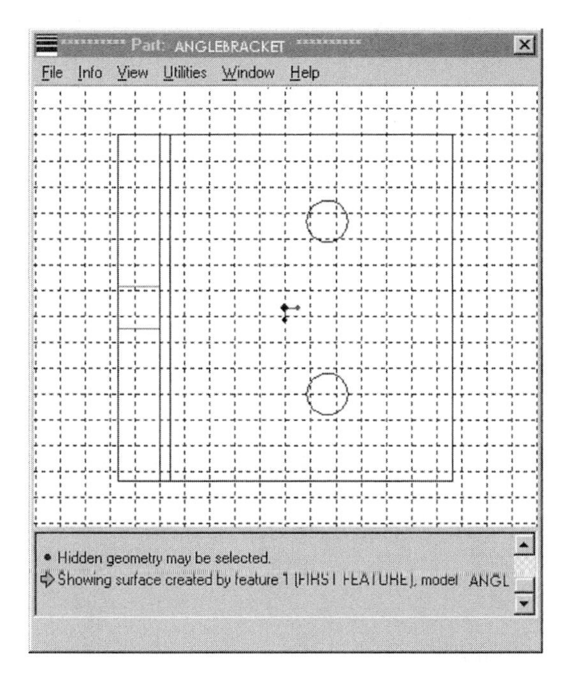


FIGURE 17.8 *The bracket reoriented in the* Sketcher.

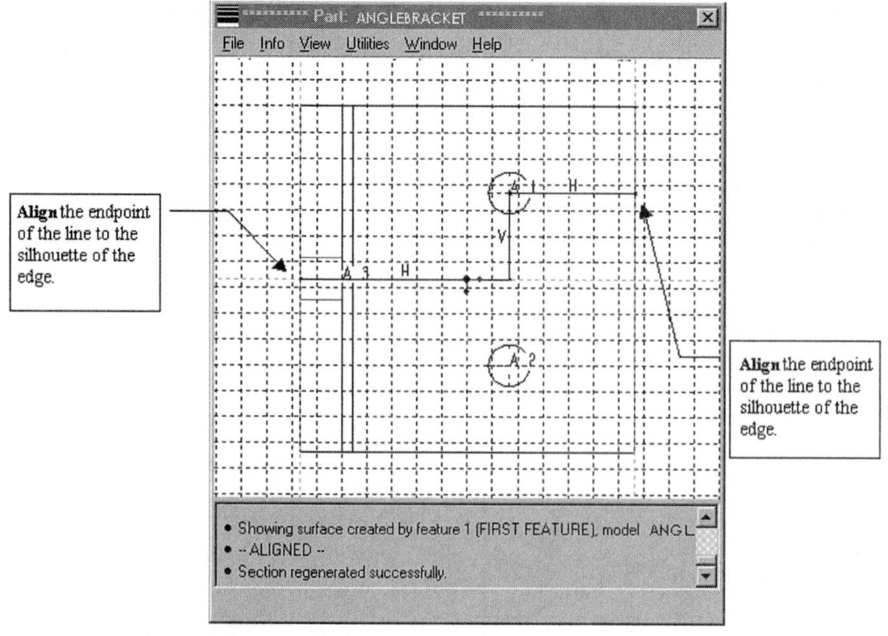


FIGURE 17.9 *The cutting plane the bracket.*

FIGURE 17.10 *An offset section generated on the general default view.*

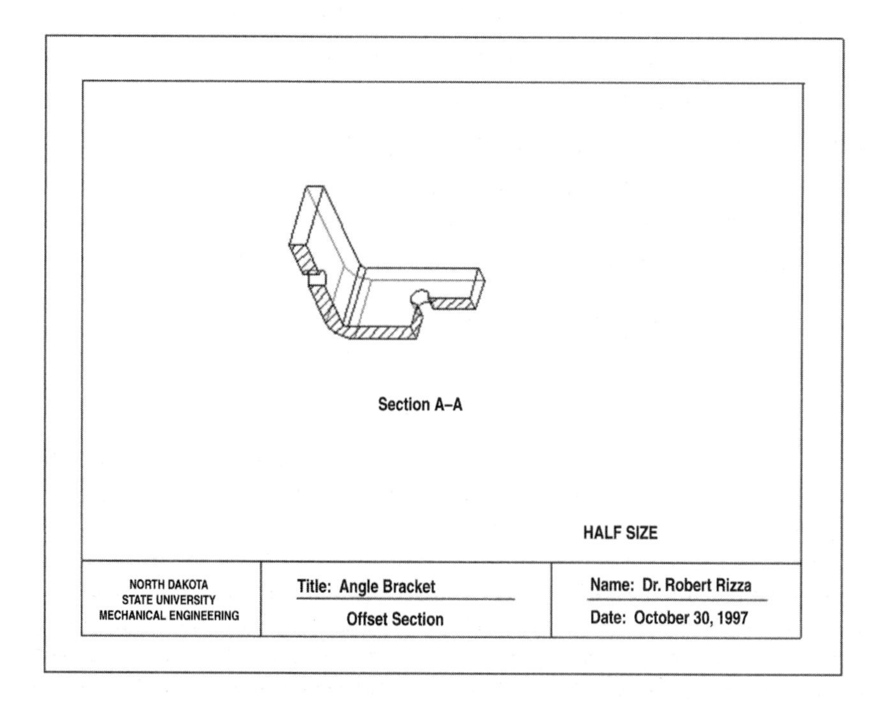

layout drawings, the section as generated is not standard practice. More often, an offset section is generated on a projection of the model. In this case, it is best to generate a multiview showing the model oriented in a top and front view. A right-hand side view should be added as well. For the sake of brevity, let us generate a top view of the model and then create a front projection on which the offset section is shown.

Add a third page to your drawing by selecting *Sheets* and then *Add*. Notice that you may move between the sheets by using the *Next* and *Previous* options.

Now, create a top view by doing the following:

1. Select *Views* and *Add View*.
2. Choose the *General, Full View, No Xsec,* and *Done*.
3. Locate the view near the top of the drawing, centered from left to right.
4. Reorient the model to create a multiview top view as shown in Figure 17.11. Notice the direction of the arrows and the location of the cutting plane line in the top view.

In this approach, the first view added is not the view that is sectioned. An additional view must be created. To add the section, follow these steps:

1. Choose *Views* and *Add View* from the *VIEWS* menu.
2. Select *Projection, Full View, Section,* and *Done*.
3. Choose *Full* and *Total Xsec* from the *XSEC TYPE* menu.
4. Then select *Create* from the *XSEC ENTER* menu.
5. Because we already created the cutting plane for this offset section, select the name "A" from the list of sections.
6. Click on the top view to locate the arrows.

Change the spacing of the crosshatch by using the *Modify* option. Enter a value of 1/16. Add the necessary text to the drawing.

### 17.1.3 Tutorial #79: A Half Section of the Angle Bracket

Let us create a half section. We will use the angle bracket as the model. The cutting plane will be defined so that the part of the model containing the large single hole will be sectioned.

In creating a *Half* section, you will need to define a reference plane that determines the extent of the section. This reference plane can be a plane or surface occurring on the model, a *Datum* plane or a plane you create using the *Make Datum* option. For this model, we are going to use the surface plane containing the double holes as the reference plane. Then, the model above this plane will be sectioned, while the part of the model below this plane will not be sectioned.

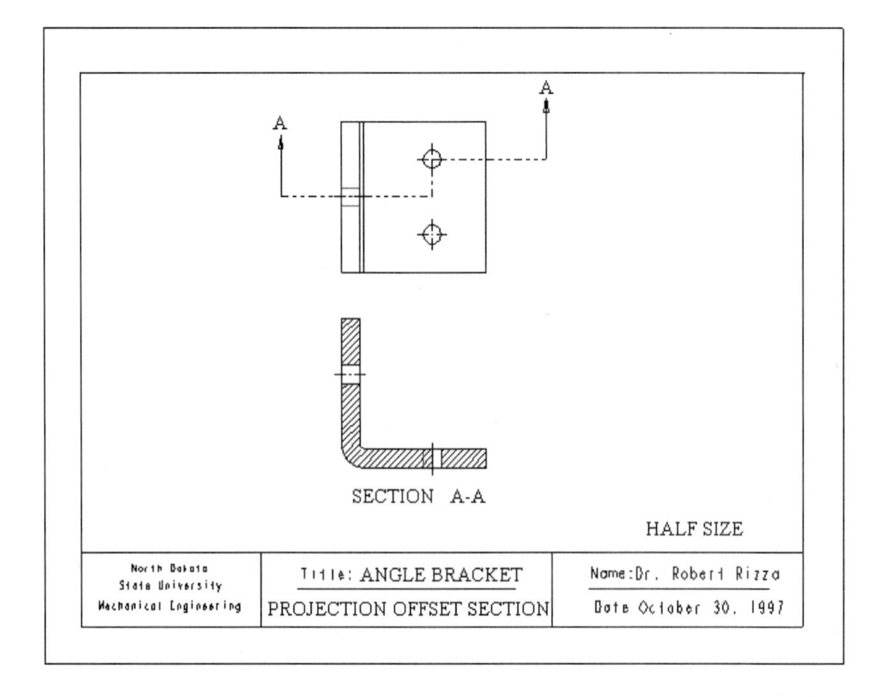


FIGURE 17.11 *The completed drawing of the offset section located in the front view of the model.*

Add a fourth sheet to your drawing by choosing *Sheets* and then *Add.* Add the general view of the model and *Reorient* as shown in the completed drawing of this section, that is, Figure 17.12.

Add the second view. Use the following sequence of steps:

1. Choose *Views* and *Add View.*
2. Select *Projection, Full View, Section,* and *Done.*
3. Choose *Half* from the *XSEC TYPE* menu. Note that the *Total Xsec* is the only option available for a *Half* section.
4. Select *Done.*
5. Locate the second view by clicking on the screen *below* the first view.
6. Now define the reference plane. In the top view, click on the surface containing the two holes as shown in Figure 17.12. A set of arrows will appear in the second view. These arrows should point upward. Notice their location. They should originate from the reference plane. The direction of the arrows indicates that all the features in that direction will be sectioned.
7. Choose *Okay.*
8. Select *Planar* and *Done.* Enter the name “C.”
9. For this model, you will need to make a datum. Use the *Make Datum* command. Choose *Through* and *Query Sel.* In the top view,

select the axis through the single hole. If necessary, cycle through the features. Then choose *Accept*. Select *Parallel*. With your mouse pointer, click on the L-shaped profile in the second view.

10. Choose *Done*.
11. Click on the first view to locate the cutting plane line.

Choose *Modify* and change the crosshatch spacing to 1/16. Your work should appear as in Figure 17.12. Add the necessary text. Save the drawing. *Erase* the file from memory.

## 17.1.4 TUTORIAL # 80: A REVOLVED SECTION OF THE CONNECTING ARM

Revolved sections are useful in visualizing thin sections such as ribs. The section is created and revolved about an axis of symmetry. Then, it is located near a reference view.

The connecting arm from chapters 6 and 8 is a perfect example where a revolved section is useful. Create a drawing called "ArmRevolvedSection" and add a general view of the arm near the top of the drawing. *Reorient* the view so that it appears as shown in Figure 17.13, which is the top view of the arm.

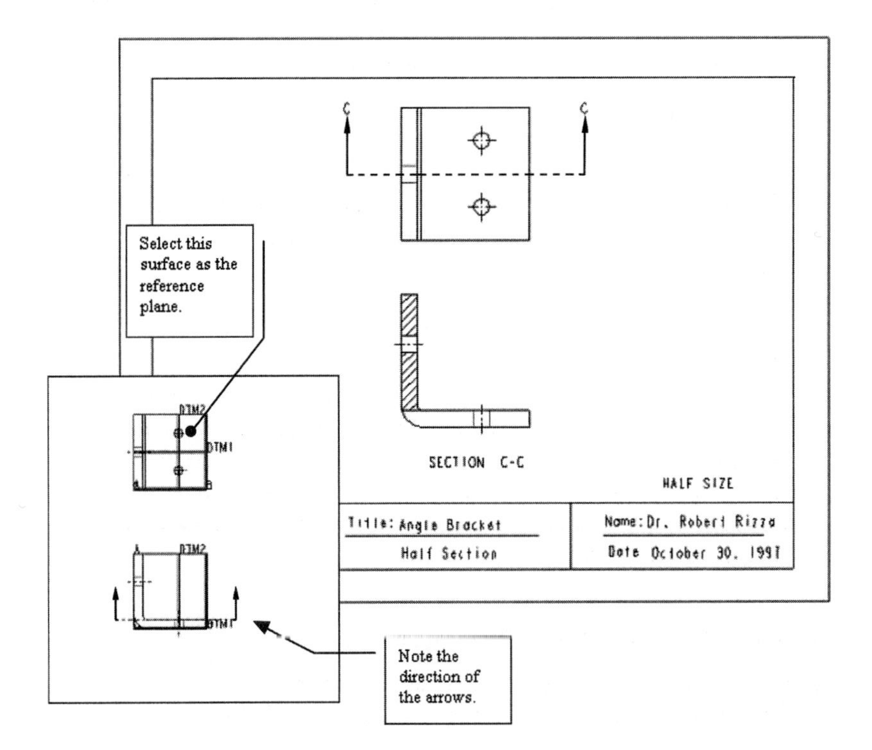


FIGURE 17.12 *The* Half *section of the angle bracket.*

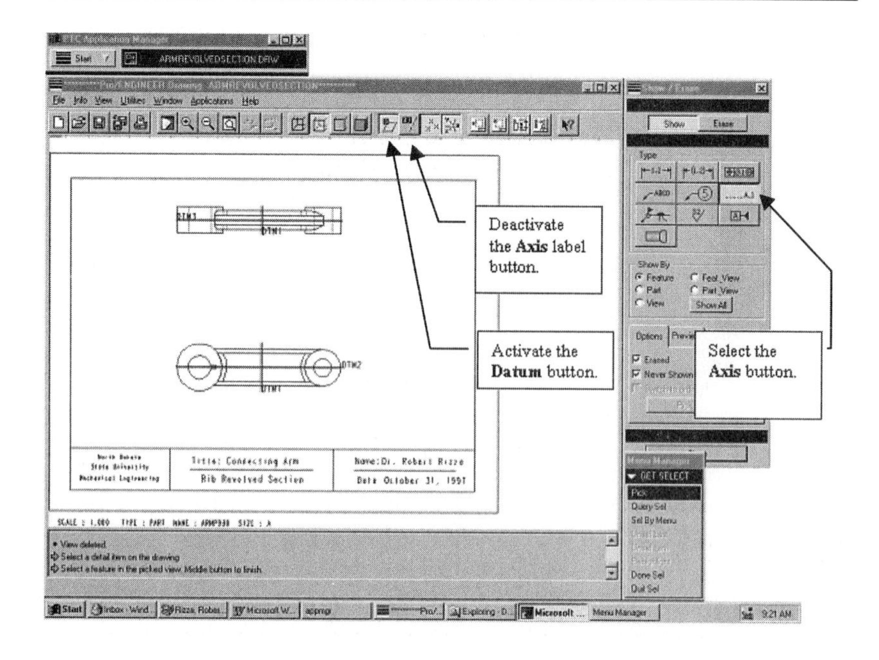


FIGURE 17.13 *A drawing showing the first and second (top and front) views of the connecting arm.*

Add a second view below the first view. This view is the front view of the model. The *Revolved* section will be created from this view.

It is advantageous to display all the datum planes and axes. Select *Show/Erase.* Choose the *Show* and *Axis* buttons. For the *Show By* option, choose *Show All.* Then select *OK. Activate* the display of the datum planes by selecting the *Datum* button on the toolbar (see Figure 1.11). *Deactivate* the display of the axis labels by selecting the *Axis* button from the toolbar (see Figure 1.11).

Now let us add the revolved section. Select *Views, Add View,* and choose *Revolved* from the *VIEW TYPE* menu (Figure 5.5). Note that the *Section* option becomes active.

For the *Center Point,* click on the screen slightly above the front view. Because we are creating a section from the front view, the *Parent View* is the *Front* view.

We can now construct the cutting plane. Luckily, for this part, the plane already exists. It is the DTM1 default datum plane. Select *Create* and then DTM1 plane. Choose *Done.*

An axis of symmetry may be defined. For our purposes, defining the axis is not necessary, so click on your middle mouse button. The revolved section will be placed. Its appearance should be similar to the one shown in Figure 17.14.

Add the appropriate text to the drawing. *Modify* the spacing of the crosshatch to 1/16". Turn off display of the datum planes to clearly see the drawing. Save the drawing.

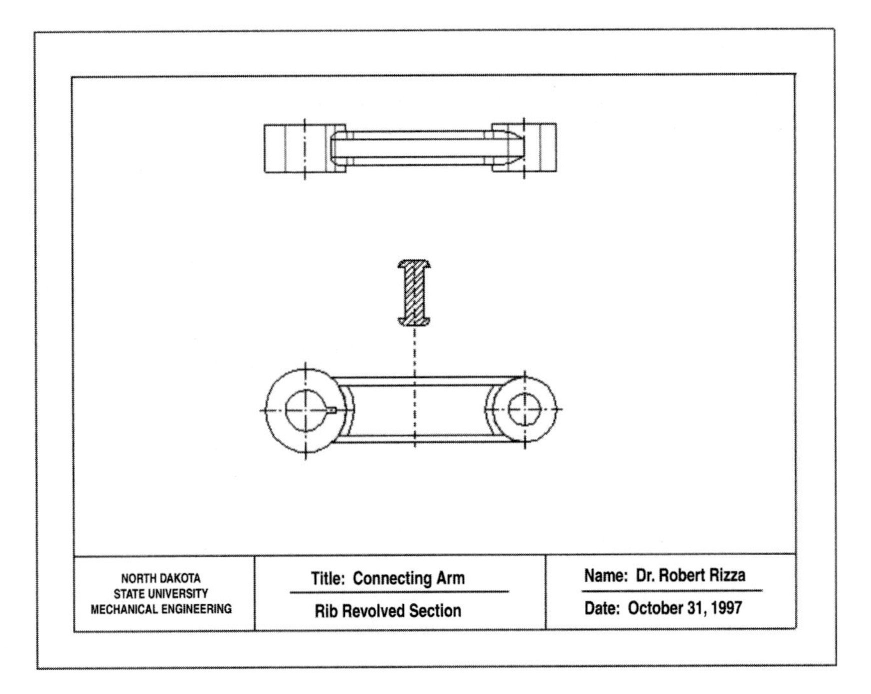


FIGURE 17.14 *A drawing of the connecting arm showing a revolved section of the rib.*

## 17.2 SUMMARY AND STEPS FOR CREATING A SECTION

A section is useful for illustrating internal features in a part. With Pro/E, the user may create several different types of sections. These sections may be full or partial sections. Standard practice is to create a section on one of the multiviews. Thus, at least two views must be added to create a drawing containing a section.

The steps for creating a section are identical for most types of sections (excluding *Revolved* sections). The procedure begins by adding the first view, which is not sectioned. Then a second view is added. This view is sectioned. The sequence of steps is as follows:

1. Select *Views* and *Add View.*
2. Choose the view type. It may be a *Projection.* However, if the drawing contains no other view, then use the *General* option.
3. Select *No Xsec* and *Done.*
4. Locate the view on the drawing.
5. Add the second view, by choosing *Add View.*
6. This view has to be a projection, so choose *Projection, Full View, Section,* and *Done.*
7. Locate the second view on the drawing.

8. Create or retrieve a section by selecting from the *XSEC ENTER* menu options *Create* and *Retrieve.*
9. If you have chosen to create the section, choose from *Planar* or *Offset.*
10. If the choice is *Planar,* choose the datum or use the *Make Datum* option and construct the datum. If the choice is the *Offset* option, select the sketching plane and the orientation plane. Then sketch the geometry of the edge of the cutting plane. *Dimension* or *Align* the sketch and *Regenerate.*
11. Select the first view as the view to locate the cutting plane line.

A revolved section requires an additional view. The procedure is the same through step 5. After this step, follow the sequence:

1. Choose *Projection, No Xsec,* and *Done.*
2. Locate the second view on the drawing.
3. The third view will be a section. Choose *Add View.*
4. Select *Revolved, Full View, Section,* and *Done.*
5. Locate the third view with respect to the second.
6. *Create* or *Retrieve* the cutting plane.
7. For symmetric sections, you may wish to select an axis of symmetry. If you wish to do so, select the appropriate axis. Otherwise, depress your middle mouse button to locate the section.

Sectioned views may be moved using the *Move View* option and dimensioned. The crosshatching may be modified by using the path *Modify* and *Xhatching.* Select *Spacing* to change the spacing of the crosshatch and *Angle* to change the orientation of the crosshatch.

Additional pages (sheets) may be added to a drawing, which is advantageous if several sections of a part are desired. Add a sheet to a drawing by choosing *Sheets* and *Add Sheet.* Use the *Next* and *Previous* options to cycle between the sheets. A sheet may be deleted from the drawing by using the *Sheets* and *Remove.*

When adding a sheet to a drawing, Pro/E automatically inserts the format used on the first page onto any additional sheet. You may remove the format by selecting *Sheets, Format,* and *Remove.* A different format may be added to a selected sheet by using *Sheets, Format,* and *Add/Replace.*

## 17.3 ADDITIONAL EXERCISES

17.1 Create a drawing showing the *Full* section of the Venturi nozzle (see also exercise 10.20) shown in Figure 17.15.

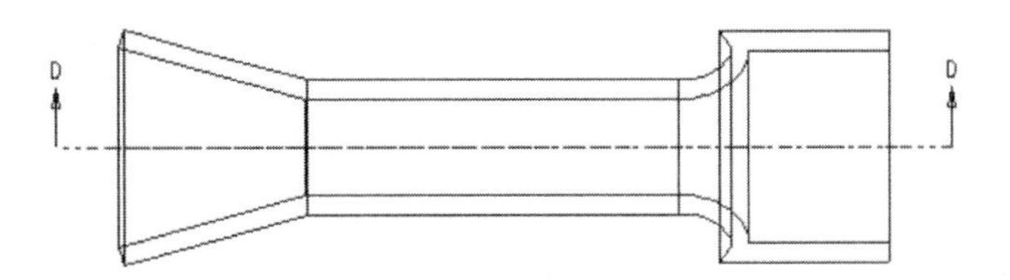


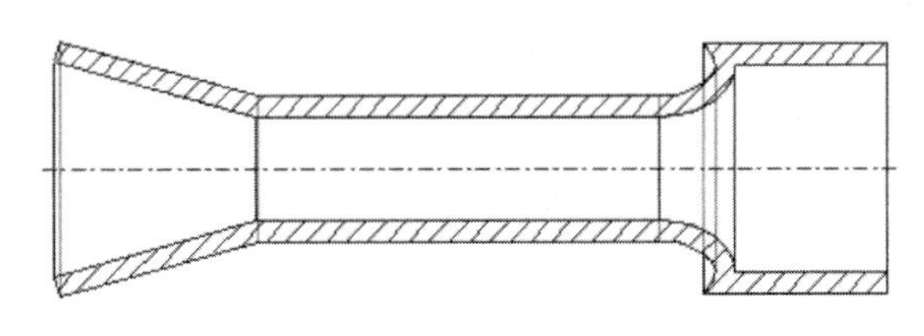


FIGURE 17.15 *Figure for exercise 17.1.*

17.2 Create a drawing of the fixed bearing cup (see also exercise 10.21) as shown in Figure 17.16. The section is a *Full* section.

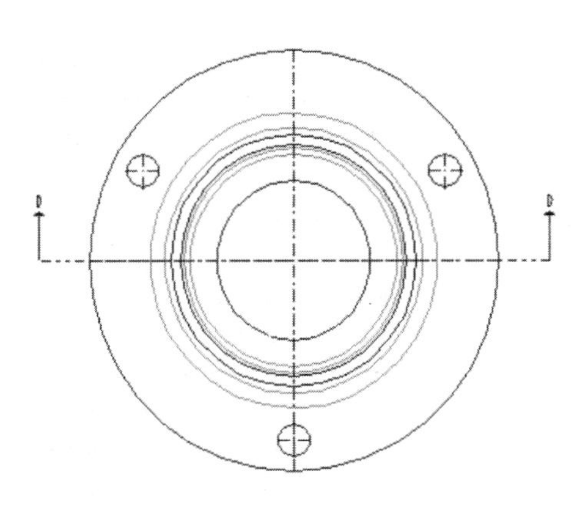

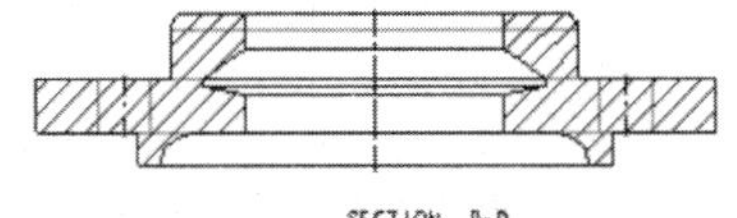


FIGURE 17.16 *Figure for exercise 17.2.*

17.3 Create a drawing of the fixed bearing cup (see also exercise 10.21) containing an *Offset* section. The section is shown in Figure 17.17.

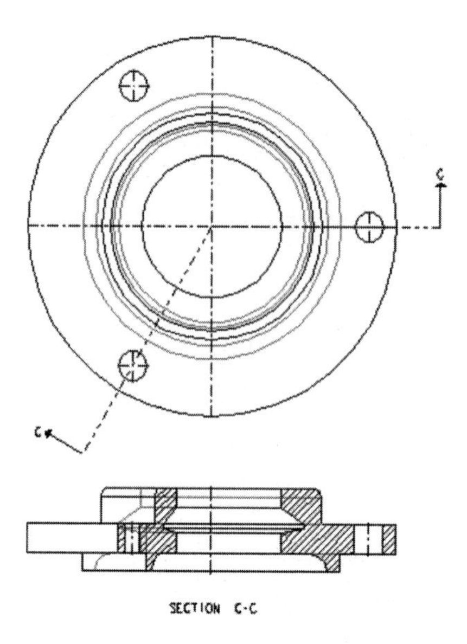


FIGURE 17.17 *Figure for exercise 17.3.*

17.4 Create a drawing with a *Full* section of the tapered collar as shown in Figure 17.18 (see also exercise 20.23).

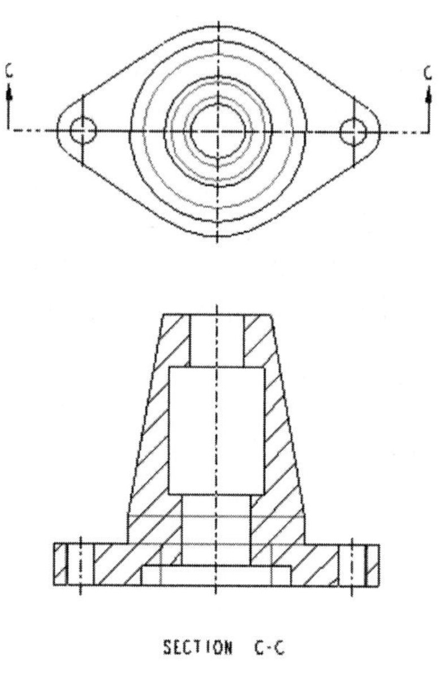


FIGURE 17.18 *Figure for exercise 17.4.*

17.5 For the model of the tapered collar (see exercise 20.23), create a drawing with the *Offset* section shown in Figure 17.19.

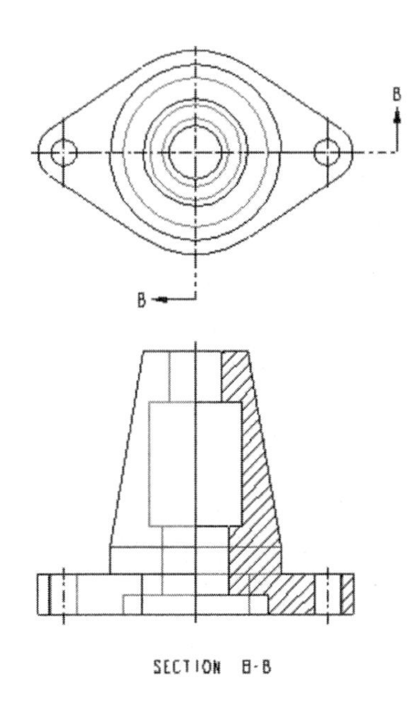


FIGURE 17.19 *Figure for exercise 17.5.*

17.6 Create a drawing with a *Full* section of the round adaptor (see also exercise 10.22) as shown in Figure 17.20.

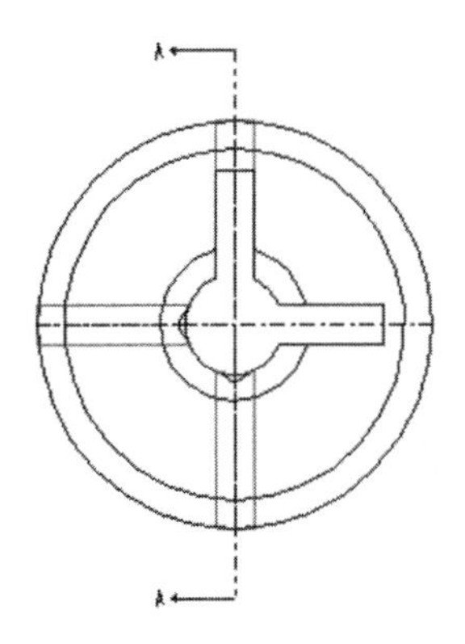


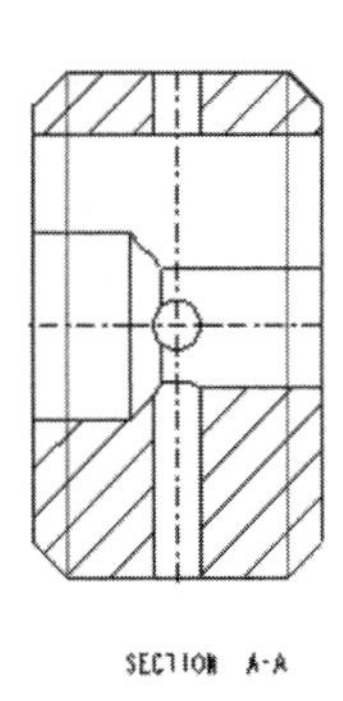


FIGURE 17.20 *Figure for exercise 17.6.*

17.7 For the round adaptor from exercise 10.22, create a drawing with the *Offset* section shown in Figure 17.21.

FIGURE 17.21 *Figure for exercise 17.7.*

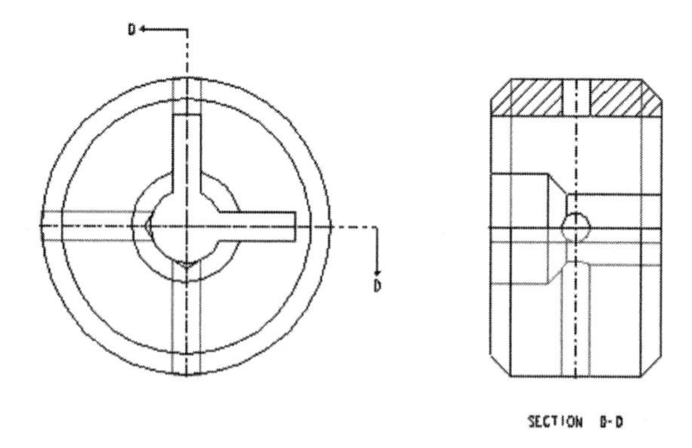


17.8 Create a drawing of the hand rail support (see also exercise 11.9) containing a *Revolved* section as illustrated in Figure 17.22.

FIGURE 17.22 *Figure for exercise 17.8.*

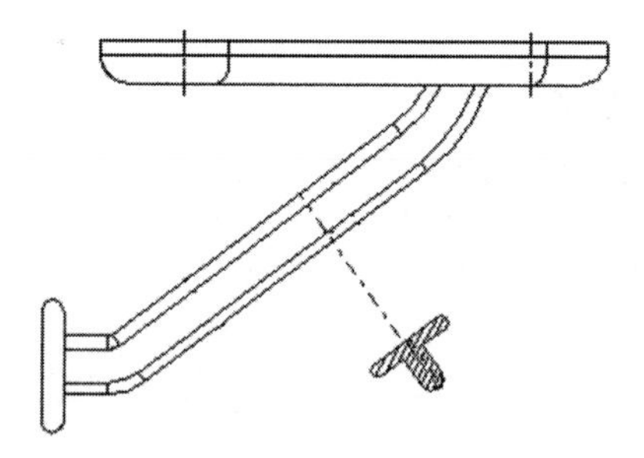

17.9 For the model of the dahpot lifter (see also exercise 11.10), create a drawing with a *Revolved* section. This section is shown in Figure 17.23.

FIGURE 17.23 *Figure for exercise 17.9.*

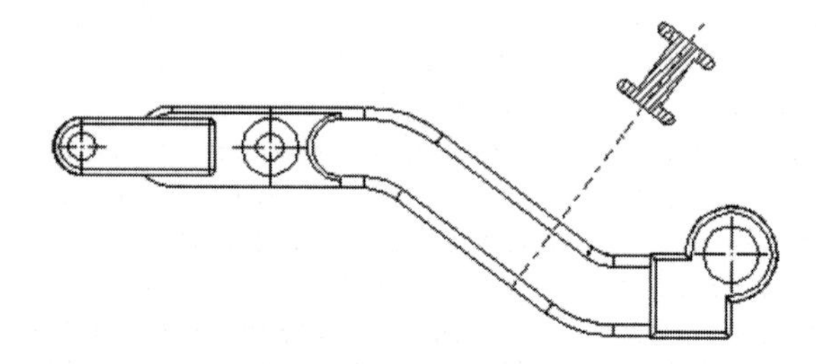

# CHAPTER 18

# ADDING TOLERANCES TO A DRAWING

## INTRODUCTION AND OBJECTIVES

In the manufacturing process, the tolerance of a feature is required. This chapter concerns specifying tolerances on drawings. Included in the methods to be discussed in this chapter is geometric dimension and tolerancing (GDT).

In Pro/E, tolerances should be prescribed when parts are created, because they determine how a feature will be constructed. If the tolerances are not prescribed by the user, then Pro/E uses default tolerance values. To see what the default values are, turn their visibility on by selecting *Utilities* and *Environment.* Then choose the *±.01 Dimensional Tolerances* option. After a model is loaded or created, Pro/E will display the default tolerance at the bottom of the drawing window as shown in Figure 18.1.

In order to change or set the bounds on any feature, select *Dim Bounds* from the *PART SETUP* menu shown in Figure 2.9. From the *DIM BOUNDS* menu reproduced in Figure 18.2, the user may set the dimension to its upper, lower, middle, or nominal value. The tolerance standard may be changed by choosing *Tol Setup* from the *DIM BOUNDS* menu. In the *TOL SETUP* menu, also shown in Figure 18.2, the user may select between *ANSI* and *ISO/DIN* standards. The default setting is *ANSI.*

A geometric tolerance may be specified by using the *Geom Tol* option in the *PART SETUP* menu. By using the options in the *GEOM TOL* menu (Figure 18.2), the user may define the reference datums, the geometric tolerance and set basic and inspection dimensions.

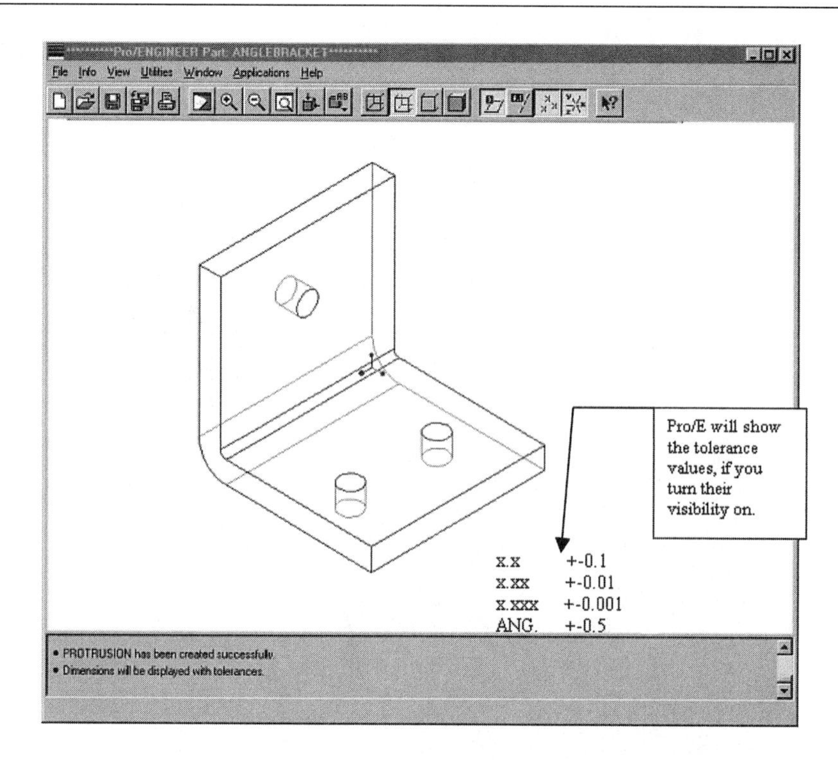


FIGURE 18.1 *An example of a part with displayed tolerances.*

These menus are useful for specifying traditional and geometric tolerances for a model in the part mode. If a tolerance is defined for a model in the part mode, the dimensions will be displayed with the tolerance in the drawing mode. In the sections that follow, we will create the tolerance within the drawing mode.

The objectives of this chapter are as follows:

1. The user will attain an ability to add a traditional or geometric tolerance to a model.
2. The reader will be able to add traditional or geometric tolerances to a multiview drawing.

## 18.1 ADDING TRADITIONAL TOLERANCES TO A DRAWING

Whether the user prescribed certain tolerances on the model when it was built is immaterial, because as was seen in the Introduction, the model was constructed with default tolerance values. Furthermore, these default values may be modified within the drawing mode, and Pro/E will update the changes in the model.

In order to add traditional tolerances to a dimension in the drawing mode, all that remains to do is to turn on the visibility of the tolerance

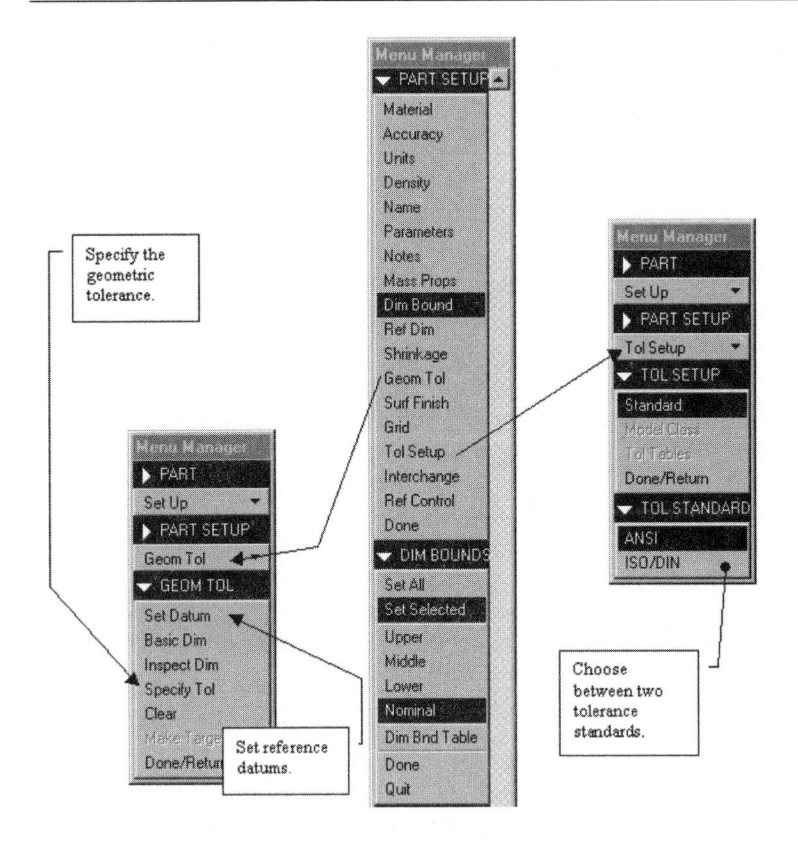


FIGURE 18.2 *Both traditional and geometric tolerances may be set for a model by using the options in the* PART SETUP *menu.*

values by loading the drawing configuration file and changing the option "tol_display" from "*NO*" to "*YES.*" The sequence of steps needed to complete this task is:

1. Select *Advanced* from the *DRAWING* menu.
2. Choose *Modify Val.*
3. Scroll down until the option "tol_display" is found.
4. Change the setting from "*NO*" to "*YES.*"
5. Select *File* and *Exit.*
6. *Done/Return.*
7. Repaint the screen.

The options *Modify* and *Dimension* may be used to change the display of the tolerances on any or all the dimensions. Pick the desired dimensions with your mouse pointer and choose *Done Sel* when finished. It will load the *Modify Dimension* dialog box shown in Figure 18.3. The options contained in the box allow the user to change the tolerance among ± *Symmetric, Limits,* and *Plus-Minus* modes. Depending on the mode desired the user might modify the upper and lower limits or the tolerances values.

### 18.1.1 TUTORIAL # 81: TRADITIONAL TOLERANCES FOR THE DRAWING "ANGLEBRACKET"

Let us display bounds on a specific part. Retrieve the drawing "Angle-Bracket" created in chapter 16. If the visibility of the tolerance values is turned off, turn it on using the procedure detailed in Section 18.1.

The software will display the tolerance values. The display should be in the ±*Symmetric* mode. If not, select *Modify* and *Dimension.* Then choose the option *Pick Many.* Drag the window across the entire drawing. Change the mode of the display.

You may need to move the views and/or dimensions in order to properly space all the elements of the drawing. Use Figure 18.4 as a reference and move the elements as necessary. Save the drawing unless told not to do so.

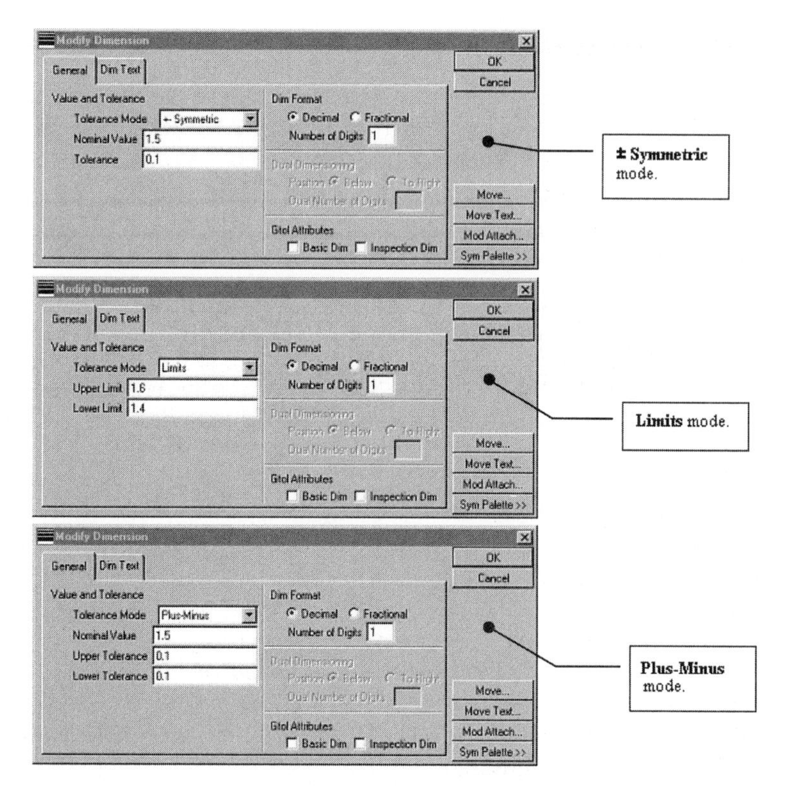


**FIGURE 18.3** *The* Modify Dimension *box. This dialog box contains options for displaying tolerances. For the first box, the* Tolerance *mode is set to* ±Symmetric. *For the second, the mode is* Limits *and for the third, the mode is* Plus-Minus.

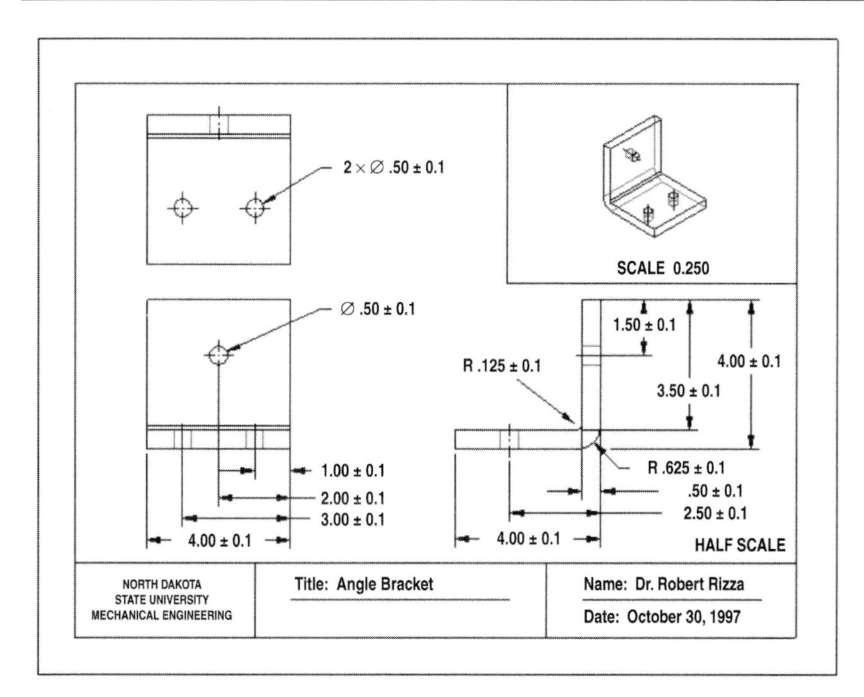


FIGURE 18.4 *The drawing of the angle bracket with added tolerance values.*

## 18.2 GEOMETRIC DIMENSIONING AND TOLERANCING

Geometric tolerances may be added to a part in the part mode or assembly mode. The user may also add the tolerances in the *Drawing* mode.

Unlike traditional tolerancing, geometric dimensioning and tolerancing (GDT) is more than just adding bounds on a dimension. It is a way of defining features based on how they are to function in engineering practice. Besides requiring a control on a feature, GDT also describes how the feature is to be controlled.

Because *Reference Datums* are used in geometric tolerancing, before adding a GDT, the user should make sure that the proper datums have been defined. Often one or more of the default datum planes can be used as a reference datum. Then, all the user needs to do is to set the desired default datum plane as a reference plane.

If none of the existing planes can be used as a reference plane, then the appropriate datum plane must be created. The plane may be created in the part mode or in the drawing mode. To create the datum in the drawing mode, use the following series of options: *Detail, Create, Datum,* and *3D Datum.*

An existing datum *must* be set as a reference datum before it can be used to define a geometric tolerance. In order to change an existing datum to a reference datum, use the *Datum/Axis* option from either the *Modify* option in the *DETAIL* menu or from *Set Datum* in the *GEOM TOL* menu. In either case, the *Datum* dialog box shown in Figure 18.5 will appear after the *Set Datum* option is chosen.

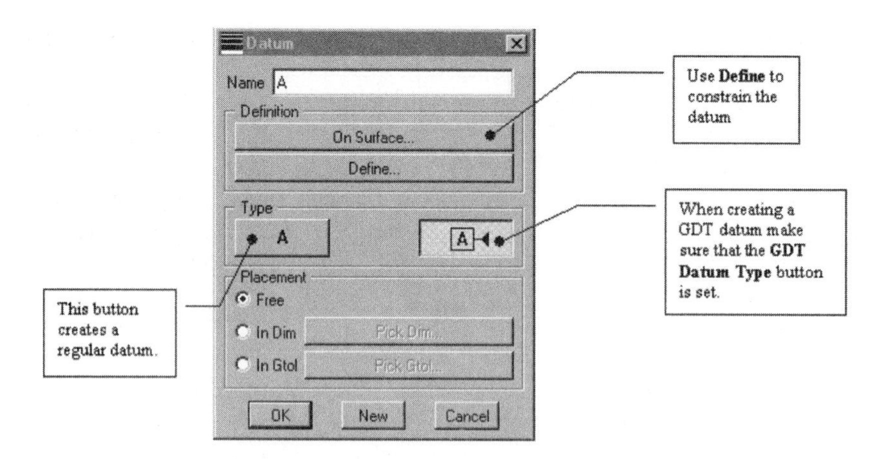


FIGURE 18.5 *The* Datum *dialog box. This box is used to create a datum.*

After reference datums have been defined, dimensions may be added to the drawing and geometric tolerances added as needed. Geometric tolerances may be created with the *Specify Tol* option from the *GEOM TOL* menu (Figure 18.2).

The *Geometric Tolerance* box shown in Figure 18.6 may be used to define the geometric tolerance. The box contains several folders with options for creating the geometric tolerance. It is possible that not all of the options will be used for creating the GDT.

In creating the GDT, the user must select the model. The name of the default model is placed in the *Model* field by Pro/E. To change the *Model,* choose the *Select Model* button.

The *Reference Entity* can be one of the following: *Edge, Axis, Surface, Feature, Datum,* and *Entity.* After one of these options has been chosen, the *Select Entity* button may be depressed and the actual edge, axis, etc. may be selected from the screen using the mouse pointer.

The geometric tolerance may be located by attaching the feature control box to a dimension (*Dimension* option). It may also be attached to another

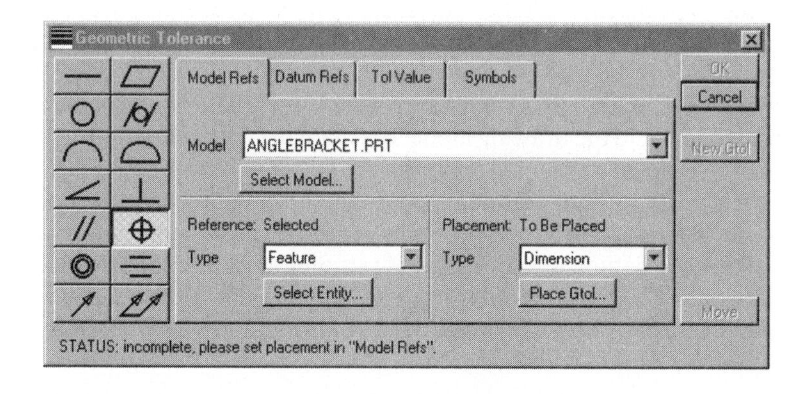


FIGURE 18.6 *The* Geometric Tolerance *box.*

control box (*gtol* option), with a leader (*Leader, Tangent Ldr,* and *Normal Ldr* options), or as a free note (*Free Note* option).

The reference datums may be placed in the feature control box by choosing the *Datum Refs* folder. The primary, secondary, and tertiary datum references are chosen by selecting from a list of available datums.

The 14 buttons along the left-hand side of the box are the available GDT control symbols. Their meaning can be accessed by placing your mouse pointer over the desired button and holding it there for a few seconds.

Notice that the number of decimal digits may also be set using the *Modify Dimension* box. Furthermore, the GDT specifications, *Basic Dimensions,* and *Inspection Dimensions* may also be set.

*Basic Dimensions* are dimensions that are theoretical values. Such dimensions do not have tolerance values associated with them. The GDT symbol for *Basic Dimensions* is a dimension enclosed in a rectangle. *Inspection Dimensions* are used for inspection. They are represented by a dimension enclosed in an oval. An example of a *Basic* and *Inspection Dimension* is given in Figure 18.7. Tolerances may be prescribed in an *Inspection Dimension.*

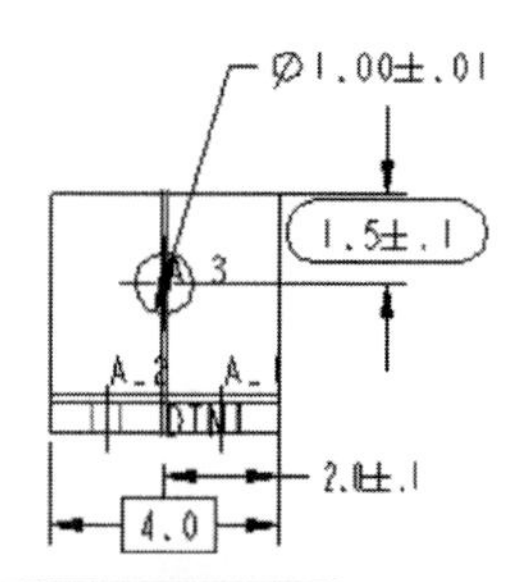


FIGURE 18.7 *An example of a* Basic Dimension *and an* Inspection Dimension. *The* Basic Dimension *is indicated with the dimensional value enclosed in a rectangle. The* Inspection Dimension *is shown with the dimensional value enclosed in an oval.*

## 18.2.1 TUTORIAL # 82: CREATING A DRAWING WITH GEOMETRIC TOLERANCES

Earlier in this chapter, a drawing of the angle bracket was used as an example for the addition of tolerances. The function of the bracket is to support a pulley wheel. In order to accomplish this task, the bracket is connected to a base plate and the wheel. In order for the connections to be performed smoothly, the locations of the hole must be properly maintained.

From the point of view of geometric tolerancing and dimensioning, proper control must be placed on the holes. In particular, the minimum size of the holes, along with their location, must be controlled. A control on the size of the holes can be placed by requiring that the minimum size of the holes be at maximum material condition.

Create a new drawing with the name "AngleBracketGDT." Load "Aformat." Place a general view of the bracket at the top left-hand side of the screen. Do not change the orientation.

Change the orientation from *Trimetric* to *Isometric,* if it is not already in *Isometric,* by using *Utilities, Environment, Isometric,* and *OK. Repaint* the screen.

The GDT datum symbol must be changed to ASME standard because the default for Pro/E is ANSI. In order to change the standard, follow these steps:

1. Select *Advanced* from the *DRAWING* menu.
2. Choose *Modify Val.*
3. Scroll down until the option "*gtol_display*" is found.
4. Change the setting from "*STD_ANSI*" to "*STD_ASME.*"
5. Select *File* and *Exit.*
6. Choose *Done/Return.*
7. Repaint the screen.

Before proceeding any further, add three GDT datum planes to the part. Create the planes using Figure 18.8 and this procedure:

1. Select *Create, Datum,* and *3D Datum.*
2. The *Datum* dialog box will appear (Figure 18.6).
3. Enter the name "A" in the cell provided. As shown in Figure 18.6, make sure that the *GDT Type Datum* button is active.
4. Select *Define.*
5. Constrain the datum. In this case, you can use the option *Through* and the L-shaped profile.
6. Select *New* from the *Datum* dialog box.
7. Enter the name "B."
8. Repeat steps 4 through 6.
9. Enter the name "C."
10. Repeat steps 4 through 6 for datum C.
11. Select *OK.*
12. Choose *Done/Return.*

Now the model may be oriented to form the top view in the drawing. Use the path *Views, Modify View,* and *Reorient* to accomplish this task. The reoriented view is shown in Figure 18.9, along with additional views. Add the additional projections by using *Views, Add View,* and the *Projection* option. After the additional views have been added, select *Modify* and change the scale to 0.4.

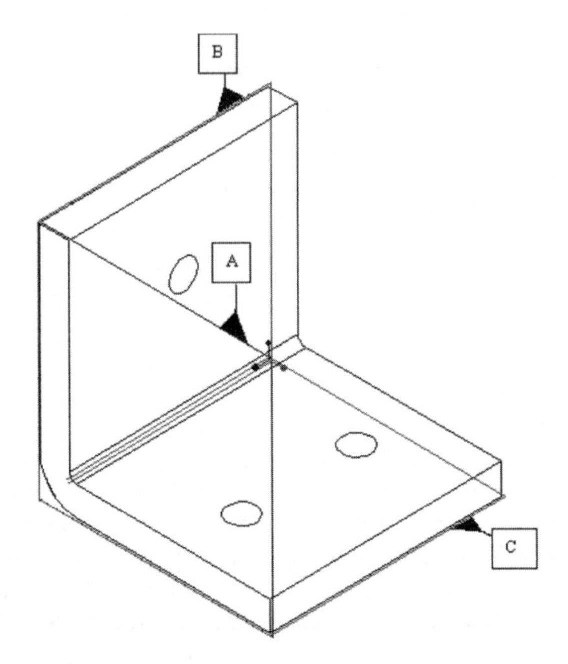


FIGURE 18.8 *The angle bracket with the GDT datums A, B, and C.*

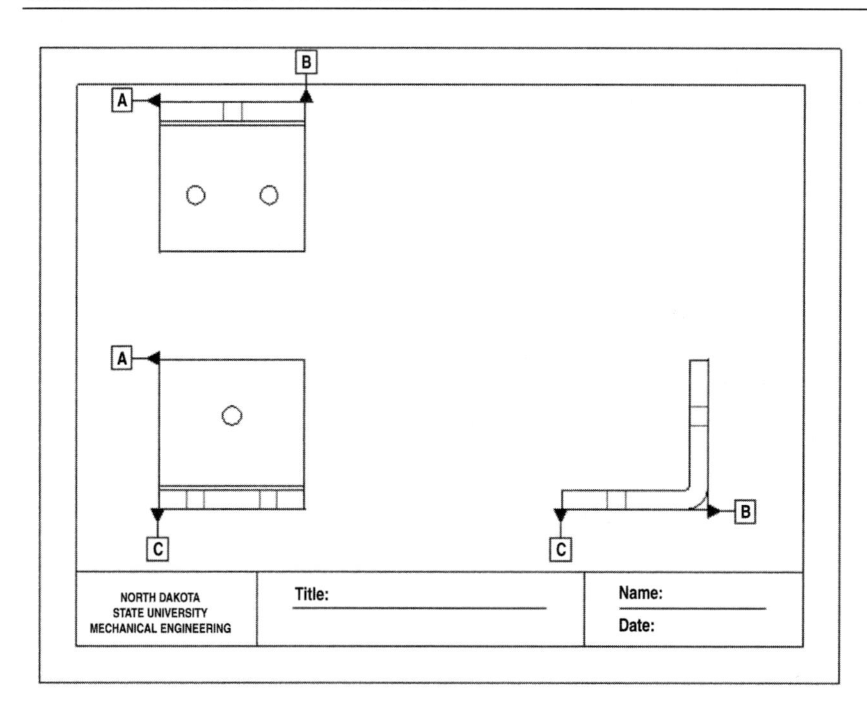


FIGURE 18.9 *Add additional views to the drawing.*

Use *Show/Erase* and turn on the axes for all the holes, but turn off their label. Display the dimensions for the features by using *Show/Erase.* Notice that because we are using geometric tolerances, the dimensions must be referenced to the GDT datum planes. Thus, you may need to erase some of the driven dimensions and create new ones. When creating dimensions from the datum planes to the center of one of the holes, use the crosshair identifying the center as the selection entity.

The labels for the GDT datums may be moved using the *Move* option. Change the arrow type from "*CLOSED*" to "*FILLED*" by selecting *Advanced, Set Up,* and *Modify Val.* Then change the setting to the option *draw_arrow_style.*

At this point, if your drawing appears as it does in Figure 18.10, then you are ready to add the geometric tolerances to the holes.

Based on how this part is to be used, we need to place a position control on the holes. Let us begin by adding a feature control box to the double 0.5 diameter holes in the top view. Select *Create, Geom Tol,* and *Specify Tol* from the *GEOM TOL* menu (Figure 18.2). The *Geometric Tolerance* box shown in Figure 18.6 will appear.

Make sure that the model is set to angle bracket. From the *Reference* field, select *Feature.* Choose *Select Entity* and click on the circle edge with your mouse pointer. Select *Dimension* for the *Placement* option. Activate the *Place Gtol* button and then click on the 0.5 dimension. Pro/E will add the feature control box. Your drawing should appear as in Figure 18.11.

Now let us add the datum reference to the control box. Select the *Datums Refs* folder and choose datum A as the primary datum. Select datum B as the

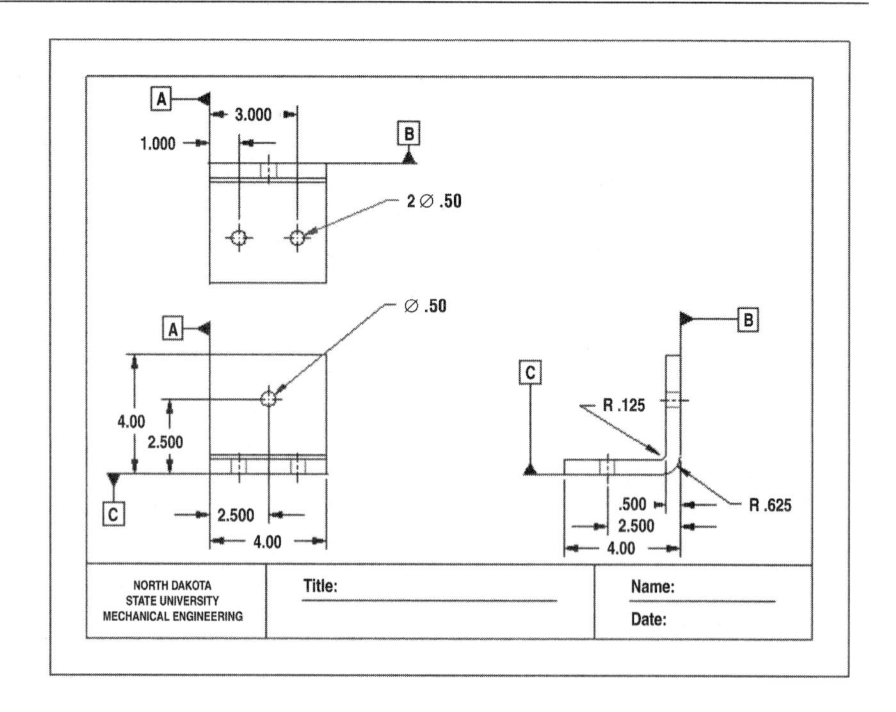


FIGURE 18.10 *Before adding any controls, add these dimensions to your drawing.*

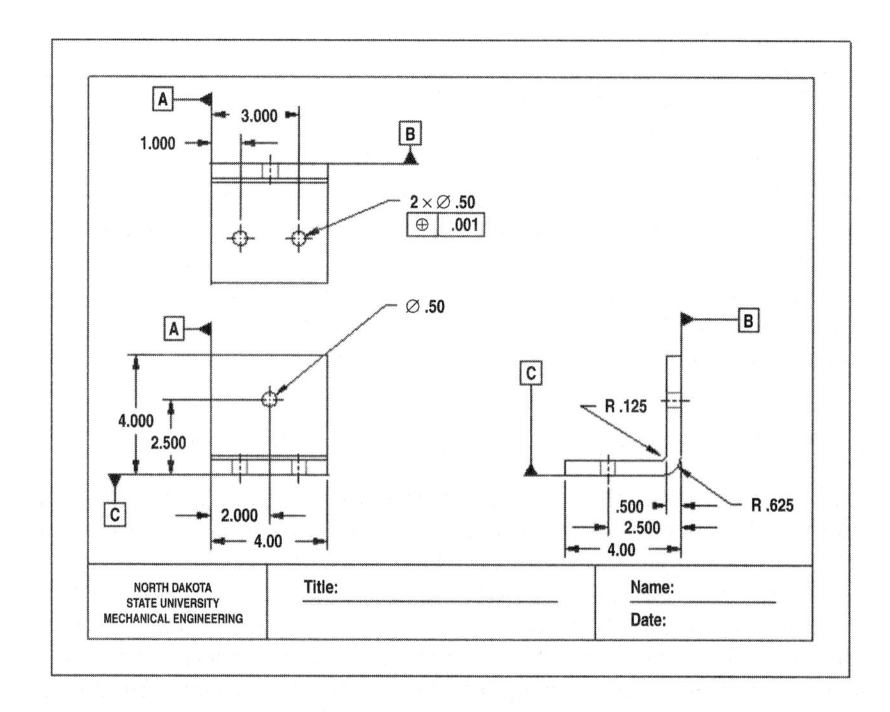


FIGURE 18.11 *A position control is added to the double holes in the top view.*

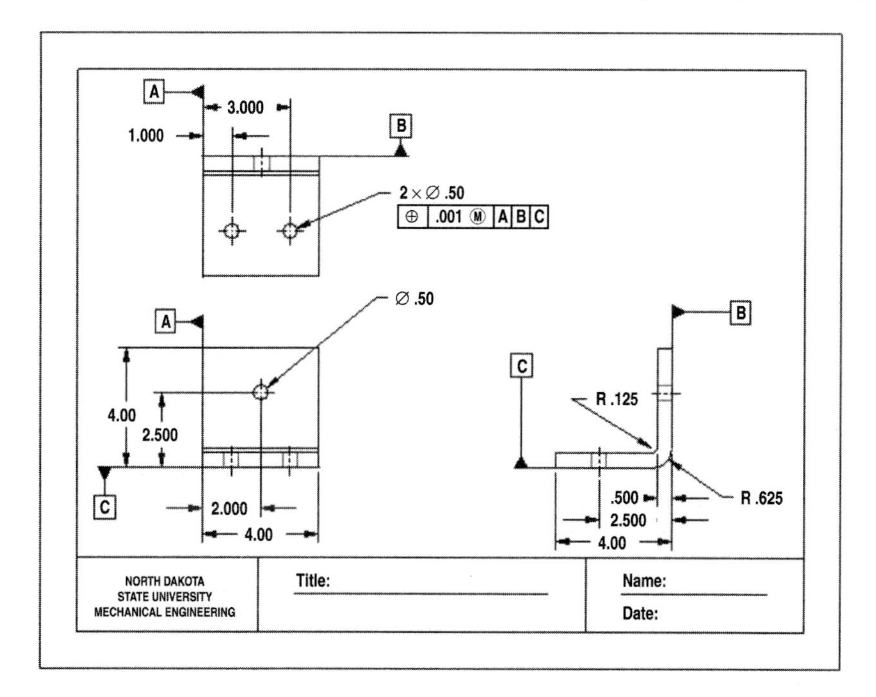


FIGURE 18.12 *The complete feature control box added to the double holes.*

secondary datum and datum C as the tertiary datum. Select the *Tol Value* folder and add maximum material condition (MMC) to the tolerance. Select *OK.* Pro/E will inquire whether it is OK to set the reference to the hole as basic. Answer yes. Your drawing should now appear as in Figure 18.12.

Repeat the process and add a similar feature control box to the 0.5 diameter hole in the front view. The feature should appear as in Figure 18.13.

Change the dimensions locating the holes to basic dimensions, because control on these holes is specified by the geometric tolerance. Change the number of decimal places and set the dimension to basic by using the options *Modify* and *Dimension.* Add the title, scale, and remaining note to your drawing as shown in Figure 18.13. Save the drawing.

## 18.3 SUMMARY AND STEPS FOR ADDING TOLERANCES TO A DRAWING

Because Pro/E assumes default tolerance values during the creation of a part, traditional tolerances may be placed in a drawing by simply turning on their visibility by simply changing the option "*tol_display*" from "*NO*" to "*YES.*"

The user may change the default tolerance values in the *Part* mode or in the *Drawing* mode. In the *Part* mode, the tolerances may be changed by using the options from the *SET UP* menu (Figure 2.9). In the drawing mode, they may be modified by using the options *Modify* and *Dimension.* With the use of the Modify Dimension box, the user may select between ±*Symmetric, Limits,* and *Plus-Minus* as the modes for displaying traditional tolerance values.

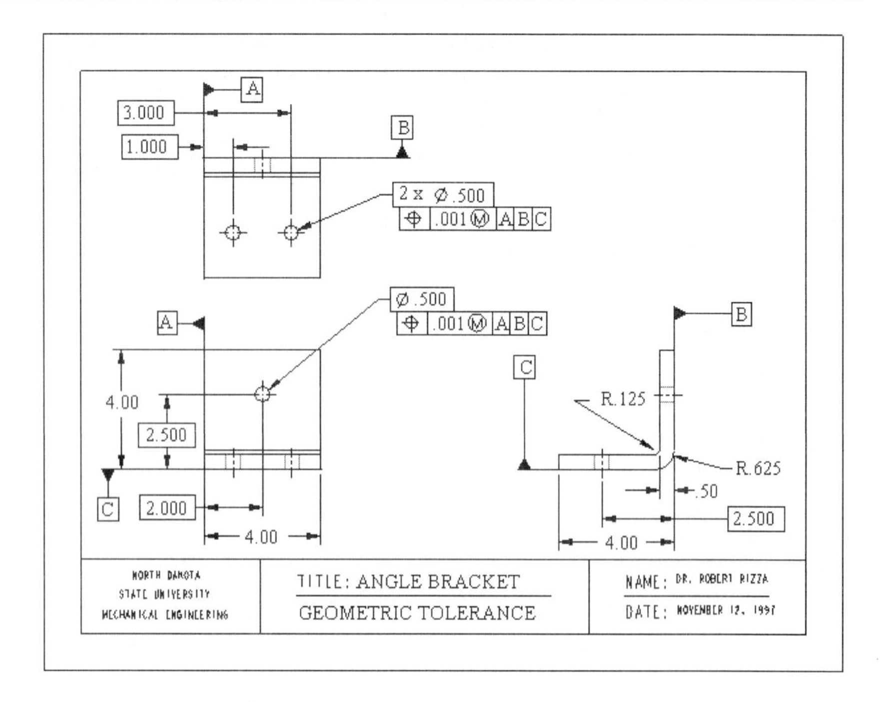


**FIGURE 18.13** *The drawing after adding the appropriate views and dimensions relative to the reference datums.*

Geometric tolerances may be added to a part as well as to a drawing. The user must define GDT datums. In the drawing mode, defining GDT datums is done by using the *Modify* option in the *DETAIL* menu or from *Set Datum* in the *GEOM TOL* menu. After a dimension has been displayed on a drawing, a feature control box may be added. The general procedure for adding a feature control box is as follows:

1. Select *Create* from the *DETAIL* menu.
2. Choose *Geom Tol* and *Specify Tol.* The Geometric Tolerance box will appear.
3. Enter the model or drawing name.
4. Select the *Reference* field, that is, feature, edge, etc. for which the geometric tolerance is to be defined.
5. Depress the *Select Entity* button and pick the appropriate entity.
6. Select the *Placement* option.
7. Depress the *Place Gtol* button and locate the geometric tolerance.
8. Choose the *Datum Refs* folder and select the datum references.
9. Select the *Tol Value* folder and enter the tolerance value. If desired, select a material condition.
10. Choose *OK.*

## 18.4 Additional Exercises

For exercises 18.1 through 18.3, create multiview drawings with traditional tolerances. For the remaining exercises, create suitable drawings with geometric dimensions and tolerances. Save your work unless instructed not to do so.

18.1 Cement trowel (see also exercise 16.1).
18.2 Control handle (see also exercise 16.3).
18.3 External lug (see also exercise 16.4).
18.4 Side guide (see also exercise 16.2).
18.5 Pipe clamp top (see also exercise 16.5).
18.6 Pipe clamp bottom (see also exercise 16.6).
18.7 Side support (see also exercise 16.7).
18.8 Guide plate (see also exercise 16.8).
18.9 Single bearing bracket (see also exercise 16.9).
18.10 Double bearing bracket (see also exercise 16.10).
18.11 Roller (see also exercise 16.11).
18.12 Baseball bat (see also exercise 16.12).

# CHAPTER 19

# ASSEMBLIES AND WORKING DRAWINGS

## INTRODUCTION AND OBJECTIVES

Individual parts can be put together to form an assembly. Therefore, complicated machines and structural components can be created by constructing the individual parts (components), independently and then assembling the parts.

In order for an assembly to be created, Pro/E must have access to the original part files. Thus, assembly construction requires large memory storage facilities. Furthermore, because assemblies are quite large, run time for regeneration of an assembly is considerable.

In Pro/E, assemblies may be created by using one of two approaches. The first approach does not use datum planes. This method is best for assembling parts with planar surfaces. The second approach uses assembly datums that are defined similar to the default datums in the *Part* mode. When using assembly datums, the user must first create the datums and then place the components. In general, it is best to use assembly datums.

Parts may be constructed in the assembly mode. Creating parts in the assembly mode has two advantages. The first advantage is that the part is automatically placed parametrically in the assembly. The second advantage is that part is constructed in a more straightforward manner, because as the part is sketched, the outline of the other components is visible.

After an assembly has been created, the user may explode the assembly. This option is desired to facilitate visualization of how the parts are combined to form the assembly. Furthermore, the user may create working drawings of the assembly containing both the exploded and unexploded states of the assembly.

In this chapter, we are concerned with the following objectives:

1. The user will be able to create an assembly with or without the use of assembly datums.
2. The reader will attain the ability to create exploded views of an assembly.
3. The Pro/E user will be able to construct parts in the assembler.
4. The reader will attain a proficiency in creating working drawings of an assembly.

## 19.1 CREATING AN ASSEMBLY FILE

The assembly of individual parts is accomplished in the *Assembly* mode. As in the creation of drawing and part files, the Pro/E user creates a file. In this case, the file has the extension ".asm." The procedure for creating an assembly file is similar to that used for part and drawing files. By using the options *File* and *New,* or the *Create New Object* icon, the user accesses the *New* dialog box shown in Figure 19.1. The assembly file is created by selecting the assembly option, entering a name and choosing the *OK* button.

After the assembly file is created, the software will load the interface containing the *ASSEMBLY* menu shown in Figure 19.2. Notice the four options: *Feature, Assemble, Create,* and *Package.* These options are used in the following manner:

1. *Feature.* The *Feature* option is used to construct features such as datum planes in the assembly mode. The features are associated with the assembly and are constructed using the same menus and procedures as in the *Part* mode.

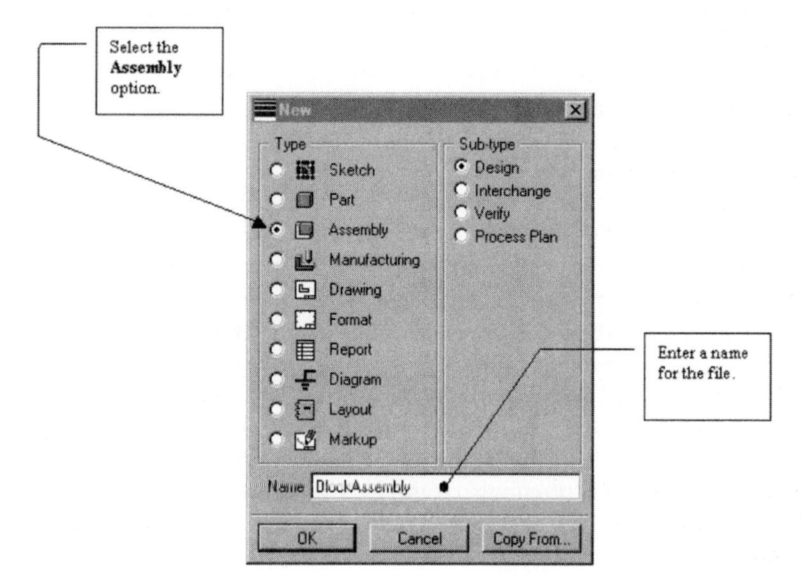


FIGURE 19.1 *An assembly file may be created by using the* New *dialog box.*

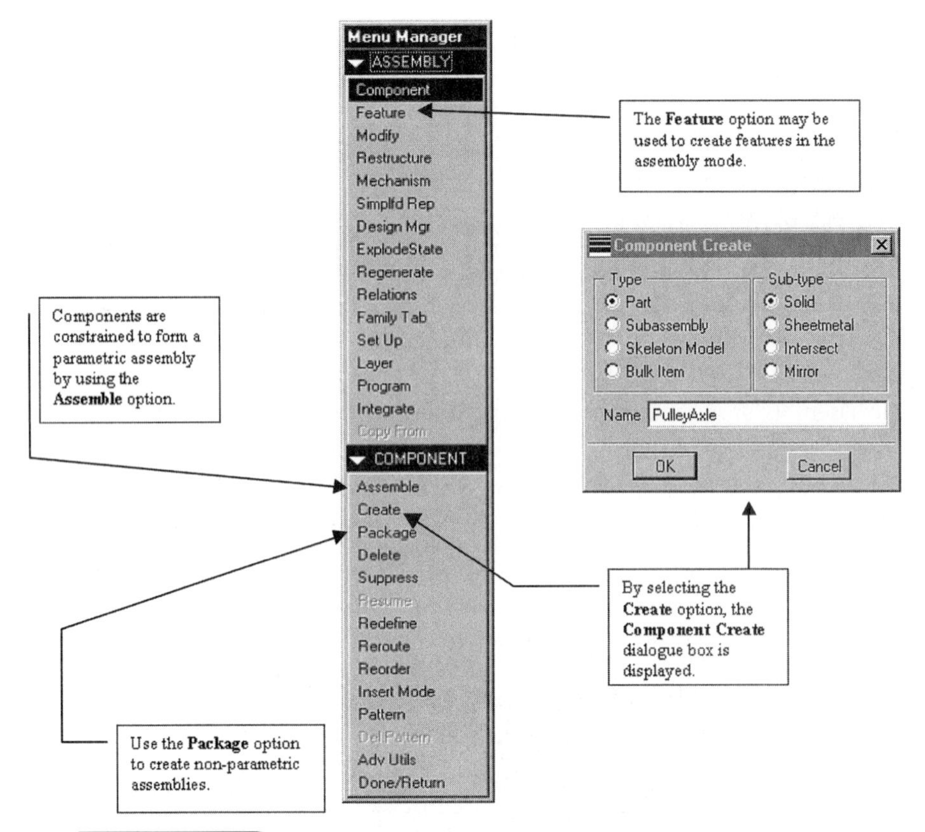


FIGURE 19.2 *The* ASSEMBLY *and* COMPONENT *menus contain options for beginning the assembly procedure.*

2. *Assemble.* This option contains the necessary menus with options for constraining components. The constrained components form a *parametric* assembly with their positions specified relative to a base feature.
3. *Create.* The *Create* option is used to construct parts or subassemblies within the assembly mode. The *Component Create* dialog box shown in Figure 19.2 allows the user to choose the desired component to be created.
4. *Package.* By using the *Package* option, components may be placed in an absolute sense; that is, the components are not placed relative to a base part. The components may be repositioned by translating or rotating the component around an axis using the mouse pointer. The *Package* option is best used when the exact placement of the components is unknown or the user does not want one or all of the components to be located with respect to another component. Components may be reassembled parametrically after

they have been placed with the *Package* option by using the *Assembly* option.

## 19.2 CREATING PARAMETRIC ASSEMBLIES

In creating parametric assemblies, the components are positioned with respect to each other. This action can be accomplished by utilizing one or more of the *Constraint Type* options in the *Place* folder illustrated in Figure 19.3. In all, 10 options are available for the placement of the components. These options are *Mate, Mate Offset, Align, Align Offset, Insert, Orient, Coordinate System, Tangent, Edge on Surface,* and *Point on Surface.* Usually more than one of these options is needed to fully constrain the component. The first four of these options are probably the most useful. The options are defined as follows:

1. The *Mate* and *Mate Offset* options are virtually the same. By selecting *Mate* or *Mate Offset,* the user is requiring that the selected planes *face each other and be coplanar.* The *Offset* option allows an offset distance to be placed between the two surfaces.

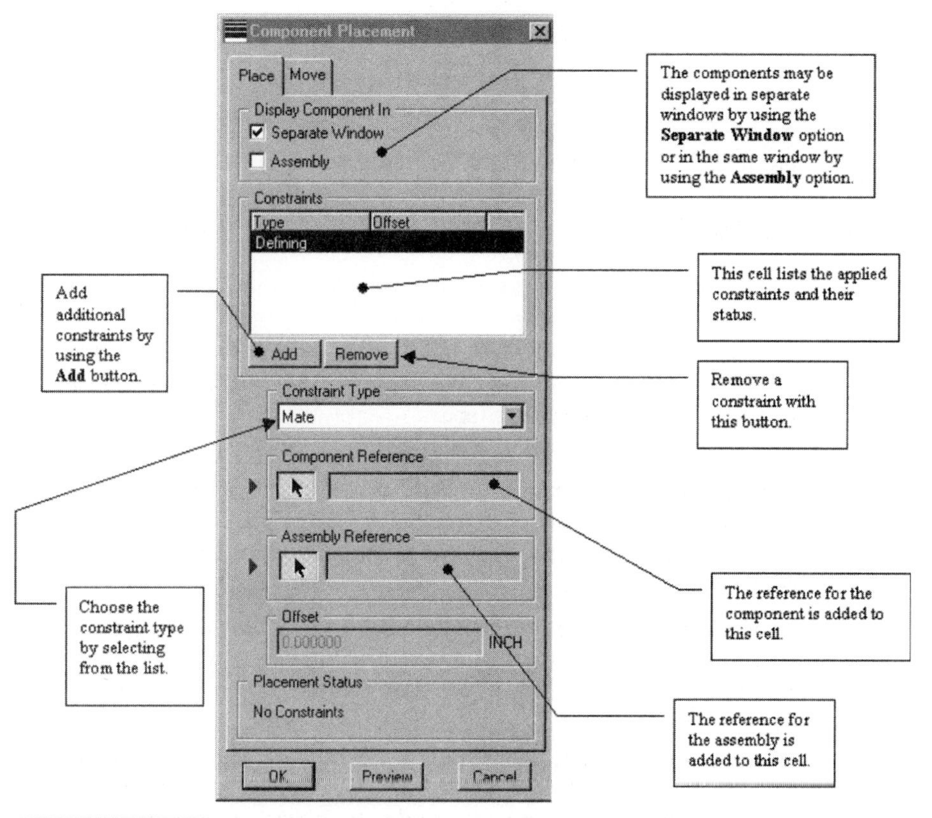


**FIGURE 19.3** *The options contained in the* Place *folder are used to fully constrain components for placement in parametric assemblies.*

2. The *Align* and *Align Offset* options require that the chosen surfaces *face the same direction and are parallel.* Again, the *Offset* option allows the user to place an offset distance between the two faces.
3. The *Insert* option can be used to place a component "inside" of another component. For example, a model of a bolt can be placed inside a hole using the *Insert* option.
4. The *Orient* option allows the user to place two planar surfaces in a parallel direction so that they are facing the same direction.
5. When two or more models contain coordinate systems, the components can be assembled by aligning the coordinate systems of one model to another by using the *Coord Sys* option. The coordinate systems are chosen by selecting with the mouse, or by selecting the names of the systems from a list.
6. The *Tangent* option is used to place two surfaces in contact. This option is similar to the *Mate* option; however, alignment does not take place.
7. The last placement option, the *Edge on Srf* option permits control of an edge with a surface. Thus, the linear edge of a feature can be placed in contact with a defined surface.
8. By using the *Pnt on Srf* option, a point may be placed in contact with a given surface. This option can also be used to control the placement of a surface so that it is at the level defined by a point.

It is worthwhile to make a few points concerning these options.

1. When using the *Offset* options, Pro/E identifies the direction of the offset with an arrow. If the opposite direction is desired, then a negative value should be entered.
2. The *Mate* and *Align* options require that the two references be of the same type. For example, if the first reference is a plane, then the second reference chosen must also be a plane. If the sections cannot be the same, consider one of the last options, *Edge on Srf* or *Orient.*
3. Because datum planes are two-sided, the side of the datum plane must be provided during the constraining process. The sides of the datum planes are indicated by yellow and red.
4. Complex assemblies should be created using datum planes. As noted in the previous point, datum planes are two-sided and hence the desired side must be indicated. During the assembly process, Pro/E indicates the yellow side with an arrow.

### 19.2.1 TUTORIAL # 83: ASSEMBLING BY USING PLACE OPTIONS

In order to get a "feel" for assembling with the options from the *PLACE* menu, we have created two parts and bundled the files with this book. These parts are called "AssembleBlockA.prt" and "AssembleBlockB.prt." Make sure that these parts are available for the software to retrieve.

Load Pro/E and create an assembly file called "BlockAssembly." Select *Component* from the *ASSEMBLY* menu (Figure 19.2). Then choose *Assemble* from the *COMPONENT* menu. Scroll down the list of available parts until you find the part "AssembleBlockA." Select the part and hit the *Open* button.

Pro/E will load the first part: "AssembleBlockA." Now let's add the second part. Choose *Assemble* again and find the second part: "AssembleBlockB."

Often it is better to display the assembly and component in different windows, because the assembly window may become too cluttered. Thus, select the *Separate Window* option in the *Component Placement* dialog box. Your interface should appear similar to the one shown in Figure 19.4.

We want to locate the parts in an assembly by mating Surfaces 6 and C, mating Surface 8 and Surface B, and aligning Surfaces 1 and A.

Choose *Mate* from the *Constraint Type* cell. Select the arrow button next to the *Component Reference* cell. Using *Query Sel,* select Surface 6. Then

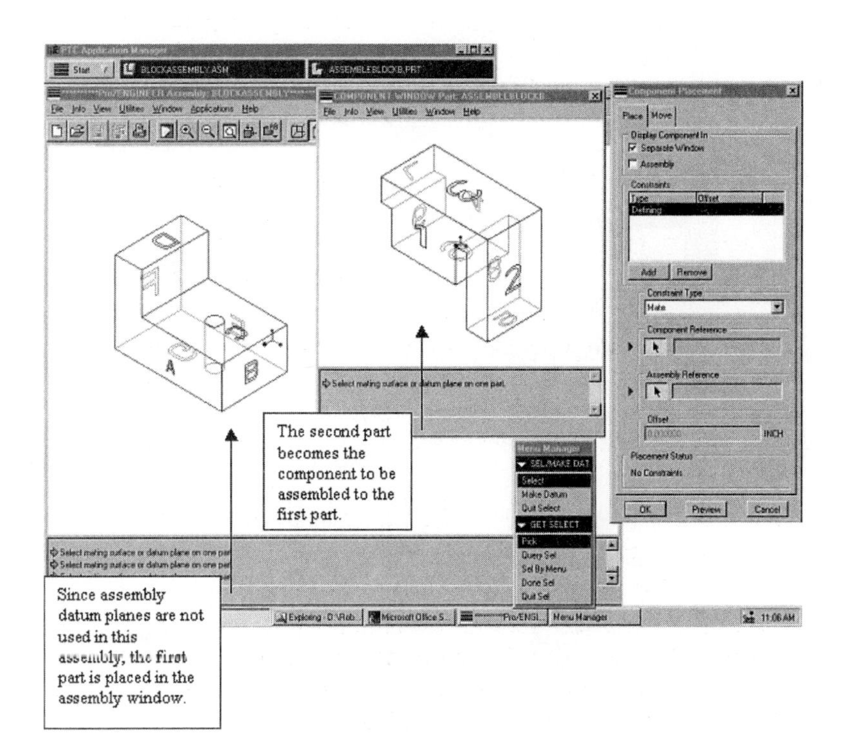


FIGURE 19.4 *The screen contains the windows for "AssembleBlockA," "AssembleBlockB," and the* Component Placement *dialog box.*

depress the arrow button for the *Assembly Reference* cell and using *Query Sel,* click on Surface C.

Notice that the *Component Placement* dialog box has a new entry in it. The box is reproduced in Figure 19.5. Notice that the status of the assembly is such that it is not fully constrained.

Now mate Surfaces 8 and B. Choose the *Mate* option. Again, choose the arrow button next to the *Component Reference* cell. Using *Query Sel,* select Surface 8. Then depress the arrow button for the *Assembly Reference* cell and using *Query Sel,* click on Surface B. The assembly is still not fully constrained.

Finally, align Surfaces 1 and A using the *Align* option. After this constraint is given, the assembly will be fully constrained. Pro/E will notify the user that the component may be placed.

Show the placement by clicking on the *Preview* button. Your assembly should appear similar to the one shown in Figure 19.6. Do the orientations of the blocks make sense? Take a few minutes and reconsider the constraints used in creating the assembly. Then, click on *OK.* Erase the assembly using the options *File* and *Erase.* In the next tutorial, we will construct a much more complicated assembly that will require the use of assembly datum planes.

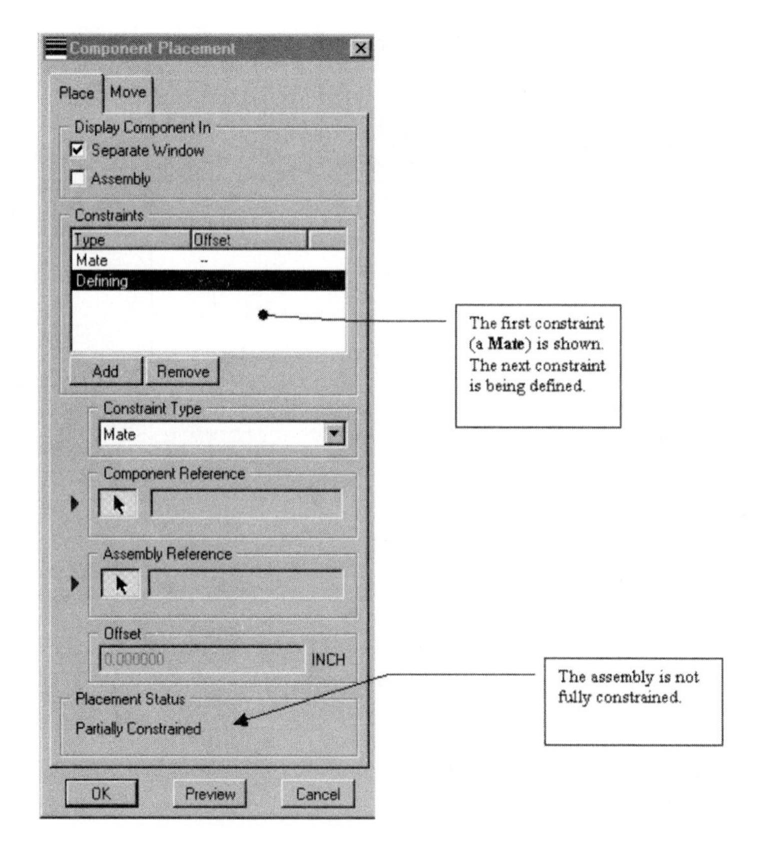


**FIGURE 19.5** *An entry has been made in the* Component Placement *dialog box after mating Surfaces 6 and C. The assembly is still not fully constrained.*

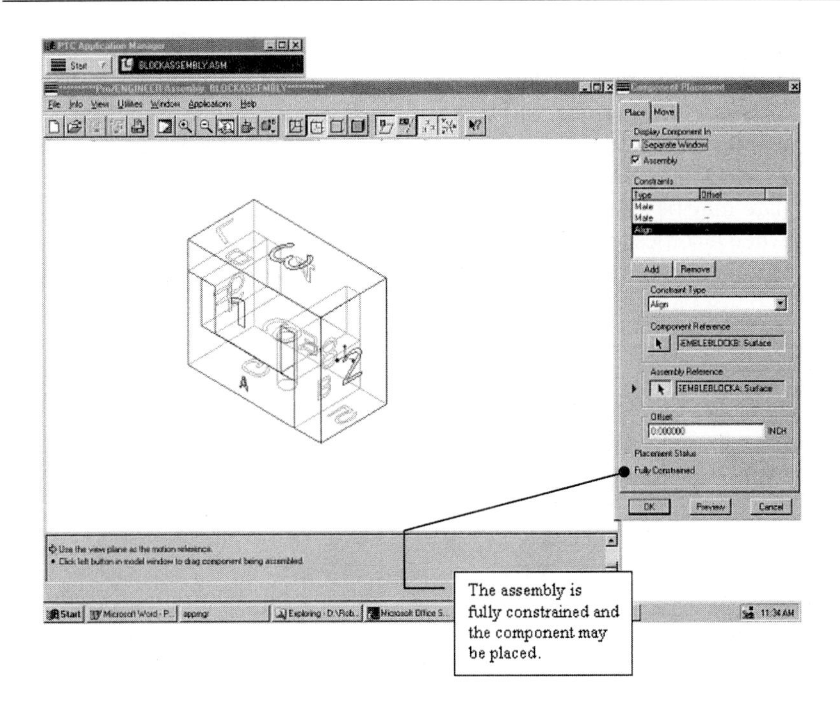


FIGURE 19.6 *The assembled blocks.*

## 19.3 ASSEMBLIES USING DATUM PLANES

The blocks used in the previous section were created to illustrate basic assembling techniques. The blocks are relatively simple to assemble because they have planar orthogonal surfaces. However, most assemblies are complicated with multiple parts. Datum planes make the assembly process easier for complicated parts.

In creating an assembly with datum planes, the set of datum planes must be created first. Then, the parts can be assembled relative to the datum planes or to each other.

The datums are created by using the *Feature* option shown in Figure 19.2. The names of datums created in the assembly mode are preceded with the letter "A" in order to indicate that they are assembly datums.

### 19.3.1 TUTORIAL # 84: AN ASSEMBLY USING DATUM PLANES

The assembly to be created in this tutorial makes use of the several parts constructed in previous chapters. These parts are "Plate_With_Sketched_Hole" (chapters 4 and 6), "AngleBracket" (chapters 2, 4, and 8), and "Pulley_Wheel" (chapter 10). In addition, we have:

1. "Half_Inch_1_25Long_Bolt"
2. "Half_Inch_Washer"
3. "Half_Inch_Hex_Nut"
4. "Wheel_Bushing"

These fasteners may be found on the disk included with this book. You will need to make sure that you have access to these parts. An additional part will be created in the assembly mode.

We approach the creation of this assembly in several phases. The first phase is to assemble the plate, angle bracket, and pulley wheel parts. The angle bracket and plate may be assembled without the use of assembly datum planes. However, when it comes to the pulley wheel, it is advantageous to use datum planes.

Create an assembly called "PulleyAssembly." The datum planes must be placed before any other component is added. The procedure to follow is:

1. Select *Feature* and *Create.*
2. Choose *Datum, Plane, Default,* and *Done.*

The software will add the assembly datum planes. Note that the names of these datums all begin with the letter "A."

The plate is the first component to be placed in the assembly. We wish to take advantage of the symmetry in the geometry of the plate. Thus, we will endeavor to place the plate equally spaced with respect to the datum planes. In the end, this placement will make it easier in locating the remaining components.

Select *Component* and *Assemble.* Scroll down the list and find the part "Plate_With_Sketched_Hole." Choose the *Separate Window* option in the *Component Placement* dialog box. Looking at the model of the plate, we note that the part was constructed so that DTM2 is in contact with the "bottom" of the plane. In addition, the datums DTM1 and DTM3 were placed symmetrically with respect to the plate. These examples use symmetry when constructing a model and make the assembly process much easier.

First, we mate datum ADTM2 and DTM2. Select the *Mate* option. Then depress the arrow button for the *Component Reference.* Choose DTM2. An arrow will appear indicating the yellow side of DTM2 as shown in Figure 19.7. Select the *red* button.

Choose the arrow button for the *Assembly Reference.* Then pick datum ADTM2. Again an arrow will appear (see Figure 19.7), indicating the yellow side of datum ADTM2. Because the *Mate* option requires that the two surfaces face each other, we must select the yellow side. Click on the yellow button.

Again, because of the symmetry, we may *Mate* ADTM3 with DTM3. The procedure is the same as the one used for mating ADTM2 and DTM2 and is as follows:

1. Select *Mate.*
2. Pick the arrow button for the *Component Reference.*
3. Select DTM3.
4. Choose the yellow side. Note the direction in Figure 19.8.
5. Pick the arrow button for the *Assembly Reference.*
6. Select ADTM3.
7. Choose the red side. The direction is shown in Figure 19.8.

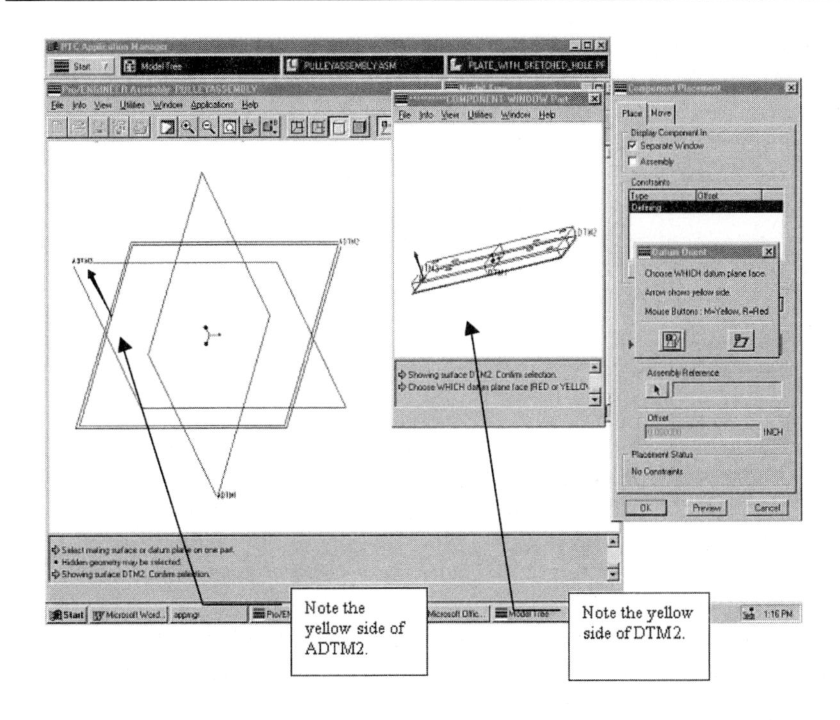


**FIGURE 19.7** *Because datum planes have two sides, an arrow is used to indicate the* yellow *side of the datum. The opposite side is the* red *side.*

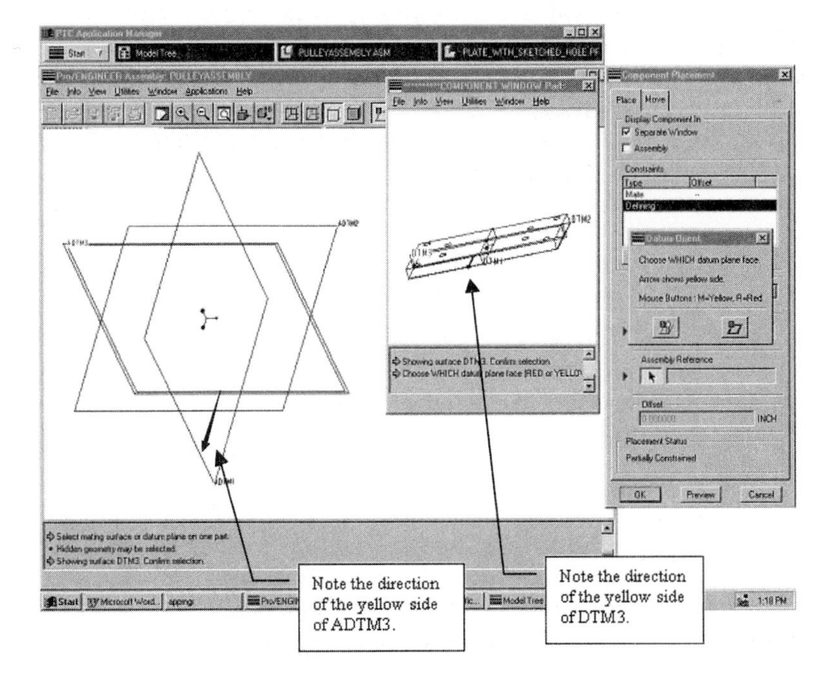


**FIGURE 19.8** *The yellow sides of the datum DTM3 and ADTM3 are shown.*

The component is not fully constrained. However, if we mate or align DTM1 and ADTM1 (taking advantage of the symmetry), the plate may be placed. The sequence of steps for doing so is as follows:

1. Select *Mate.*
2. Pick the arrow button for the *Component Reference.*
3. Choose datum DTM1.
4. Choose the red side. Note the direction in Figure 19.9.
5. Pick the arrow button for the *Assembly Reference.*
6. Select ADTM1.
7. Choose the yellow side using Figure 19.9.

The component can now be placed. Select *OK* from the *Component Placement* box. Your model should appear as in Figure 19.10.

Of course, other ways can be used to constrain the plate. This approach used the symmetry of the models and the location of the datum planes. Notice that the three controls were chosen to constrain the model along the three isometric directions, as illustrated in Figure 19.10.

The two angle brackets are placed next because their location may be used to locate the pulley wheel. Unfortunately, the default datum planes were not used in the construction of the angle bracket. Although, GDT datum planes exist on the model, because of their location, they are not useful

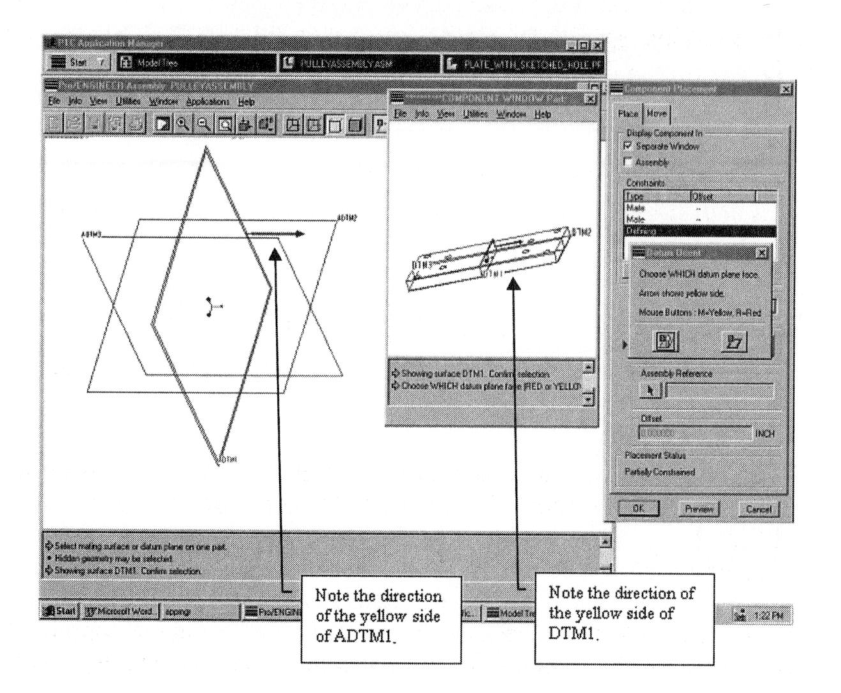


**FIGURE 19.9** *The figure shows the yellow directions of the datum planes DTM1 and ADTM1.*

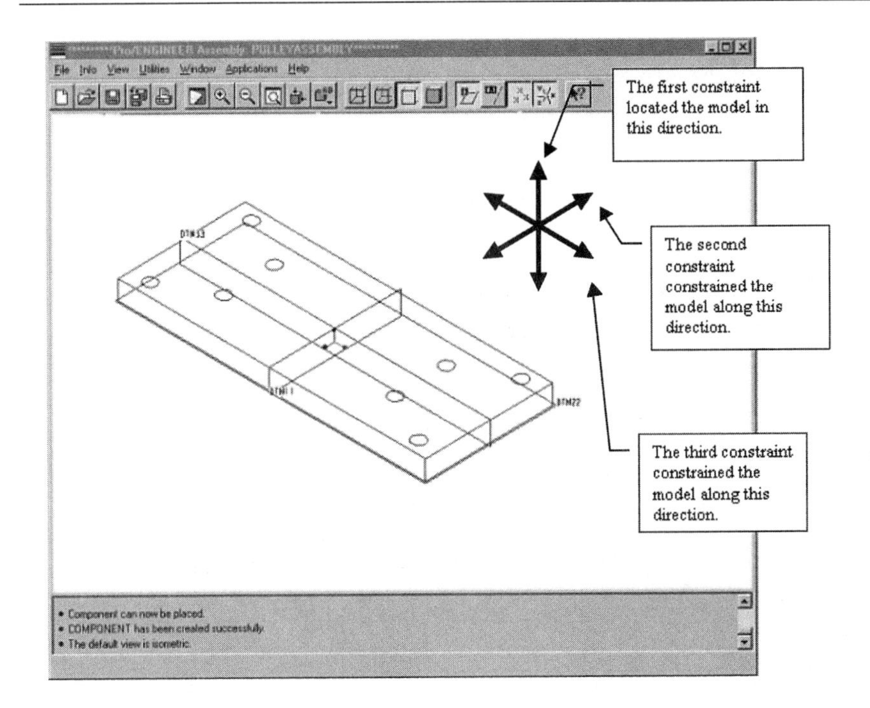


FIGURE 19.10 *The plate is placed in the assembly. Note the locations of the datum planes.*

for assembling the component. Their location makes the constraining of this component a bit more difficult. Luckily, the double holes in the angle bracket must line up with the double holes in the plate to help to constrain the part.

In order to add the bracket to the assembly, follow this procedure:

1. Select *Component* and then *Assemble.*
2. Scroll through the list of available parts and choose "Angle-Bracket."
3. Choose *Mate.*
4. Pick the two surfaces shown in Figure 19.11.
5. Select *Align* and then the two axes A_2 and A_5 shown in Figure 19.12.

Because of the orientation and location of the two axes, the bracket will be fully constrained. Choose *OK.* The assembly at this point is shown in Figure 19.13.

A second bracket must be added to the assembly. The second bracket is not quite as easy to place as the first. Aligning the appropriate axes will not achieve the result, because the bracket must face in the opposite direction. Because the direction must be specified, two alignments must be

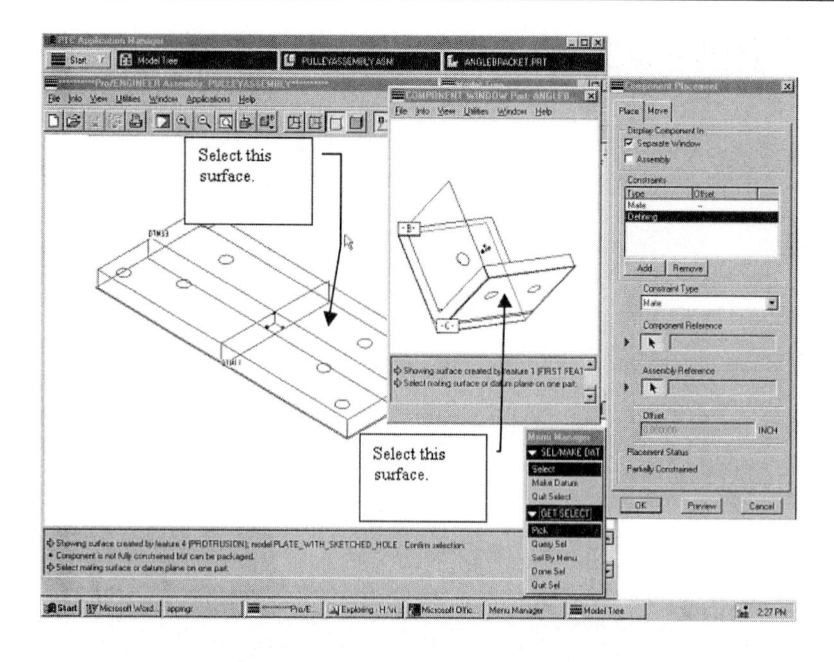


FIGURE 19.11 *Pick the shown surfaces when using the* Mate *option.*

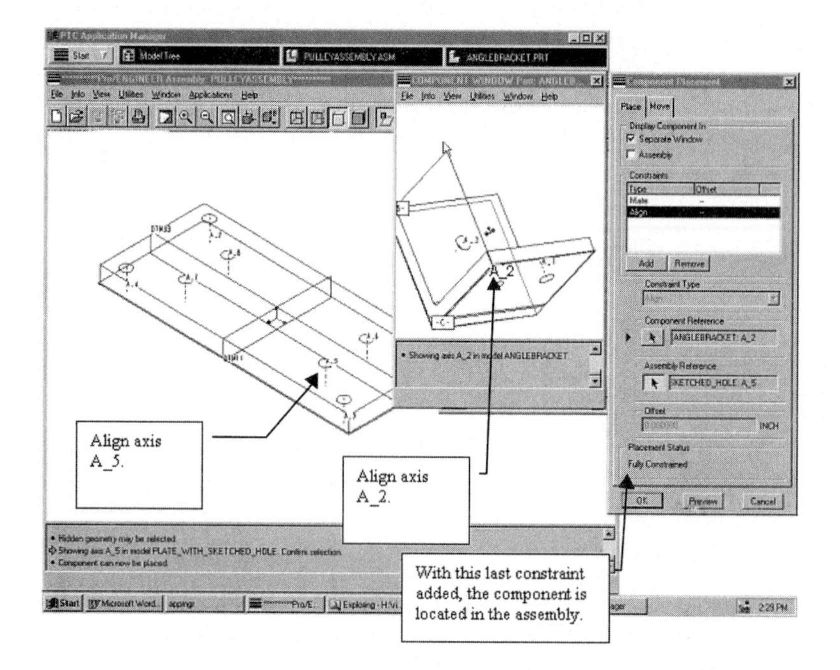


FIGURE 19.12 *Select the axes A_2 and A_5 to constrain the model.*

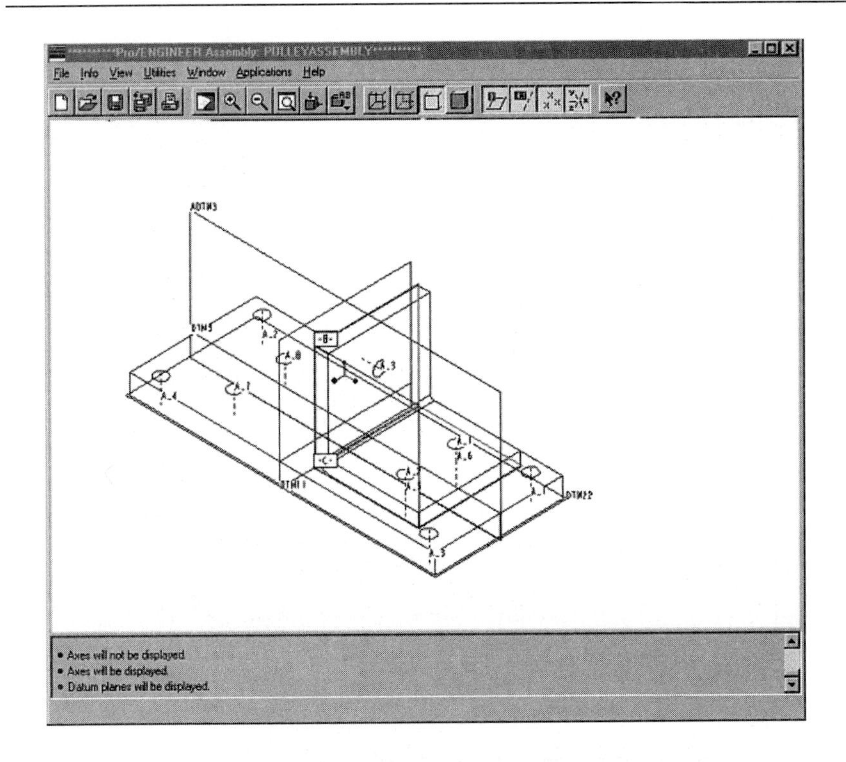


**FIGURE 19.13** *The assembly after placing the first bracket.*

used. The second bracket may be added to the assembly by using the following procedure:

1. Select *Component* and *Assemble.*
2. Scroll through the list of available parts and retrieve "Angle-Bracket."
3. Choose *Align.*
4. Select axis A_2 and axis A_8 as shown in Figure 19.14.
5. Again, choose *Align.*
6. Pick axis A_1 and A_7 as shown in Figure 19.14.
7. These selections will constrain the component along two of the three directions. In order to constrain the model along the third direction, select *Mate.*
8. Choose the surfaces shown in Figure 19.15.
9. The model will be fully constrained, so select *OK*. The assembly at this point is shown in Figure 19.16.

With the brackets in place, we can now add the pulley wheel. The wheel is constrained by aligning the center of the hub with the single hole on the brackets. Because default datum planes were used in the construction of

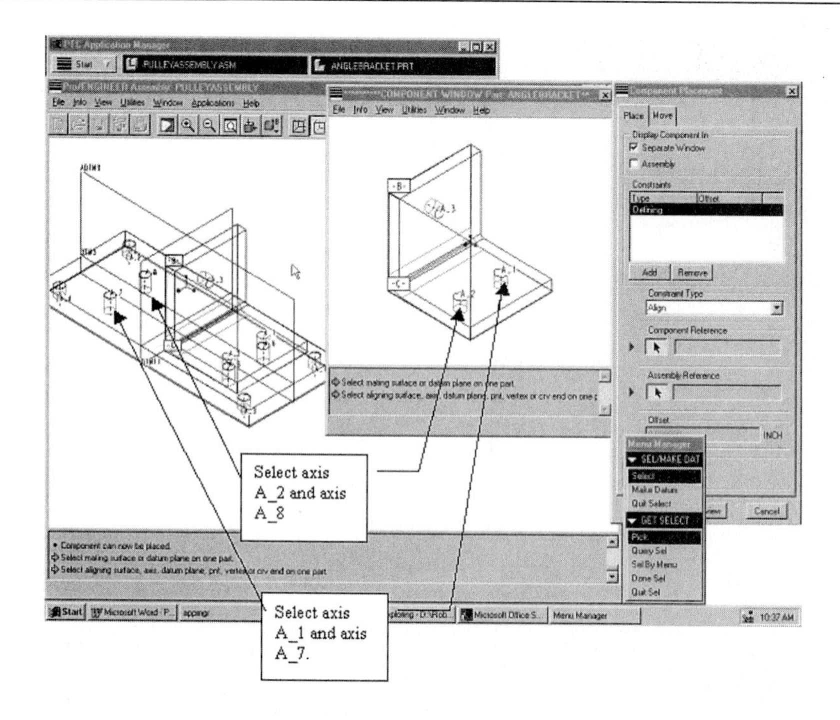


**FIGURE 19.14** *Use this figure to select the proper axes.*

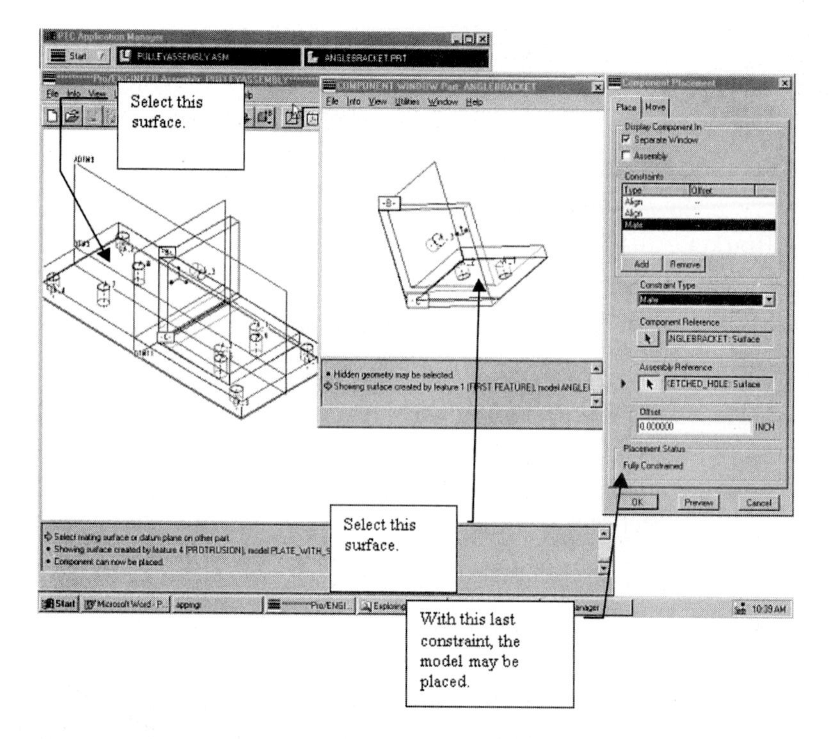


**FIGURE 19.15** Mate *the given surfaces.*

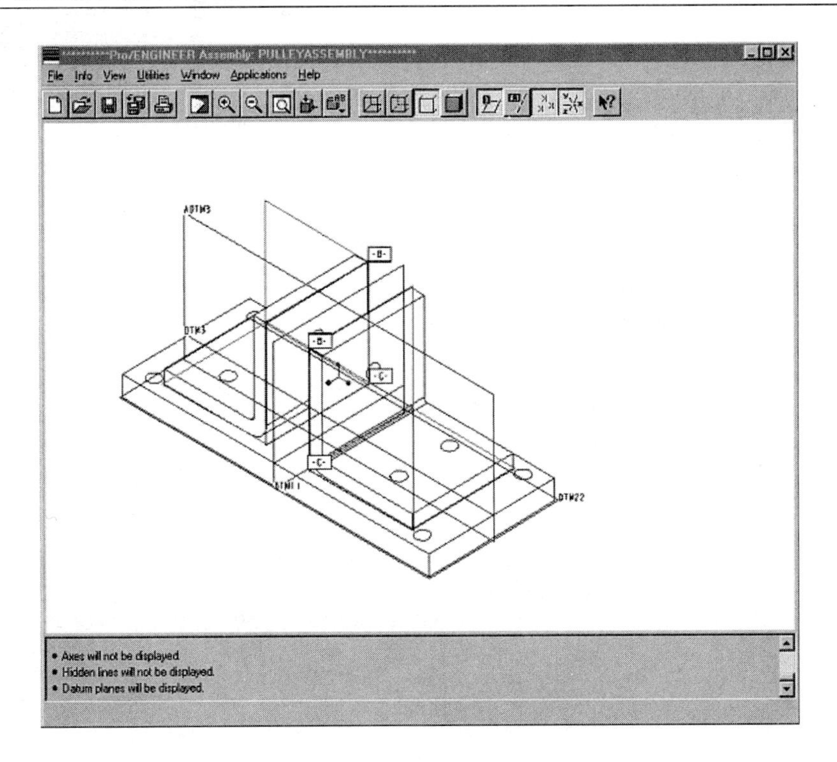


FIGURE 19.16 *The assembly after the addition of the second angle bracket.*

the model, and planes were placed symmetrically with respect to the model, the rest of the constraints follow easily. Depending on which sides of the planes are selected, we can either *Mate* or *Align* the planes. The *Align* option will be used. Add the pulley wheel using the following sequence of steps:

1. Choose *Component* and *Assemble.*
2. From the list select "Pulley_Wheel."
3. Choose *Align.*
4. Select DTM1 and the *yellow* direction as shown in Figure 19.17.
5. Choose ADTM1 and the *yellow* direction as shown in Figure 19.17.
6. Again, choose *Align.*
7. Pick DTM3 and the *yellow* direction shown in Figure 19.18.
8. Select ADTM3 and the *yellow* direction as shown in Figure 19.18.
9. Choose *Align.*
10. Using Figure 19.19, pick axis A_1 and axis A_3.
11. Choose *OK.* The assembly with the wheel added is shown in Figure 19.20.

As you can see, the placement of the models in this tutorial required, for the most part, the use of datum planes. The use of datum planes made the assembly process relatively simple. The difficulty arose in selecting the proper

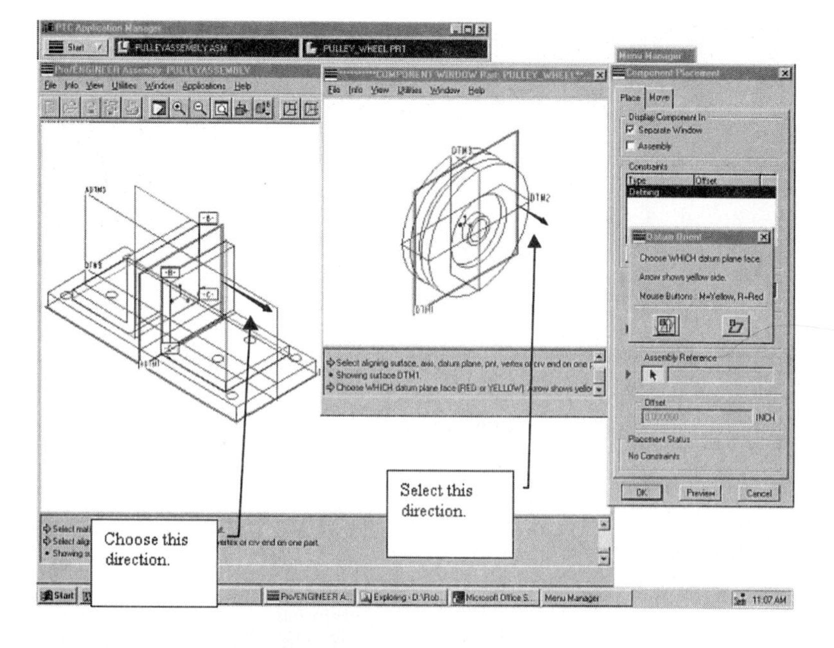


FIGURE 19.17 Align *the* yellow *sides of the planes DTM1 and ADTM1.*

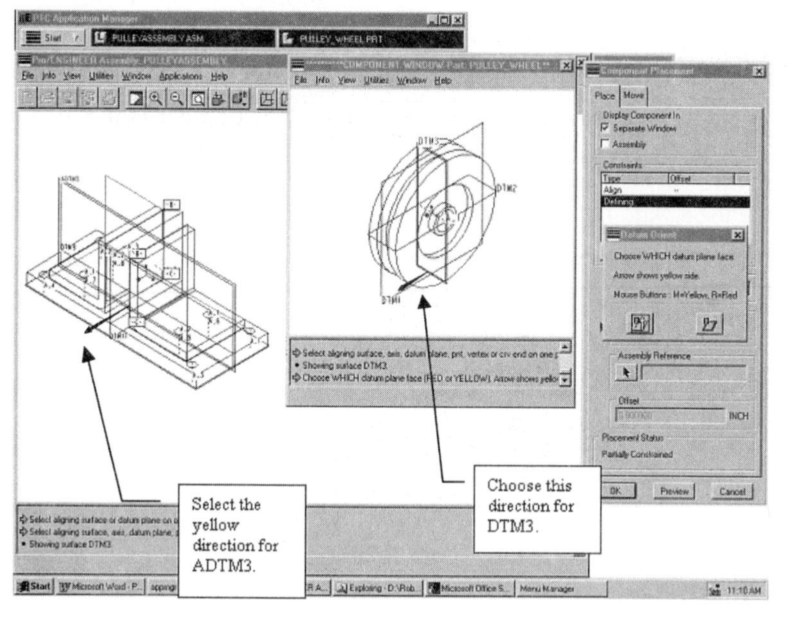


FIGURE 19.18 Align *the* yellow *sides of the planes DTM3 and ADTM3.*

side of the datum plane. The selection of the proper side is easy to determine if one remembers the difference between the *Align* and *Mate* option. The *Mate* option requires that the two surfaces face each other, whereas the *Align* option requires that the two surfaces face the same direction. Thus, when deciding which side of the datum to pick, the user should be asking: "Which sides face each other and which sides face the same direction?"

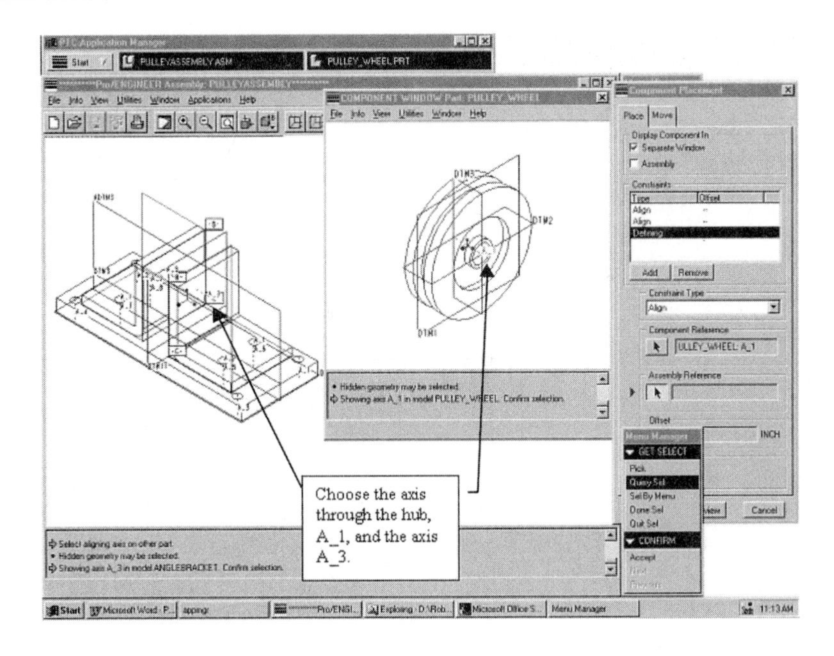


FIGURE 19.19 *Align the axes A_1 and A_3.*

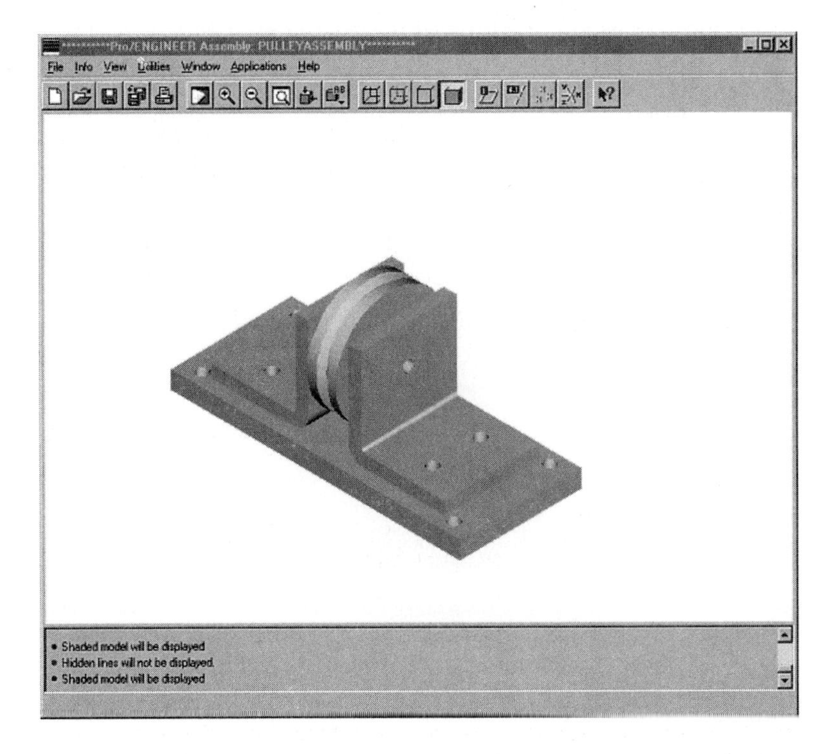


FIGURE 19.20 *With the pulley wheel fully constrained, the model may be added to the assembly.*

## 19.4 CREATING PARTS IN THE ASSEMBLER

Parts created in the assembler may be treated as parts constructed in the part mode. Creating a part in the Assembly mode has the advantage that the assembly may be used as a reference for the part. Thus, the size and position of the part with respect to the assembly is readily visible. This procedure saves time in the end.

Model construction in the assembler proceeds along a sequence of three steps. First, a base feature is created and assembled. The easiest approach is to load a part containing default datums (this you may recall, is a Start Part). Secondly, the datum planes are assembled. Lastly, using the *Modify* option, additional features are added to the assembled part.

Figure 19.21 shows the *Creation Options* box. This box contains options for creating the base feature. These options may be used as follows:

1. *Copy From Existing.* Load and assemble an existing Start Part. The assembly of the component may be redefined. Another advantage of using this option is that if the same Start Part is used for all the parts in the assembly, then all the parts including the one created in the assembler will have the same standards.
2. *Locate Default Datums.* Create the component datums and assemble. The assembly of the component may be redefined; however, the component may not have the same standards as the rest of the parts.
3. *Empty.* Construct the component without any initial geometry. Geometry may be added from existing parts using the *Copy From* option.
4. *Create First Feature.* Using existing assembly references, create the geometry. The assembly of the component cannot be redefined. However, the feature may be modified.

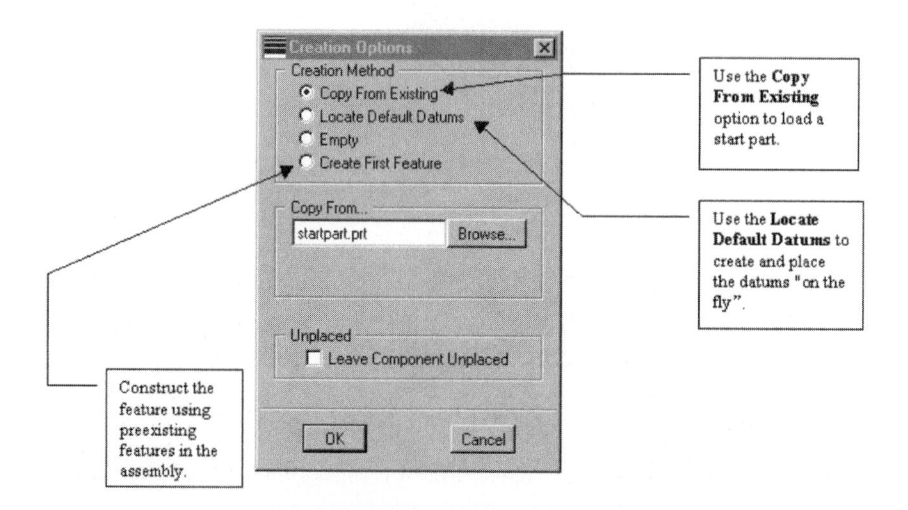


FIGURE 19.21 *The* Creation Options *box is used to initialize the creation of a component in the assembler.*

### 19.4.1 TUTORIAL # 85: AN AXLE FOR THE PULLEY ASSEMBLY

Let us create an axle for the pulley assembly in the assembler. The *Copy From Existing* option is the best way to create such parts. However, for the parts in this assembly, the default standards were used, so it really doesn't matter if a Start Part is used. Furthermore, we may use ADTM1, ADTM2, and ADTM3 to create the part. The software will automatically constrain the component to these datums. If we use the option *Copy From Existing,* we would have to constrain the Start Part to these datums. Therefore, in this case, not much time is saved in using a Start Part.

Select *Component* and *Create.* Choose the *Part* and *Solid* options from the *Component Create* box (Figure 19.2). Enter the name "PulleyAxle." Then select *Locate Default Datum, Three Planes,* and *OK.*

The axle is constructed from two protrusions. For the first protrusion, follow this sequence:

1. Select *Protrusion, Extrude, Solid,* and *Done.*
2. Choose *Both Sides* and *Done.*
3. In the *Sketcher,* draw a circle that is aligned to the edges of the datum planes ADTM3 and DTM2. See Figure 19.22 for the geometry.
4. *Dimension* and *Modify* to a diameter of 0.75 inches.
5. Extrude the geometry using the *Blind* option to a depth of 1.700, the distance between the angle brackets.
6. Select *Done/Return.*

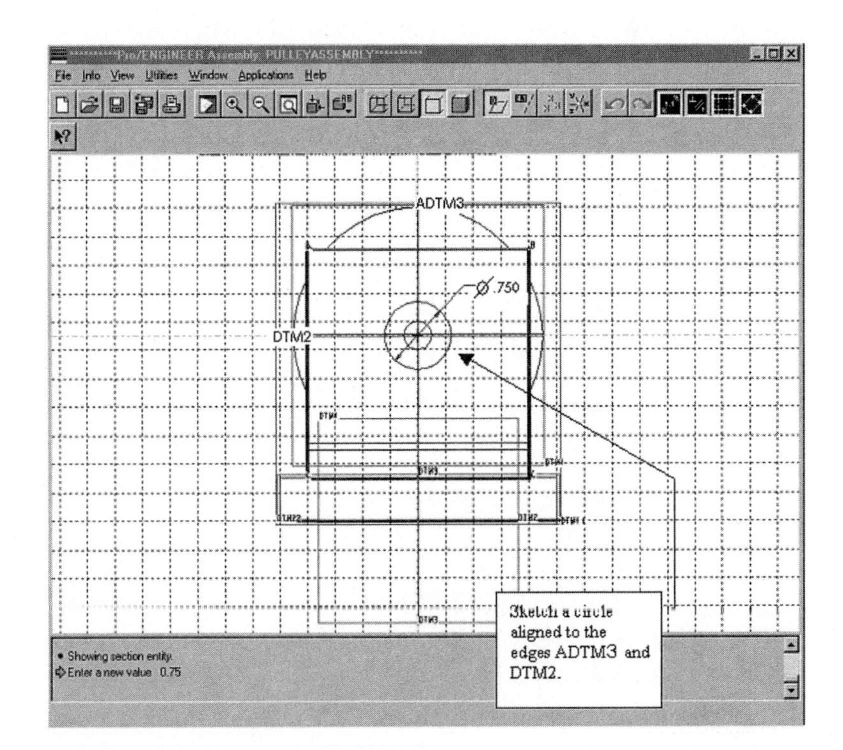


FIGURE 19.22 *The cross section of the axle is a circular section.*

You may recall that the size of the hole in the hub of the pulley wheel is 0.75. Therefore, a clearance fit must be specified on the hole and the diameter of the axle. An RC5 class fit, for example, would have limits of 1.6 and 3.6 thousandths of an inch. Certainly, the axle and the hole can be modified to have these limits.

Additional features may be added to the component by using the *Modify* option. The procedure is as follows:

1. Select *Modify* and *Mod Part.*
2. Using your mouse pointer, select the axle component.
3. Choose *Feature* and *Create.*
4. Select *Protrusion, Extrude, Solid,* and *Done.*
5. Choose *Both Sides* and *Done.*
6. Select ADTM1 (or DTM1 on the component) as the sketching plane.
7. Choose ADTM2 (or DTM2 on the component) as "*Top.*"
8. In the *Sketcher,* draw a circle. Align the center to the circle to the ADTM3 and DTM2 edges as shown in Figure 19.23.
9. *Dimension* and *Modify* the circle so that it has a value of 0.5.
10. Use the *Blind* option and extrude the feature to a depth of 4.00.

The axle is shown in Figure 19.24. We have enlarged the area around the axle.

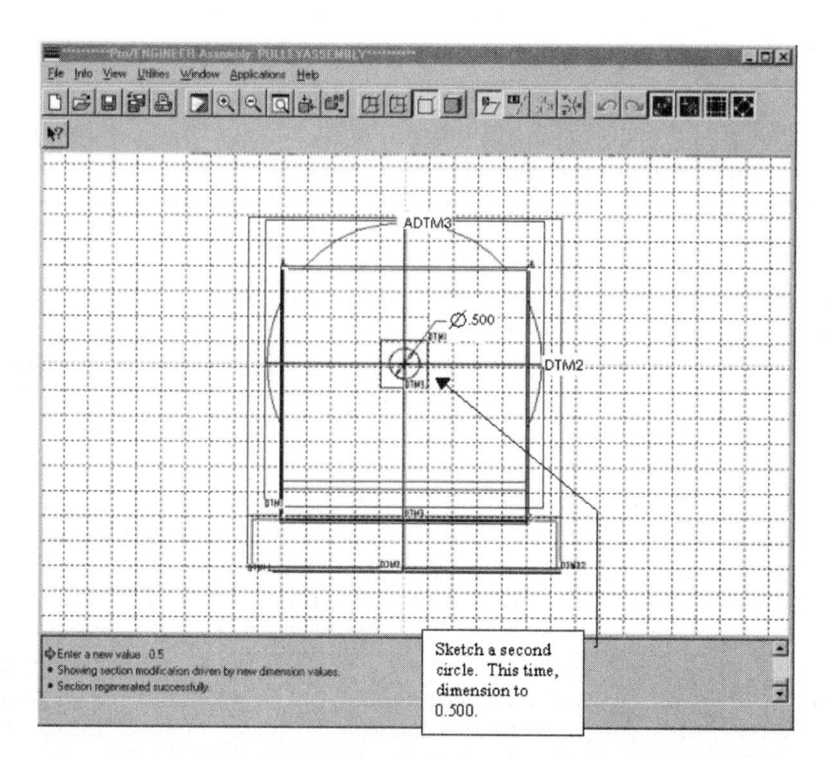


FIGURE 19.23 *The cross section of the second protrusion is also a circular section.*

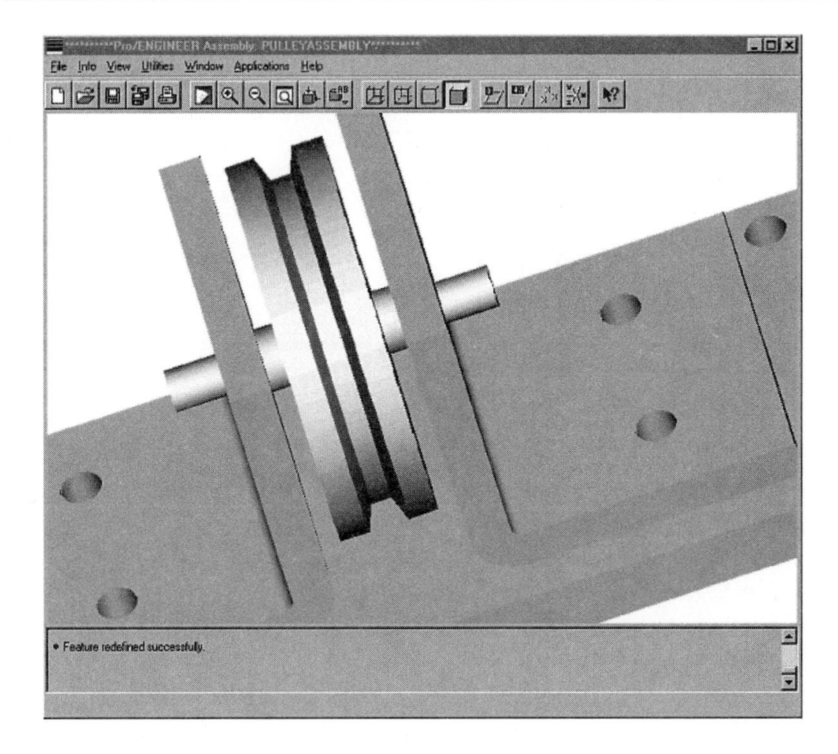


FIGURE 19.24 *The axle with the area near it enlarged.*

## 19.5 ADDITIONAL CONSTRAINTS

Like *Mate,* the *Mate Offset* option assumes that the chosen surfaces face each other. However, an offset distance may be prescribed. Likewise, *Align* and *Align Offset* assume that the chosen entities face the same direction. However, the *Align Offset* option allows for an offset distance between the two entities.

The *Insert* constraint is useful for placing fasteners into an assembly. In general, the threaded surface of the screw, nut, and so on, is chosen, along with the threaded surface of the hole.

Because both a hole and cylinders have two surfaces, the selection of the appropriate surface may be required. However, because fasteners are symmetric more often than not, it is not important which surface is chosen.

### 19.5.1 TUTORIAL # 86: COMPLETING THE PULLEY ASSEMBLY

The pulley assembly may be completed by adding the fasteners. These fasteners are assembled by using primarily the options *Align* and *Mate.*

As shown in Figure 19.25, the bushing is required on either side of the pulley wheel. This bushing prevents the pulley from sliding from side to side along the axle. The available space between the pulley and one of the angle brackets is 0.25 inches. The bushing is 0.20 inches. Therefore, the bushing must be offset from the angle bracket and the pulley by 0.025 inches (if

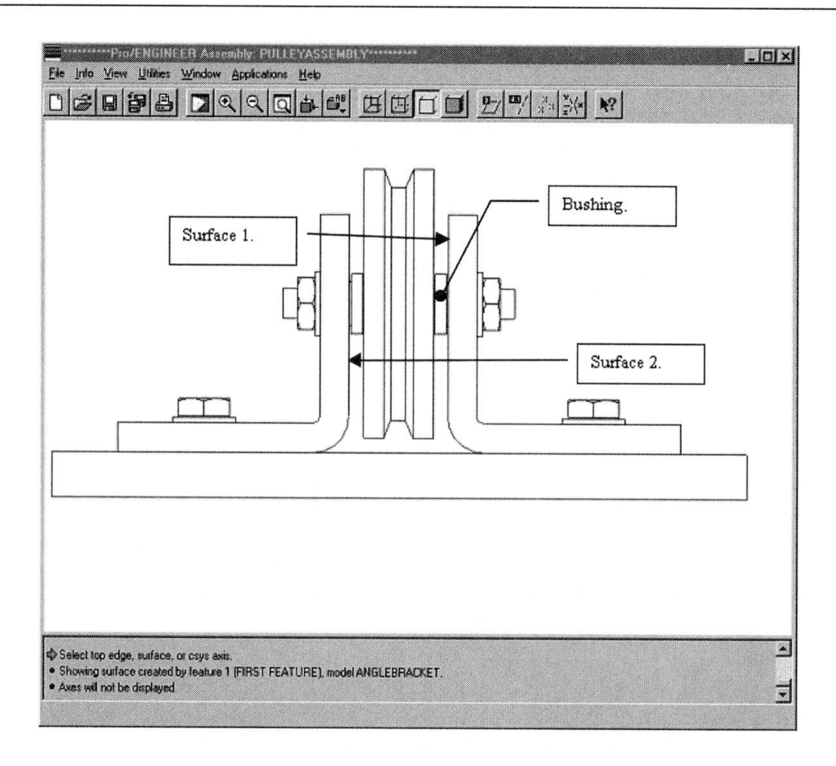


**FIGURE 19.25** *Front view of the assembly with bushings to be added.*

equally spaced). The *Mate Offset* option may be used in this case with a value of 0.025.

The procedure for adding the two bushings is as follows:

1. Select *Component* and *Assemble.*
2. Retrieve the part "Wheel_Bushing."
3. Select *Align.*
4. Choose axis A_1 on the component and axis A_3 on the assembly. See Figure 19.25 for these entities.
5. Select *Mate Offset.*
6. Choose Surface 1 in Figure 19.25.
7. Select Surface 3 in Figure 19.26.
8. Enter the value 0.025.
9. Choose *OK.*
10. Repeat for the second bushing using Surfaces 2 and 3.

The washer may be added to the assembly using the *Align* and *Mate* options. The procedure to be used is as follows:

1. Select *Component* and *Assemble.*
2. Scroll to find the part "Half_Inch_Washer." Double click.

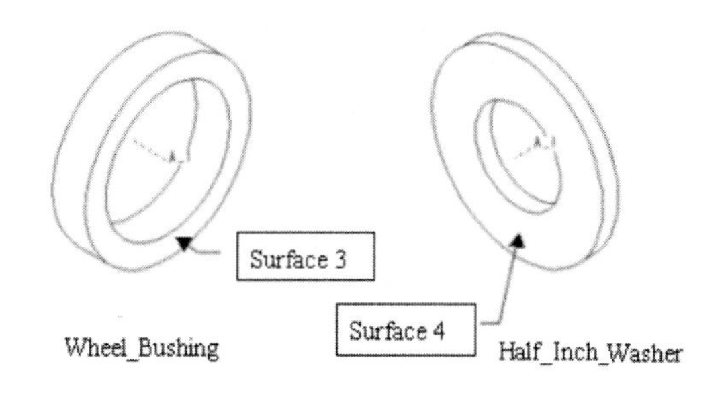


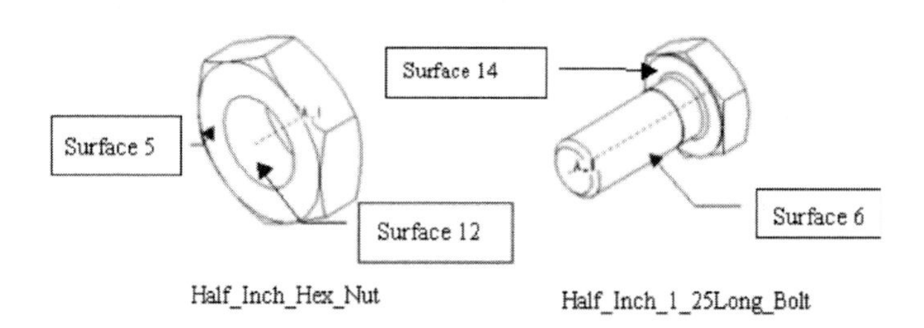


FIGURE 19.26 *Additional fasteners to be added to the assembly.*

3. Choose *Align.*
4. Select axis A_1 on the washer model in Figure 19.26.
5. Choose axis A_3 in Figure 19.27.
6. Select *Mate.*
7. Choose Surface 4 in Figure 19.26.

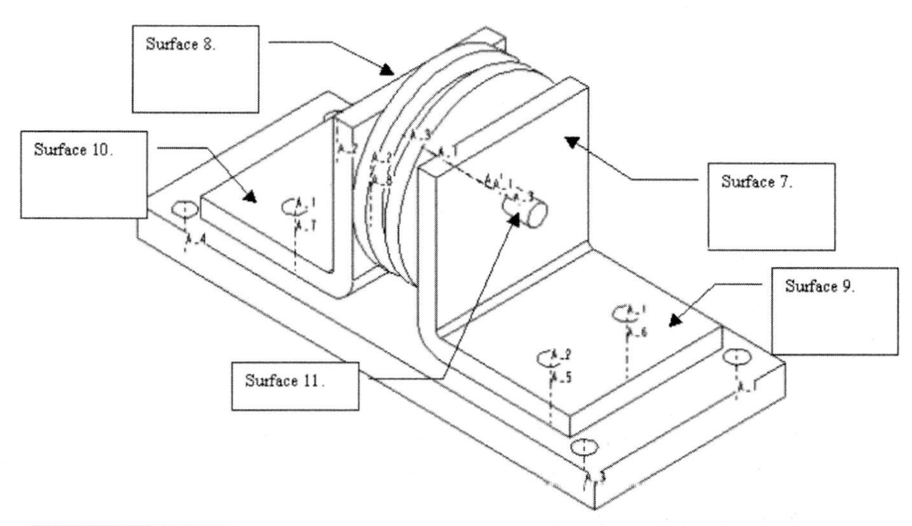


FIGURE 19.27 *The different surfaces to be used in placing the washers in the assembly.*

8. Select Surface 7 in Figure 19.27.
9. Choose *OK*.
10. Repeat steps 1 through 8 for the other angle bracket, using Surface 8 in step 8.
11. Repeat steps 1 through 4.
12. Choose axis A_5 in Figure 19.27.
13. Select *Mate*.
14. Choose Surface 4 in Figure 19.26.
15. Select Surface 9 in Figure 19.27.
16. Choose *OK*.
17. Repeat steps 11 through 16, except for step 12; use axis A_6.
18. Repeat steps 11 through 16 using axis A_8 for step 12 and Surface 10 for step 15.
19. Use steps 11 through 16 one last time. Use axis A_8 for step 12 and Surface 10 for step 15.

The hex nuts may be added to the assembly by using the *Align* and *Insert* options. The following sequence of steps may be used to place the two nuts:

1. Select *Component* and *Assemble*.
2. Retrieve part "Half_Inch_Hex_Nut."
3. Select *Mate*.
4. Choose Surface 5 in Figure 19.26.
5. Select Surface 7 in Figure 19.27.
6. Choose *Insert*.
7. Select Surface 12 in Figure 19.26.
8. Choose Surface 11 in Figure 19.27.
9. Choose *OK*.
10. Repeat steps 1 through 9 for the other side of the axle.

Likewise, the cap screws may be added to the assembly using the *Align* and *Insert* options. The procedure for doing so is as follows:

1. Select *Component* and *Assemble*.
2. Load the part "Half_Inch_1_25Long_Bolt."
3. Select *Mate*.
4. Choose Surface 14 in Figure 19.26.

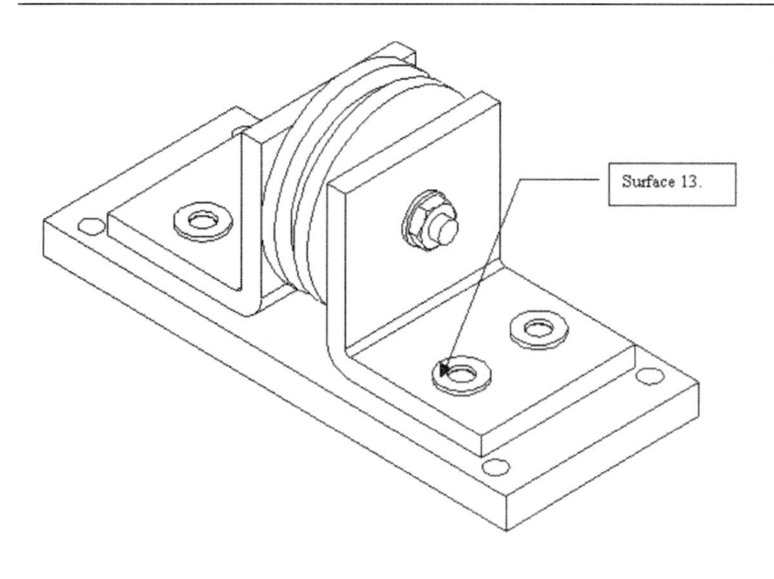


FIGURE 19.28 *Surface 13 is the top of the washer.*

5. Pick Surface 13 in Figure 19.28.
6. Choose *Insert.*
7. Select Surface 6 in Figure 19.26.
8. Choose the hole corresponding to the location of the washer.
9. Choose *OK.*
10. Repeat for the remaining three holes and washers.

The completed assembly is given in Figure 19.29. Save the assembly.

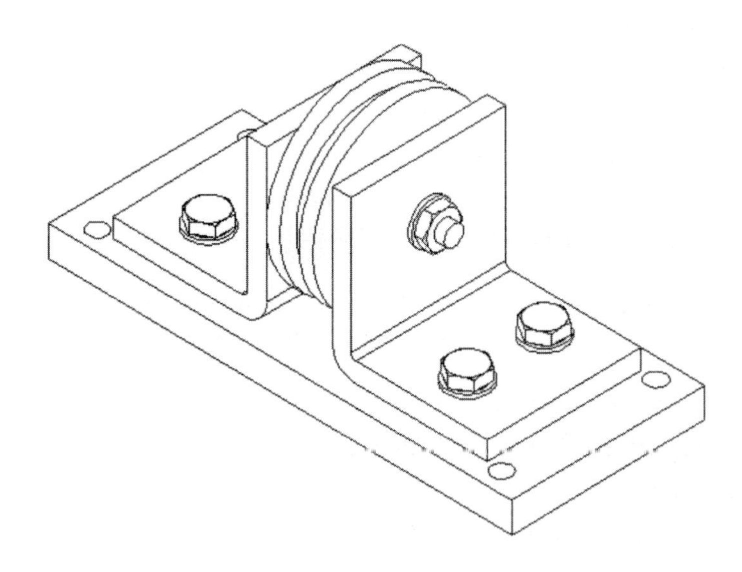

FIGURE 19.29 *The completed assembly.*

## 19.6 EXPLODED ASSEMBLIES

Assemblies are often created with exploded views; that is, the components are shown as to how they are to be put together. Once an assembly has been created, it can be exploded by choosing the *ExplodeState* option in the *ASSEMBLY* menu (see Figure 19.2).

After selecting the *ExplodeState* option, the exploded state must be given a name. Multiple exploded states may be constructed. The software automatically generates a default-exploded state. This state may be modified.

The options shown in Figure 19.30 are used as follows:

FIGURE 19.30 *The* EXPLD STATE *menu is used to define or redefine an exploded state.*

1. *Create.* This option creates a new exploded state. The name of the state must be given.
2. *Set Current.* The current state may be set from those available with this option by clicking on the desired state.
3. *Copy.* The contents of one exploded state may be copied into another.
4. *Redefine.* A user is able to modify the current state.
5. *Delete.* An exploded state may be permanently removed.
6. *List.* The different exploded states may be listed by name by using this option.

After an exploded state has been created and named, the *MTNPREF* menu appears. This menu is reproduced in Figure 19.31.

An assembly is exploded by selecting a reference entity, along whose direction one or more components is moved. The component(s) may be

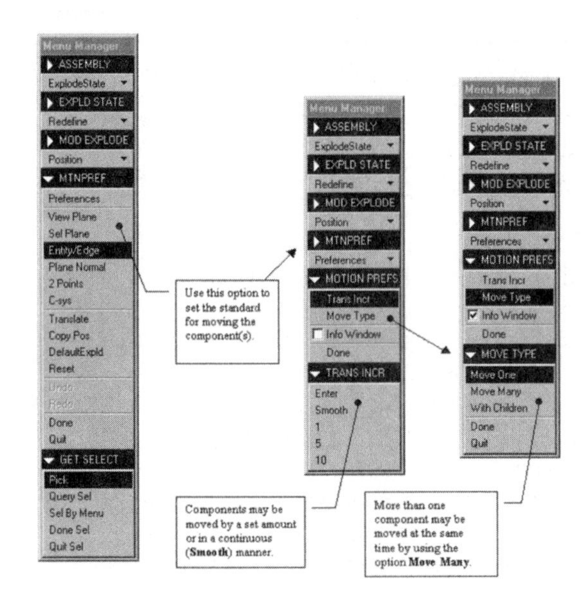


FIGURE 19.31 *The* MTNPREF *menu contains options for defining the direction and amount of translation in an explosion. By using the* TRANS INCR *menu, the translation increment can be set by selecting one of the preset values (1, 5, or 10) or by entering a value.*

moved incrementally by a set amount or in a continuous (*Smooth*) manner. The way in which a component is moved is set by using the *Preference*s option in the *MTNPREF* menu. The *Trans Incr* option is used to set the incremental value.

The *MOVE TYPE* menu (Figure 19.31) may be used to select whether to move one or more components.

The entity that is used as the reference for the move may be selected from the *MTNPREF* menu using one of the following options:

1. The *View Plane* option uses the viewing plane as a reference to position the component. If a plane other than the *Viewing Plane* is to be chosen, then the *Sel Plane* option is helpful. In either case, a plane parallel to the selected plane is constructed, which is used to locate the component.
2. The *Entity/Edge* option is used to pick an entity, axis, edge, or datum plane, which can be used to define the direction of the explosion. The component will be moved so that it is in a line parallel to the feature.
3. The *Plane Normal* option is used to select a plane as the reference plane. The component is placed along a line normal to the plane.
4. The *2 Points* option may be used to define a line between two points or vertices. The component is positioned along a line connecting the two points.
5. A coordinate system may be used to reposition a component by using the *C-sys* option. The component is placed in the direction of the coordinate system.
6. The *Translate* option translates a component along a given increment determined by using the *TRANS INCR* menu. The *Smooth* option in the *TRANS INCR* menu allows the component to be placed without incrementing.
7. To use the position instruction from one component to another, use the *Copy Pos* command.

A default-exploded state is always created by Pro/E using the *Default-Expld* option. The exploded positions of the component are determined by default values.

The last three options in the *MTNPREF* menu, *Reset, Undo,* and *Redo,* are used to manipulate the exploded state. The *Reset* option removes all explode instructions and returns the assembly to its original state. The *Undo* command undoes the last performed step, while the *Redo* performs the last step over again.

The *Dynamic Explode Component* window contains the current instructions for the creation of the exploded assembly. See Figure 19.32. Notice that the *Translate* and *Smooth* options are defaults.

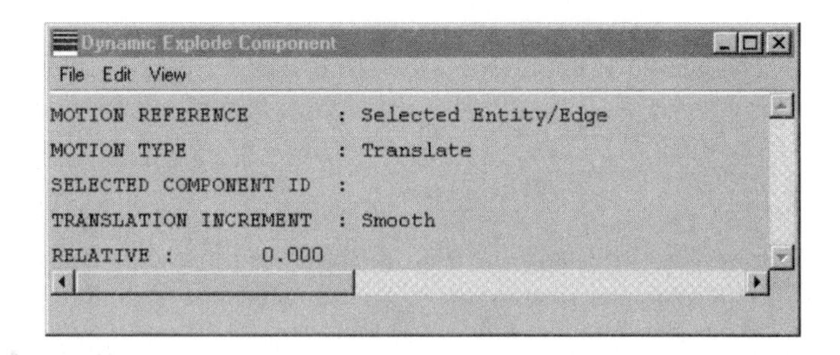


FIGURE 19.32 *The* Dynamic Explode Component *window lists the preference for the move. This particular example shows that the move will be done in a* Smooth *fashion.*

## 19.6.1 TUTORIAL # 87: AN EXPLODED STATE FOR THE PULLEY ASSEMBLY

It is best to create an exploded state in *Isometric* orientation for two reasons. The first reason is that most engineers are more familiar with *Isometric* orientation than *Trimetric.* The second reason is that quite often an entity coincident with one of the isometric axes may be used as the reference for the move.

Retrieve the pulley assembly. Then, select *Utilities, Environment,* and change the orientation to *Isometric.* Choose *OK.* Repaint the screen.

In addition, it is more practical to minimize the size of the model with respect to the screen. The reason to minimize is that as components are moved away from one another, they begin to take up more and more room on the screen.

Select *Explodestate* from the *ASSEMBLY* menu (Figure 19.2). Enter "Exploded" for the name of the state.

Choose *Preferences* from the *MTNPREF* menu. We want to change the move type to incremental with a preset value. With multiple components, it is easier to keep track of each component if they are moved by a certain amount. It is also a good idea to turn off the display of datum features, such as datum planes.

Select the increment value 10 from the *TRANS INCR* menu (Figure 19.31). Also, turn the *Dynamic Explode Component* window on by clicking on the *Info Window* option. Choose *Done.* Notice the changes in the setting of the *Dynamic Explode Component* window.

Because the plate forms the base of the assembly, we are going to move just the plate. It is easier than keeping the base stationary and moving all the other components. After the plate has been moved, we will move the rest of the components away from each other.

Select the line shown in Figure 19.33 as the reference entity. Note that this line is parallel to one of the isometric axes. Then click on the plate and move it away from the rest of the components. Because we selected an incremental move, the plate will only move an increment of ten units each time you move the mouse. Use only one move, so that the plate is moved only ten units from the rest of the assembly.

The next step is to move the rest of the components away from each other. Because the pulley wheel offers some symmetry, we can keep the

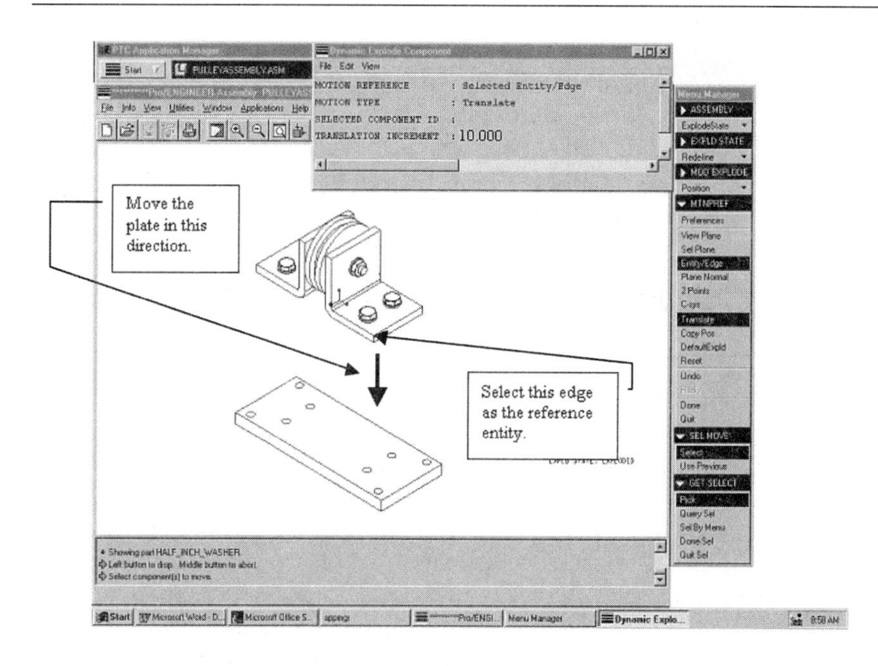


FIGURE 19.33 *Select the line forming one edge of the angle bracket as the reference entity.*

wheel stationary and move the rest of the components away from the pulley. A feature on the wheel, such as the axis through the pulley hub, can be used as the reference entity. In order to select this entity, the axis display must be turned back on.

Turn on the display of the datum axes. Because the assembly has more than one component to move, select *Preferences, Move Type, Move Many,* and *Done.* Then choose *Entity/Edge* from the *MTNPREF* menu. The best way to select the proper axis is to use the *Sel By Menu* option. Choose *Sel By Menu.* Then from the list, choose "Pulley_Wheel." Select *Axis.* The software will list the axes contained in the model. This case has only one, called A_1. Choose the axis from the list.

We will work on each side of the pulley wheel, one side at a time, as shown in Figure 19.34. This process will involve moving more than one component at a time. In order to select all the proper components to move, it is best to use the *Sel By Menu* option.

Choose the *Sel By Menu* option. The procedure to be used in selecting the proper components is as follows:

1. From the list, choose "AngleBracket."
2. The software will highlight one of the angle brackets. Choose the bracket required for the move shown in Figure 19.34 by using the *Next* and *Accept* options.
3. Again, choose *Sel By Menu* option and repeat steps 1 through 2 for the "Half_Inch_Hex_Nut."

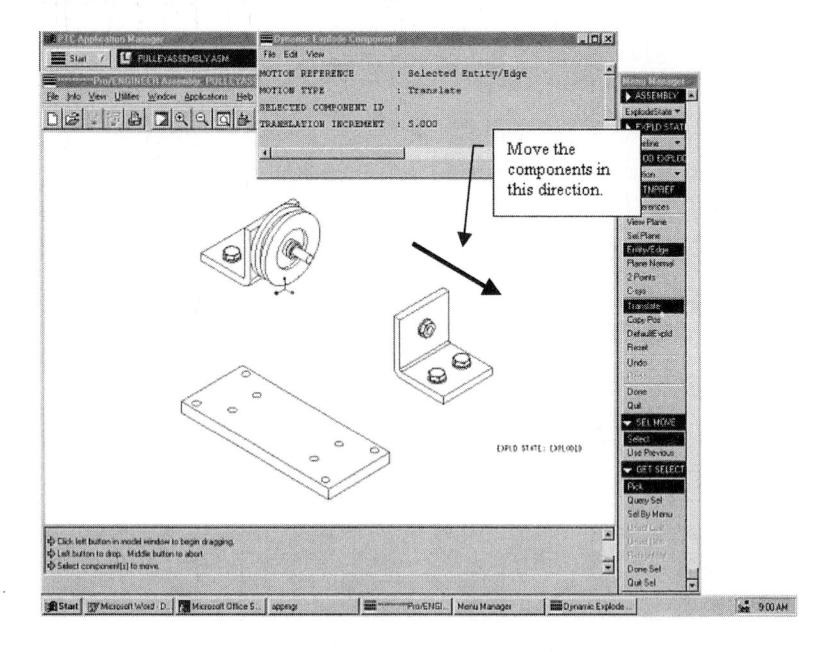


**FIGURE 19.34** *Move the components 10 units with respect to the pulley wheel.*

4. After choosing *Sel By Menu* again, repeat steps 1 through 2 for the "Half_Inch_Washer." Note that three washers means you will have to do this step three times.
5. Select *Sel By Menu* one last time and pick "Half_Inch_1_20Long_Bolt." Repeat steps 1 through 2 twice.
6. Choose *Done Sel.*

Move the components to the right as shown in Figure 19.34. Repeat steps 1 through 6 for the components on the other side of the pulley wheel. The assembly, after these components have been moved, is shown in Figure 19.35.

Now, select *Preferences* and change the increment to seven by using the *Enter* option. Using *Query Sel,* select one of the bushings and move it 7 units. Use *Query Sel* again and move the other bushing.

Choose *Preferences* one more time along with the *Enter* option. Enter a value of 2. Then, using *Query Sel,* select the axle and move it. The positions of the axle and bushings are shown in Figure 19.36.

The fasteners may be moved using a similar approach. The bolts and associated washers are moved using the axis of their corresponding holes as the reference entity. Because the axes of the two holes are parallel, it does not matter which axis is used for the move. The hex nut and associated washer are moved using the axis of the single hole as the reference entity. Figure 19.37 may be used as a reference in moving these components.

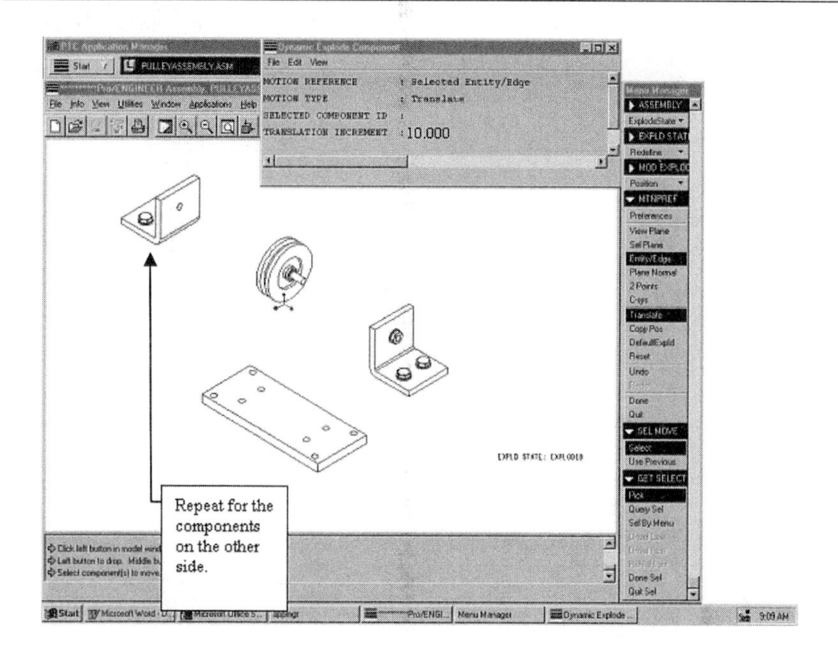


**FIGURE 19.35** *Move the components on the other side of the pulley wheel using the same approach.*

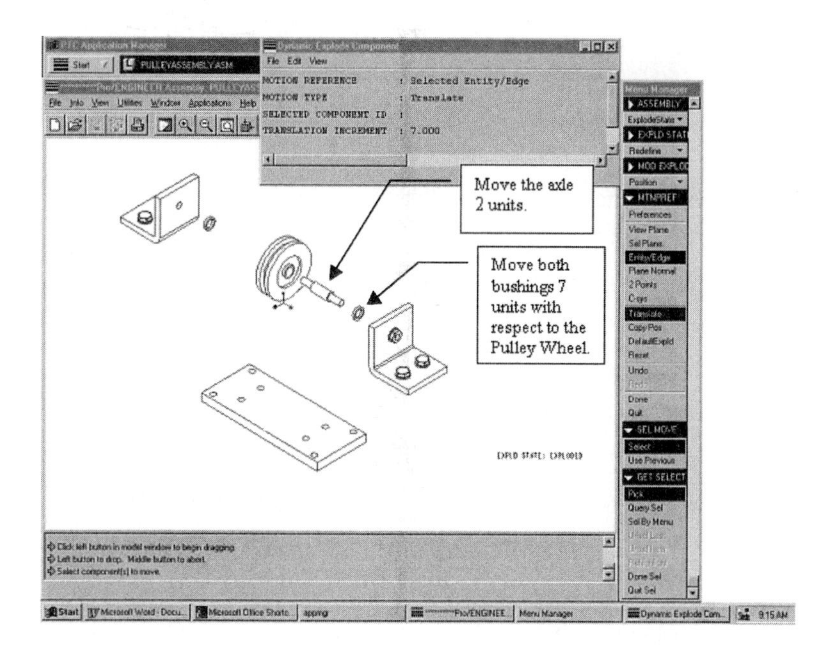


**FIGURE 19.36** *Move the axle 2 units and the bushings 7 units from the pulley wheel.*

FIGURE 19.37 *The fasteners may be moved by using the appropriate reference entity.*

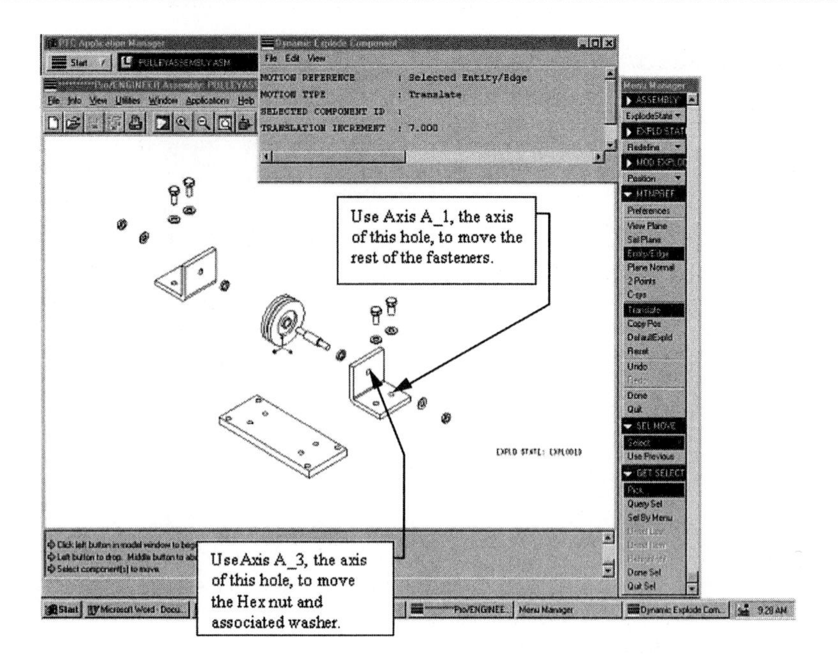


The procedure for moving the bolts and washers is as follows:

1. Select *Preferences* and *Enter.*
2. Enter the value 10.
3. Choose *Entity/Edge* and *Sel By Menu.*
4. Pick "AngleBracket" from the list. Then choose *Axis.*
5. Select A_1 or A_2.
6. Using *Query Sel,* pick all four of the bolts.
7. Choose *Done Sel.*
8. Move the bolts by 7 units.
9. Select *Preferences* and *Enter.*
10. Enter a value of 7.
11. Using *Query Sel,* pick all four of the washers associated with the bolts.
12. Choose *Done Sel.*
13. Move the washers 7 units.

The hex nuts and the associated washers may be moved by using the same procedure. However, replace the reference entity with axis A_3. After all the fasteners have been moved, select *Done* and *Done/Return.* Save the assembly.

## 19.7 ADDING OFFSET LINES TO ASSEMBLIES

Offset lines are added to exploded assemblies to show how the parts come together by using the *Offset Lines* option. The offset lines can be *Hidden, Centerline, Phantom Line, Cut Plane,* and *Geometry* type. The default is *Hidden.* Select the *Set Def Style* option and the options in the *Line Style* box to set the style of the line to use. The *Line Style* box is shown in Figure 19.38.

After changing the line style, select *Create.* The offset lines are drawn between two entities selected by the user. The type of entity, as shown in Figure 19.38, can be an *Axis,* normal surface (*Surface Norm* option), or an edge or curve (*Edge/Curve* option). Select the appropriate option from the menu when drawing the offset lines.

### 19.7.1 TUTORIAL # 88: OFFSET LINES FOR THE PULLEY ASSEMBLY

If the pulley assembly is not already in memory, *Open* the file. Then select *ExplodeState* and *Redefine* from the *EXPLD STATE* menu. Choose the state "Exploded" and *Done.* Select *Offset Line* and *Set Def Style.* For the *Style* option cell choose *Hidden.* For the *Line Font* cell choose *Dashfont.* Select *Apply* and close the window.

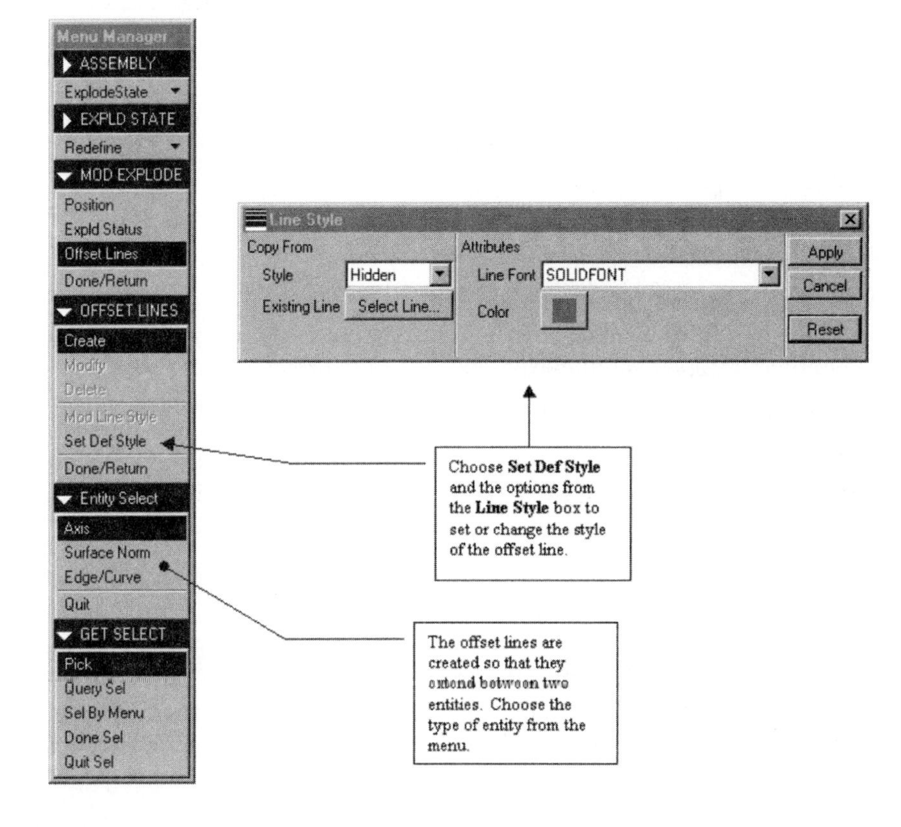


FIGURE 19.38 *Options for creating and defining the style of an offset line.*

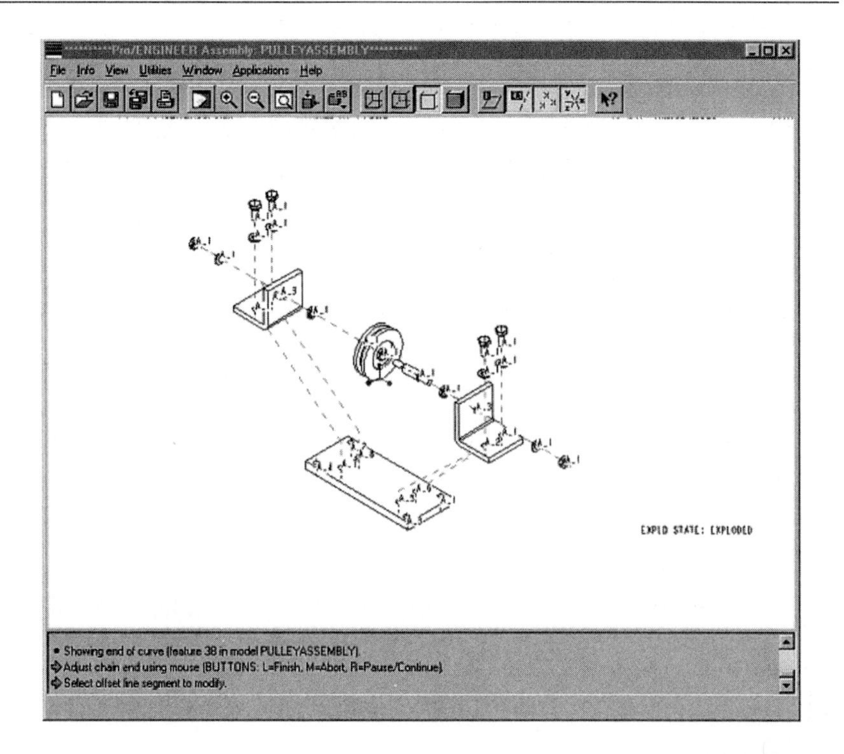


FIGURE 19.39 *The exploded assembly of the pulley with offset lines.*

Choose *Create* and *Axis.* For this assembly all the offset lines may be added by creating lines between the various axes. Create the offset lines as shown in Figure 19.39. Use the *Modify* option to change the position of the offset lines. Save the assembly.

## 19.8 ADDING AN ASSEMBLY TO A DRAWING

An assembly may be added to a drawing by simply specifying the "*.asm" file as the model when adding the view to the drawing. It is advantageous to add both an exploded and unexploded assembly to the drawing. When doing so, the scale of the unexploded assembly should be properly reduced from the scale of the exploded assembly.

Recall that Pro/E identifies the exploded state of the assembly with a name. Thus, when loading the exploded assembly, the name of the assembly must be given by choosing the correct exploded state.

### 19.8.1 TUTORIAL # 89: A WORKING DRAWING OF THE PULLEY ASSEMBLY

Create a drawing called PulleyAssembly.drw. Retrieve the format "Aformat." Select *Views.* Find the file "PulleyAssembly.asm" in the list.

Either the exploded or the unexploded model may be loaded into the drawing. It is customary to make the unexploded assembly larger scale than the exploded assembly. In order to properly size the drawing, let us add the exploded assembly first.

Select *Exploded, No Scale,* and *Done.* Notice that we have chosen the *No Scale* option, because we don't really know how the exploded assembly will fit in the drawing. It is best to place the assembly and change the scale afterward if it is unacceptable.

Click on the screen with your mouse pointer somewhere near the middle of the available space to place the exploded assembly. Select the "*Exploded*" state not the *Default* state. Do not reorient the assembly.

Now that we have placed the assembly, a decision regarding its scale can be made. Select *Modify* and click on *Scale.* Enter a value of 0.2.

We can add a second view containing a model of the unexploded assembly. It turns out that for this case, the same scale may be used for both views. Again, choose *Views* from the *DRAWING* menu. Then choose *Add View, General, Unexploded, No Scale,* and *Done.*

The *General* option is chosen so that the new view is independent of the first view. Furthermore, because we have some idea of the available space, we can choose the *Scale* option.

Click somewhere near the lower left-hand side of the screen to locate the new view. After the model is placed, do not change its orientation.

Move the views so that your drawing appears as in Figure 19.40. Save the drawing.

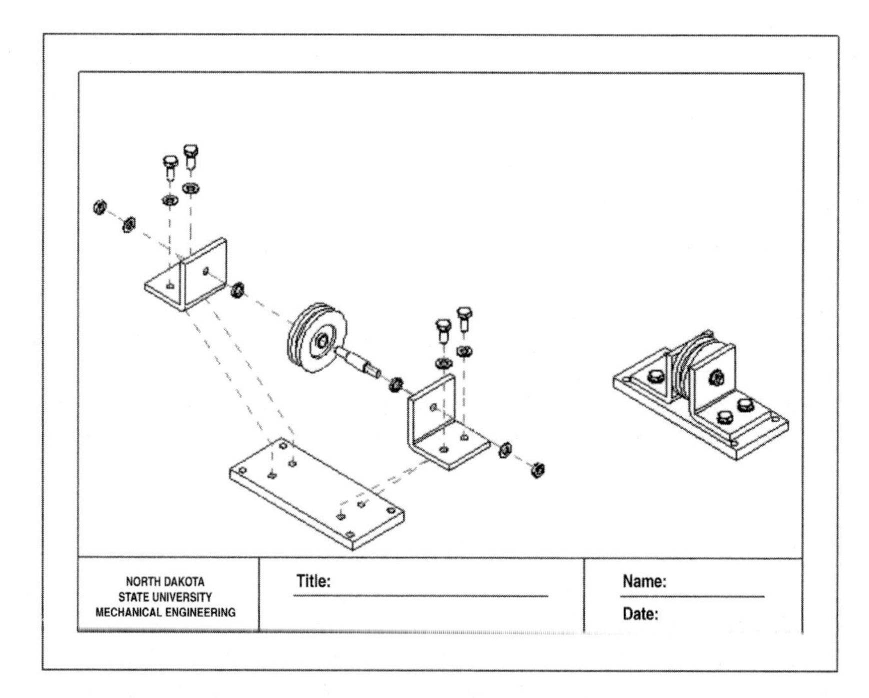


FIGURE 19.40 *Exploded and unexploded views.*

## 19.9 BILL OF MATERIAL (BOM) AND BALLOONS

A working drawing is incomplete without a bill of material (BOM). A BOM may contain the part or component name, the quantity of each component, the material from which it is to be made, and any required finish. If a component is to be obtained from a subcontractor or external supplier, then the supplier is listed, along with the part.

Pro/E has commands for creating a bill of material tailored to the user. A default BOM can be generated by selecting the *BOM* option from the *Info* option in the toolbar. The information may be sent to the screen, file, or both.

All results sent to a file are saved in a file with the extension .BOM. This file can be loaded into a drawing. The text may be modified using the path *Modify, Text,* and *Full Note.*

Balloons are used to reference the parts in the assembly to the bill of material. A balloon may be created by using the path *Create, Balloon,* and *Make Note.*

### 19.9.1 TUTORIAL # 90: ADDING A BOM TO THE WORKING DRAWING

Let us generate a simple BOM of the pulley assembly. Select *Info* from the toolbar and then *BOM.* A bill of materials listing the components in each assembly will be created, as well as a summary of all the assemblies. Because the pulley assembly is a single assembly, the information contained in the summary will be identical to the information contained in the assembly list. Choose the *Top Level* option. Then *OK.* Save the information to the disk by selecting *File.*

The BOM can be placed on the drawing. Select *Detail, Create,* and *Note.* For the placement, use the *Default* or *Justify Left;* otherwise, the text may not fit on the screen.

Select *File* and *Make Note.* Click somewhere near the upper right-hand side of the screen.

We need to enter the name of the file containing the text to be displayed. This file is, of course, the file "PulleyAssembly.BOM." In order to enter the name of the file, choose the *Enter Name:* option. Then enter the name of the file (PulleyAssembly.BOM). Select *Done/Return.*

Use *Move Text* from the *DETAIL* menu to properly place the text. Then create a border around the BOM by using the *Sketch* option.

Edit the note by using *Modify, Text,* and *Full Note.* Select the text. After the software has loaded the page editor, add the additional text and spaces to produce the BOM shown in Figure 19.41.

Balloons may be added to the exploded assembly to help refer to the bill of material. Add a balloon for each part in the exploded assembly. In some cases, the same balloon may be used for multiple parts by selecting both parts. Use the following procedure to place the balloons:

1. Select *Create* and *Balloon.*
2. Choose *Leader, Enter,* and *Make Note.* We suggest using the default orientation.

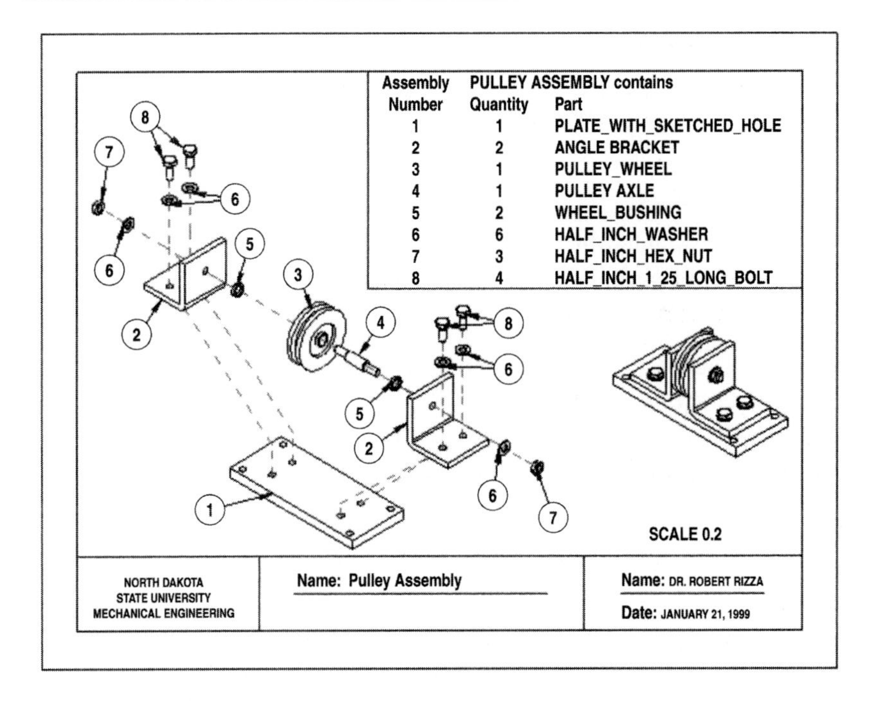


FIGURE 19.41 *A working drawing of the pulley assembly.*

3. Select the entity on the part. If you want the balloon to reference several parts, pick an entity on each part.
4. Choose *Done Sel.*
5. Select the type of leader. We suggest the default (*ArrowHead*).
6. *Done.*
7. Pick the location for the note.
8. Select *Make Note.*
9. Enter the text for the balloon.
10. Hit *Return* twice.

Move the balloons as necessary, using the *Move* option. The arrowheads should be of "*Filled*" type. If they are not, select *Advanced* and *Modify Val.* Change the arrowhead style from "*Closed*" to "*Filled.*"

Add the scale size, your name, title, and date to the format. Save the drawing.

## 19.10 SUMMARY AND STEPS FOR DEALING WITH ASSEMBLIES

Parts may be put together to form an assembly. Each part in the assembly is called a component. In order for the component to be placed in the assembly, the software must have access to the part.

Assemblies are created as assembly files using the method analogous for part and drawing files. After an assembly file has been created, components may be added to the assembly either parametrically, using the *Assemble* option, or nonparametrically, using the *Package* option. Components placed by using the *Package* option are located in an absolute sense. They are not placed relative to the base part, which is helpful if the exact placement of the part is unknown.

In this chapter, we concentrated on placing the components in a parametric fashion. When placing components using the *Assemble* option, the relative positions of the components must be known. Because the assembly process is parametric, the location of the components may be modified.

An assembly may be created with or without assembly datum planes. If the datum planes are not used, then the first part added to the assembly becomes the base part in the assembly. Assembly datums are useful in locating components that do not have planar surfaces.

Assembly datum planes may be added to an assembly using the following:

1. Select *Feature* and *Create.*
2. Choose *Datum, Plane, Default,* and *Done.*

The sequence of steps for adding a part to an assembly is:

1. Select *Component* and *Assemble.*
2. Retrieve the desired part.
3. Choose the required constraint.
4. Select the component reference.
5. Choose the assembly reference.
6. Add additional constraints, as needed.
7. When the component is fully constrained, select *OK.*

A part may be constructed in the assembler. In order to ensure that the part has the same standards as the components, it is best to start with a Start Part containing the standards and the default datum planes. This option is *Copy From Existing.* Then, the datum planes may be assembled and the new part created using the datum planes as sketching and orientation planes.

If all the parts contain the default standard, the datum planes may be created in the assembler using the option *Locate Default Datums.* The datums are assembled and the part constructed using the assembled datums.

In order to create a part in the *Assembly* mode, use the following sequence of steps:

1. Choose *Component* and *Create.*
2. Enter a name for the part and select *OK.*
3. Select from *Copy From Existing, Locate Default Datums, Empty,* or *Create First Feature.*
4. Choose *OK.*

5. Assemble the part based on your choice in step (3).
6. Construct the part using the appropriate option and sketching and orientation planes.

After a part has been created in the assembler, it may be modified. Additional features may be added to the part by doing the following:

1. Select *Modify* and *MOD Part.*
2. Choose the part.
3. Select *Feature.*
4. Choose the appropriate option and construct the feature.

In order to see how the components come together to form the assembly, an exploded state of the assembly may be created. The exploded state is created by moving the components apart. Moving components can be done in an incremental fashion or continuously in a smooth manner. The components are moved with respect to a reference. In general, an exploded state may be created by using the following procedure:

1. Select *ExplodeState* and *Create.*
2. Enter a name for the state.
3. If you wish to move the component or components incrementally, select *Preferences* and change the move type.
4. Select the reference.
5. Choose the component or components to move.
6. Click the left mouse button to begin the move.
7. Move the component(s) by moving your mouse.
8. Choose *Done* and *Done/Return* twice.

After an exploded state is defined, offset lines may be added to the state. The line type may be changed using the options in the *Line Style* box. The lines are created by defining the feature to which they are attached. This feature may be an axis, normal surface, edge, or curve. The sequence of steps to follow for adding offset lines to an assembly is:

1. Select *Offset Lines* from the *EXPLD STATE* menu (Figure 19.38).
2. Choose *Create.*
3. Select *Mod Line Style* to set the line style.
4. Choose the type of entity.
5. Select the entity on the part.
6. After you are finished adding all the offset lines, select *Done/Return.*

Assemblies may be added to a drawing by selecting *Views* and *Add View*. Because the assembly may contain an exploded state, choose between the *Exploded* and *Unexploded* option to place the desired state.

## 19.11 ADDITIONAL EXERCISES

19.1 Using the parts "AssembleBlockA" and "AssembleBlockB," obtain the assembly shown in Figure 19.42.

FIGURE 19.42 *Figure for exercise 19.1.*

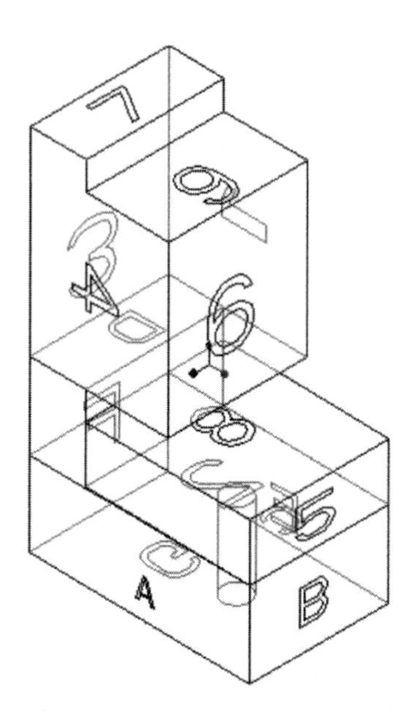


19.2 Assemble the parts "AssembleBlockA" and "AssembleBlockB" to obtain the assembly shown in Figure 19.43.

FIGURE 19.43 *Figure for exercise 19.2.*

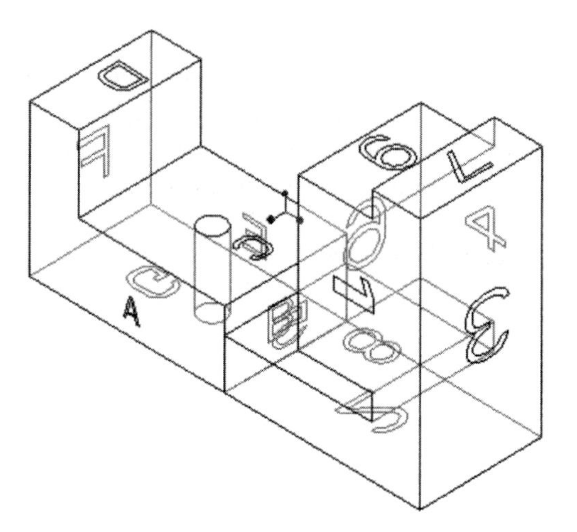

19.3 Use the parts "AssembleBlockA" and "AssembleBlockB" to create the assembly shown in Figure 19.44.

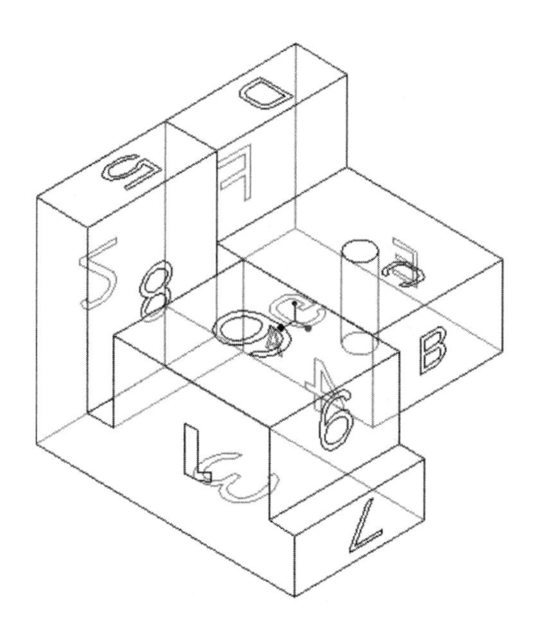


FIGURE 19.44 *Figure for exercise 19.3.*

For exercises 19.4 through 19.9, create the exploded assemblies using the given figures. Some of the parts have already been constructed in the tutorial sections of this book. Additional parts may be required. Consult the BOM associated with each assembly for the parts and the corresponding reference figure, tutorial, or exercise. Create a working drawing if assigned.

19.4 Propeller assembly. Consult Figures 19.45–46.
19.5 Linkage assembly. Use Figures 19.47–51 as a reference.
19.6 Flood light subassembly. Use Figures 19.52–54 as a reference.
19.7 Steady rest assembly. Consult Figures 19.55–57 for the components.
19.8 Blower assembly. Consult Figures 19.58–62 for the parts in this assembly.
19.9 Emergency light assembly. Use Figures 19.63–65 for this assembly.

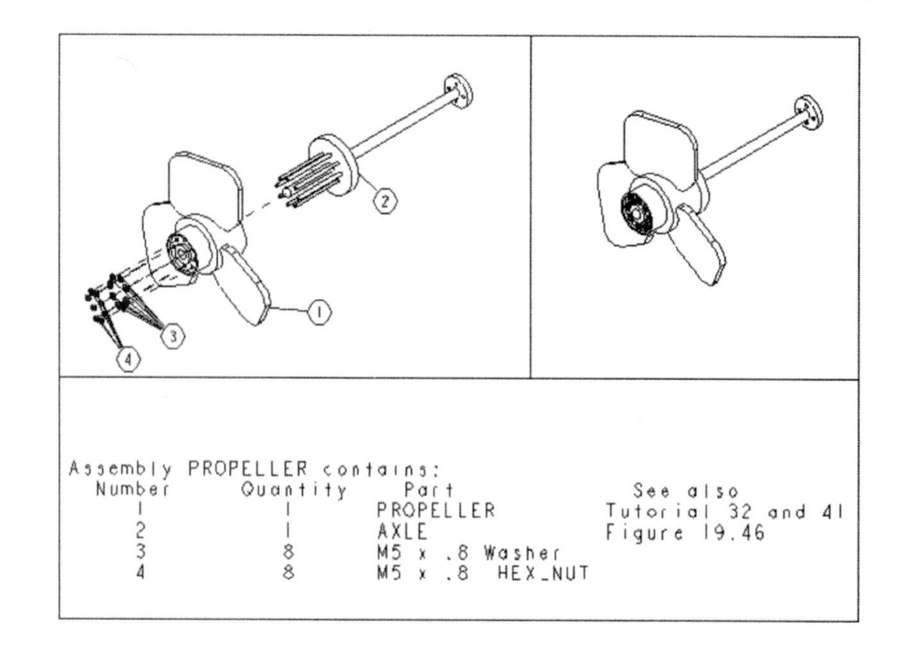


Assembly PROPELLER contains:

| Number | Quantity | Part | See also |
|---|---|---|---|
| 1 | 1 | PROPELLER | Tutorial 32 and 41 |
| 2 | 1 | AXLE | Figure 19.46 |
| 3 | 8 | M5 x .8 Washer | |
| 4 | 8 | M5 x .8 HEX_NUT | |

**FIGURE 19.45** *Figure for exercise 19.4.*

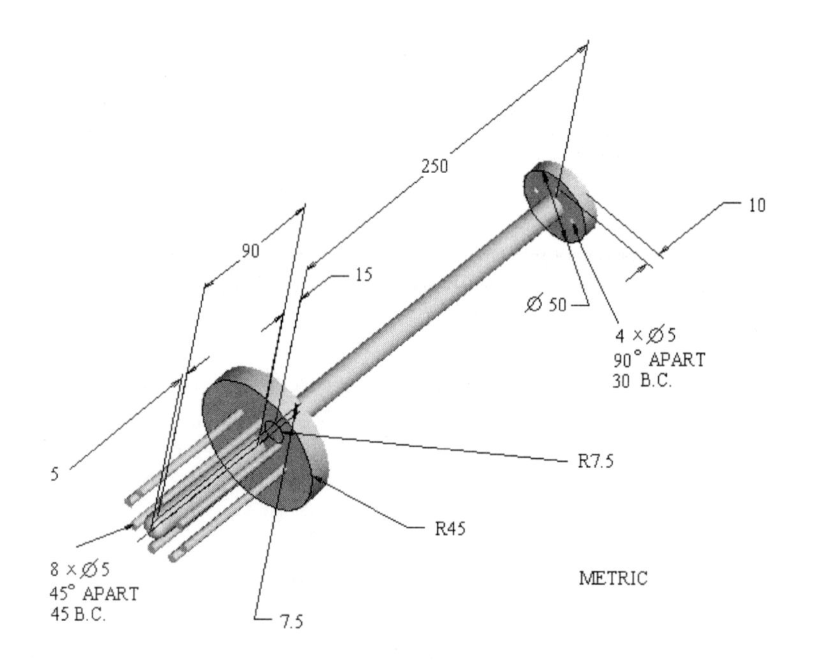


**FIGURE 19.46** *Geometry of the axle.*

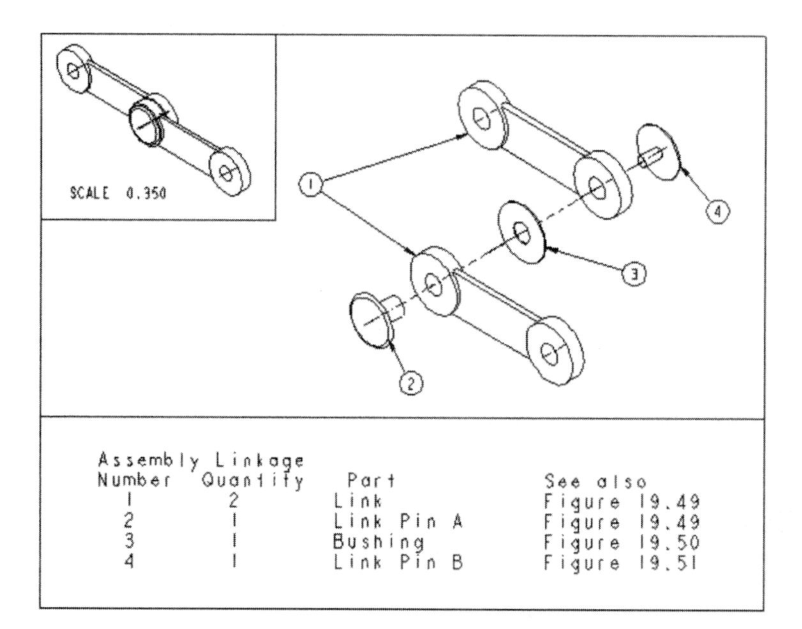


| Assembly Linkage Number | Quantity | Part | See also |
|---|---|---|---|
| 1 | 2 | Link | Figure 19.49 |
| 2 | 1 | Link Pin A | Figure 19.49 |
| 3 | 1 | Bushing | Figure 19.50 |
| 4 | 1 | Link Pin B | Figure 19.51 |

FIGURE 19.47 *Figure for exercise 19.5.*

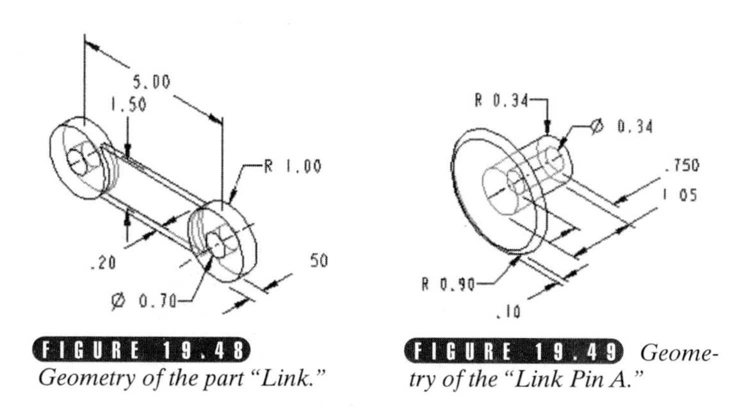


FIGURE 19.48 *Geometry of the part "Link."*

FIGURE 19.49 *Geometry of the "Link Pin A."*

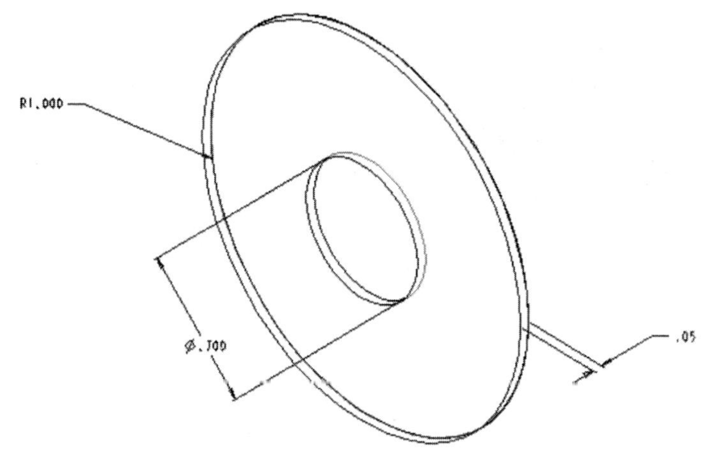


FIGURE 19.50 *Geometry of the part "Bushing."*

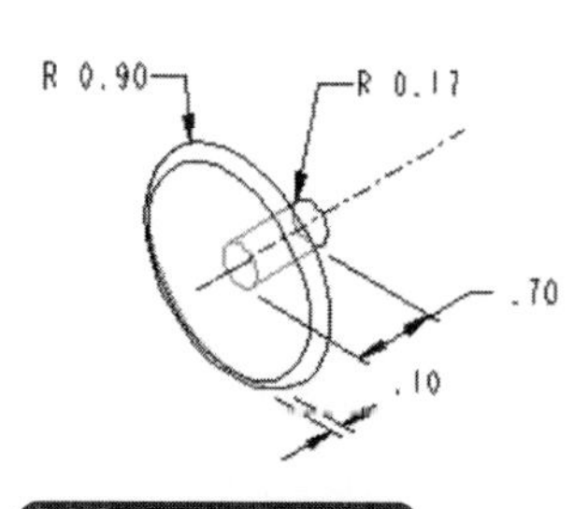


FIGURE 19.51 *Geometry of "Link Pin B."*

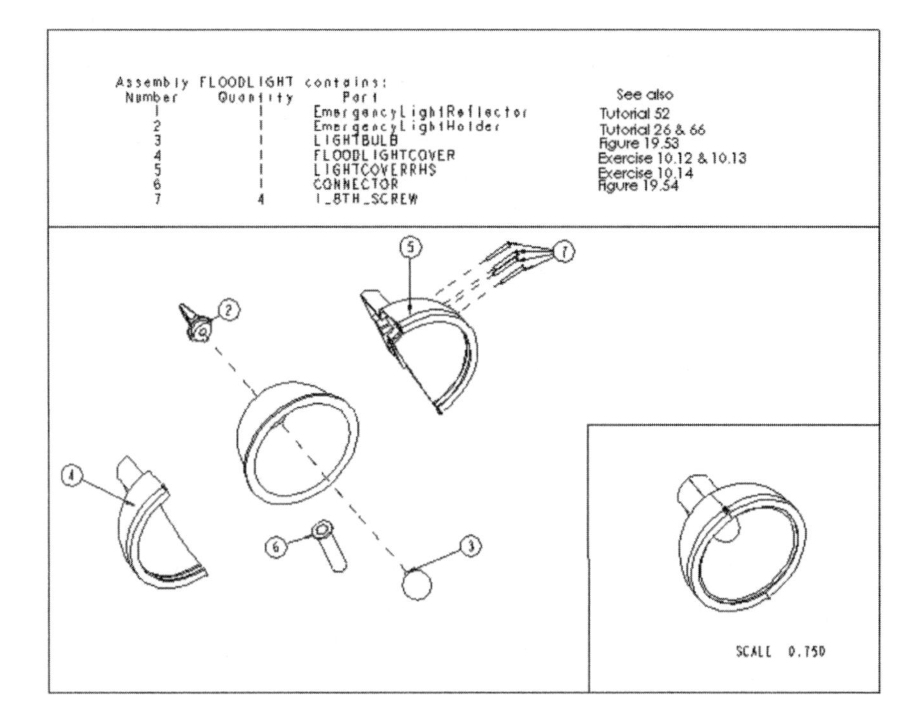

Assembly FLOODLIGHT contains:

| Number | Quantity | Part | See also |
|---|---|---|---|
| 1 | 1 | EmergencyLightReflector | Tutorial 52 |
| 2 | 1 | EmergencyLightHolder | Tutorial 26 & 66 |
| 3 | 1 | LIGHTBULB | Figure 19.53 |
| 4 | 1 | FLOODLIGHTCOVER | Exercise 10.12 & 10.13 |
| 5 | 1 | LIGHTCOVERRHS | Exercise 10.14 |
| 6 | 1 | CONNECTOR | Figure 19.54 |
| 7 | 4 | 1_8TH_SCREW | |



FIGURE 19.52 *Figure for exercise 19.6.*

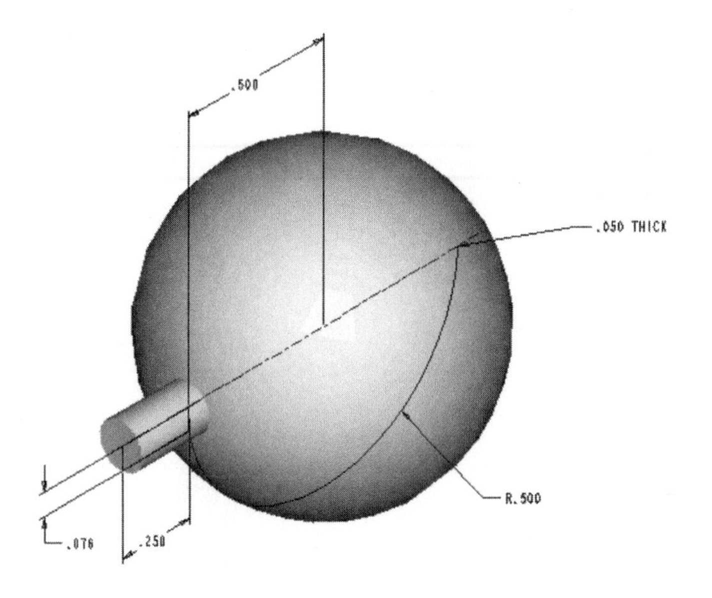


FIGURE 19.53 *Geometry of the part "LightBulb."*

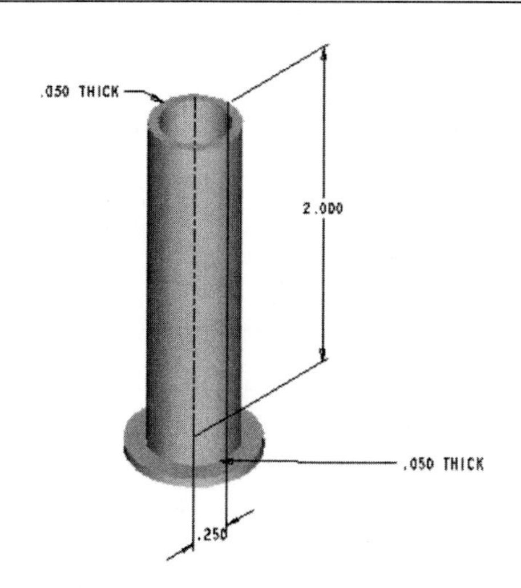


**FIGURE 19.54** *Geometry of part "Connector."*

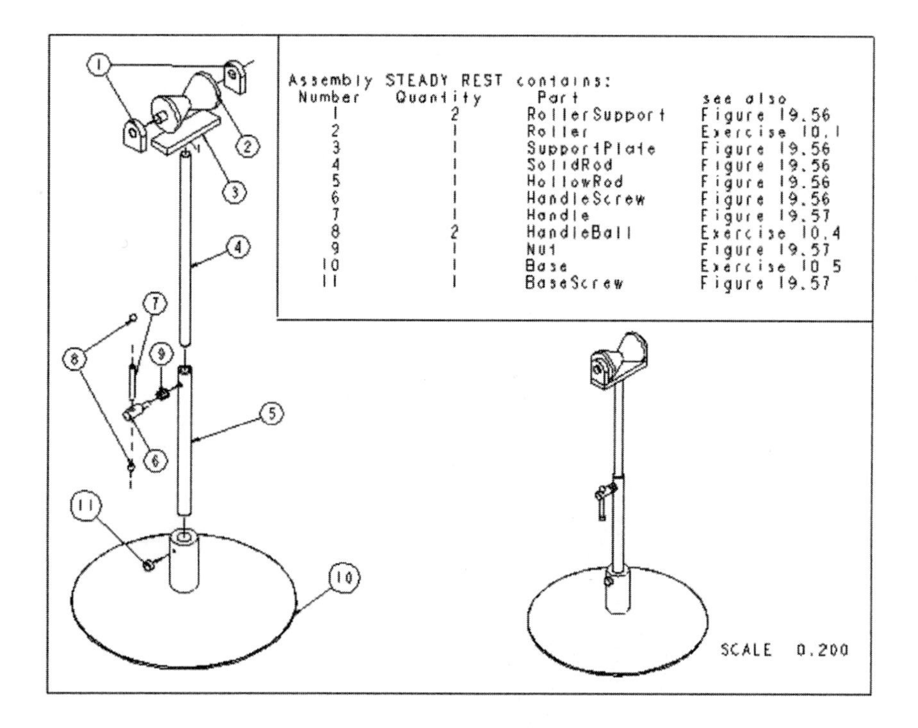
Assembly STEADY REST contains:

| Number | Quantity | Part | see also |
|---|---|---|---|
| 1 | 2 | RollerSupport | Figure 19.56 |
| 2 | 1 | Roller | Exercise 10.1 |
| 3 | 1 | SupportPlate | Figure 19.56 |
| 4 | 1 | SolidRod | Figure 19.56 |
| 5 | 1 | HollowRod | Figure 19.56 |
| 6 | 1 | HandleScrew | Figure 19.56 |
| 7 | 1 | Handle | Figure 19.57 |
| 8 | 2 | HandleBall | Exercise 10.4 |
| 9 | 1 | Nut | Figure 19.57 |
| 10 | 1 | Base | Exercise 10.5 |
| 11 | 1 | BaseScrew | Figure 19.57 |



**FIGURE 19.55** *Figure for exercise 19.7.*

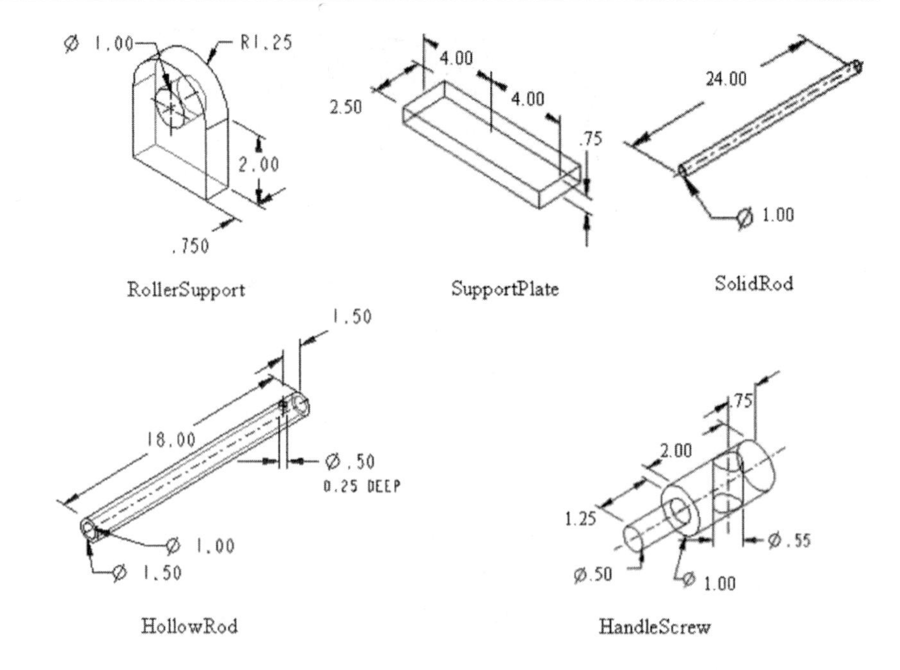


**FIGURE 19.56** *Some parts in the steady rest assembly.*

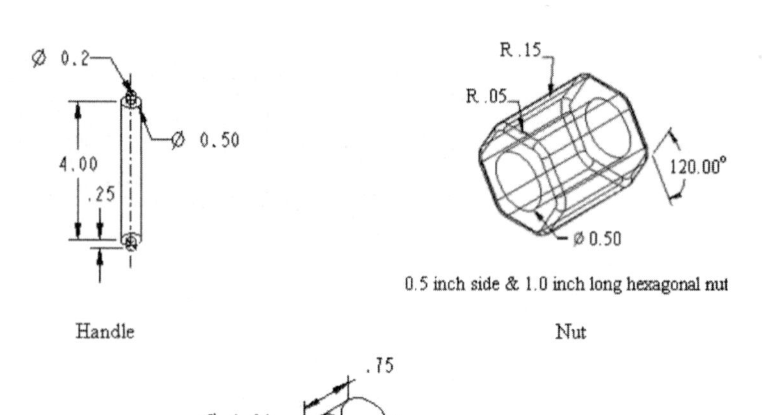


**FIGURE 19.57** *More parts for the steady rest assembly.*

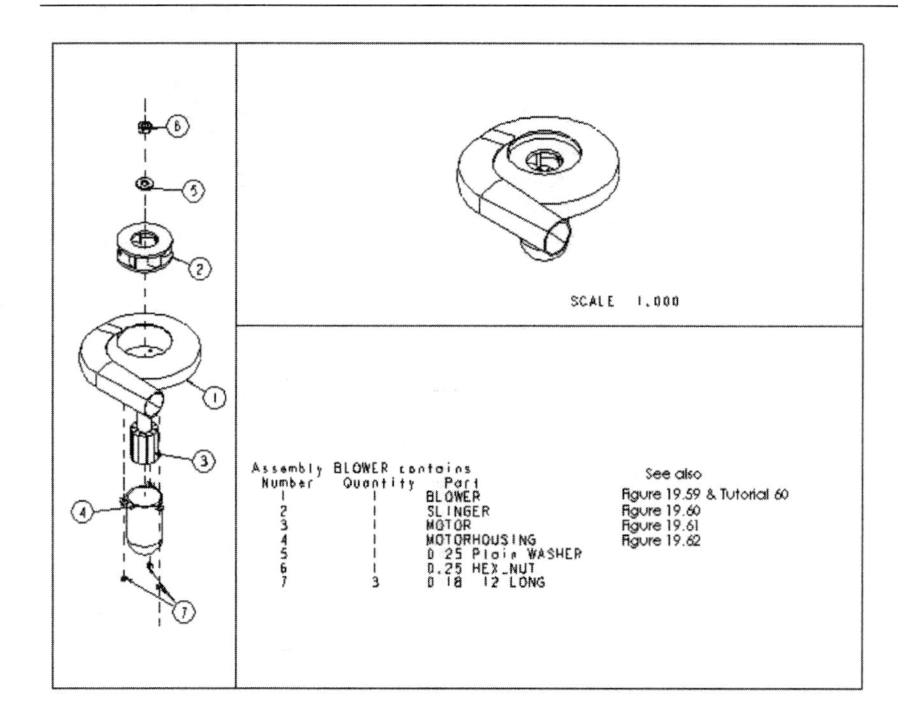


Assembly BLOWER contains

| Number | Quantity | Part | See also |
|---|---|---|---|
| 1 | 1 | BLOWER | Figure 19.59 & Tutorial 60 |
| 2 | 1 | SLINGER | Figure 19.60 |
| 3 | 1 | MOTOR | Figure 19.61 |
| 4 | 1 | MOTORHOUSING | Figure 19.62 |
| 5 | 1 | D 25 Plain WASHER | |
| 6 | 1 | D.25 HEX_NUT | |
| 7 | 3 | D 18 12 LONG | |

**FIGURE 19.58** *Figure for exercise 19.8.*

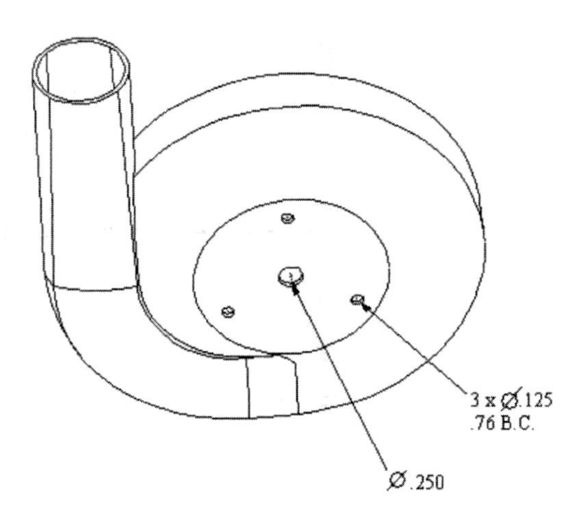


**FIGURE 19.59** *Add the holes to the "Blower" from Tutorial # 60.*

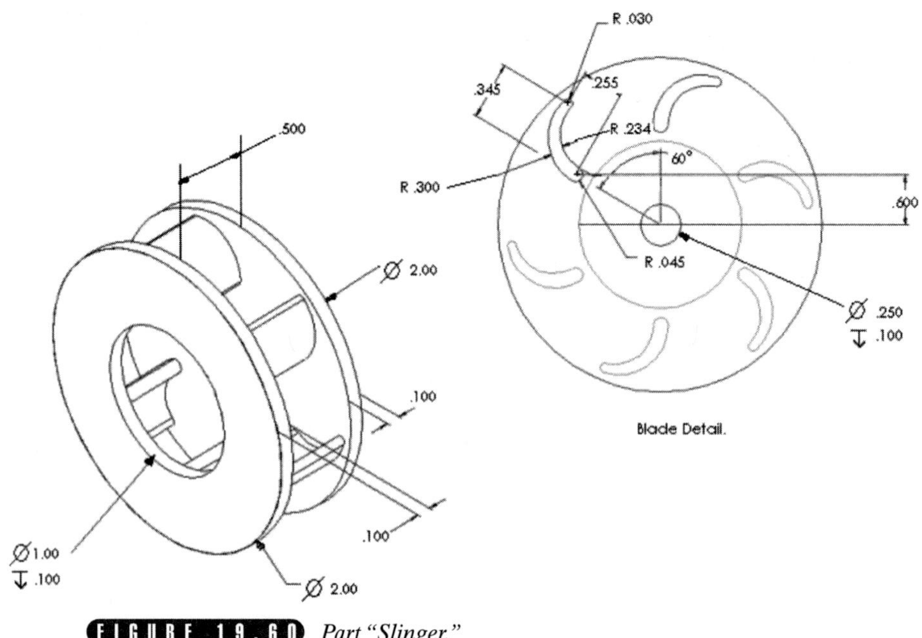


FIGURE 19.60 *Part "Slinger."*

FIGURE 19.61 *Part "Motor" for the assembly "Blower."*

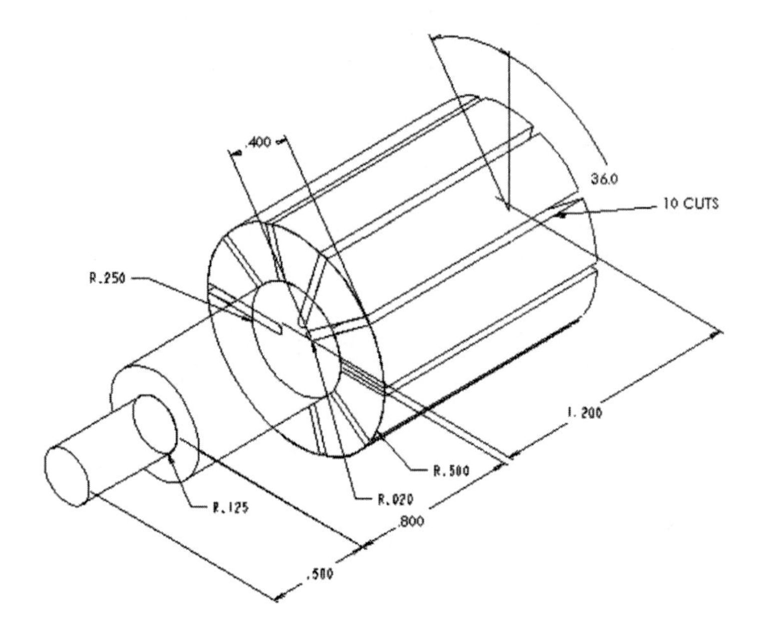

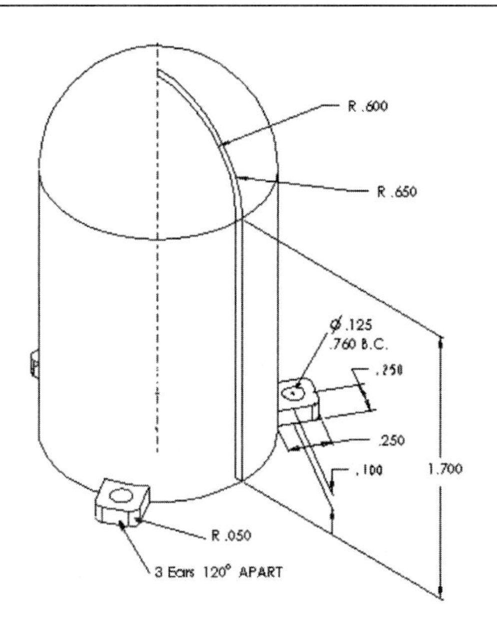


**FIGURE 19.62** *Part "MotorHousing" for the motor in Figure 19.61.*

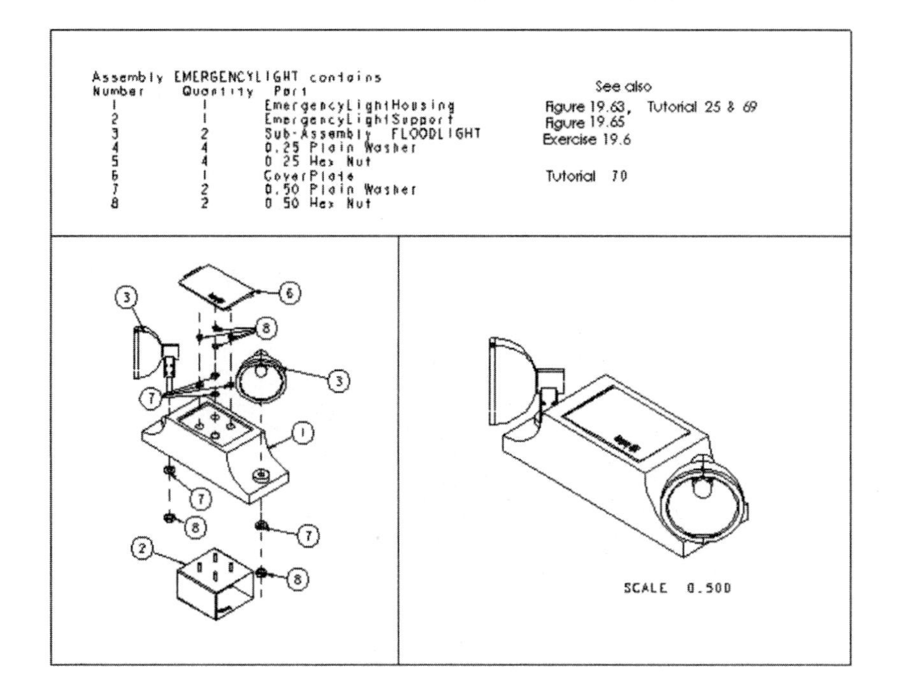


**FIGURE 19.63** *Figure for exercise 9.9.*

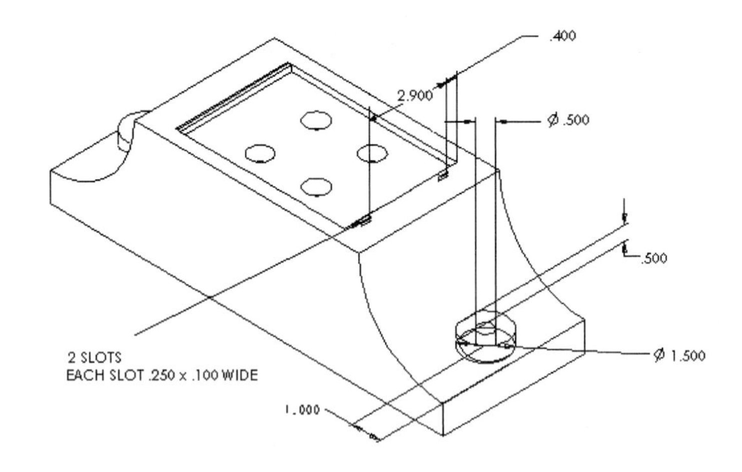


FIGURE 19.64 *Additional features for the "EmergencyLightHousing."*

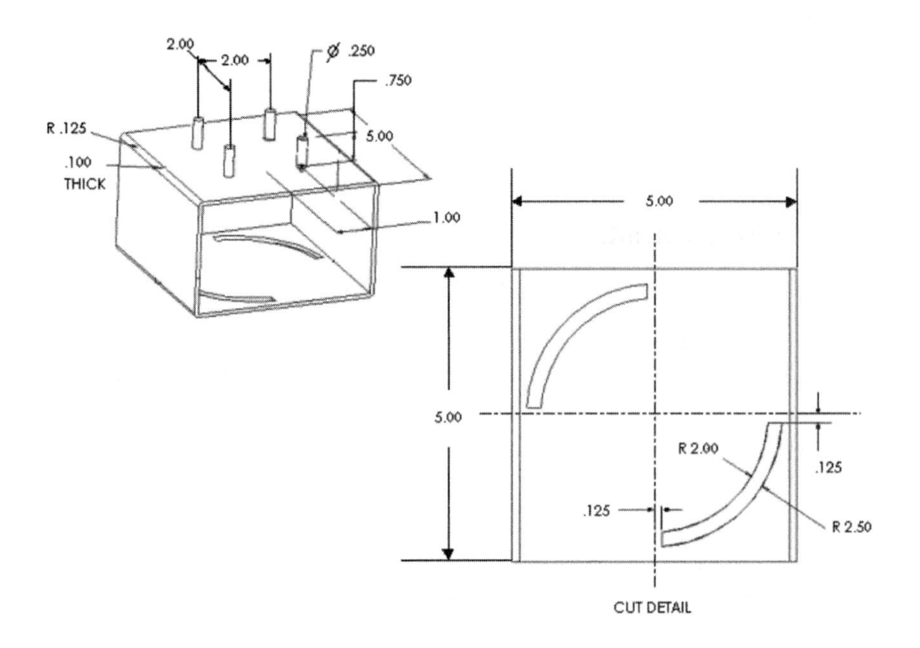


FIGURE 19.65 *Part "EmergencyLightSupport."*

# CHAPTER 20

# ENGINEERING INFORMATION AND FILE TRANSFER

## INTRODUCTION AND OBJECTIVES

Pro/E provides several options for obtaining engineering information, such as mass properties, lengths of a feature, interference, and clearance checks. The options for obtaining such information may be found as suboptions in the *Info* command. For revision 21, the options *Measure* and *Model Analysis* are located under the *Analysis* command as shown in Figure 20.1.

Many of these options require properties of engineering materials. These properties may be entered into the informational database using the *Set Up* option during model construction. If the properties have not been assigned, then the software will query the user for the required property during the computational phase.

A coordinate system is required for the computation of many of these engineering properties. A coordinate system may be created using the methods outlined in chapter 9 during model construction, or "on the fly" during calculation of the engineering property.

The initial graphics exchange specification (IGES) is used to transfer graphics and text to another CADD or CAM system. Models and some sections may be imported or exported as IGES files. When exporting a model as an IGES file, the features that the receiving system may receive must be taken into consideration. The proper options must be used in order to achieve the desired results. The software will create two files. The first file is the actual IGES transfer file with the extension ".igs." The second file, with extension "out.log," contains information about the processor, type, and quantity of each entity.

The *Geom Check* option may be used to correct geometry errors. This option should always be gray. If it is not, then the suggested steps should be undertaken and the geometry fixed.

The objectives of this chapter are as follows:

1. The user will be able to assign material properties to a model.
2. The software user will be able to obtain engineering properties and dimensions of a model.
3. The reader will be able to find the clearance or interference between two parts.
4. The user will be able to export a part as an IGES file.

## 20.1 DEFINING AND ASSIGNING MATERIAL PROPERTIES

Material properties may be assigned to a model by using the options in the *PART SETUP* menu. This menu is shown in Figure 20.1. The *Material* option may be used to define and assign material properties. The *Define* option allows the user to enter data about a new material. This material is given a

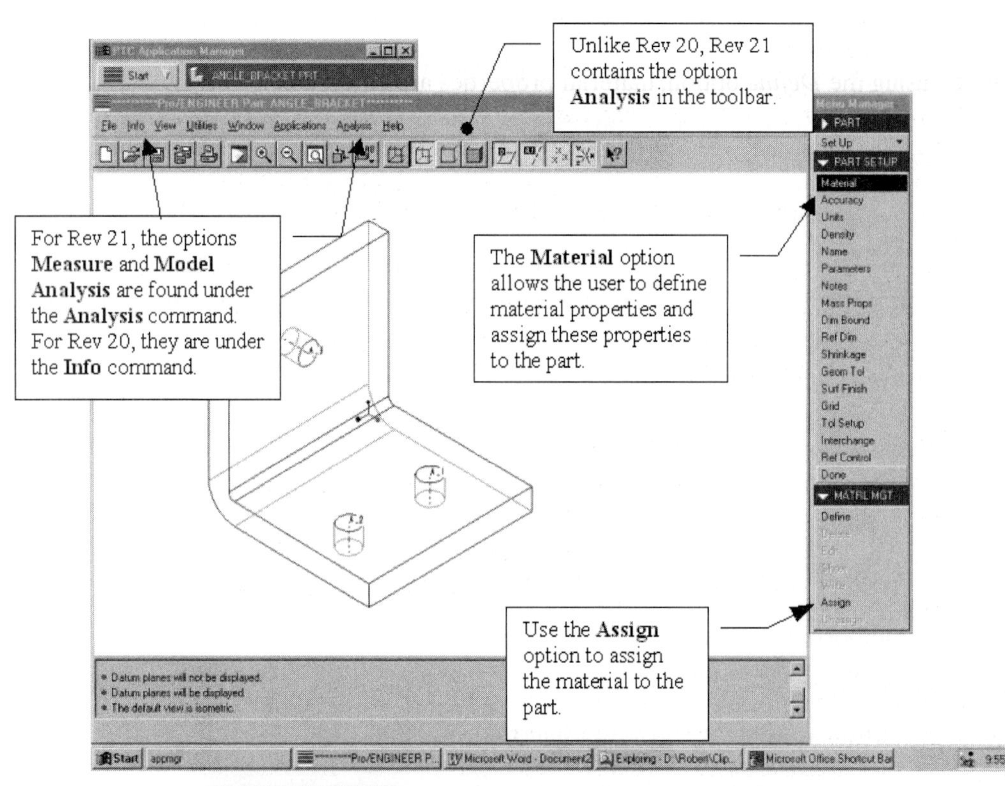


FIGURE 20.1 *Use the* Material *option to define and set material properties for a part.*

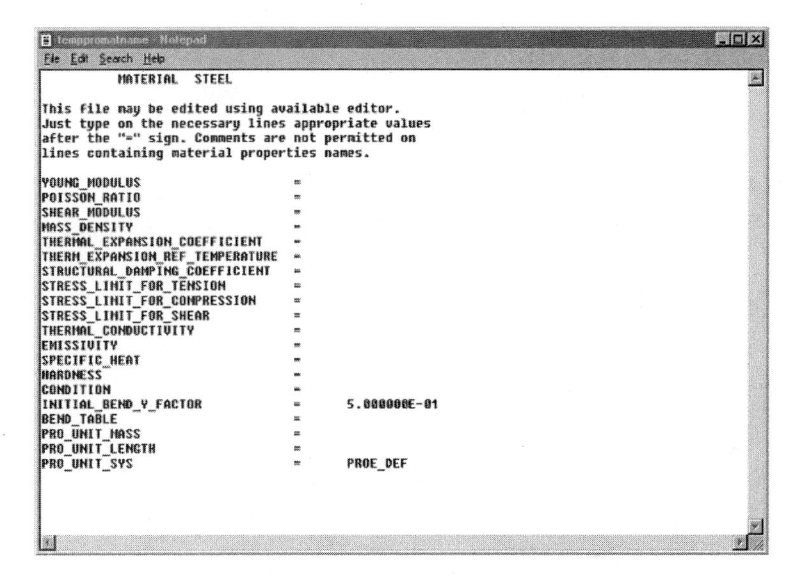


**FIGURE 20.2** *The material properties may be defined by using the* Define *option and by entering the values in the appropriate spaces.*

name by the user. After the material is defined, it may be assigned to a part by using the *Assign* option.

When using the *Define* option, material properties are entered in a window. This window is reproduced in Figure 20.2. In order to define a property, the user simply types the value in the appropriate space. Notice that the density may be entered directly by choosing the *Density* option from *PART SETUP* menu. This approach is used when the density is the only material property required.

When entering material properties, enter the values with the correct units. Also, notice that the computation of engineering properties is dependent on the accuracy of the model. The accuracy may be set by using the *Accuracy* option in the *PART SETUP* menu.

### 20.1.1 TUTORIAL # 91: ENTERING THE DENSITY FOR A MODEL

Retrieve the part "AngleBracket." For our needs in this chapter, we only need to define the density of the material. Therefore, select *Set Up* and *Density.* The units of the model are inches and the material is aluminum. For aluminum, the density is 170 lb/ft$^3$ (.09848 lb/in$^3$). Enter the density in lb/in$^3$. Choose *Done.*

## 20.2 MODEL ANALYSIS

Model analysis may be carried out by using the options *Info* and *Model Analysis* (*Analysis* and *Model Analysis* for Rev 21). The *Model Analysis* box, shown in Figure 20.3, will become available. The type of analysis may be one of the following types:

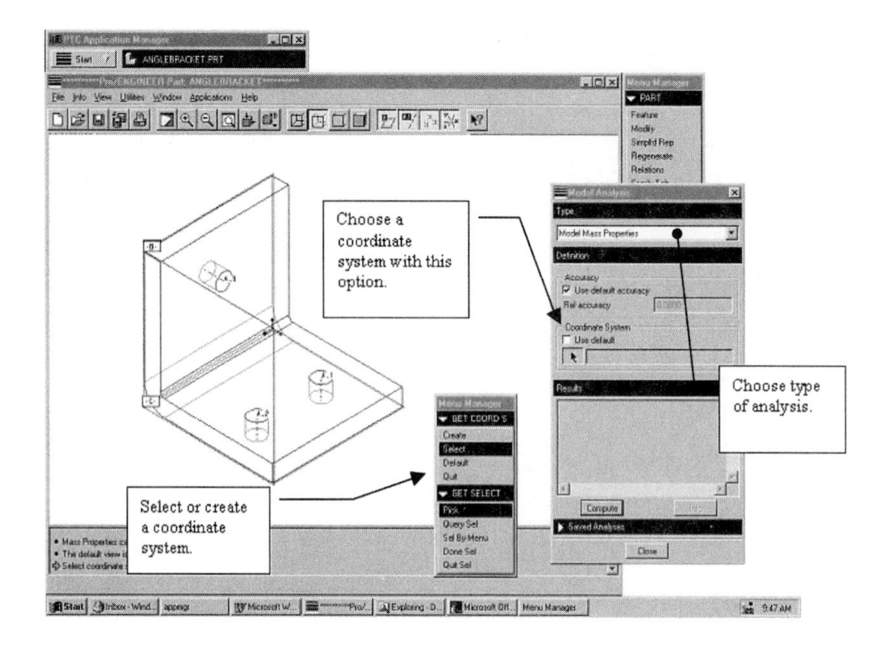


**FIGURE 20.3** *Use the* Model Analysis *box to perform an analysis on the part or assembly.*

1. *Model Mass Properties.* The software calculates mass properties of the part or assembly that include:
   a. The *VOLUME* of the model.
   b. The total *SURFACE AREA* of the model.
   c. The total *MASS* of the model.
   d. The *CENTER OF GRAVITY,* with respect to the chosen coordinate system. The center of gravity is shown on the model.
   e. The *INERTIA TENSOR* for the specified coordinate system.
   f. The *INERTIA TENSOR* at the center of gravity.
   g. The *PRINCIPAL MOMENTS OF INERTIA* at the center of gravity and the principal axes. The principal moments are denoted as 1, 2, and 3.
   h. The *ROTATION MATRIX* for rotation from the chosen coordinate axes to the principal axes. The angle $\phi$ is the angle of rotation about the axis 1, $\theta$ is the angle about axis 2, and $\psi$ is the angle about the axis 3.
   i. The *RADII OF GYRATION* at the center of gravity.
   j. For an assembly, a summary is provided detailing the mass properties of all the components.
2. *X-Section Mass Properties.* The mass properties are calculated for a chosen cross section.

3. *One-Sided Volume.* The volume on one side of a chosen datum is calculated.
4. *Pairs Clearance.* The software calculates the clearance distance or the amount of interference between any two parts, subassemblies, entities, or surfaces. In the case of interference, the software highlights the interfering parts, subassemblies, entities, or surfaces.
5. *Volume Interference.* The software checks to see whether keepin/keepout regions have been violated.
6. *Short Edge.* The software highlights all the edges that are shorter than a prescribed value.
7. *Edge Type.* For a selected edge, Pro/E lists the edge type.
8. *Thickness.* The thickness of a part is checked to see whether it is beyond a given bound. The user may select the lower and/or upper bound for the thickness.

A few words must be said about the analysis of parts with suppressed or blanked features. Because suppressed features are removed from the model generation list, they are not used when performing an analysis. Blanked features are still in the generation list; their visibility has simply been turned off. Therefore, blanked features are used when carrying out model analysis.

### 20.2.1 TUTORIAL # 92: MASS PROPERTIES OF THE ANGLE BRACKET

With the density of the angle bracket entered into the database, the mass properties of the part may be calculated. Choose *Info* and *Model Analysis* (*Analysis* and *Model Analysis* for Rev 21). Then select *Model Mass Properties* for the analysis type. This model does not contain any additional coordinate system. So, use the default coordinate system. Choose the *Compute* button. The software will calculate the properties and load them in the *Results* cell. Select *Done.*

The software will also save the results in a file called "AngleBracket_m_p." This file is a text file and may be opened and printed. Figure 20.4 shows the mass properties for the angle bracket.

### 20.2.2 TUTORIAL # 93: INTERFERENCE BETWEEN TWO PARTS

Whether two parts interfere with each other is important in design. You may recall that in the pulley wheel assembly in chapter 19, an axle was created with an RC5 fit. This tolerance was not added to the model. The diameter of the hole and the axle were set the same. As created, no interference nor a clearance occurs between the two parts.

We can check to see whether this lack of clearance is the case by using the *Model Analysis* and *Pairs Clearance* options. Retrieve the assembly called "PulleyAssembly." Then, choose *Info* and *Model Analysis* (*Analysis* and *Model Analysis* for Rev 21). The *Model Analysis* box will appear. Select the *Two Pairs* type. As shown in Figure 20.5, you need to select the parts in the

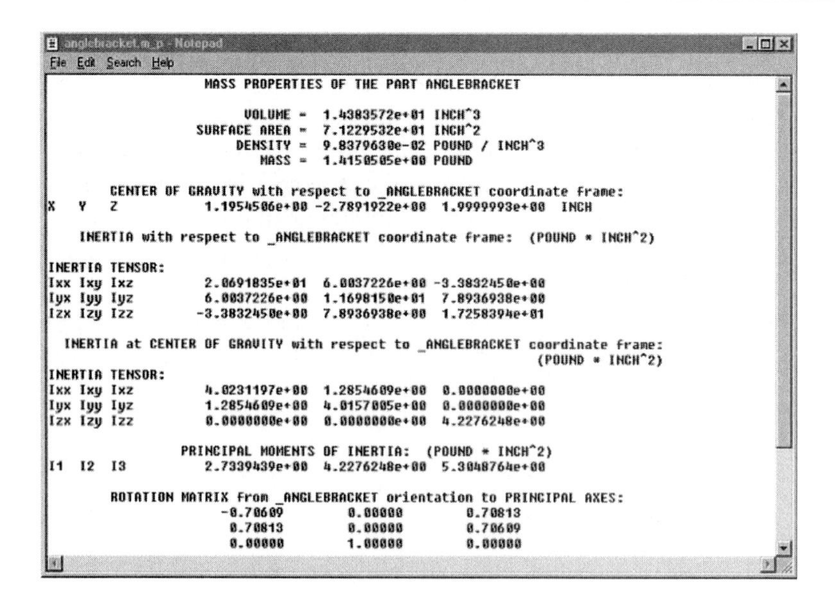

```
anglebracket.m_p - Notepad
File Edit Search Help
                         MASS PROPERTIES OF THE PART ANGLEBRACKET

                             VOLUME =  1.4383572e+01 INCH^3
                       SURFACE AREA =  7.1229532e+01 INCH^2
                            DENSITY =  9.8379630e-02 POUND / INCH^3
                               MASS =  1.4150505e+00 POUND

        CENTER OF GRAVITY with respect to _ANGLEBRACKET coordinate frame:
X   Y   Z            1.1954506e+00 -2.7891922e+00  1.9999993e+00  INCH

    INERTIA with respect to _ANGLEBRACKET coordinate frame:  (POUND * INCH^2)

INERTIA TENSOR:
Ixx Ixy Ixz          2.0691835e+01  6.0037226e+00 -3.3832450e+00
Iyx Iyy Iyz          6.0037226e+00  1.1698150e+01  7.8936938e+00
Izx Izy Izz         -3.3832450e+00  7.8936938e+00  1.7258394e+01

  INERTIA at CENTER OF GRAVITY with respect to _ANGLEBRACKET coordinate frame:
                                                              (POUND * INCH^2)
INERTIA TENSOR:
Ixx Ixy Ixz          4.0231197e+00  1.2854609e+00  0.0000000e+00
Iyx Iyy Iyz          1.2854609e+00  4.0157005e+00  0.0000000e+00
Izx Izy Izz          0.0000000e+00  0.0000000e+00  4.2276248e+00

                 PRINCIPAL MOMENTS OF INERTIA:  (POUND * INCH^2)
I1  I2  I3           2.7339439e+00  4.2276248e+00  5.3048764e+00

        ROTATION MATRIX from _ANGLEBRACKET orientation to PRINCIPAL AXES:
                          -0.70609        0.00000        0.70813
                           0.70813        0.00000        0.70609
                           0.00000        1.00000        0.00000
```

**FIGURE 20.4** *The mass properties are saved in a text file. This file may be printed.*

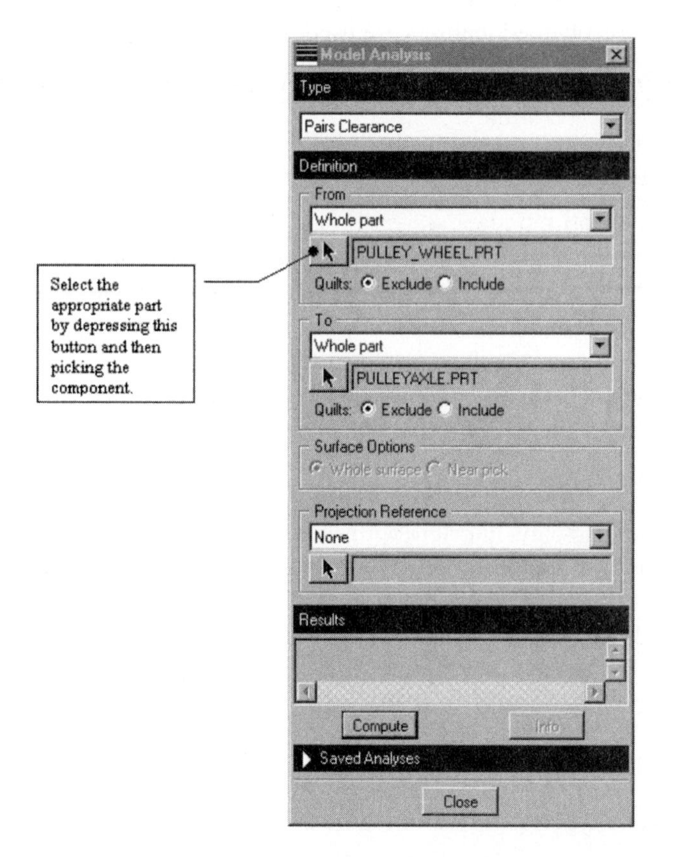


**FIGURE 20.5** *Obtain the clearance between the axle and pulley wheel by selecting the two components.*

assembly. Choose the pulley wheel and axle. Then depress the *Compute* button. The software should respond with the statement:

```
"Zero clearance. No interference was detected."
```

Choose the *Close* button. Note that if the upper and lower limits of the respective parts were changed by using the clearance fit, and the analysis redone, the clearance between the two parts would no longer be zero.

## 20.3 The Measure Option

The *Measure* option may be used to analyze the geometry model or assembly. The *Measure* box shown in Figure 20.6 contains options for carrying out the measurement. The types of measurements that are available are as follows:

1. Curve *Length.* Calculate the length of a curve or edge.
2. *Distance.* Measure the distance between two entities, parts, or subassemblies.
3. *Angle.* Calculate the angle between two entities. The entities may be axes, planar curves, or planar nonlinear curves.
4. *Area.* Obtain the surface area of a chosen entity.
5. *Diameter.* Calculate the diameter of any revolved surface.
6. *Transform.* For two given coordinate systems, the software obtains the transformation matrix.

### 20.3.1 Tutorial # 94: Use of the Measure Option

If the assembly "PulleyAssembly" is not already in memory, retrieve it. Turn off the display of the datum planes and axes. Reorient the model as shown in Figure 20.7. Use either the *Orient the Model* button in the toolbar or the path *View* and *Orient.*

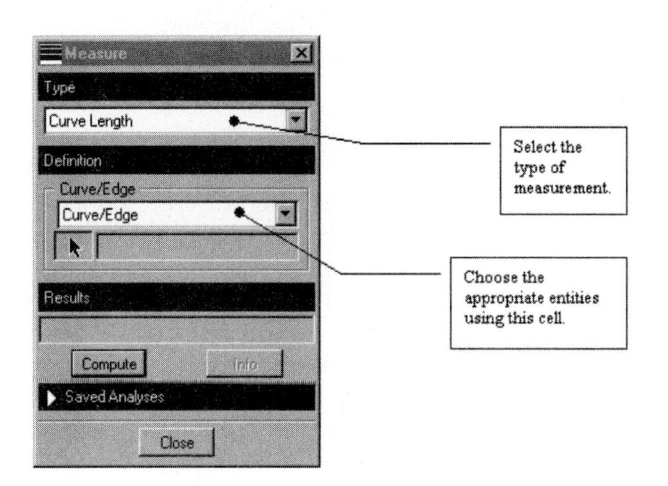


FIGURE 20.6 *The options in the* Measure *box may be used to analyze the geometry of a model or assembly.*

**FIGURE 20.7** Reorient *the model as shown. Also, notice the entities to choose when using the* Measure *option.*

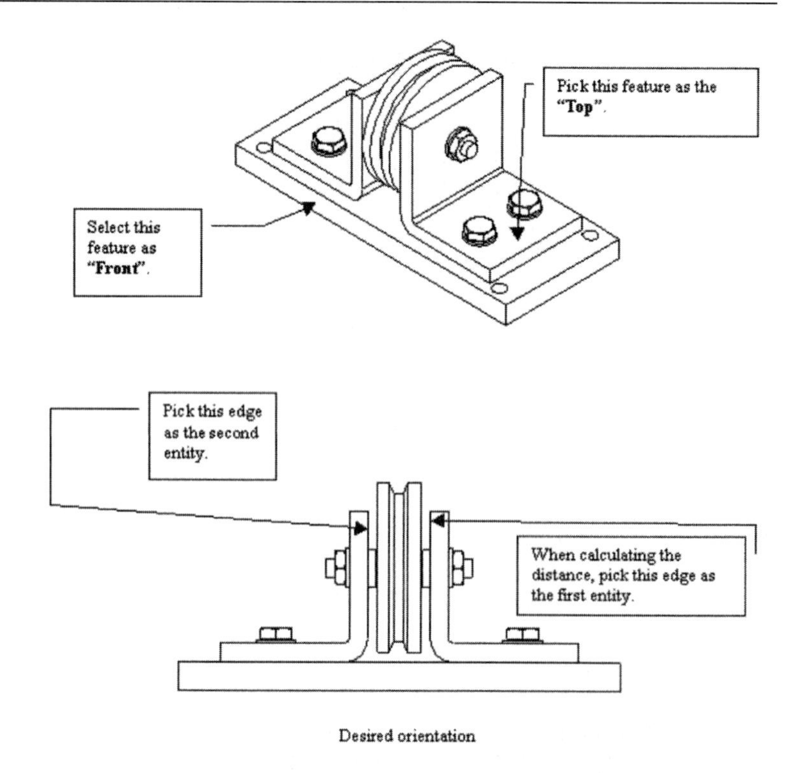


As an example, let us measure the distance between the two edges shown in Figure 20.7. Choose *Info* and *Measure* (*Analysis* and *Measure* for Rev 21). Select *Distance* as the measurement type. Then choose *Line/Axis* for the *Definition.* Pick the two lines shown in Figure 20.7. The linear distance between the two lines is 1.700 inches. Choose *Close.*

Now restore the model to its default *Isometric* orientation. Select *Info* and *Measure* (*Analysis* and *Measure* for Rev 21). For the measurement type, choose *Area* and *Surface* as the definition type. Pick the "*Top*" of the plate as the area as shown in Figure 20.7. The surface is 58.4292 in$^2$. Choose *Close.* Erase the model.

## 20.4 IGES FILE TRANSFER

A part or an assembly may be transferred using an IGES format. The creation of IGES files from an assembly is more complicated because of the number of parts.

The *Export IGES* box shown in Figure 20.8 contains the options for exporting a model in an IGES format. The part may be exported as a wireframe or as a surface model. In a wireframe model, the part is exported with edge information only. The *Surfaces* option, on the other hand, exports the part in a format containing information on the surface as well as the edges. Notice that the *Customize Layers* button may be used to select the appropriate layers.

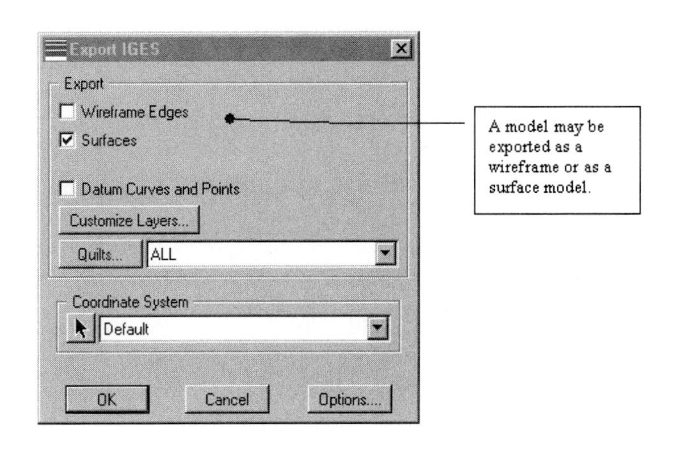


FIGURE 20.8 *The* Export IGES *box is used to export a file.*

When exporting an assembly in IGES format, additional information must be provided concerning the treatment of the individual components in the assembly. This information is provided by using the *File Structure* option in Figure 20.9.

The possible choices for the *File Structure* option are as follows:

1. *Flat.* Output the assembly to a single file. The assembly is treated as a single part. When using this option, it is recommended that each component be placed on a separate layer in order to discriminate among the parts.
2. *One Level.* The assembly is exported to a file with references to the IGES files of the components. The IGES assembly file only contains information regarding the assembly of the components.

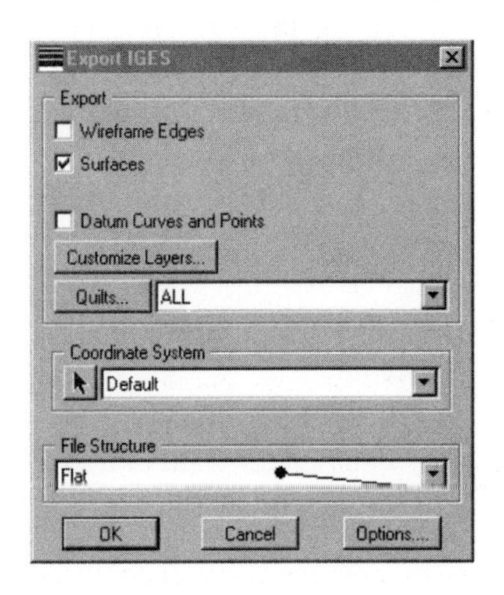


FIGURE 20.9 *Additional information must be provided for an IGES transfer of an assembly. Use the* File Structure *option to provide this information.*

3. *All Levels.* The assembly is exported with references to all the components and component IGES files.
4. *All Parts.* Multiple files are created. These files contain information on each part and assembly. If a component is found more than once in the assembly, a file is created for each copy of the component. The format of the file name is "_cpy_#.igs," where # is the numbered copy of the part.

### 20.4.1 TUTORIAL # 95: AN IGES FILE OF THE ANGLE BRACKET

Retrieve the part "AngleBracket." Then, select *File, Export, Model,* and *IGES.* Accept the default name of the file by hitting the *Return* key. Keep the default settings in the *Export IGES* box. Select *OK.*

The software will create two files. The first of these is "AngleBracket.igs" and is the IGES file of the model. The second file called "AngleBracket_out.log" is a log file containing information on the IGES file and processor. Check to see whether these files exist in your directory.

## 20.5 SUMMARY AND THE STEPS FOR THE OPTIONS

Parts and assemblies may be analyzed by using various options in the *Info* (*Analysis* for Rev 21) command. These options include *Measure* and *Model Analysis.* The *Geom Check* option is an option that should be gray. If it is not, then the model contains geometric error. The software will list suggestions for correcting the errors.

The *Model Analysis* option may be used to obtain mass properties of a model or a cross section. It may also be used to find parts or subassemblies that interfere with one another.

The calculation of mass properties requires the definition of the material of the model. In general, a material may be created, properties assigned, and the material assigned to a part using the following approach:

1. Select *Set Up* from the *PART* menu.
2. Choose *Material* and *Define.*
3. Enter the material properties. Exit.
4. Select *Assign.*
5. Choose the material.

The general steps for obtaining an analysis of a model or assembly are as follows:

1. Select *Info* and *Model Analysis* (*Analysis* and *Measure* for Rev 21).
2. Choose the analysis type.
3. If required, select the desired coordinate system.
4. Enter any additional information required by the type of analysis.

5. Choose *Compute.*
6. Select *Done.*

The *Measure* option may be used to calculate such quantities as distances, surface area, and curve lengths. The following sequence of steps may be used to analyze a part or assembly using the *Measure* option:

1. Select *Info* and *Measure* (*Analysis* and *Measure* for Rev 21).
2. Choose the measurement type.
3. Select the *Definition* type.
4. Choose the appropriate feature or features.
5. If required, select *Compute.*

A part or assembly may be exported in IGES format. The geometry may be exported as a wireframe or surface model. When exporting assemblies, information must be provided as to how to treat the individual components. Options are provided for exporting a single file representing the entire assembly or multiple files each containing information on the components and the assembly. In general, a part or assembly may be exported by using the following procedure:

1. Select *File, Export, Model,* and *IGES.*
2. Enter a name for the IGES file or accept the default name.
3. Pick the Export type, coordinate system, and, for an assembly, the *File Structure.* If desired, select the appropriate layer or layers.
4. Choose *OK.*

## 20.6 ADDITIONAL EXERCISES

For exercises 20.1 through 20.12, obtain mass properties of the given parts. Use the default coordinate system unless instructed otherwise.

20.1 Cement trowel (see also exercise 8.1).
20.2 Side guide (see also exercise 8.2).
20.3 Control handle (see also exercise 8.3).
20.4 External lug (see also exercise 8.4).
20.5 Pipe clamp top (see also exercise 8.5).
20.6 Pipe clamp bottom (see also exercise 8.6).
20.7 Side support (see also exercise 8.7).
20.8 Guide plate (see also exercise 8.8).
20.9 Single bearing bracket (see also exercise 8.9).
20.10 Double bearing bracket (see also exercise 8.10).
20.11 Roller (see also exercise 10.1).
20.12 Baseball bat (see also exercise 10.2).
20.13 Using the *Distance* option, check the dimensions of the angle bracket. Do these values conform to Figure 2.20?

20.14 Show that in the subassembly "Flood Light" (exercise 19.6), interference exists between the parts "FloodLightCover" and "EmergencyLightReflector." Redesign one or more of the parts to eliminate the interference.

20.15 Check the assembly "Steady Rest" (exercise 19.7) for interference. If interference exists, redesign the conflicting parts.

20.16 Check the assembly "Blower" (exercise 19.8) for any possible interference. If interference exists, redesign the appropriate parts.

# BIBLIOGRAPHY


Abbott, I. H. and A. E. Von Doenhoff, 1959. *Theory of Wing Sections,* Dover, NY.

Giesecke, F. E., A. Mitchell, H. A. Spencer, I. L. Hill, J. T. Dygdon, J. E. Novak, and S. Lockhart, 1997. *Technical Drawing* 5th ed., Prentice-Hall, Upper Saddle River, NJ.

Parametric Technology Corporation, 1997.
*—Pro/Engineer Assembly Modeling User's Guide*
*—Pro/Engineer Drawing User's Guide*
*—Pro/Engineer Fundamentals*
*—Pro/Engineer Interface Guide*
*—Pro/Engineer Part Modeling User's Guide*

Parametric Technology Corporation, 1998. *Pro/Engineer 20 Release Notes.*

Parametric Technology Corporation, 1999. *Pro/Engineer 21 Release Notes.*

# INDEX

*Italics* indicate options, *CAP ITALICS* indicate menus, **boldface numbers** indicate tutorials. *See also* suggests related topics or more detailed breakdowns.


Adaptors, round, 389
*Add,* 17
Adding material, 123–143. *See also* Material addition
*Add Inn Fcs,* 261
*Add Items,* 299, 300, 307
*Add Point,* 211
*Add View,* 381. *See also View* options
Adjustment screws, **108–111**
*ADVANCED FEATURE OPTIONS,* 255
*ADVANCED GEOMETRY, 41,* 41–42, 231
Aerator, 293
Aircraft wing, **334–340**
Airfoil section, **215–217**
*Align,* 408, 425–429
  vs. *Mate,* 419–420
*ALIGNMENT, 57*
*Alignment,* 56
*Align Offset,* 408, 425
*Align* options, 373–374
*All Levels,* 466
*All Parts,* 466
Analysis, 455, *456*
*Analysis,* 457–459, 464–466. *See also Info*
*ANGLE,* 110, *111*
Angle brackets, **39–41, 81–84,** 142, **187–188, 351–356, 359–364**
  full section, **375–378**
  half section, **381–383**
  mass properties, **459**
  offset section, **378–381**
  in pulley assembly, 414–417
  revolved section, **383–384**
  tolerances, **394, 397–401**
Angled blocks, 320
Angular dimensions, 31
ANSI standards, 391
*Appearance Editor,* 17–18, *18*
*Applications,* 6
Arc, centering, 37
*ARCTYPE,* 133
Arms, connecting, **130–134, 188, 383–384**
Arrowhead style, 359
Arrows
  flipping of, 360
  reference plane, 382
  style of, 349
ASME standard, 397
*Assemble,* 406, 428–429
AssembleBlockA/AssembleBlockB, 409–410
Assemblies, 404–454
  adding to drawing, 438, **438–439,** 444
  additional constraints, 425, **425–429**
  bill of material (BOM) and balloons, 440, **440–441**
  creating file, 404–407, 441–442
  datum planes for, 411, **411–421,** 442
  exploded, 430–431, **432–436,** 443
  offset lines in, 437, **437–438,** 443
  parametric, 407–409, **409–410**
  parts creation, 422, **423–424,** 442–443
  summary and steps, 441–444
*ASSEMBLY, 405,* 405–406, 432
*Assembly Reference,* 410
*At Center,* 208
*ATTRIBUTES,* 124, *247,* 278, *279*
Auxiliary views, 364–365, **365–367**
*2Axes,* 203
*Axis,* 384
Axis
  datum, 200–203, **201–202,** 218
  radial dimension and, 90
  reference, 90
  of symmetry, 384
Axles
  for pulley assembly, **423–424**
  slotted, **107–109**

Ball, handle, 238
*Balloon,* 440–441
Baseball bat, 237
Base feature(s). *See also* individual subtopics
  adding draft to, **32,** 324–325, **325.** *See also Tweak*
  adding holes, 80–100
    to angle bracket, 82–84
    bolt circle in pipe flange, 87–92
    coaxial for bushing, 84
    general procedure, 80–81
    on inclined plane on rod support, 92–93
    sketched in plate, 85–87
    summary of steps, 93–94
  datum planes for, 53–79. *See also* Datum planes
  material removal, 101–122
  for *Shell,* **111–115**
BaseOrient
  printing, 21
  reorienting, 15–16, *16*
  retrieving and saving, 11–14, *13*
Bases, clamp, 371
*Basic,* 17
Bat, baseball, 237
Bearing cups, 240, 387, 388
Bill of material (BOM) and balloons, 440, **440–441**
Blades, propeller. *See* Propellers
Blanking
  analysis and, 459
  with *Blank,* 297
  with *Set Display,* 297
  solid features *(Suppress),* 298, 307
*Blank* UDF, 168

*Blend,* 277
HVAC takeoff, **277–281**
*BLEND OPTS, 282,* 284
Blends, 276–292
basic principles, 276–281
general, 276–277
parallel, 276–277
rotational, 276–277
summary and steps, 288–289
swept, 276, 282, **282–288**
twisted, 277, 288
*Blend Section,* 67
*Blind,* 34, 61, 63, 81
Blocks, angled, 320
Blowers, centrifugal, **282–287**
*Bndry Chain,* 261
Bolt circles, **87–92**
Bolts, moving of, 434–436
BOM (bill of material), 440, **440–441**
BOM files, 440
Borders, 349–350, 356
Bosses, 243
Boss-hole combinations, 161–162
*Both Sides,* 61, 63
Brackets
angle. *See* Angle brackets
double bearing, 142
single bearing, 142
U, **102–103, 134–135**
Bridge truss, 225, 270. *See also* Trusses
Browser, with *Help,* 6
Bushings, **54–59, 84**

CADD
information transfer to, 455–466
traditional vs. Pro/E, 1–4
CAM systems, information transfer to, 455–466
*Cancel,* 8
Cartesian coordinates, 205–206. *See also* Coordinate systems
Center, of circle, 56
Centering, 355
pf arc, 37
*CenterLine,* 129
*Centerline,* 87
*Center Point,* 384
Centrifugal blower, **282–287**
*CHAIN/CHAIN OPT,* 261, *261*
Chamfers, 185, 191–193
on adjustment screw and nut, **191**
Circles. *See also* Pulleys; Wheels
bolt, 87–92, **87–92**
center, 56
diameter, 56
washers, **103–106**
Circular cavity, 152–154
Circular edge, **188**
Circular holes, 282–284
Circular protrusion, 310
Circular/rectangular blends, 276–277, 280
Clamp, pipe, 179, 319
Clamp bases, 371
Clearance, **459–460**
pairs, 459
Clip
paper, 269
spring, 269
*Coaxial,* 81
Coaxial holes, 84, 299–300
Collars, tapered, **201–203,** 241, 388, 389
Color
of datum planes in assemblies, 408
for undimensioned element, 29
*Color Editor,* 17
*Colors,* 7–8
Color scheme, 17–18
#command, 19
Command options, 6–11. *See also* specific commands
*COMPONENT,* 406, *406*
*Component,* 428–429
*Component Placement,* 409, *409,* 410
*Component Reference,* 409
Components, list of (bill of material), 440–441
Cone head rivet, 139
Configuration file (Config.pro), 9, 18–19
Conic sections, 41–42
Connecting arms, **130–134, 188, 383–384**
*CONNECT TYPE,* 211, *212*
Constraints, plane, 66–67
*Constraint Type,* 407–408, 409–410
Coordinates, for grid, 350
*COORDINATE SYSTEM, 205*
Coordinate systems, 67
assembly by, 408
for blends, 284
in component repositioning, 431
datum, 203–207, **204–207,** 218, 219–219
COORDINATE SYSTEMS OPTIONS, *203,* 203–204
*Copy,* 162–163, 177
with copy mirror, **163–165**
*Copy From Existing,* 423–424
Copy mirror, **163–165**
*Copy Pos,* 431
Cosmetic features, 311–321
options for, 311–313
summary and steps, 318
text for cover plate, **316–317**
threads on nut, **314–316**
Counterbored holes, 85–87, 99
Couplers, 141, 319
Cover plates, **325–328**
cosmetic text on, **316–317**
Covers, 345
electric motor, 321
emergency light, **325**
floodlight, 243
light shell, **111–115**
*Create,* 37, 39, 43
assembly, 406
*Create New Object,* 347, *347*
*Creation Options,* 422
Crosshatch, modifying, 375
Cross sections. *See* Sections; *Sketcher*
*CRV OPTIONS,* 210, *211*
*Crv X Crv,* 209
cs[two]2.pts file, 336
Cups
bearing, 387, 388
fixed bearing, 240
*Curve, UpTo,* 82
*Curve Chain,* 257, 261
Curves
datum, 210–219. *See also* Datum curves
neutral, 324
*Quilt* and airplane wing, **334–340**
*Curve X Surf,* 208
*Customize Screen,* 9–10
*Cut,* 63, 157
*Cut/Slot,* 101–109, 115–116
Cylinders, 58–60, 123–124. *See also* Flanges
collar, **201–203**
Cylindrical coordinates, 205–206

Dash pot lifter, 274, 390
*DATUM,* 201, *201*
Datum, reference, 395–397
Datum axis, 200–203, **201–202,** 218
*Datum/Axis,* 395
Datum coordinate systems, 203–207, **204–207,** 218, 218–219
Datum curves, 210–211, 219
  adding to two-dimensional frame, **211–212**
  airfoil section, **215–218**
  king post truss, **213–214**
  sweep along, **261–262**
*DATUM* menu, *56*
*DATUM PLANE,* 107, 201, *205*
Datum planes, 53–79, 375, 381, 384
  assembly, 408, 411, **411–421,** 442
  for base features, 53–79
  constraining, 66–67
  default, 53–65, *55,* 74–76, 103
  GDT, 399–400
  start part and, 69–73
  steps for adding, 73–74
  turning off, 7
  user-defined, 65–69, 76–77
*DATUM POINT,* 207, *208,* 209
Datum points, 200, 206–209, **209,** 219
Decimal digits setting, 397
*Default,* 16
*Default-Expld,* 431
Default icons, *10,* 10–11
*Default Orientation,* 9–10
Defaults
  changing, 32
  coordinate system, 204
  datum planes, 53–65, *55,* 74–76
  exploded state, 431
  for measures, 31
  sections, 373
  standards, 397
  tolerances, 391
  tolerance values, 401–402
Default settings, 7, 9
*Define,* 27, 28, 456–457
*Defined/Defining,* 28
*Delete,* 36
  all or old versions, 11
*Density,* 456–457
*Dependent/independent,* 163
*Depth,* 311
Descriptive text. *See* Text
*Destination,* 21
*DETAIL,* 349, *349,* 358, 395. *See also* Drawings
*Detailed,* 348
Dialog boxes. *See Command options* and specific boxes
Diameter, of circle, 56
Digits, decimal setting, 397
*DIM BOUNDS,* 391, *393*
*Dimension,* 25, 42, 43, 60, 61, 393–394, 396–397. *See also* Tolerances
Dimensioning, 29
Dimensions, on drawing, 356–358, **359–364**
*DIM PARAMS,* 358, *359*
*DIMS,* 164
Display, turning off, 7
*Divide,* 36
*Done,* 39, 43
*Done/Return,* 45
Doorknob, 237
Double bearing bracket, 142
Double-sided supports, 180
*Draft,* 324–325, 340–341
*DRAFT OPTS,* *324,* 325
Drafts, adding to base feature, **32,** 324–325
*DRAWING,* 378
Drawing. *See also Sketcher;* Sketching
  adding dimensions and text, 356–358, **359–364,** 369
  adding tolerances, 391–403. *See also* Tolerances
  auxiliary views, 364–365, **365–367**
  creating drawing, **348–356,** 368–369
  creating format, 346–348, **348–352,** 368
  sections
    full section, **375–378**
    half section, **381–383**
    offset section, **378–381**
    revolved section, **383–384**
    summary and steps, 385–386
Drawings, adding assemblies to, 438, **438–439,** 444
*Dynamic Explode Component* window, 431–432, *432*

*Ear,* 323, 325–327, **327–330,** 341
Edge
  circular, **188**
  for propeller blades, **189**
  short, 459
*Edge Chain,* 186, 188, 190
*Edge on Srf,* 408
*Edge-Surf,* 186
Edge type, 459
*Edit Config,* 9, 18–19
Editing, relation, 45
Editors
  *Appearance Editor,* 17–18, *18*
  *Color,* 17
*Edit Points,* 209
*Edit Rel,* 31
Electric motor covers, 321
Emergency light cover, **111–115, 325**
Emergency light holder, **123–129,** 308
Endpoints
  aligning, 131
  defining, 284
Engineering information, 455–466
  defining and assigning material properties, 456–457, **457**
  IGES file transfer, 462–464, **464**
  *Measure* option, 461, **461–462**
  model analysis, 457–459, 464–465
    interference between parts, **459–461**
    mass properties, **459**
  summary and steps, 464–465
*Enter Points,* 209
Entities
  number in blend, 276–277, 278
  reference, 396, 432
*Entity,* 8
*Entity Colors,* 8, *9*
*Entity/Edge,* 431
*ENVIRONMENT,* 7
*Environment,* 7, *8,* 21
*Erase,* 357, *357,* 366, 384, 399
Erasing files, 11
*EXPLD STATE,* 430, *430,* 437–438
Exploded assemblies, 430–431, **432–436,** 443
*ExplodeState,* 430–431, 432, 437–438
*Export IGES,* 462–464, **464**
*Extrude,* 37, 39, 43, 51, 52

*Failure Diagnostics,* 302, *303,* 304
*False,* 315
*Family Table,* 167
*Family Tables,* 171

Fasteners, moving of, 434–436
*Fcs* options, 261
*FEAT,* 145, *145,* 157, 293–310. *See also* Feature management; Speed-up options
*FEAT PARAM,* 311, *311*
*FEATURE,* 55
*Feature,* 37, 39, 43
  in *ASSEMBLY,* 405, 411
Feature-based modeling, 1–2
Feature Creation box, *294,* 294–295
*Feature* creation box, 27
Feature management, 293–310. *See also Specific options*
  with *LAYERS,* 297–300
  with *Redefine,* 294–296
  with *Reorder* and *Reroute,* 296–297
  with *Resolve,* 303–306
  summary and steps, 306–308
  with *Suppress* and *Resume,* 300–302
Feature order, 165–166
Features
  base. *See* Base features
  blanked, 459
  cosmetic, 311–321. *See also* Cosmetic features
  suppressed, 459
*File,* 6–7, 21–22
Files
  assembly, 404–407, 441–442
  configuration, 9
  creating new, 43
  importing datum curve, 210
  opening, 11
  part, 11–14
  transfer of, 462–464, **464.** *See also* Engineering information
*File Structure,* 465–466
Fillets, 185–187, 193. *See also* Rounds
  adding draft to, 324–325
  with round on angle bracket, **187–188**
Fink truss, 224, 270
*First,* 147
Fixed bearing cup, 240
*Fix Model,* 302
Flanges, 123–124, **124–129,** 136
  pipe, **87–92**
  radial hole placement, 147–148
*Flat* file structure, 465
Flat roof truss, 221–222
Flat top countersunk rivet, 140
*Flip Arrows,* 360
Floodlight cover, 243–244
Forks, shifter, **59–65, 299–300,** 309
*Format,* 347–348
Format creation, for multiviews, 346–348, **348–352,** 368
Frame
  highway sign, 271
  two-dimensional, **204–206**
*Free Note,* 397
*FromDifModel,* 163
*FromDifVers,* 163
*From File,* 204, 210
*From-To,* 261
Full sections, **375–378**
Function keys, 19

Gaskets, 181, **249–256**
*General,* 355, 439
General blends, 276–277
Geometric dimension and tolerances (GDT), 395–401. *See also* Tolerances
Geometric tolerances, 395–397, **397–401**
*GEOMETRY,* 41–42, 133, 150, 231
*Geometry,* 8
*GEOM TOL,* 395–399, *396,* 402–403
*GEOM TOOLS,* 46
*Geom Tools,* 36–39, 46, 280, 284–287
*Grid,* 349
Grid coordinates, 350
*Groove,* 311
*GROUP,* 159
*Group,* 157–161, 177
  vibration isolator pad, **160–162**
*Group elements,* 164
Groups, UDF-driven, 167
Gyration, radii of, 458

Half-sections, **381–383**
*Half View,* 366
Handle ball, 238
Handles, 268, 318
  ladle, 275
  vise, **25–29**
Handrail supports, 274, 390
Hanger, shaft, 273
*Helical Sweep,* 255–261, 268–269
  screw threads, **258–261**
  spring, **256–258**
Helical sweeps, **256–261,** 294
*Help,* 6
  browser used for, 6
Highway sign frame, 271
Hinges, **32–35,** 184
  completion with *Mirror Geom,* **165–166**
Holders, lamp, **123–129**
*Hole,* 252
Holes, 80–100
  to angle bracket, 82–84
  bolt circle in pipe flange, 87–92
  circular, 282–284
  clearance for, 424
  coaxial, 84, 299–300
  counterbore, 85–87, 99
  on inclined plane on rod support, 92–93
  Pro/E features for, 80–81
  radial, 87–91
  radial repeated, 147–148
  *Sketched,* 80–81, 85–87
  sketched in plate, 85–87
  *Straight,* 80
  summary of steps, 93–94
  tolerances and, 397–401
Hook spanner, 272
Horizontality, 31
Houses, 343–344
Housing, robot arm, 321
Howe roof truss, 224, 342
HVAC takeoff, **277–281**

Icons
  customizing, 9–10
  default, *10,* 10–11
  grayed, 9–10
*Identical,* 147
Inclined planes, 65–69
  holes in, 92–93
  oblique, 365
*Independent/Dependent,* 163
*Independent Groups,* 167
Inertia, principal moments of, 458
Inertia tensor, 458
*Info,* 6, 27, 455, 457–459, 464–466
*Info Window,* 432
Initial graphics exchange specification (IGES), 455, 462–464, **464.** *See also* Engineering information; Files, transfer of
*Insert,* 408

*Insert Point,* 211
Interface. *See* Pro/E interface
Interference between parts, **459–461**
Interlocking parts, **330–333**
*Intersect,* 36
*Intersection,* 276, 280
*Intr Surfs,* 210
*Investigate,* 302
ISO/DIN standards, 391
*Isometric,* as default, 7
Isometric orientation, 20
*Items* options, 299, 300, 307

*Justify,* 440

Keys
  function, 19
  hot *(Mapkeys),* 18–20
King Post Truss, **213–215**

Labels, moving of, 399
Ladle, 242, 275
Lamp holders, **123–129**
*Landscape,* 347, 349
Latches, 244, 328–329
*Layer,* 297–299, **299–300**
*Layer Items,* 300
*LAYERS, 299*
*Leader,* 397
Levels, multiple. *See* Layers
Lever, lift, 142
Lifter, dash pot, 138, 274, 380
Lift lever, 142
Lightholder, emergency, 308
Light shell covers, **111–115**
Light source, 16–17
*Linear,* 81, 83
Linear increment, *Pattern* with, **146–147**
*Line Style,* 437, *437*
*LINE TYPE,* 86, *86*
*Lip,* 330, **330–333,** 341–342
*Load Config,* 9
*Local section,* 373
Lock plunger, 138
Lugs, 224

Machine wheel, 242
Main drawing window, 5
*Major Diam,* 311
*MAKE DATUM,* 66
*Make Datum,* 66, 148–150, *149, 150,* 154, 375, 381. *See also* Datum planes
  reference axis, 90–91
*MapKey,* 7, 18–20, *19*
Mass properties, **459**
  model, 458
  section, 458–459
*Mate,* 407, 412, 425–429
  vs. *Align,* 419–420
*Mate Offset,* 407, 425, 426
*Material,* 456–457
Material addition, 123–143
  flanges, 123–124, **124–129,** 136
  ribs, 130, **130–134,** 136
  shafts, 134, **134–135,** 136–137
Material removal, 101–122
  cutting/slotting, 101–109
  necks, 109–111
  *Shell* option, 111–115
  summary steps, 115–116
*Measure,* 455, *456,* 461–462
Measure, units of, 31, 50, 249
Memory, erasing to clean up, 11
Menus
  pop-up, 6
  pull-down, 6
*Merge Ends,* 247, 266
Message window, 5
Metric. *See* Units of measure
Metric start part construction, 71–83
Minimization, 432
Mirror, copy, **163–165**
*Mirror,* vs. *Mirror Geom,* 162
*Mirror Geom,* 165–166, 178, 213, 254
  hinge completion, **165–166**
*Model Analysis,* 455, *456*
Model analysis, 457–459, 464–465
  interference between parts, **459–461**
  mass properties, **459**
Modeling
  constraint-based, 3–4
  feature-based, 1–2
  non-constraint-based (primitive), 2, *3, 4*
  parametric, 1, *2*
  traditional CADD, 1, *2, 3*
Model mass properties, 458
Models
  solid, 16–18
  three-dimensional. *see* Drawing; *Sketcher*
*Model Setup,* 18
*Model Tree,* 11–14, *12*
  copying and, 163
  order of features in, 165–166
  for vise handle, *26,* 27
Model tree, reordering, 111
*Modify,* 19, 43, 56, 349
*Modify Dimension,* 364
*MODIFY PROMPT, 170,* 171
*MODIFY SEC TEXT2,* 313, *313*
*MODIFY TEXT,* 357
Mouse
  button features, *14,* 14–15
  defined for *Sketcher,* 25
  orienting with, 15
*Move,* 163
*Move Text,* 357
*MOVE TYPE, 430,* 431
Moving components, 432–433, 434
*MTNPRF, 430,* 430–431
Mug (*Thin* option), **228–229**
Multiple levels. *See* Layers
*Multiple Rad,* 211
Multiview drawings, 346–371, 379–381. *See also* Drawing

*Name,* 22
*Neck,* 110, 119
Necks, 108–111, 115–116
Neutral curves, 324
Neutral planes, 324
*New,* 20
*New Drawing,* 347
*New Refs,* 162
*No Inn Fcs,* 261
Non-constraint-based (primitive) modeling, 2, *3, 4*
*Normal Pin,* 201
*Normal* UDF, 168
*Note* (text), 351–352
*NOTE TYPES,* 351–352
Nozzle, Venturi, 239
*Nrm to Spine,* 282
Number of entities, in blend, 276–277, 278
Nuts
  cosmetic threads on, **314–316**
  with rotational pattern, **148–150**

Oblique planes, 65, 365. *See also* Inclined planes
*Offs By View,* 204

*Offset,* 107, 204, 407, 408
*Offset/Coord Sys,* 67
*Offset Csys,* 208
Offset cutting planes, 374
*OffsetFromSrf,* 210
*OffsetInSrf,* 210
*Offset Lines,* 437, *437*
Offset lines, in assemblies, 437, **437–438,** 443
*Offset/Plane,* 66–67
*Offset Point,* 209
Offset sections, **378–381**
Offset supports, 319
*Offset Surf,* 208
Oil pans, 345
*OK,* 27
*On Curve,* 208
*One By One,* 186, 261
One-level file structure, 465
*One Side,* 61, 83, 84, 86
One-sided volume, 459
*On Point,* 81
*On Surface,* 208
*On Vertex,* 208
*Open,* 11, 21–22
*OPTIONS,* 128, *203,* 203–204
Options, command, 6–11. *See also* Command options *and specific commands*
*Orient,* 15–16, 22, 408
*Orientation,* 15–16, 22
Orientation, *Mapkeys* for, 20
Orientation/reorientation
  BaseOrient, 15–16, *16*
  model, 14–15
*Origin Point,* 43
*Orig + Zaxis,* 204
Orthographic views, 15

*Package,* 406–407
Pages (sheets), adding, 378
Pairs clearance, 459
*Pan,* 14–15
Pans, oil, 345
Paper clip, 269
Parabolic reflector, **229–236.** *See also Sketcher*
Parallel blends, 276–277
Parameters, setting up in *Sketcher,* 31–35
Parametric assemblies, 407–409, **409–410**
Parametric modeling, 1, *2*
*PART,* 12, 22, 152, 293
*Part,* 25–29, 43
Parts
  adding lip to, **330–333**
  construction with *Sketcher,* 24–53
  creation in *Assembly* mode, 422, **423–424,** 442–443
  datum planes used for, 54. *See also* Datum planes
  interference between, **459–461**
  start, 69–73, **71–72,** 422
*PART SETUP,* 31, 32, 391, *393, 456,* 456–457
*PAT OPTIONS,* 146, *146*
*PATTERN,* 150
*Pattern,* 145–157, 175–177, 325–326
  in creating UDF, 169
  within *GROUP,* 159
  with linear increment, **146–147**
  nut with rotational pattern, **148–150**
  propeller, **150–157**
  with radial increment, **147–148**
*PhotoRender,* 16–17
*Pin + 2Axes,* 204
Pins, 139
  detent, 141
Pipe clamp, 179, 319
Pipe flanges, **87–92**
*Place,* 407–408
*PLACEMENT,* 81, 83
Placement planes, 87
*Place Sec,* 43, 44, 46
Planar sections, 374–375
*Plane,* 55. *See also* Datum planes
*Plane Normal,* 431
*Plane* options, 431
*Planes,* 203
Planes
  BaseOrient, 15–16
  datum, 53–79, 375, 381, 384, 399–400, 408, 411, **411–421,** 442. *See also* Datum planes
  inclined, 65–69
    holes in, 92–93
  inclined oblique, 365
  neutral, 324
  oblique, 65–67
  offset cutting, 374
  placement, 87
  reference, 382
  sketching, 151–152, 155, 378
Plates, **85–87**
  cover, **316–317, 325–328**
  skew, 182
  tension, **36–39**
  wing, 182
*Plus-Minus* options, 401–402
*Pnt + 2Axes,* 203
*Pnt Normal Pin,* 201
*Pnt on Srf,* 408
*Pnt on Surf,* 201
*POINT ARRAY,* 207, *207,* 209
*Point* options, 211, 336–338, 384
Points, datum. *See* Datum points
Pop-up menus, 6
*Preferences,* 7, 9, 18–19
  in *MTNPREF,* 432
*Preview,* 27–28
  of assembly, 410
Primitive (non-constraint-based) modeling, 2, *3, 4*
Principal moments of inertia, 458
*Principal Sys,* 45
*Print,* 20–21
Pro/E
  general characteristics, 1–4
  interface, *5,* 5–11, *7–9*
*Projected,* 210
*Projection,* 348
*PROMPTS,* 169, *170*
Propellers, **150–157, 189**
*Properties,* print, 21
PROTAB editor, 359
*Protrusion,* 37, 39, 43, 102, 121
Protrusions. *See also* Ears
  as base feature, 58–65
  circular, 310
  U-shaped, 58–60
.prt extension, 11, 21
Pull-down menus, 6
Pulley wheels, **227–228**
  as assembly, **411–422**
  axles for, **423–424**
  interference between parts, **459–460**
  measuring, **461–462**

*Query Sel,* 188, 189, 190, 232, 314–315, 317, 325, 409–410, 434
*Query Set,* 83

*Quick Fix,* 302
Quilts, 322, 334–340, **335–340,** 342. *See also Tweak*
*QUILT SURF, 323,* 334

*Rad,* 211, 221
*Radial,* 81
Radial holes, 81, 87–91
Radial increment, *Pattern* with, **147–148**
Radii of gyration, 458
Ratchets, 183
*Read-Only* UDF, 168
*Rectangle,* 150
*Redefine,* 294, **294–296,** 306–307
*Redo,* 431
Reference datums, 395–397
Reference entity, 396, 432
Reference plane, 382
*Refit,* 15
Reflector, parabolic, **229–236.** *See also Sketcher*
*Refs,* 27
*Regenerate,* 12, 22, 25, 34, 39, 42–44, 56
  with blends, 287
*Regeneration*
  in *Make Datum,* 148
  *Resolve* and, **301–306**
  steps, 45
  *Suppress* and *Resume* and, 300
Regeneration, failed, 300, 301, **301–306**
Regeneration symbols, 29–31, 30t
Relation
  editing, 45
  to reference axis, 90
  steps, 45
*Relations,* 31, 278
Removing material, 101–122. *See also* Material removal and specific operations
*Reorder,* 111, 296, **296–297,** 307
*Repaint,* 349
*Reroute,* 296, 307
*Reset,* 431
*Resolve,* 185
*RESOLVE FEAT,* 302, *303*
*Resume,* 299, 308
*Revolve,* 226–243
*Revolved,* 348
Revolved sections, 227–243, **383–384**
  mug (*Thin* option), **228–229**
  parabolic reflector, **229–236.** *See also Sketcher*
  pulley wheel (*Solid* option), **227–228**
*REV TO,* 227, *227,* 236
Rho method, 43, 231–232
*Rib,* 130
Ribs, 130, **130–134,** 136
Rivet
  cone head, 139
  flat top countersunk, 140
RND SET ATTR, 186, *186*
Robot arm housing, 321
Rod supports, **67–69, 92–93**
Roller, 236
Roof truss, Howe, 342
*Rotate,* 14–15
Rotation, 36
Rotational blends, 276–277
Rotation matrix, 458
*Round,* 229
Round adaptors, 389, 390
Rounds, 185–187, 193. *See also* Fillets
  adding, **303–306**
  for angle bracket, **187–188**
  on circular edge, **188**
  for connecting arm, **188**
  edge for propeller blades, **189**
  surface-to-surface, **189–190**

*Same Dims,* 174
*Save As,* 12
*Saved View List,* 16
*Save Views,* 16, 22
Saving updated model, 14
Screws, adjustment, **108–111**
Screw threads, **258–261**
*SEC ENVIRON* menu, *35*
*SECTION, 294*
Section mass properties, 458–459
Sections, **42–44,** 372–390
  airfoil, **215–217**
  conic, 41–42
  creating and using, 42–44
  full section, **375–378**
  half section, **381–383**
  local, 373
  offset section, **378–381**
  parabolic (conic), 226. *See also Sketcher*
  placing in *Sketcher,* 46
  planar, 374–375
  revolved section, **383–384**
  steps for drawing, 45
  summary and steps, 385–386
  thin, 39–41
  three-dimensional, 24–53. *See also Sketcher*
  two-dimensional, 43–44
*SEC TOOLS,* 246, 277
*Sec Tools,* 35, 44
*Sel By Menu,* 433–434
*Select All,* 261
*SELECTION,* 146
*Select Working Directory,* 6–7
*Set* appearance, 18
*Set Def Style,* 437
*Set Display,* 297
*Set Items,* 299, 300, 307
*Set Plane,* 431
Setting up parameters, 31–35
*Set to Initial,* 8
*Setup,* 17, 31, 45
*SETUP SK PLAN,* 66, *66. See also MAKE DATUM*
*Shade,* 16–17
*Shading,* 16
Shading, 18
Shaft hanger, 273
Shafts, 134, **134–135,** 136–137
*Sheets,* 378, 381, 382–383, 386
*Shell,* 111–115, 116, 121, 332
  order of feature creation, 115
Shells, 111–115, 116
Shifter fork, **59–65, 299–300,** 309
Short edge, 459
*Show/Erase,* 300, 359, 366, 384, 399
*SHOW ERRORS,* 304, *304*
*Show* options, 356–357
*Side Blind,* 61, 81
Sides, for hole placement, 81–82
Side supports, 320
Side views, 61
Single bearing bracket, 142
*Single Point,* 336–338
*Single Rad,* 211, 221

Single-sided supports, 180
*Sketch,* 46
  cosmetic uses, 311
  in *DETAIL,* 349
*Sketched* holes, 80–81
*SKETCHER,* 246
*Sketcher,* 24–52, 229, 266
  conic sections, 41–42
  creating and using section, 42–44
  datum planes and, 55, 56
  *Geom Tools* option, 36–39
  interface, 24–28, *26*
  regeneration symbols, 29–31, 30t
  *Relations* option, 31
  *Sec Tools* option, 35
  setting up parameters, 31–35
  *Sketch View* option, 36
  summary and steps, 44–52
  thin sections, 39–41
Sketching
  for blends, 277, 279–280
  cutouts, 106
  holes, 85–87
  of profile, 135
  for U bracket cut, 102–103
Sketching planes, 151–152, 155, 378
*SKETCH VIEW,* 56, *56*
Skew plates, 182
Slotted axles, **107–109**
*Smooth,* 431
*SOLID,* 123. *See also* Material addition
  *Cut/Slot,* 101–109, 115–116
  necks, 109–111, 116
  shells, 111–115, 116
*Solid,* 25, 37
  revolving pulley wheel section, **227–228**
Solid features, blanking of, 298, 307
*SOLID OPTS,* 284
Spanner, hook, 272
*SPEC TO,* 81, *81,* 84
Speed-up options, 144–184. *See also* individual subtopics
  *Copy,* 162–163, **163–164,** 177–178
  *Group,* 157–160, **160–161,** 177
  *Mirror Geom,* 165–166, 178
  *Pattern,* 145–146, **146–157,** 175–177
  review and steps, 175–179
  user-defined features (UDFs), 166–168, **168–174,** 178–179
Spherical coordinates, 205–206
*Spline,* 211, 213, 262, 336–338
Spring clip, 269
Springs, **256–258**
*Srfs* (surface) options, 208–210
Standalone UDFs, 167
Standards, changing, 397–398
Start parts, 69–73, **71–72,** 422
  datum planes and, 69–73
*Start Point,* 250
Start points, for blends, 277, 280, 287
Status changes, 28
*Straight,* 80, 83
Subordinate UDFs, 167
Subsections, 276–277. *See also* Blends
Supports
  double-sided, 180
  handrail, 274, 390
  offset, 319
  rod, **67–69, 92–93**
  side, 320
  single-sided, 180
*Suppress,* 298–299, 300, 307
  features, **301–302**
  features in layer, **302**
Suppressed features, 459
*Surface, UpTo,* 82
Surface area, of model, 458
Surface-to-surface rounds, **189–190**
*Surf Chain,* 187, 261
*SURF OPTIONS,* 324, *324*
*Surf-Surf,* 186, 190
*Sweep,* 245–275
  along datum curve, **261–262**
  gasket, **249–256**
  helical sweep, **256–261**
  King post truss, **262–267**
  simple sweep, **246–249**
  summary and steps, 267–268
Sweeps, 209, 213
  helical, 294
*SWEEP TRAJ,* 246, *246*
Swept blends, 282, **282–288**
  centrifugal blower, **282–287**
*Switch View,* 360
Symbols, regeneration, 29–31, 30t
Symmetry, 145. *See also* Speed-up options
  axis of, 384
  *Mate* and, 412–413
*System,* colors, 8
Takeoff, HVAC, **277–281**
*Tan Curv,* 201
*Tangent,* 408
Tangential planes, 201
*Tangnt Chain,* 187, 261
Tapered collar, **201–203,** 241, 388, 389
Tension plates, **36–39**
Text *(Note),* 169, 351–352
  adding to multiviews, 356–358, **359–364,** 369. *See also* Drawing
  bill of material (BOM), 440–441
  cosmetic for cover plate, **316–317**
  descriptive, 169
  location of, 352
*Text Style,* 313, *313,* 357
Thickness, 459
*Thin,* 25, 39–41, 43, 49, 50, 52, 102
  cut/slot, 101–109
  revolved, **228–229**
Thin sections, 39–41
*Thread,* 311–313, **314–316**
Threads
  cosmetic on nut, **314–316**
  screw, **258–261**
  suppressing, **301–302**
Three-dimensional sections, 24–53. *See also Sketcher*
*Three Srf,* 208
*Through/Plane,* 66–67
*Thru All,* 83, 85, 103
*Thru Cyl,* 201
*Thru Edge,* 201, 222–223
*Thru Next,* 82
*Thru* options, 82
*Thru Points,* 210, 211, *211*
*Thru Until,* 82
Toggling, 7, 277, 278
tol_display, 393
Tolerances, 391–403
  default, 391, 401–402
  geometric, 395–397, **397–401**
  summary and steps, 401–402
  traditional, 392–394, **394**
*TOL SETUP,* 391, *393*
Toolbar, *5,* 6
*Toolbars,* 9
*Top,* 56, 58
*Total Align,* 373–374
*Total Unfold,* 374
Traditional CADD vs. Pro/E, 1–4

Traditional tolerances, 392–394, **394**
Trajectory. *See also Sweep*
defining, 245–246, 257
for swept blends, 282
Trajectory sweeps, 245–275. *See also Sweep*
*TRANS INCR,* 431
*Translate,* 431
Transparency, 18
*Trim,* 36
*Trimetric,* as default, 7
Trusses
bridge, 225, 270
fink, 224, 270
flat roof, 221–222
Howe roof, 224, 342
King post, **262–267**
*TWEAK,* 322, *323*
*Tweak*
adding draft to base feature, 324–325, **325**
ears, 325–327, **327–330**
general features, 322–324
lips, 330, **330–333**
quilts, 334, **335–340**
summary and steps, 340–342
Twisted blends, 277, 288
*2Axes,* 203
Two-dimensional frame, **204–206**
Two-dimensional sections, 43–44
*Two Planes,* 201, 203
*Two Pnt/Vtx,* 201
*2 Points,* 431

U Brackets, **102–103, 134–135**
UDF-driven groups, 167
*UDF FEATS,* 169, *170*
*UDF Library,* 167
Underdimensioning, 29
*Undo,* 431
*Undo Changes,* 302
*Unfold Xsec,* 374
*Ungroup,* 159
*UNIT MGR,* 31
*Units,* 31, 32, 45
Units of measure, 31, 50, 249
*Unset,* 18
Updating, model, 12–14
*UpTo*
*Curve,* 82
*Pnt/Vtx,* 82
*Surface,* 82
*Upto, Plane,* 227
*Up To Pnt/Vts,* 227
User-defined datum planes, 65–69
User-defined features (UDFs), 166–174, 178–179
adding to part, **172–174**
creating, **168–172**
retrieval and placement, 172–174
*Utilities,* 6, 7, 19, 21. *See also specific utilities*

Valve poppet, 140
*Var Dims,* 164, 167, 171–172
*Variable,* 325
*Varying,* 150
Venturi nozzle, 239
Vertex, 82
Verticality, 30–31
Vibration isolator pad, **160–162**
*View,* 6, 9–10, 17, 22, 366, 367, 381
solid model, 16
*View Plane,* 431
*VIEWS,* 375
Views
auxiliary, 364–365, **365–367**
centering of, 355
default, 16
multi-, 346–361. *See also* Drawing
orientation of, 353
orthographic, 15
saving, 16
types of, 348
for working drawing, 439
Vise handles, **25–29**
Volume
of model, 458
one-sided, 459
Volume interference, 459
V symbol, 30

Walls, 343–344
Washers, **103–107**
moving of, 434–436
Wheels. *See also* Pulley wheels
machine, 242
*Whole Array,* 211
*Window,* 6
Windows
main drawing, 5
message, 5
Wing, aircraft, **334–340**
Wing plates, 182
*With/Without Dims,* 207
*Work Directory,* 21–22
*Working Directory,* 6–7
Working drawings, 404–454. *See also* Assemblies

*X&Y Spacing,* 349
X-axis, 205
*Xhatching,* 375, 377
*XSEC CREATE,* 378
*XSEC ENTER, 374,* 377, 378
*X-section mass properties,* 458
*XSEC TYPE,* 373, *373*

Y-axis, 205

Z-axis, 205
*Zoom in/Zoom out,* 14–15, 217